Integrative Strategies for Bioremediation of Environmental Contaminants, Volume Two

Developments in Applied Microbiology and Biotechnology

Integrative Strategies for Bioremediation of Environmental Contaminants, Volume Two

Avenues to a Cleaner Society

Edited by

Riti Thapar Kapoor
Amity Institute of Biotechnology
Amity University Uttar Pradesh,
Noida, Uttar Pradesh, India

Maulin P. Shah
Environmental Microbiology Lab, Bharuch,
Gujarat, India

ACADEMIC PRESS
An imprint of Elsevier
ELSEVIER

Academic Press is an imprint of Elsevier
125 London Wall, London EC2Y 5AS, United Kingdom
525 B Street, Suite 1650, San Diego, CA 92101, United States
50 Hampshire Street, 5th Floor, Cambridge, MA 02139, United States
The Boulevard, Langford Lane, Kidlington, Oxford OX5 1GB, United Kingdom

ISBN: 978-0-443-14013-6

For information on all Academic Press publications visit our website at https://www.elsevier.com/books-and-journals

Publisher: Stacy Masucci
Acquisitions Editor: Linda Versteeg-Buschman
Editorial Project Manager: Billie Jean Fernandez
Production Project Manager: Sajana Devasi P K
Cover Designer: Christian J. Bilbow

Typeset by TNQ Technologies

Contents

List of contributors

Uzma Afaq
Department of Biosciences, Integral University, Lucknow, Uttar Pradesh, India

Syed Nikhat Ahmed
P.G. Department of Environmental Sciences, Sambalpur University, Burla, Odisha, India

Iswar Baitharu
P.G. Department of Environmental Sciences, Sambalpur University, Burla, Odisha, India

Sneha T. Bankar
Department of Chemistry, Rashtrasant Tukadoji Maharaj Nagpur University, Nagpur, Maharashtra, India

Lashika Batra
Amity Institute of Biotechnology, Amity University, Noida, Uttar Pradesh, India

Pranoti Belapurkar
Department of Biosciences, Acropolis Institute, Indore, Madhya Pradesh, India

Vivek P. Bhange
Department of Biotechnology, Priyadarshini College of Engineering, Nagpur, Maharashtra, India

Avinash D. Bholay
Department of Microbiology, K.T.H.M. College, Nashik, Maharashtra, India

P.K. Chauhan
Shoolini University, Solan, Himachal Pradesh, India

Parul Chugh
Amity Institute of Biotechnology, Amity University, Noida, Uttar Pradesh, India

Shivani Dubey
Department of Biosciences, Acropolis Institute, Indore, Madhya Pradesh, India

Jagdish Gabhane
Chintamani College of Art and Science, Gondwana University, Gondpipri, Maharashtra, India

Sharmila S. Ghangale
Department of Biotechnology, C. K. Thakur Arts, Commerce and Science College, Mumbai, Maharashtra, India

Piyush Gupta
Department of Chemistry, Faculty of Engineering and Technology, SRM Institute of Science and Technology, Ghaziabad, Uttar Pradesh, India

Vishakha Gupta
Amity Institute of Biotechnology, Amity University, Noida, Uttar Pradesh, India

Anita V. Handore
Research and Development Department, Phytoelixir Pvt. Ltd., Nashik, Maharashtra, India

Dilip V. Handore
Research and Development Department, Sigma Wineries Pvt. Ltd., Nashik, Maharashtra, India

Syed Khalida Izhar
Department of Biosciences, Integral University, Lucknow, Uttar Pradesh, India

Vinita S. Jagtap
Medical Oncology Molecular Laboratory, Tata Memorial Hospital, Mumbai, Maharashtra, India

Anshika Jain
Department of Biotechnology, Dr. Ambedkar Institute of Technology for Handicapped, Kanpur, Uttar Pradesh, India

Riti Thapar Kapoor
Amity Institute of Biotechnology, Amity University, Noida, Uttar Pradesh, India

Ritika Kapoor
Amity Institute of Biotechnology, Amity University, Noida, Uttar Pradesh, India

Rajib Karmakar
Department of Agricultural Chemicals, B.C.K.V, Directorate of Research, Nadia, West Bengal, India

Kalpana Katiyar
Department of Biotechnology, Dr. Ambedkar Institute of Technology for Handicapped, Kanpur, Uttar Pradesh, India

Seema Kelotra
Department of Biosciences, Acropolis Institute, Indore, Madhya Pradesh, India

Sharad R. Khandelwal
H.A.L. College of Science & Commerce, Nashik, Maharashtra, India

Pooja Kumari
Department of Biotechnology, Dr. Ambedkar Institute of Technology for Handicapped, Kanpur, Uttar Pradesh, India

Jitendra Kumar
Department of Biosciences, UIBT, Chandigarh University, Mohali, Punjab, India

Virendra Kumar
Department of Biotechnology, Dr. Ambedkar Institute of Technology for Handicapped, Kanpur, Uttar Pradesh, India

Kumari Manorma
Dr. Yashwant Singh Parmar University of Horticulture and Forestry, Solan, Himachal Pradesh, India

Vaishali Mishra
Department of Biotechnology, Dr. Ambedkar Institute of Technology for Handicapped, Kanpur, Uttar Pradesh, India

Jyoti Pandey
Department of Biotechnology, APS University, Rewa, Madhya Pradesh, India

Kamal Prasad
Centre for Advanced Agriculture Research, Absolute Bioscience, Delhi NCR, Gurugram, Haryana, India

Neha Rana
Department of Pharmaceutical Sciences, KIET School of Pharmacy, KIET Group of Institutions, Ghaziabad, Uttar Pradesh, India

Arya Sahu
Department of Biotechnology, Dr. Ambedkar Institute of Technology for Handicapped, Kanpur, Uttar Pradesh, India

Shweta Sharma
ICAR-Directorate of Mushroom Research, Solan, Himachal Pradesh, India

Abhishek Sharma
CSIR-Imtech Chandigarh, Chandigarh, Punjab, India

Nitya Singh
Department of Biotechnology, Dr. Ambedkar Institute of Technology for Handicapped, Kanpur, Uttar Pradesh, India

Rachana Singh
Amity Institute of Biotechnology, Amity University, Noida, Uttar Pradesh, India

Gagandeep Singh
Central Ayurveda Research Institute, Jhansi, Uttar Pradesh, India

Charu Vyas
Department of Biotechnology, Faculty of Life Science and Technology, AKS University, Satna, Madhya Pradesh, India

Ashwini A. Waoo
Department of Biotechnology, Faculty of Life Science and Technology, AKS University, Satna, Madhya Pradesh, India

Ranjana Yadav
Indian Plywood Industries Research & Training Institute, Mohali, Punjab, India

Kavita Yadav
Centre for Advanced Agriculture Research, Absolute Bioscience, Delhi NCR, Gurugram, Haryana, India

CHAPTER

1

Bioaugmentation: an emerging ecofriendly technology for wastewater treatment

Anita V. Handore[1], Vinita S. Jagtap[2], Sharmila S. Ghangale[3], Sharad R. Khandelwal[4], Avinash D. Bholay[5], Rajib Karmakar[6] and Dilip V. Handore[7]

[1]Research and Development Department, Phytoelixir Pvt. Ltd., Nashik, Maharashtra, India; [2]Medical Oncology Molecular Laboratory, Tata Memorial Hospital, Mumbai, Maharashtra, India; [3]Department of Biotechnology, C. K. Thakur Arts, Commerce and Science College, Mumbai, Maharashtra, India; [4]H.A.L. College of Science & Commerce, Nashik, Maharashtra, India; [5]Department of Microbiology, K.T.H.M. College, Nashik, Maharashtra, India; [6]Department of Agricultural Chemicals, B.C.K.V, Directorate of Research, Nadia, West Bengal, India; [7]Research and Development Department, Sigma Wineries Pvt. Ltd., Nashik, Maharashtra, India

1. Introduction

Since last few decades, there is drastic progression in technological development. The resulted vigorous civilization together with industrialization, exhaustive usage of synthetic xenobiotics, heavy metals, typical pollutants including pesticides PAHs, that is, polycyclic aromatic hydrocarbons, petroleum products, organic dyes, chlorophenols, nitrophenols and their derivatives, etc., releases highly polluted wastewater causing environmental pollution and health hazards.

There is requirement to treat such polluted wastewater prior to its reuse or discharge into environment. Globally, legislations on pollutants discharge are being confined due to the awareness of pollutants and its consequences on human health and environment resulted in the development of different strategies for improving the efficiency of treatment plants.

In this context, there is a need of ecosustainable solution, leading to bioremediation of heavily polluted environment. These days application of several beneficial microbes has been emphasized as a workable biotechnological remedy, alleviating various environmental hazards and contamination due to their ability to biodegrade most of the hazardous and man-made pollutants (Charles and Anani, 2021).

In most of the biological and physicochemical wastewater treatments, extremely complex chemicals are not efficiently biodegraded by microbes as they are recalcitrant to biodegradation and subsequently persist in the wastewater compromising the water quality. To overcome such challenges, bioaugmentation strategy could be used as an effective, ecofriendly, and cost-effective solution. This process is the addition of specific and efficient pollutant biodegrading microorganisms in the polluted environment. In this, specific microbes having distinct ability to break down specific pollutants are used in place of the indigenous microbes which are not available in sufficient amount or unable to

Integrative Strategies for Bioremediation of Environmental Contaminants, Volume Two. https://doi.org/10.1016/B978-0-443-14013-6.00004-4

break down the chemicals. In bioaugmentation, microorganisms should not only possess degradation capacity w.r.t. particular pollutants of mobilized/immobilized inoculum state but also could survive in adverse environmental condition (Simarro et al., 2013). This process is influenced by the relationship between indigenous and exogenous microbial population (Charles and Anani, 2021).

Following are some important characters of microorganisms necessary for bioaugmentation.

1.1 Need of bioaugmentation

With the intention of establishment of bioreactors for meritorious treatment of chemical mixed wastes, it is necessary to maintain the important consortia of microbes having sufficient activity w.r.t degradation of pollutants in waste. Microbial populations which could be able to colonize such tough niches with preferred catabolic characters and support to develop particular inocula exploiting by process of bioaugmentation are required. In bioaugmentation, microbial strains could fail to grow or to be active in the bioreactor due to predation or competition with indigenous microbes/bacteriophages or due to failure to acclimatize environmental conditions. In such a dynamic context, the introduced microbes may not colonize or show any population dominance w.r.t variation in pH, temperatures, starvation periods, etc. Microbial degradation of pollutants could be increased by bio stimulation.

1.2 Significance of bioaugmentation

Basically, bioaugmentation works by increasing the level of active microorganisms within the treatment environment. These microbes work to transform the pollutants into less harmful chemicals, but the microbes could handle such transformation at certain level. If the ratio of active microbes to detrimental pollutant is not optimal, then the process efficiency will be prominently reduced resulting in the leftover of harmful compounds even after the treatment phase.

Bioaugmentation process is the targeted process in which measured as well as scientific approach could give opportunity for getting more from pushing the effectiveness w.r.t. cleaning of pollutant even optimizing the costs and results. Wastewater-activated sludge contains naturally existing microbes that could biodegrade numerous pollutants, but due to high toxicity, low water solubility, low bioavailability, high stability, and low biodegradability, some microbes show resistance toward biodegradation. In this situation, microbial metabolic enzymes may not be efficiently used as substrates. However, consortia of different microorganisms could help for their appropriate biodegradation. In numerous cases, recalcitrant chemicals might be new; therefore, microorganisms couldn't adapt themselves to use them as a substrate (Providenti et al., 1993). Bioaugmentation process can overcome such challenges, as this treatment could be tailored w.r.t. certain pollutant showing dominance in environment. Accordingly, this approach is effective to address both the increasing number of new pollutants and contaminant which are available at very high concentrations.

1.3 Factors affecting bioaugmentation efficiency

1.3.1 Selection of microorganisms for bioaugmentation consortium

In bioaugmentation process, it is reported that the microbial strain selection is an important aspect as successful bioaugmentation is predominantly influenced by the behavior of bioaugmentation enrichment culture, comprising metabolic adaptation, good survival, and retention in biosystems

(Singer et al., 2005). In this way, effective application of bioaugmentation depends on the acclimatization of coexistence of bioaugmentation microbes to indigenous microbial community. The interactions of bioaugmentation microbes with operation conditions are very critical w.r.t. development and survival of bioaugmentation microorganisms. Also, the interactions of bioaugmentation microbes with native predating/competing microbes are significant factors affecting bioaugmentation.

1.3.2 Bioaugmentation dosage

Bioaugmentation plays a key role w.r.t. improving overall operational performance in biosystems. Mostly, amplified inoculum quantity could be used to avoid washout from the reactors. However, excessive inoculation of specific strains could break down the structure and function of indigenous microbial community. The selection of bioaugmentation dose is restricted by certain operational factors, such as seeding source, yield of parent reactor supplying bioaugmentation consortium, possibility of transferring of bioaugmentation biomass, as well as cost–benefit ratio. Usually, higher dose of bioaugmentation shows advantages to operation performance. However, regulated bioaugmentation dosage has been desired for economic benefit.

1.3.3 Bioaugmentation time

Bioaugmentation effects are found to be determined by timely bioaugmentation. It can be explained with the example that with the sudden oxytetracycline shock treatment, operation performance of five reactors, that is, R0, R1, to R4, in which, R1 has been bioaugmented before shock, R2 has been bioaugmented during the shock, R3 has been bioaugmented after shock, and R4 has been bioaugmented during and after shock. During the performance recovery tests, it was observed that R0, R1, R2, R3, and R4 required 45, 38, 38, 60, and 38 h, respectively, indicating that effects of bioaugmentation get affected by bioaugmentation time (Jin et al., 2014).

1.3.4 Bioaugmentation times and frequency

In spite of significant increase in operation performance, positive impacts of bioaugmentation could last for a less period. Consequently, quantification of bioaugmentation time has to be implemented for ensuring its success under severe stresses. It is reported that there is substantial reduction in the removal of *p-nitrophenol* on the initial day of bioaugmentation; however, similar shock intensity couldn't influence the removal performance in successive transient shock. Thus, bioaugmentation times and frequency should be considered. It is reported that the performance couldn't get influenced by any organic matter shock during long-term bioaugmentation application in Anammox sequencing batch reactor (SBR) (Martin-Hernandez et al., 2012; Tang et al., 2014).

1.3.5 Bioaugmentation interval

Optimal bioaugmentation interval between two bioaugmentation is very important w.r.t. economic cost. It is reported that the utmost effect of one-time bioaugmentation depends on dosage of bioaugmentation and decays due to losing of bioaugmented nitrifiers. Besides, the addition of bioaugmentation dosage by every day could result in long-term satisfactory performance. However, sufficient large bioaugmentation dosage is necessary. Besides successful bioaugmentation, total system management including the controlling the crucial factors like oxygen levels, pH, temperature, etc., is required.

1.4 Advantages of bioaugmentation

In several areas, bioaugmentation has been proved as beneficial. It increases the diversity and numbers of microbes for achieving the expected results. Numerous industrial waste plants face complications to achieve nitrification/toxic shocks. However, by regular addition of nitrifying microbes, appropriate population w.r.t. removal of ammonia could be maintained. Moreover, it could help to reduce odor. Due to strict environmental restrictions, various waste treatment plants face various compliances and challenge the capabilities of existing wastewater treatment plants. Sometime, bioaugmentation could also be the long-term solution due to the lack of capital investments of mechanical solutions. Moreover, other areas where bioaugmentation could be beneficial include odor reduction, grease and oil removal, rapid startup, and amended tolerance to toxic shocks. Besides, extensive research could endure exploration of new application areas.

Several areas where bioaugmentation has prevent beneficial are discussed below.

1.4.1 Amended solids settling

Solid removal is an important step in biological waste treatment commonly by settling in a clarifier or pond lagoon. Bacteria or other microbes naturally form a biopolymer to support in settling. Poisonous shocks and system alterations could be resulted in microbial population with minute biopolymer and poor settling properties. Although, the traditional method of addition of biological polymer or coagulants as settling aids is effective. Demand of polymer could be reduced greatly or eliminated by inoculating the system with organisms recognized as resistant to toxicity and outstanding flock formers.

1.4.2 Degradation of particular compounds

Low levels of specific compounds could be achieved by the use of particular microbes. Compounds like chlorinated aromatic hydrocarbons, phenols, etc., could also be reduced with bioaugmentation.

1.4.3 Improved nitrification

Numerous industrial waste plants have problems to achieve nitrification/toxic shock. However, due to frequent addition of nitrifying bacteria, proper microbial population for removal of ammonia could be retained.

1.4.4 Controlling level of fecal coliform

There are few microbial accelerators (MAs) which could control the *Escherichia coli* level in effluent or sewage without using bleaching powder or chlorine. There are some specific microorganisms that could synthesize normal strains to harmless strains.

1.4.5 Benefits to businesses

1.4.5.1 To optimize cost

Bioaugmentation is a cost-effective process than other wastewater treating methods. It costs only for sourcing/cultivating particular microbial culture.

1.4.5.2 Regulatory compliances

Bioaugmentation process is normally reinforced by regulators across different authorities. Regulators could apply various metrics as they could control viability of treatment process, including the level of risk which the harmful chemicals may be left over, stability and sustainability toward the environment, and the level of disruption which the treatment causes in the local community.

1.4.5.3 Social acceptance

In any businesses, there is a need to maintain a good association with local public by decreasing noise and environmental distraction along with the confirmation of safe and accountable wastewater disposal. Therefore, bioaugmentation could help to achieve the positive relationship.

1.4.5.4 Carbon footprint reduction

During bioaugmentation, there is no need to waste energy for heating the reactor or to give any treatment against environment nor the related businesses need any high levels of pressure for achieving the desired treatment results. It helps to optimize the cost as well as secure ongoing environmental sustainability and thereby reduces carbon footprint.

1.4.5.5 Efficient contaminant treatment

Bioaugmentation treatment could certainly play a key role in ecosustainable businesses. It is a very effective contaminant treatment w.r.t. scientific approach which could efficiently support the businesses for targeting environmental sustainability and stability in economic sense. However, it needs appropriate research and understanding.

1.4.5.6 Microbial selection criteria

The application of commercially useful products, potent microbial culture, indigenous-exogenous strains/tailor-made consortia have been proved as useful bioaugmentation options in which the main strategy has based on the isolation of natural environmental samples, selection of efficient microbial strains, selection followed by the use of mutagenic agents, and fermentation tests for single strains using traditional enrichment techniques. Adaptation w.r.t. scaling up is an extremely important phase for effective bioaugmentation. At this key stage, complexity of engineered wastewater biotreatment needs specific efforts.

Priming is predisposing the microbial population to future conditions carried out to perform a specific function. Accordingly, acclimation of microbial consortia must be considered in the process flow chart for predisposing them for better physicochemical conditions as well as to achieve efficient performance.

It is reported that microbes used in this process should be catabolically able to degrade the pollutant, even in the presence of other potentially inhibitory pollutants; they must survive and remain competitive after their introduction into the biosystem, and they should be compatible with the indigenous microbial populations. Therefore, the selection of microorganisms should be carefully carried out, because very few organisms may be appropriate for bioaugmentation, and it is not necessarily from the community needing bioaugmentation (Yu and Mohn, 2002). Besides, the selected microbes should not be closely related to human pathogens while using in field operation. It has been proposed that significant progress should be carried out by developing current information w.r.t. recent techniques in molecular biology, worldwide access to culture collections, etc., simultaneously, updating the information from accessible research supporting the effective way of overcoming barriers in bioaugmentation ((Singer et al., 2005) (Table 1.1).

Table 1.1 Microorganisms used for bioaugmentation.

Microorganisms	Degraded substrates or substances
P. aeruginosa, strain WatG	Diesel
Pseudomonas sp., strain ST41	Marine gas oil
A. cylindrospora	Fluorene
A. chlorophenolicus, strain A6L	4-Chlorophenol
C. testosterone, strain BR60	PAHs and crude oils
P. putida, strain pDH5/Paw 340	4-Chlorobenzoic
C. necator, strain RW112	Chlorobenzoates aroclor 1221 and 1232
B. xenovorans, strain LB400/ohb	Aroclor 1242
P. fluorescens, strain RE	2,4-Dinitrotoluene
P. fluorescens, strain MP	2,4-Dinitrotoluene
P. fluorescens, strain KT2442	Naphthalene
P. fluorescens, strain F113rifpcbrrnBP1: gfpmut3	Naphthalene and PCBs
Strain RHAI, *Rhodococcus* sp.	4-Chlorobenzoate
Streptomyces spp.	Naphthalene
Phenol-degrading bacteria	Phenol
S. xenophaga	Bromoamine
C. humicolus and unidentified cyanide-degrading microbes	Cyanide
E. coli, strain AtzA	Atrazine
Rhodococcus sp., strain F92	Various petroleum products
P. putida, strain BCRc14349	Trichloroethane and phenol
P. fluorescens, strain CS2	Ethylbenzene and biphenyl
P. funiculosum and *Rhizopus* spp.	Petroleum hydrocarbons
P. aeruginosa, strains NM and M, *B. subtilis*, strain DM-04	Crude petroleum oil hydrocarbon
Pseudomonas spp., *Acinetobacter* spp., and *Rhodococcus* spp.	PAH congeners (pyrene, phenanthrene, and fluorine)
Acremonium spp., *Verticillium* spp., *Aspergillus* spp., *T. canadense*, and *F. oxysporum*	HMW-PAHs (4−7 rings)
S. chlorophenoticum, strain ATCC 39723	Pentachlorophenol
Immobilized phenol-utilizing *Arthrobacter* spp.	Phenol, naphthalene, carbazole, dibenzofuran, and dibenzothiophene
Desulfitobacterium spp.	2,4,6-Trichloro-phenol
Consortium of bacteria	2,4-Dichlorophenol(2,4-DCP)
Comamonas testosterone	3-Chloroaniline
Acinetobacter spp.	4-Fluoroaniline
S. xenophaga	Bromoamine
Sphingomonas spp.	Bromoamine
Serratia spp.	Diethylene glycol butyl ether
Streptomyces spp.	Naphthalene
Acinetobacter spp.	Nicotine

Table 1.1 Microorganisms used for bioaugmentation.—cont'd

Microorganisms	Degraded substrates or substances
Pseudomonas spp.	Nicotine
Phenol-degrading bacterial species	Phenol
Immobilized arthrobacter spp. phenol-utilizer	Phenol (PH) and naphthalene (NAP) along with carbazole (CA), dibenzofuran (DBF), and dibenzothiophene
Rhizobium spp./*P. denitrificans*	Pyridine
Bacillus spp./*B. pickettii*	Quinoline

Adapted from Charles, O.A., Anani, A., 2021. Bio augmentation-A powerful biotechnological techniques for sustainable eco restoration of soil and groundwater contaminants. Microorg. Sustain. 1, 373–398. ISBN: 978-981-15-7447-4 (eBook). https://doi.org/10.1007/978-981-15-7447-4 and Alexis, N., Shaikh, A.R., Jesse, Z., 2016. Bioaugmentation-an emerging strategy of industrial wastewater treatment for reuse and discharge. Int. J. Environ. Res. Publ. Health. 13 (9), 1–20.

1.5 Bioaugmentation products

These days, various ready-to-use bioaugmentation products made up with MAs (bacterial/fungal strains) are available in market. These products have diverse applications in industrial, agricultural, and consumer field wastewater treatment and do not require any special storage. As soon as these are introduced into a polluted area, the microbes could immediately receive and being feeding and reproducing while attacking the material in the water causing the pollution.

Basically, the microorganisms are isolated from nature and mostly selected w.r.t. their accelerated reproduction rates as well as potential to perform particular function, for example, good floc forming abilities to augment settling/capability to degrade particular compound, etc. These products have been sold in various forms, for example, bran carrier and liquid products, dried organisms, etc. These natural products are particularly nonpathogenic and nontoxic that are pure and safe for plant, animal, human, aquatic, and marine life. These are compatible with the indigenous microbial populations which could work synergistically with microbes, increases their ability by creating a competitive environment. These products could efficiently stabilize the complex processes in effluent treatment plants (ETPs) and sewage treatment plants (STPs) and optimize efficacy in special reactors with/without activated sludge.

These products could be used in the fermenting and anoxidial process. It works aerobically/anaerobically with or without light and works in a wide range of temperature and pH value tolerance.

These could be efficiently used for food processing industries, fruit and pulp processors, beverages and bottling plants, milk, dairy and cheese processing plants, ice-cream plants, pharmaceutical industries, meat processing industries, paper and pulp mills, sugar industries, distilleries, wine and alcohol producing units, solvent extraction plants, paints and resin industries, town and municipalities, STPs, lagoons, lakes and ponds, city drains, zoo, textile units (using organic dyes), military bases, refineries, petrochemical industries, hotels and resorts, golf clubs (for their ponds), fish farming and aquaculture units, and any industry that produces organic wastewater/effluent.

1.6 Reasons for failures/limitations of bioaugmentation technologies

Although the concept of bioaugmentation w.r.t. wastewater treatment has been extensively studied with encouraging results at laboratory scale, till date success has not been turned to full-scale wastewater treatment. It was found that due to various biotic and abiotic stresses, number of exogenous microorganisms decreases very shortly after addition to a site. The reasons include deficient substrates, variations in temperature, pH, nutrient limitation, struggle between introduced microbes and indigenous microbes, phase infections, pollutant load shock, protozoa grazing, and some factors allied with quorum sensing (Alexis et al., 2016).

Till date, various reasons have been proposed by different researchers w.r.t. the fiasco of inocula used for bioaugmentation strategy (Alexis et al., 2016). The success of such strategy is influenced by the interactions between the inoculated microbes and new biological as well as nonbiological environments w.r.t. their survival, activity. During bioaugmentation, in spite of substantial increase in bioreactor performance, positive effects are found for a short duration after the inoculation of microorganisms.

According to few reports, the failures might be due to selecting bioaugmentation in the initial place, irrespective to the fact that indigenous microbial community may possess suitable catabolic genes for the degradation of targeted compounds. Therefore, the effectiveness of reported results w.r.t. bioaugmentation might frequently be questionable. As the microbes of similar genres are not equally fit for specific chores, and few might be competitive under a broad range of conditions, whereas remaining microbes may be suited for particular conditions.

In this context, species-level variation, even among the microbes exhibiting the identical catabolic characters, has been stated as a failure reason, alongside alterations at strain level w.r.t. tenacity in system followed by microbial inoculation leading to efficient bioaugmentation designs. Most often, the problems come across with regard to the limiting growth conditions resulted because of the low concentration of substrate; occurrence of inhibitory substances in the stream and even secreted by other microbes showing antagonistic effects like bacteriocin, antibiotics, poor biofilm formation, occurrence of bacteriophages, etc. (Fu et al., 2009). Generally, bioaugmentation has been correlated with population of consortia of bioaugmentation and disturbance of ecosystem equilibrium. In this context, hypotheses for bioaugmentation failure have been proposed as:

(i) Concentration of pollutant in targeted environment may be too low to support the metabolism of bioaugmentation bacteria.
(ii) Microbial inhibitors in environment may control the growth of bioaugmentation consortium.
(iii) Growth rate of bioaugmentation microbes may be slower than out rate due to predation/washout.

In fact, the failure rate of bioaugmentation may be high due to the complicated interactions of microbial inoculants with their growth environment (Alexis et al., 2016; Liu et al., 2015; Wu et al., 2005).

1.7 Promising strategies—mechanisms for efficient bioaugmentation

Before the application of bioaugmentation under actual conditions, study relating to the ability of bioaugmented microorganisms w.r.t. their growth efficiency in tested environment, inoculum size should be considered. During the bioaugmentation of groundwater, an inoculum of 10^6–10^7 cells/mL bacterial concentration is recommended. It is reported that the infections of bacteriophages could significantly affect the bioaugmented bacterial growth. Also, till date no sustainable strategies exist w.r.t. removal of bacteriophages from wastewater. However, before initiating a bioaugmentation, the ability of bacteria to grow in the new environment should be recognized. Various methods may be used to monitor microbial growth, including plating, the most probable number, polymerase chain reaction, that is, PCR, quantitative PCR, that is, qPCR, and microarray. During monitoring of groundwater treatment, if bacterial concentration falls down below 10^6 cells/mL, its necessary to add new inocula for maintaining the bioaugmentation efficiency.

1.7.1 Improvement of detrimental conditions

In the anammox, that is, anaerobic ammonium oxidation system, bioaugmentation support increased in biomass with high speed by creating favorable environment. It also increased the amount of required microbes by exhibiting few growth factors necessary for functional microbes. In case of shock stress bioaugmentation-activated sludge system with particular strains of microbes, there is rapid reduction in toxicity on microbial community.

The rate of degradation toxic pollutants gets augmented by bioaugmentation in biosystems exposed to the shock of toxic substance. Consequently, indigenous bacterial culture along with other microbes get protected from toxic shocks, resulted in fast recovery of microbial activity. Bioaugmentation not only ensures optimum growth and activity of various microorganisms but also resulted in better sludge characteristics and better nitrification/denitrification (Zhang et al., 2017).

1.7.2 Bioaugmentation consortium and their retention, adaptation, and reactivation

During bioaugmentation, bacteria/other microbes go through significant criteria like persistent activity and compatibility before reflecting effects of bioaugmentation. Thereafter, steady treatment competency could be attained over long-term addition of bioaugmentation consortium. Successively, amount and activity of these microbes are found to enhance. Thus, the amplification of bioaugmentation consortium with improved activity is the principal cause behind better performance (Zhang et al., 2017).

1.7.3 Quorum-sensing effect

Biofilms, that is, structures where bacterial communities are attached through self-produced hydrated polymeric matrix formation is mediated by a quorum sensing. During this process, microbes like bacteria release chemical signals for bacteria—bacteria communication which leads to biofilm formation.

Such variation of whole bacterial community structures might be resulted from quorum-sensing ability of bioaugmentation consortium. Accordingly, in the biological systems, the bioaugmentation bacteria predominantly change the structure of entire bacterial community (Zhang et al., 2017). Besides, modulation of quorum-sensing could support the advancement of bioaugmentation w.r.t. wastewater treatment by regulating the microbial population dynamics in the bioreactors and help to open up the route for modified efficiency of bioaugmentation in the treatment of wastewater (Alexis et al., 2016).

1.7.4 Assessment of risk

It is found that sometime, stress is caused due to lack of nutrients, adverse environmental conditions such as competition between microbes, very less temperature, shock by toxic chemicals, etc. In this situation, it's become very hard to the bioaugmentation microbes for keeping a foothold, retentiveness, growth, and become dominant. In this context, it is required to carry out risk assessment with its optimization strategies. Thus, the detail assessment needs to be performed in order to reduce the bioaugmentation risk for optimization of bioaugmentation effects in future (Zhang et al., 2017).

1.7.5 Bioaugmentation and cell immobilization

Immobilization is the encapsulation of microorganisms allied with higher concentration of biomass. Their boosted cell survival could be used to overcome limitations/failure of bioaugmentation. It helps to protect against protozoa grazing, bacteriophage infections; increases biological/physical stability by reducing the challenges w.r.t. sudden and brief variations of temperature/pH; and protects from inhibitory effect of toxic substances including heavy metals and increase of shear stress (Alexis et al., 2016).

1.7.6 Utilization of gene transfer/genetically modified microorganisms

Bioaugmentation advancement could be carried out by using genetically modified microbes. Although, this method might be risky for environmental equilibrium of microorganisms. In this context, several countries are approving restrictive legislations against their widespread use in the environment (Alexis et al., 2016).

1.7.7 Plasmid-mediated bioaugmentation

In nature, microbes/bacteria could acquire new-fangled catabolic function by getting gene encoding catabolic enzyme from closely/distant-related microorganism or bacteria, through plasmids or transposons. In this context, bioaugmentation improvement could be carried out by horizontal gene transfer among microbes/bacteria. To carry out this process, there is need to use donor microbe or bacteria having plasmids of interest which could be mixed and cultured with recipient microbes like bacteria. Once the transfer occurs, the recipient microorganisms/bacteria become trans-conjugants by acquiring new catabolic biochemical processes.

However, sometime, this method fails due to incompetence of donors for persisting into the environment, incompetence of plasmid transfer in the recipient microbes like bacteria, less number of donor/recipient microbes, and decreased stability of plasmids once in the recipient microbe/bacteria. Therefore, understanding the genetic characters of recipient microbes including bacteria w.r.t. said biological events is essential to improve such plasmid-bioaugmentation approach for wastewater treatment. In this context, new tools such as metabolomics, metagenomics, transcriptomics, proteomics, etc., are needed for successful bioaugmentation (Alexis et al., 2016; Blakely, 2015; Thomas and Nielsen, 2005; McClure et al., 1989; Hausner et al., 2011; Top et al., 2002) (Table 1.2).

1.8 Applications of bioaugmentation

1.8.1 Removal of chlorinated/fluorinated compounds

Globally, there is extensive use of fluorinated as well as chlorinated solvents in both industry and home causing water contamination. In this regard, bioaugmentation has been proved as an important strategy for their removal. At laboratory scale, *Comamonas testosteroni, Acinetobacter* spp., etc., show the ability to biodegrade 4-fluoroaniline and 3-chloroaniline in synthetic wastewater medium added with activated sludge (Wang et al., 2013b).

Table 1.2 Monitoring techniques for bioaugmentation.

Sr. No.	Monitoring techniques
1	Quantitative competitive polymerase chain reaction (PCR)
2	Quantitative competitive polymerase chain reaction-denaturing gradient gel electrophoresis (PCR–DGGE) biolog system
3	Terminal restriction fragment length polymorphism (T-RFLP)
4	Quantitative competitive PCR and reverse transcription polymerase chain reaction (RT-PCR)
5	Ribosomal intergenic spacer analysis (RISA)
6	PCR-temperature gradient gel electrophoresis (TGGE)
7	Autofluorescence, PCR–DGGE
8	Competitive PCR and RT-PCR
9	Fluorescent in situ hybridization (FISH)

1.8.2 Lignin removal

Paper and pulp industry produces large amount of wastewater, that is, black liquor containing polysaccharides, lignin, resinous compounds, etc. During the wastewater treatment, natural biological treatment with activated sludge fails to reduce/eliminate these compounds efficiently due to the absence of microbes having ability to degrade lignin (Wu et al., 2005). Accordingly, selection and supplementation of such lignin-biodegraders into wastewater could provide remarkable strategy for the removal of specific pollutants originated from black liquor. It is reported that in a laboratory set-up consisting of an SBR, consortia of such lingo-cellulose-biodegrading microbes isolated from activated sludge in the SBR could enhance the reduction of lignin (>50%) (Zheng et al., 2013).

1.8.3 Pyridine and quinoline

Pyridine and quinoline solvents are used for paints, dyes, and wood treatment chemicals, thereby found in the industrial wastewater. It is reported that there is increase in biodegradation of quinoline and pyridine by using *Bacillus* spp., *Burkholderia pickettii*, *Paracoccus* spp., *Pseudomonas* spp., respectively (Tuo et al., 2012; Jianlong et al., 2002; Bai et al., 2009).

1.8.4 Synthetic dyes

Man-made dyes, which consisting azo-/anthraquinone-based molecules, are extensively used in textile/cosmetics industries. Basically, azo dyes are found to be the largest diverse group of dyes and are generally resistant to biodegradation with conventional activated sludge treatment (Pandey et al., 2007). It is reported that azo dye/Acid Orange-7 can be successfully removed with *Shewanella* sp. (Wang et al., 2012). Also at laboratory scale, it was found that the strain of the *Sphingomonas* spp., *Sphingomonas xenophaga* could successfully remove Bromoamine acid in bioaugmentation studies with synthetic wastewater medium (Qu et al., 2006, 2009).

1.8.5 Cyanides

It is reported that the use of *Cryptococcus humicolus*, that is, cyanide-degrading yeast and other unidentified cyanide degraders could remove ferric cyanide in wastewater at laboratory scale (Park et al., 2008).

1.8.6 Nicotine

Pollutants including nicotine compounds could be efficiently removed by using bioaugmentation strategy. Numerous bacterial species are showing ability to degrade nicotine, for example, *Acinetobacter* spp., *Sphingomonas* spp., *Acinetobacter* spp., etc. Besides, strain, *Pseudomonas* spp. HF-1, in an SBR could be used for the treatment of tobacco wastewater (Wang et al., 2011; Wang et al., 2013a; Wang et al., 2009).

1.8.7 Glycol ethers

Glycol ethers like ethylene glycol monobutyl ether and diethylene glycol monobutyl ether are polar solvents which are easily miscible with organic chemicals as well as water, thereby used in paints and cleaners. These compounds show toxicity in animal models. As they are recalcitrant to biodegradation, they are found to be accumulated in the environment. Strain of *Serratia* spp. shows potential to remove diethylene glycol monobutyl ether during bioaugmentation of contaminated wastewater both at laboratory as well as full-scale (Sitarek et al., 2012; Chen et al., 2016).

1.8.8 PAHs—polycyclic aromatic hydrocarbons

PAHs are primarily found in petroleum products and in the waste streams from various industrial processes like coal conversion and synthesis of organic chemicals. It is reported that the use of strains of *Streptomyces* spp. in a membrane bioreactor showed a noteworthy removal of naphthalene. Besides phenol, pyridine, quinoline, carbazole, dibenzofuran, dibenzothiophene, etc., present in the coking wastewater could be effectively removed by bioaugmentation with consortia of *Paracoccus denitrificans*, *Pseudomonas* sp, and *Arthrobacter* sp. both free and immobilized (Xu et al., 2015; Zhu et al., 2015; Shi et al., 2015).

Besides these applications, bioaugmentation is beneficial for the removal of enhanced biochemical oxygen demand (BOD), and it can be effectively used for the improvement of solid setting as it provides an overall healthier biomass. It can efficiently improve nitrification by regularly adding nitrifying bacteria. Moreover, other diverse areas where bioaugmentation strategy offers benefits include reduction of odor, removal of oil and grease, rapid system start-up, and improved tolerance to toxic shock. Moreover, extensive research has been continued for the exploration of new application areas for this progressing technology (Yanbo et al., 2021).

1.9 Future prospects of bioaugmentation for wastewater treatment

In upcoming era, it's necessary to fill up the crucial gaps in the addressed limited research for successful application of bioaugmentation. There is need to operate parallel studies with control and test process streams.

In this context, following are the future prospects:

- Two stages bioaugmentation reactor with first stage parent reactor for supplying bioaugmentation bacteria with timeliness, promptness, automaticity, etc., are necessary.
- Studies w.r.t. optimization of bioaugmentation parameters, for example, consortium, time of dosage, frequency, interval, interaction of bioaugmentation microbes with the indigenous species, quality of wastewater, configuration of reactor, and operation conditions are very critical for appropriate bioprocessing.

- Exact quantification of beneficial or adverse impacts of bioaugmentation on the bioprocess are still discreet. Hence, suitable model to evaluate its outcome should be developed.
- Further study related to real-time tracking molecular technique to monitor structure of microbial community, for example, analyzing the dominant microbes like bacteria in the bioaugmentation systems is needed.
- Technology w.r.t. sidestepping the loss of biomass during bioaugmentation process is urgently needed.
- Further studies related to the selection of proper bioaugmentation consortia, appropriate duration, and in what manner to employ bioaugmentation for successful adaptation under treatment stress are required.
- Progress related to understanding the degradation pathways and finding a suitable microbial species for the reduction of particular compound is needed.
- The synergistic effect of consortium as well as the collective action of a biosurfactant and a preadapted consortia could support the degradation of wastewaters having PAHs.
- Progresses in genetical study w.r.t. the development of suitable strains for targeting the xenobiotic compounds having limitation for elimination are needed.
- Detail understanding w.r.t. treatment process conditions, environmental factors, and location for introduction of the microbial strain for reducing its exposure to undesirable conditions might help to provide possible solutions.
- A close interaction between engineering and biotechnology approaches is needed for promoting the toxicity removal efficiency at higher level. These days, advent of membrane reactors in wastewater treatment plants and the capacity to keep the consortia in bioreactor have been considerably increased leading to improved performance as well as comparatively better quality of effluents. In this context, emerging areas of research will be helpful for developing innovative technologies for novel applications in bioaugmented bioreactors.
- Protein engineering has a wide potential for technological applications.
- Nanotechnology is also a promising area of research with considerable potential. Besides, reverse micellar extraction has the benefits w.r.t. simplification of scaling-up as well as continuous operation.
- Single cell analysis will be essential for revealing the cellular multiplicity and heterogeneity. In this regard, in-depth understanding at single cell level accompanied at the community level is required for understanding the key controlling factors of bioaugmentation.

2. Conclusions

Bioaugmentation is an outstanding strategy to remove the contaminant from wastewater. It can be effectively used in conventional biological waste management system to reduce different parameters such as biochemical oxygen demand, chemical oxygen demand, total suspended solids from the wastewater, etc. It can be efficiently applied in wastewater treatment plant of various industries during the system reseeding, startup operation of a biotreatment unit. Although, this methodology shows success at laboratory scale, few challenges are still existing. There is need to give attention toward the parameters and limitations described in chapter. Emerging research areas can be helpful for the development of appropriate bioaugmented bioreactors for successful application of this emerging ecofriendly technology.

Acknowledgment

The authors are grateful to the Research and Development Department, Phytoelixir Pvt Ltd. Nashik. Special thanks to Mr. V.C. Handore and Mrs. Hira V. Handore for valuable support.

References

Alexis, N., Shaikh, A.R., Jesse, Z., 2016. Bioaugmentation-an emerging strategy of industrial wastewater treatment for reuse and discharge. Int. J. Environ. Res. Publ. Health 13 (9), 1–20.

Bai, Y., Sun, Q., Zhao, C., Wen, D., Tang, X., 2009. Simultaneous biodegradation of pyridine and quinoline by two mixed bacterial strains. Appl. Microbiol. Biotechnol. 82, 963–973.

Blakely, G.W., 2015. Mechanisms of horizontal gene transfer and DNA recombination. In: Molecular Medical Microbiology, second ed., vols. 1–3. Elsevier, pp. 291–302.

Charles, O.A., Anani, A., 2021. Bio augmentation-A powerful biotechnological techniques for sustainable eco restoration of soil and groundwater contaminants. Microorg. Sustain. 1, 373–398. https://doi.org/10.1007/978-981-15-7447-4. ISBN 978-981-15-7447-4.

Chen, M., Fan, R., Zou, W., Zhou, H., Tan, Z., Li, X., 2016. Bioaugmentation for treatment of full-scale diethylene glycol monobutyl ether (DGBE) wastewater by *Serratia* sp. BDG-2. J. Hazard Mater. 309, 20–26.

Fu, S., Fan, H., Liu, S., Liu, Y., Liu, Z., 2009. A bioaugmentation failure caused by phage infection and weak biofilm formation ability. J. Environ. Sci. 21, 1153–1161.

Hausner, M., Starek, M., Bathe, S., 2011. Bioremediation- plasmid-mediated bioaugmentation of microbial communities-experience from laboratory-scale bioreactors. In: Treatise on Water Science. Elsevier, pp. 369–376.

Jianlong, W., Xiangchun, Q., Libo, W., Yi, Q., Hegemann, W., 2002. Bioaugmentation as a tool to enhance the removal of refractory compound in coke plant wastewater. Process Biochem. 38, 777–781.

Jin, R.C., Zhang, Q.Q., Zhang, Z.Z., Liu, J.H., Yang, B.E., Guo, L.X., Wang, H.Z., 2014. Bioaugmentation for mitigating the impact of transient oxytetracycline shock on anaerobic ammonium oxidation (ANAMMOX) performance. Bioresour. Technol. 163, 244–253.

Liu, X., Chen, Y., Zhang, X., Jiang, X., Wu, S., Shen, J., Sun, X., Li, J., Lu, L., Wang, L., 2015. Aerobic granulation strategy for bioaugmentation of a sequencing batch reactor (SBR) treating high strength pyridine wastewater. NJ. Hazard. Mater. 295, 153–160.

Martin-Hernandez, M., Suarez-Ojeda, M.E., Carrera, J., 2012. Bioaugmentation for treating transient or continuous p-nitrophenol shock loads in an aerobic sequencing batch reactor. Bioresour. Technol. 123, 150–156. https://doi.org/10.1016/j.biortech.2012.07.014.

McClure, N.C., Weightman, A.J., Fry, J.C., 1989. Survival of Pseudomonas putida UWC1 containing cloned catabolic genes in a model activated-sludge unit. Appl. Environ. Microbiol. 55, 2627–2634.

Pandey, A., Singh, P., Iyengar, L., 2007. Bacterial decolorization and degradation of azo dyes. Int. Biodeterior. Biodegrad. 59, 73–84.

Park, D., Lee, D.S., Kim, Y.M., Park, J.M., 2008. Bioaugmentation of cyanide-degrading microorganisms in a full-scale cokes wastewater treatment facility. Bioresour. Technol. 99, 2092–2096.

Providenti, M.A., Lee, H., Trevors, J.T., 1993. Selected factors limiting the microbial degradation of recalcitrant compounds. J. Ind. Microbiol. 12, 379–395.

Qu, Y., Zhou, J., Wang, J., Song, Z., Xing, L., Xing, F., 2006. Bioaugmentation of bromoamine acid degradation with Sphingomonas xenophaga QYY and DNA fingerprint analysis of augmented systems. Biodegradation 17, 83–91. https://doi.org/10.1007/s10532-005-3544-0.

Qu, Y., Zhou, J., Wang, J., Xing, L., Jiang, N., Gou, M., Salah Uddin, M., 2009. Population dynamics in bioaugmented membrane bioreactor for treatment of bromoamine acid wastewater. Bioresour. Technol. 100, 244–248.

Shi, S., Qu, Y., Ma, Q., Zhang, X., Zhou, J., Ma, F., 2015. Performance and microbial community dynamics in bioaugmented aerated filter reactor treating with coking wastewater. Bioresour. Technol. 190, 159–166.

Simarro, R., González, N., Bautista, L.F., Molina, M.C., 2013. Assessment of the efficiency of in situ bioremediation techniques in a creosote polluted soil: change in bacterial community. J. Hazard Mater. 262, 158–167.

Singer, A.C, Van Der Gast, C.J, Thompson, I.P., 2005. Perspectives and vision for strain selection in bioaugmentation. *Trends Biotechnol.* 23: 74–77.

Sitarek, K., Gromadzińska, J., Lutz, P., Stetkiewicz, J., Świercz, R., Wąsowicz, W., 2012. Fertility and developmental toxicity studies of diethylene glycol monobutyl ether (DGBE) in rats. Int. J. Occup. Med. Environ. Health 25, 404–417.

Tang, C.J., Zheng, P., Ding, S., Lu, H.F., 2014. Enhanced nitrogen removal from ammonium- rich wastewater containing high organic contents by coupling with novel high-rate ANAMMOX granules addition. Chem. Eng. J. 240, 454–461.

Thomas, C.M., Nielsen, K.M., 2005. Mechanisms of and barriers to horizontal gene transfer between bacteria. Nat. Rev. Microbiol. 3, 711–721.

Top, E.M., Springael, D., Boon, N., 2002. Catabolic mobile genetic elements and their potential use in bioaugmentation of polluted soils and waters. FEMS Microbiol. Ecol. 42, 199–208.

Tuo, B.H., Yan, J.B., Fan, B.A., Yang, Z.H., Liu, J.Z., 2012. Biodegradation characteristics and bioaugmentation potential of a novel quinoline-degrading strain of *Bacillus* sp. isolated from petroleum-contaminated soil. Bioresour. Technol. 107, 55–60.

Wang, J., Liu, G.F., Lu, H., Jin, R.F., Zhou, J.T., Lei, T.M., 2012. Biodegradation of Acid Orange 7 and itsauto-oxidative decolorization product in membrane-aerated biofilm reactor. Int. Biodeterior. Biodegrad. 67, 73–77.

Wang, J.H., He, H.Z., Wang, M.Z., Wang, S., Zhang, J., Wei, W., Xu, H.X., Lv, Z.M., Shen, D.S., 2013a. Bioaugmentation of activated sludge with *Acinetobacter* sp. TW enhances nicotine degradation in a synthetic tobacco wastewater treatment system. Bioresour. Technol. 142C, 445–453.

Wang, M., Xu, J., Wang, J., Wang, S., Feng, H., Shentu, J., Shen, D.S., 2013b. Differences between 4-fluoroaniline degradation and autoinducer release by *Acinetobacter* sp. TW: implications for operating conditions in bacterial bioaugmentation. Environ. Sci. Pollut. Res. Int. 20, 6201–6209.

Wang, M., Yang, G., Min, H., Lv, Z., Jia, X., 2009. Bioaugmentation with the nicotine-degrading bacterium Pseudomonas sp. HF-1 in a sequencing batch reactor treating tobacco wastewater: degradation study and analysis of its mechanisms. Water Res. 43, 4187–4196.

Wang, M.Z., Yang, G.Q., Wang, X., Yao, Y.L., Min, H., Lv, Z.M., 2011. Nicotine degradation by two novel bacterial isolates of *Acinetobacter* sp. TW and *Sphingomonas* sp. TY and their responses in the presence of neonicotinoid insecticides. World J. Microbiol. Biotechnol. 27, 1633–1640.

Wu, J., Xiao, Y.Z., Yu, H.Q., 2005. Degradation of lignin in pulp mill wastewaters by white-rot fungi on biofilm. Bioresour. Technol. 96, 1357–1363.

Xu, P., Ma, W., Han, H., Jia, S., Hou, B., 2015. Isolation of a naphthalene-degrading strain from activated sludge and bioaugmentation with it in a MBR treating coal gasification wastewater. Bull. Environ. Contam. Toxicol. 94, 358–364.

Yanbo, J., Jianyi, J., Xiandong, W., Wei, L., Lincheng, J., 2021. Bioaugmentation technology for treatment of toxic and refractory organic waste water based on artificial intelligence. Front. Bioeng. Biotechnol. 9, 1–12.

Yu, Z.T., Mohn, W.W., 2002. Bioaugmentation with the resin-acid degrading bacteria *Zoogloea resiniphila* DhA-35 to counteract pH stress in an aerated lagoon treating pulp and paper mill effluent. Water Res. 36, 2793–2801.

Zhang, Q.Q., Yang, G.F., Zhang, L., Zhang, Z.Z., GuangMing, T., Jin, R.C., 2017. Bioaugmentation as a useful strategy for performance enhancement in biological wastewater treatment undergoing different stresses-application and mechanisms. Crit. Rev. Environ. Sci. Technol. 47 (19), 1877–1899.

Zheng, Y., Chai, L.Y., Yang, Z.H., Tang, C.J., Chen, Y.H., Shi, Y., 2013. Enhanced remediation of black liquor by activated sludge bioaugmented with a novel exogenous microorganism culture. Appl. Microbiol. Biotechnol. 97, 6525–6535.

Zhu, X., Liu, R., Liu, C., Chen, L., 2015. Bioaugmentation with isolated strains for the removal of toxic and refractory organics from coking wastewater in a membrane bioreactor. Biodegradation 26, 465–474.

CHAPTER

Role of biochar and modified biochar for effective removal of heavy metals

2

Parul Chugh, Ritika Kapoor, Lashika Batra and Rachana Singh
Amity Institute of Biotechnology, Amity University, Noida, Uttar Pradesh, India

1. Introduction

With the development of industry and advancements in the anthropogenic activities, soil pollution is increasing day by day and has become a serious problem. Heavy metals are among the most harmful present in the soil. These contaminants are not only produced by the natural activities, but the major source of these are the potential human activities. Humans for their own benefit have destroyed the environment with activities like mining, refining, overuse of pesticides, etc. The excessive use of land and natural resources has resulted in a tremendous pressure on the ecological environment; consequently, these toxicants have been distributed even in the nonpolluted areas as heavy metals tend to spread via dust particles and also through sewage sludge contaminating the ecosystem. These heavy metals include selenium (Se), zinc (Zn), copper (Cu), arsenic (As), and lead (Pb). Heavy metals adversely affect the microbiota of soil and plants and also have severe effects on the health of human beings and animals (Bisaria et al., 2021; Christopher et al., 2012).

Owing to the harmful impacts of these heavy metals, it has become an important task for the researchers around the world to remove them by implying a variety of techniques including physical methods such as thermal extraction and steam leaching and chemical methods such as immobilization-stabilization techniques, redox reactions, organic matter improvement, etc., but the major problem with the use of chemicals include harming the environment in one way or the other. Therefore, it led to the discovery of techniques that involve the use of various biological agents such as plants, microbes, their constituents, their products, and the by-products for the removal of heavy metals. Bioremediation is one such technique which includes the removal of harmful pollutants from the environment by the use of biological agents such as plants and microbes (Singh and Sinha, 2013).

With the discovery of different remediation techniques involving the use of biological agents and the advancement in the study of other abiotic factors including soil properties, especially the effective properties of Amazon black soil as proposed by Lehmann (2007), it gave the researchers the idea of using biochar. It is an environment-friendly soil improver along with higher ecological and economic benefits. Biochar is a by-product of various biological wastes obtained by pyrolysis of the materials including agricultural products, their wastes, domestic wastes, and many other biological products found in our environment (Singh, 2012). Biochar is generally considered as the heterogeneous mixture

Integrative Strategies for Bioremediation of Environmental Contaminants, Volume Two. https://doi.org/10.1016/B978-0-443-14013-6.00006-8

of all the by-products left after the pyrolysis including char, charcoal, soot, graphite, and ash. Biochar contains a high content of recalcitrant aromatic forms of the organic carbon which has the irregularly arranged chains of six carbon bonded together having no oxygen or hydrogen in its structure. Biochar has a great adsorbing potential property because of its high surface area, increased capacity of volume holding by the pores, its functional groups, and also the high contents of the ash constituents.

Biochar has a great capacity for the extraction of the heavy metals from the soil without even degrading the quality of the soil. Heavy metals like copper, zinc, arsenic, lead, and mercury can be removed by the use of biochar developed by various different biomass such as hardwood, orange peel, municipal wastes, etc., and also it has the ability of removing various other toxicants including pesticides, dyes like methylene blue, etc., thereby reducing the toxicant content in the environment leading to a clean environment (Mehrotra et al., 2021). New techniques have been developed using the biochar by enhancing its properties through different methods like chemical oxidation, chemical reduction, or metal impregnation for the effective removal of the harmful toxicants from the environment (Fig. 2.1).

This chapter deals with various biochar materials being used as the technique for the effective removal of the heavy metals. It also focuses on various modifications done on biochar in order to enhance its adsorption capacity against various contaminants.

2. Heavy metal pollution in soil

Heavy metals are those metals whose density is relatively higher and have atomic number more than 20 and are very toxic even though their concentration is much less in the environment (Bisht et al., 2021).

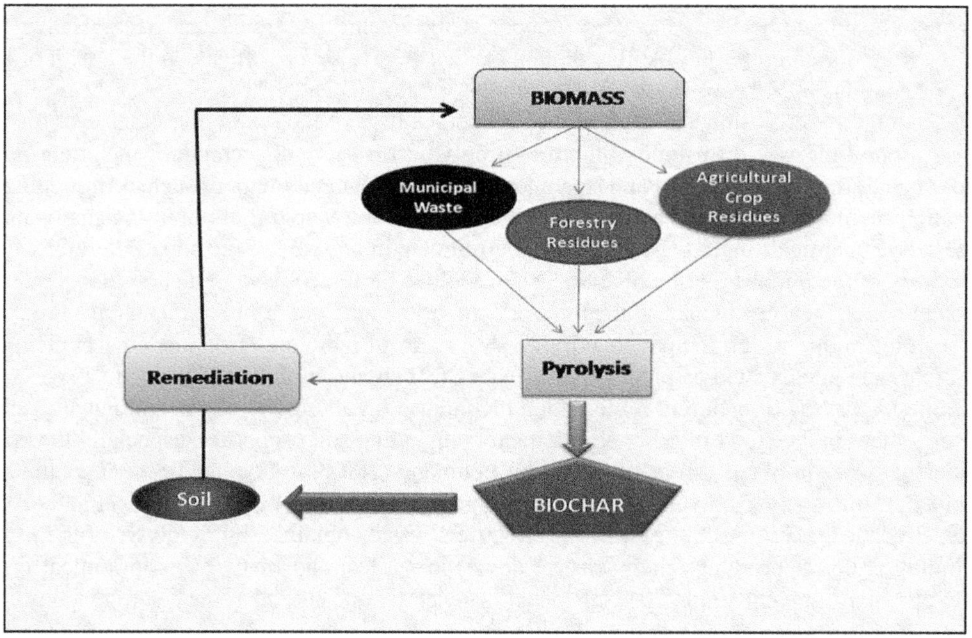

FIGURE 2.1

Biochar and soil remediation.

The heavy metals like copper (Cu), cadmium (Cd), zinc (Zn), lead (Pb), etc., are present in the soils. Their mobility is affected by the various factors including what complexes they form, in what ionic state they are present, and their oxidation-reduction properties (Uchimiya et al., 2010). These heavy metals cause a great damage to our environment including changes in the soil properties like fertility, leaching, etc., thereby affecting plant growth and crop yield. They also enter in the food chain severely disrupting other ecosystems. There are various health risks associated with human exposure to the soil contaminated with heavy metals.

Metals like copper and cadmium form different organic complexes with different organic nutrients found in the soil. The contamination of copper in the soil is the result of using sewage, municipal wastes manure, and different copper-containing fungicides which results in the deterioration of the soil quality and reduces the microflora population of the soil (Uchimiya et al., 2011). Similarly, cadmium also causes various hazardous effects on the environment leading to reduction in the plant nutrient content in the soil. Higher concentration of cadmium is the result of application of sewage sludge and pesticides on agricultural soils (Hayyat et al., 2016). Bioavailability of zinc and lead in the soils is also due to the extensive use of chemical pesticides and certain other activities like combustion of coal (Hayyat et al., 2016). Arsenic is one of the most toxic heavy metals that is found in our environment. Arsenic found in the soil causes damage to the soil biota and also affects the oxidative cycles of the cells and growth of the plants (Abdelhafez et al., 2014; Gregory et al., 2014). Nickel contamination in the soil is the result of sewage sludge and other agricultural activities.

These metals cause a great amount of damage to our environment, and it has become so important to minimize the accessibility and phytoavailability of heavy metals to plants along with the restoration of contaminated soil.

3. Biochar

Biochar is a by-product of the biomass material including different biological products including agricultural, domestic, and forestry waste, which is obtained as the result of pyrolysis. It contains high content of organic carbon in the form of charcoal, graphite, and ashes. The biochar in its chemical structure has different functional groups, which enhances its property of adsorption. Adsorption of particles is the reason for the biochar being one of the major adaptive methods for remediation of different harmful toxicants present in our environment. The biochar has an advantage because of the fact that it can form good interactions with the soil and water molecules and also remains in the environment for longer duration, as it cannot be easily degraded, which helps in the carbon dioxide efflux from the environment, thereby reducing earth warming (Lehmann and Joseph, 2015; Sánchez et al., 2009).

3.1 Preparation of Biochar

The preparation of biochar involves different processes including pyrolysis, hydrothermal carbonization (HTC), and microwave carbonization (Van Zwieten et al., 2009; Liu et al., 2013; Bisaria et al., 2022). These processes involve the use of heat in degrading the biomass materials like crop straw, poultry manure, and municipal sludge at high temperature and under anoxic environment so as to obtain high content of organic carbon (Zhang et al., 2021).

3.1.1 Pyrolysis

Pyrolysis is the main mechanism which is generally adapted for the production of biochar. This involves burning of the biomass at high temperatures above 300°C under oxygen-free conditions. It has some disadvantages such as the heat/energy losses, the uncontrolled nature of the process, unwanted side reactions leading to lower yield, superficial/nonselective heating, and longer residence times (Meyer et al., 2011; Yin, 2012; Tripathi et al., 2015; Tsubaki et al., 2012).

3.1.2 Hydrothermal carbonization

HTC is a thermochemical process that converts biomass into HTC-coal, a coal-like substance having a higher carbon content, generated by heating biomass at high temperature (180–220°C), suspended in water under saturated pressure for several hours. It has high conversion efficiency and can be operated at relatively low temperature. Hydrothermal treatment of biomass generates liquid (biooil), gaseous (mainly carbon dioxide), aqueous, and solid products (biochar). In a study conducted by Liu et al. (2013), coconut fiber and dead eucalyptus leaves biomass were used to generate biochar using HTC. Biomass was crushed to form particles of size less than 5 mm and then was dried at 105°C for 24 h. The dried biomass was loaded with deionized water into an autoclave reactor, and the autoclave was heated to reach the desired temperature (150–375°C). Once it reached the final temperature, it was held for 30 min at the same temperature and then was quickly cooled down to room temperature. The biochar can be easily recovered as a solid residue by vacuum filtration and then dried in an oven at 105°C for 24 h (Liu et al., 2013).

3.1.3 Microwave carbonization

In microwave heating, heat is generated in the material by the molecular interaction with electromagnetic waves (Guiotoku et al., 2014). Microwave heating is simpler as compared to the conventional method of heating as done in pyrolysis. It is cheaper, faster, and a more energy-efficient method for biochar preparation. Here, heat is produced throughout bulk of material in a way that microwave penetrates the material resulting in the conversion of microwave energy into internal energy (Mubarak et al., 2016). Moreover, it provides homogenous heating to the biomass, thereby reducing the reaction time as well (Nizamuddin et al., 2018).

3.2 Properties of Biochar

Biochar being a carbon-rich material is now being popularized for its advantageous use in the agricultural sector and in the restoration of the natural environment of an area. Its properties majorly include a highly porous structure having a good amount of carbon and other organic content. Depending upon the pore size, biochar can be divided into micropores (<2 nm), mesopores (2–50 nm), and macropores (>50 nm). Micropores and mesopores are located on the inner side of biochar and play a major role in adsorption, while macropores are located on the outer surface of biochar and act as a channel for transporting heavy metals to the inner pores since the adsorption takes place at the walls of biochar (Fahmi et al., 2018; Mandal et al., 2020). It has a large surface area which results in the good adsorptive capacity. The surface area increases with the increase in pyrolysis temperature (Li et al., 2018a,b). Biochar has the ability for reducing the carbon dioxide emission and also has the nutrient and pesticide uptake capacity, thus making this material efficient in variety of biological processes (Hayyat et al., 2016).

3.2.1 Factors affecting the properties of biochar

The major two factors that affect the biochar material and its properties are the feedstock used for its production and the methodology applied for its production. Biochar prepared from plant waste is enriched in elements like carbon, hydrogen, oxygen, nitrogen, phosphorus, sulfur, potassium, calcium, and magnesium. Pyrolyzed biochar is also rich in nitrogen and potassium. The increased pyrolysis temperature also increases the carbon content of biochar prepared from municipal waste (Zhang et al., 2021). Due to its enriched elemental content, it is very beneficial for promoting microbial growth. The major changes that are observed in the properties include the amount of organic and inorganic components, thereby changing the effective adsorptive nature of the biochar material. Other factors such as the quantity of biomass material taken for the production, the alteration in the pyrolysis conditions, the type of metal taken, or the type of soil used are the factors taken into the account for the biochar material and its effectiveness.

4. Role of biochar in effective removal of heavy metals

Biochar has been reported to have high pH value and contents of organic carbon, which has led the researchers to use biochar for remediation purpose. The content of different minerals including calcium, phosphorus, and magnesium has helped in the enhancement of the remediation properties. Some reports have also stated that biochar helps in immobilizing the heavy metal ions in the environment and transforming them into less toxic and bioaccessible forms (Sizmur et al., 2017). Various mechanisms have been reported by which biochar can be effectively used in eradicating the toxicants including ion exchange, precipitation, and complexation but majorly due to adsorption (Ahmad et al., 2014). The biochar fabricated from different sources utilized in the removal of various heavy metals through different mechanisms is summarized in Table 2.1.

Table 2.1 Different biochar sources for removal of heavy metals.

Biochar source	Metal removed	Mechanism involved	Remarks	References
Dairy manure	As^{5+}, Pb^{2+}, Cu^{2+}, and Zn^{2+}	Electrostatic adsorption, precipitation with PO_4^{3-} and CO_3^{2-}	Dairy manure adsorbs great amount of As^{5+}, and among the other three metals, Pb^{2+} was adsorbed easily.	Guan et al. (2013), Xu (2015)
Sludge biochar	Pb^{2+}	Precipitation and complexation	The rate of adsorption is quite high around 60%.	Lu et al. (2012)
Rice husk	Pb^{2+}, Cu^{2+}, Zn^{2+}, and Cd^{2+}	Formation of complex on surface with the phenolic hydroxyl group	Among all the metals, Pb^{2+} was adsorbed easily.	Xu (2015)

Continued

Table 2.1 Different biochar sources for removal of heavy metals.—cont'd

Biochar source	Metal removed	Mechanism involved	Remarks	References
Stinging nettle	As^{5+} and Cu^{2+}	Biochar sorption	Cu^{2+} leaching was reduced.	Sneath et al. (2013)
Poultry litter	Pb^{2+}, Cu^{2+}, and Cd^{2+}	Precipitation and chemiosorption	Pb^{2+} and Cd^{2+} were adsorbed easily.	Park et al. (2011)
Sewage sludge	Cd^{2+}	Precipitation	Conversion to less toxic state of Cd.	Zhang et al. (2016)
Grain husk	Zn^{2+}	Precipitation and surface sorption	The amount of zinc is reduced.	Kumar et al. (2018)
Maize straw	As^{5+}, Cd^{2+}, and As^{3+}	Electrostatic adsorption and precipitation	Maize straw adsorbs least amount of As^{5+}.	Guan et al. (2013), Zhang et al. (2016), Huang et al. (2014)
Orchard prunings	As^{3+}, Pb^{2+}, Cu^{2+}, Zn^{2+}, and Cd^{2+}	Metal mobility by altering soil pH	As^{3+} and Cu^{2+} adsorbed more.	Beesley et al. (2010)
Water hyacinth hardwood	Zn^{2+}	Electrostatic adsorption	The adsorption was good.	Wang et al. (2017)

4.1 Ion exchange

Cation exchange capacity (CEC) is the total number of cations that can be adsorbed on the surface of the biochar (Zhang et al., 2021). The positive charge on the soil surface and the negative charge on the surface of biochar material result in the electrostatic interaction which further result in the adsorption of the heavy metal cations from the soil. The retention power of the heavy metal becomes more when the number of cation exchange is more (Yang et al., 2019; Rees et al., 2013). The CEC of biochar is generally affected by the preparation temperature and biochar raw materials. The temperature also affects the integrity of functional groups ultimately affecting CEC. In a study conducted by Li et al. (2018b), it was reported that the straw biochar prepared at low temperature had higher reactive functional groups and hence showed higher CEC (Li et al., 2018b).

4.2 Precipitation

It is one of the primary mechanisms for removing inorganic pollutants from biochar. It entails the formation of mineral precipitates in the solution or on the surface of the sorbing material, particularly for biochar, which is produced by the pyrolysis of cellulose and hemicelluloses at temperatures greater than 300°C and has an alkaline property. Researchers discovered that biochar made from sugarcane and straw dust can improve Cd and Zn precipitation. However, they claimed that the efficiency of surface precipitation of biochar is dependent on the pyrolysis temperature, implying that more research in the optimization of the pyrolysis temperature is required in the future (Chang et al., 2019).

4.3 Complexation

This metal complexation mechanism includes the formation of multiatom complexes through the interaction of specific metal ligands. Because of the functional groups that contain oxygen in their structure, such as phenolic, lactonic, and carboxyl, biochar produced at low temperatures can bind with heavy metals. This oxygen content can increase the surface oxidation of biochar, resulting in increased metal complexation. When compared to biochar prepared from animals such as dairy manure and poultry litter, biochar prepared from vegetal biomass has a high efficiency in binding potentially metals such as Cu, Cd, Ni, and Pb to form metal complexes with carboxylic and phenolic functional groups. Scientists came to the conclusion that plant-derived biochar has a high surface complexation and ion exchange capacity (Xia et al., 2020).

4.4 Adsorption

The adsorption process is the most important phenomenon that occurs during remediation. The large pore size of the biochar material results in the adsorption of heavy metals with smaller pore sizes. The adsorption process is caused by the basic characteristic feature of biochar, that is, porosity, and a large surface area. Pore filling, hydrophobic interaction, partitioning, electrostatic interaction, and electron donor—acceptor (EDA) interaction are the various mechanisms involved in organic pollutant adsorption (Tomczyk et al., 2020).

4.4.1 Partitioning

The adsorbate material diffuses into the pores of the noncarbonized portion of the biochar during this process. This portion is easily able to interact with the organic adsorbate, resulting in its sorption (Pap et al., 2018).

4.4.2 Electron donor and acceptor interaction

The electron donor and acceptor interaction mechanism is commonly used in the adsorption of aromatic compounds on graphene-like biochar. A temperature of more than 1100°C should be reached during the biochar preparation to achieve complete graphitization (Inyang and Dickenson, 2015).

4.4.3 Hydrophobic interaction

Through partitioning and hydrophobic interaction processes, this mechanism can be used for the adsorption of hydrophobic and neutral organic compounds. The hydrophobic interaction mechanism requires less energy than the partitioning process (Stevens-Garmon et al., 2011).

5. Modifications of biochar in effective removal of heavy metals

With the advent of technology, the researchers began to modify the original biochar material with different chemical groups including oxygen containing functional groups such as −COOH and −OH (Li et al., 2018a,b), treating the biochar with alkali bases such as sodium hydroxide (NaOH) and potassium hydroxide (KOH) (Lyu et al., 2018), and also by the adsorption of certain metals such as iron (Fe) and silver (Hg) (Kong et al., 2021), so as to enhance the properties of the biochar material and to achieve higher adsorption rates. The modification property of biochar is affected by various factors including the type of the activator used, the time required for the activation, the time required for soaking the biochar material into the modification medium, and the activation temperature. The commonly used biochar modification mechanisms are physical, chemical, and biological. Physical modification includes grinding of the biochar to decrease the size, which increases the surface area and, hence, adsorption capacity of biochar. Biological methods of modification of biochar involve the biomass feedstock to undergo pretreatment biologically through aerobic or anaerobic digestion or by bacterial conversion, but so far, it has been reported using sugar beet tailings, bagasse, and corn straw silage (Tao et al., 2019a,b). Chemical methods involved for the modification of biochar include chemical reduction, chemical oxidation, and metal impregnation (Sun et al., 2020).

5.1 Chemical oxidation

The oxygen-containing functional groups like hydroxide (-OH) and carboxyl (-COOH) on the surface of biochar get oxidized with oxidants such as hydrochloric acid (HCL), thereby increasing the hydrophilicity of the biochar material, which results in the enhanced adsorption property of that material (Sajjadi et al., 2019).

5.2 Chemical reduction

In order to enhance the nonpolarity of the biochar material, certain reducing agents like NaOH and KOH are used to reduce the functional groups present on the biochar surface. This reaction helps in enhancing the adsorption capacity of the biochar material and also in increasing the porosity and surface area. This method is also known as the alkali modification method (Shafawi et al., 2021).

5.3 Metal impregnation

Certain metal ions such as iron (Fe) and metal heteroatoms get adsorbed on the surface of biochar, which results in the increase of the surface area and also in the enhancement of the adsorption capability of that biochar material (Othmani et al., 2021). Table 2.2 depicts the different biochar sources along with their modification reagents and mechanisms and the results of how effective they are in the heavy metal remediation.

Table 2.2 Different biochar modified with chemical reagents for removal of heavy metals.

Biochar source	Modification Reagent and mechanism	Metal removed	Modification effects	References
Dairy manure	Treatment with NaOH; chemical reduction	Pb and Cd	The effects include increased adsorption of the contaminants by the biochar.	Chen et al. (2019)
Municipal solid wastes	Treatment with KOH; chemical reduction	As	The surface area of the biochar increased which in turn increased adsorption.	Jin et al. (2014)
Bamboo hardwood	Treatment with NaOH and CS_2; chemical reduction	Cd	The sulfur modified biochar had more brittle structure.	Wu et al. (2019)

Table 2.2 Different biochar modified with chemical reagents for removal of heavy metals.—cont'd

Biochar source	Modification Reagent and mechanism	Metal removed	Modification effects	References
Coconut shell	Treatment with HCl; chemical oxidation	Cd, Ni, and Zn	The modified coconut biochar had superior qualities.	Liu et al. (2018)
Corn straw	Treatment with Na_2S and KOH; chemical reduction	Hg	The adsorption capacity of the modified biochar increased around 20%.	Tan et al. (2016)
Thalia dealbata	Treatment with $MgCl_2$	Cd	The adsorption capacity of the modified biochar increased around 1% and also the mobility of Cd decreased.	Tao et al. (2019a,b)
Hickory wood	Treatment with $KMnO_4$; oxidizing agent modification	Cd, Pb, and Cu	The adsorption capacity was enhanced.	Wang et al. (2015)
Swine manure	Treatment with MnO_2; oxidizing agent modification	Cd and Pb	The removal capacity increased on modification.	Liang et al. (2017).
Rice husk	Treatment with Fe(III); metal impregnation	As(III) and As(V)	The removal capacity increased.	Samsuri et al. (2013)
Auricularia auricular dreg (AAD)	Treatment with cetyltrimethyl ammonium bromide (CTAB)	Cr	The adsorption capacity of the modified biochar was increased.	Li et al. (2018a,b)

6. Biochar for soil improvement

Biochar being an organic material has number of applications in different fields including agriculture, environment sustainability, etc. (Novak and Busscher, 2012). With the increase in population, at a rapid rate, it has become a task to feed the increasing population, provide a roof over their head, and also give them a healthy lifestyle along with maintaining a proper and natural environment around the world. There are various benefits associated with biochar in maintaining a good and healthy environment in all spheres of nature including soil.

The use of biochar in the field of agriculture has helped in the enhancement of various properties of soil including fertility, nutrient content, pH of the soil, the water holding capacity, etc., and has remarkably helped in restoring the natural microenvironment of soil making it suitable for the growth of microorganisms. Biochar in its natural form is alkaline as it contains basic functional groups because due to pyrolysis, the temperature increases, thereby increasing the ash content and decreasing the elements like H and O. This results in an increase in its aromaticity, thereby enhancing the pH, making it alkaline. The alkaline nature of biochar can be used to neutralize the soil acidification. But, when the biochar is prepared at a lower temperature around 400°C, it is slightly acidic (Zhang et al., 2021). It was observed that on adding maize residue biochars to sandy loam and clayey texture calcareous soils, various attributes of soil like pH, electrical conductivity, total nitrogen content, plant-available potassium, and the fluorescein diacetate hydrolysis activity were enhanced significantly, while the plant-available phosphorus was decreased and these factors are largely influenced by soil texture, pyrolysis temperature, and application rate (Khadem et al., 2021). These properties further enhance the crop productivity, the rate of germination, and the other properties of soil and the crops. The biochar material has helped a lot in the removal of heavy metals and other harmful biological agents including various pesticides which destroys the crop and also affects the soil. The biochar material helps in enhancing the cation exchange property along with maintaining a high pH of the soil, which results in the high adsorption capacity of the soil. Biochar also aids in the reduction of atmospheric greenhouse gas (GHG) emissions, which is heavily influenced by feedstock particle size (Deng el al., 2021). The biochar material aids in the reduction of waste, both domestic and agricultural, because it can now be used in the treatment of pollution and other harmful toxicants found in the environment. Biochar has also aided in the reduction of various GHGs such as carbon dioxide, nitrogen dioxide, and others, as well as demonstrating its economic benefits due to its low production and handling costs.

7. Conclusion and future perspectives

A significant number of research have been carried out to look at the possibility of utilizing biochar to treat soils polluted by heavy metals and other organic contaminants. Biochar amendment can stabilize heavy metals and organic compounds in soil and attenuate soil pollution; however, performance varies by biochar source, amendment dose, soil type, pollutant species, and even biochar location in soil. Biochars from various sources have a wide range of physical and chemical quality features, as well as different capabilities and efficiencies for fixing heavy metals and organic pollutants in soil. Biochar does not eliminate heavy metals from soil; instead, soil-incorporated biochar converts the water-soluble and bioaccessible part of heavy metals into immobilized forms, lowering the hazardous elements' bioavailability and ecotoxicity. This is accomplished largely by raising the pH of the soil, adding carbonates and phosphates, and increasing surface sorption. Electrostatic attraction, ion exchange-based surface adsorption, surface complexation, and coprecipitation are the basis of detoxification by biochar.

Biochar is more economically accessible and can provide additional environmental advantages such as carbon sequestration and soil health improvement when properly applied, even though its stabilization effectiveness may not be comparable to other agents such as lime, phosphate salts, and activated carbon. Biochar amendments might be used in mildly contaminated farmland to ensure safe food production, in mine land restoration initiatives to encourage vegetation, and in soil bioremediation and phytoextraction projects to speed up the process. However, the pollutant-stabilizing impact of biochar in agricultural soils may wane over time. Research at higher scales is definitely required to investigate the long-term impact of biochar amendment on soil pollution mitigation in real-world situations.

References

Abdelhafez, A.A., Li, J., Abbas, M.H.H., 2014. Feasibility of biochar manufactured from organic wastes on the stabilization of heavy metals in a metal smelter contaminated soil. Chemosphere.

Ahmad, M., Rajapaksha, A.U., Lim, J.E., Zhang, M., Bolan, N., Mohan, D., Vithanage, M., Lee, S.S., Ok, Y.S., 2014. Biochar as a sorbent for contaminant management in soil and water: a review. Chemosphere.

Beesley, L., Moreno-Jiménez, E., Gomez-Eyles, J.L., 2010. Effects of biochar and greenwaste compost amendments on mobility, bioavailability and toxicity of inorganic and organic contaminants in a multi-element polluted soil. Environ. Pollut. 158.

Bisaria, K., Singh, R., Gupta, M, Mathur, A., Dixit, A., 2022. Novel acoustic-activated alkali-functionalized *Trapa bispinosa* peel biochar for green immobilization of chlorpyrifos from wastewater: artificial intelligence modelling and experimental validation. Biomass Conv. Bioref.

Bisaria, K., Sinha, S., Singh, R., Iqbal, H.M., 2021. Recent advances in structural modifications of photo-catalysts for organic pollutants degradation—a comprehensive review. Chemosphere 131263.

Bisht, D., Sinha, S., Nigam, S., Bisaria, K., Mehrotra, T., Singh, R., 2021. Adsorptive decontamination of paper mill effluent by nano fly ash: response surface methodology, adsorption isotherm and reusability studies. Water Sci. Technol. 83 (7), 1662—1676.

Chang, R., Sohi, S.P., Jing, F., Liu, Y., Chen, J., 2019. A comparative study on biochar properties and Cd adsorption behavior under effects of ageing processes of leaching, acidification and oxidation. Environ. Pollut. 254, 113123.

Chen, Z.-L., Zhang, J.-Q., Huang, L., Yuan, Z.-H., Li, Z.-J., Liu, M.C., 2019. Removal of Cd and Pb with biochar made from dairy manure at low temperature. J. Integr. Agric. 18 (1). https://doi.org/10.1016/S2095-3119(18)61987-2.

Christopher, D., Kaur, S., Singh, R., 2012. Water quality status of river Yamuna in Delhi with reference to presence of heavy metals : A review. J. Pharm. Med. Bio. Sci. 1.

Deng, B., Yuan, X., Siemann, E., Wang, S., Fang, H., Wang, B., Gao, Y., Shad, N., Liu, X., Zhang, W., Guo, X., Zhang, L., 2021. Feedstock particle size and pyrolysis temperature regulate effects of biochar on soil nitrous oxide and carbon dioxide emissions. Waste Manag. 120, 33—40.

Fahmi, A.H., Samsuri, A.W., Jol, H., Singh, D., 2018. Physical modification of biochar to expose the inner pores and their functional groups to enhance lead adsorption. RSC Adv. 8 (67), 38270—38280.

Gregory, S.J., Anderson, C.W.N., Arbestain, M.C., McManus, M.T., 2014. Response of plant and soil microbes to biochar amendment of an arsenic-contaminated soil. Agric. Eco. Environ. 191, 133—141. https://doi.org/10.1016/j.agee.2014.03.035.

Guan, L.Z., Zhou, J.J., Zhang, J., et al., 2013. Effects of biochar produced from different sources on arsenic adsorption and desorption in soil. Chin. J. Appl. Ecol. 24 (10), 2941—2946 (Chinese). PMID: 24483091.

Guiotoku, Marcela, Rambo, Hatzo, et al., 2014. Charcoal produced from cellulosic raw materials by microwave-assisted hydrothermal carbonization. J. Therm. Anal. Calorim. https://doi.org/10.1007/s10973-014-3676-8. In press.

Hayyat, A., Javed, M., Rasheed, I., Ali, S., Shahid, M.J., Rizwan, M., Javed, M.T., Ali, Q., 2016. Role of biochar in remediating heavy metals in soil. Phytoremediation 421—437.

Huang, D.K., Li, X.Q., Dong, Z.Q., et al., 2014. Soil environmental influence of biochar and its application in soil heavy metal restoration. Gui Zhou Agric. Sci. 42 (11), 159—165. http://www.gznykx.org.cn/Jweb_gznykx.

Inyang, M., Dickenson, E., 2015. The potential role of biochar in the removal of organic and microbial contaminants from potable and reuse water: a review. Chemosphere 134, 232—240.

Jin, H., Capareda, S., Chang, Z., Gao, J., Xu, Y., Zhang, J., 2014. Biochar pyrolytically produced from municipal solid wastes for aqueous As(V) removal: adsorption property and its improvement with KOH activation. Bioresour. Technol.

Khadem, A., Raiesi, F., Besharati, H., et al., 2021. The effects of biochar on soil nutrients status, microbial activity and carbon sequestration potential in two calcareous soils. Biochar 3, 105–116. https://doi.org/10.1007/s42773-020-00076-w.

Kong, S., Tang, J., Ouyang, F., Chen, M., 2021. Research on the treatment of heavy metal pollution in urban soil based on biochar technology. Environ. Technol. Innovat. 23, 101670.

Kumar, A., Joseph, S., Tsechansky, L., Privat, K., Schreiter, I.J., Schüth, C., et al., 2018. Biochar aging in contaminated soil promotes Zn immobilization due to changes in biochar surface structural and chemical properties. Sci. Total Environ. 626, 953–961. ISSN 0048-9697. https://doi.org/10.1016/j.scitotenv.2018.01.157.

Lehmann, J., 2007. A handful of carbon. Nature 447, 143–144. https://doi.org/10.1038/447143a.

Lehmann, J., Joseph, S., 2015. Biochar for Environmental Management: Science, Technology and Implementation, 8 (9).

Li, H., Cao, Y., Zhang, D., Pan, B., 2018a. pH-dependent KOW provides new insights into understanding the adsorption mechanism of ionizable organic chemicals on carbonaceous materials. Sci. Total Environ. 618, 269–275.

Li, Y., Wei, Y., Huang, S., Liu, X., Jin, Z., Zhang, M., Qu, J., Jin, Y., 2018b. Biosorption of Cr(VI) onto *Auricularia auricula* dreg biochar modified by cationic surfactant: characteristics and mechanism. J. Mol. Liq. 269, 824–832. https://doi.org/10.1016/j.molliq.2018.08.060.

Liang, J., Li, X., Yu, Z., Zeng, G., Luo, Y., Jiang, L., Yang, Z., Qian, Y., Wu, H., 2017. Amorphous MnO_2 modified biochar derived from aerobically composted swine manure for adsorption of Pb (II) and Cd (II). ACS Sustain. Chem. & Eng.

Liu, Z., Quek, A., Hoekman, S.K., Balasubramanian, R., 2013. Production of solid biochar fuel from waste biomass by hydrothermal carbonization. Fuel 103, 943–949.

Liu, H., Xu, F., Xie, Y., Wang, C., Zhang, A., Li, L., Xu, H., 2018. Effect of modified coconut shell biochar on availability of heavy metals and biochemical characteristics of soil in multiple heavy metals contaminated soil. Sci. Total Environ.

Lu, H., Zhang, W., Yang, Y., 2012. Relative distribution of Pb2+ sorption mechanisms by sludge-derived biochar. Water Res.

Lyu, H., Gao, B., He, F., Zimmerman, A.R., Ding, C., Tang, J., Crittenden, J.C., 2018. Experimental and modeling investigations of ball-milled biochar for the removal of aqueous methylene blue. Chem. Eng. J. 335, 110–119.

Mandal, Pu, He, Ma, Hou, et al., 2020. Biochar induced modification of graphene oxide & nZVI and its impact on immobilization of toxic copper in soil. Environmental Pollution 259. https://doi.org/10.1016/j.envpol.2019.113851.

Mehrotra, T., Dev, S., Banerjee, A., Chatterjee, A., Singh, R., Aggarwal, S., 2021. Use of immobilized bacteria for environmental bioremediation: a review. J. Environ. Chem. Eng. 105920.

Meyer, S., Glaser, B., Quicker, P., 2011. Technical, economical,and climate-related aspects of biochar production tech-nologies: a literature review. Environ. Sci. Technol. 45, 9473–9483.

Mubarak, N.M., Nizamuddin, S., Tiripathi, M., Jayakumar, S.N., Sahu, J.N., Ganesan, P., 2016. Chemical, dielectric and structural characterization of optimized hydrochar produced from hydrothermal carbonization of palm shell. Fuel 163, 88–97. https://doi.org/10.1016/j.fuel.2015.08.057.

Nizamuddin, S., Baloch, H.A., Siddiqui, M.T.H., Mubarak, N.M., Tunio, M.M., Bhutto, A.W., Jatoi, A.S., Griffin, G.J., Srinivasan, M.P., 2018. An overview of microwave hydrothermal carbonization and microwave pyrolysis of biomass. Rev. Environ. Sci. Biotechnol. 17 (4), 813–837.

Novak, J.M., Busscher, W.J., 2012. Selection and use of designer biochars to improve characteristics of southeastern USA Coastal Plain degraded soils. In: Lee, J.E. (Ed.), Advanced Biofuels and Bioproducts.

Othmani, A., John, J., Rajendran, H., Mansouri, A., Sillanpää, M., Chellam, P.V., 2021. Biochar and activated carbon derivatives of lignocellulosic fibers towards adsorptive removal of pollutants from aqueous systems: critical study and future insight. Separ. Purif. Technol. 274, 119062.

Pap, S., Bezanovic, V., Radonic, J., Babic, A., Saric, S., Adamovic, D., Sekulic, M.T., 2018. Synthesis of highly-efficient functionalized biochars from fruit industry waste biomass for the removal of chromium and lead. J. Mol. Liq. 268, 315–325.

Park, J.H., Choppala, G.K., Bolan, N.S., Chung, J.W., Cuasavathi, T., 2011. Biochar reduces the bioavailability and phytotoxicity of heavy metals. Plant Soil 348.

Rees, F., Simonnot, M.O., Morel, J.L., 2013. Short-term effects of biochar on soil heavy metal mobility are controlled by intra-particle diffusion and soil pH increase. Eur. J. Soil Sci. 65 (1), 149–161.

Sajjadi, B., Zubatiuk, T., Leszczynska, D., Leszczynski, J., Chen, W.Y., 2019. Chemical activation of biochar for energy and environmental applications: a comprehensive review. Rev. Chem. Eng. 35 (7), 777–815.

Samsuri, A.W., Sadegh-Zadeh, F., Seh-Bardan, B.J., 2013. Adsorption of As(III) and As(V) by Fe coated biochars and biochars produced from empty fruit bunch and rice husk. J. Environ. Chem. Eng.

Sánchez, M.E., Lindao, E., Margaleff, D., Martínez, O., Morán, A., 2009. Pyrolysis of agricultural residues from rape and sunflowers: production and characterization of bio-fuels and biochar soil management. J. Anal. Appl. Pyrolysis 85, 142–144.

Shafawi, A.N., Mohamed, A.R., Lahijani, P., Mohammadi, M., 2021. Recent advances in developing engineered biochar for CO_2 capture: an insight into the biochar modification approaches. J. Environ. Chem. Eng. 106869.

Singh, R., 2012. Selective separation of Lead ions from waste water by immobilized Chlorella based biosorbents. Int. J. Chem. Environ. Eng. 3 (2), 108–114. ISSN 2078-0737.

Singh, R., Sinha, S., 2013. Bioremediation of heavy metals by algae: a review on evaluation of low cost and high performance biosorbents. Indian J. Agric. Biochem. 26 (1), 1–9. ISSN 0970-6399.

Sizmur, T., Fresno, T., Akgül, G., Frost, H., Moreno-Jiménez, E., 2017. Biochar modification to enhance sorption of inorganics from water. Bioresour. Technol. 246, 34–47.

Sneath, H.E., Hutchings, T.R., de Leij, F.A., 2013. Assessment of biochar and iron filing amendments for the remediation of a metal, arsenic and phenanthrene co-contaminated soil. Environ. Pollut. 178, 361–366. https://doi.org/10.1016/j.envpol.2013.03.009.

Stevens-Garmon, J., Drewes, J.E., Khan, S.J., McDonald, J.A., Dickenson, E.R., 2011. Sorption of emerging trace organic compounds onto wastewater sludge solids. Water Res. 45 (11), 3417–3426.

Sun, Y., Chen, S.S., Lau, A.Y., Tsang, D.C., Mohanty, S.K., Bhatnagar, A., et al., 2020. Waste-derived compost and biochar amendments for stormwater treatment in bioretention column: co-transport of metals and colloids. J. Hazard Mater. 383, 121243.

Tan, G., Sun, W., Xu, Y., Wang, H., Xu, N., 2016. Sorption of mercury (II) and atrazine by biochar, modified biochars and biochar based activated carbon in aqueous solution. Bioresour. Technol. 211, 727–735. https://doi.org/10.1016/j.biortech.2016.03.147.

Tao, Q., Chen, Y., Zhao, J., Li, B., Li, Y., Tao, S., Li, M., Li, Q., Xu, Q., Li, Y., Li, H., Li, B., Chen, Y., Wang, C., 2019a. Enhanced Cd removal from aqueous solution by biologically modified biochar derived from digestion residue of corn straw silage. Sci. Total Environ. 674, 213–222.

Tao, Q., Li, B., Li, Q., Han, X., Jiang, Y., Jupa, R., Wang, C., Li, T., 2019b. Simultaneous remediation of sediments contaminated with sulfamethoxazole and cadmium using magnesium-modified biochar derived from Thalia dealbata. Sci. Total Environ. 659, 1448–1456. https://doi.org/10.1016/j.scitotenv.2018.12.361.

Tomczyk, A., Sokołowska, Z., Boguta, P., 2020. Biochar physicochemical properties: pyrolysis temperature and feedstock kind effects. Rev. Environ. Sci. Biotechnol. 19 (1), 191–215.

Tripathi, M., Sahu, J., Ganesan, P., Dey, T., 2015. Effect of tem-perature on dielectric properties and penetration depth ofoil palm shell (OPS) and OPS char synthesized by micro-wave pyrolysis of OPS. Fuel 153, 257–266.

Tsubaki, S., Oono, K., Onda, A., Yanagisawa, K., Azuma, J-i, 2012. Microwave-assisted hydrothermal hydrolysis of cellobioseand effects of additions of halide salts. Bio Technol. 123, 703–706.

Uchimiya, M., Lima, I.M., Klasson, K.T., Wartelle, L.H., 2010. Contaminant immobilization and nutrient release by biochar soil amendment: roles of natural organic matter. Chemosphere.

Uchimiya, M., Klasson, K.T., Wartelle, L.H., Lima, I.M., 2011. Influence of soil properties on heavy metal sequestration by biochar amendment, copper desorption isotherms. Chemosphere.

Van Zwieten, L., Kimber, S., Morris, S., Chan, K.Y., Downie, A., Rust, J., Joseph, S., Cowie, A., 2009. Effects of biochar from slow pyrolysis of papermill waste on agronomic performance and soil fertility. Plant Soil 327 (1–2), 235–246.

Wang, H., Gao, B., Wang, S., Fang, J., Xue, Y., Yang, K., 2015. Removal of Pb(II), Cu(II), and Cd(II) from aqueous solutions by biochar derived from $KMnO_4$ treated hickory wood. Bioresour. Technol. 197, 356–362.

Wang, H., Xia, W., Lu, P., et al., 2017. Adsorption characteristics of biochar on heavy metals of Pb and Zn in Soil. Environ. Sci. J. Integr. Environ. Res. 38 (9), 3944–3952.

Wu, C., Shi, L., Xue, S., Li, W., Jiang, X., Rajendran, M., Qian, Z., 2019. Effect of sulfur-iron modified biochar on the available cadmium and bacterial community structure in contaminated soils. Sci. Total Environ. 647, 1158–1168.

Xia, S., Song, Z., Jeyakumar, P., Bolan, N., Wang, H., 2020. Characteristics and applications of biochar for remediating Cr (VI)-contaminated soils and wastewater. Environ. Geochem. Health 42 (6), 1543–1567.

Xu, X.Y., 2015. The Sorption and Transformation of Inorganic Contaminants by Biochar and the Underlying Mechanisms. Shanghai Jiao Tong University.

Yang, X., Zhang, S., Ju, M., Liu, L., 2019. Preparation and modification of biochar materials and their application in soil remediation. Appl. Sci. 9 (7), 1365. Available at: https://www.mdpi.com/2076-3417/9/7/1365/pdf. (Accessed 24 April 2020).

Yin, C., 2012. Microwave-assisted pyrolysis of biomass for liq-uid biofuels production. Biores. Technol. 120, 273–284.

Zhang, Y., Chen, T., Liao, Y., Reid, B.J., Chi, H., Hou, Y., et al., 2016. Modest amendment of sewage sludge biochar to reduce the accumulation of cadmium into rice (*Oryza sativa* L.): a field study. Environ. Pollut. 216.

Zhang, Y., Wang, J., Feng, Y., 2021. The effects of biochar addition on soil physicochemical properties: a review. Catena 202, 105284.

CHAPTER

3 Potential of microbes for the remediation of heavy metals—contaminated soil

Syed Nikhat Ahmed and Iswar Baitharu
P.G. Department of Environmental Sciences, Sambalpur University, Burla, Odisha, India

1. Introduction

Due to rapid urbanization, continuous expansion of mining and mineral industries, and the application of fertilizers and pesticides in agricultural farms mostly in developing nations, numerous harmful persistent organic compounds and heavy metals (HMs) are getting added to the soil system. As heavy metals are elemental in nature that do not undergo further degradation and can easily move from one environmental compartment, pollution due to heavy metals in water and soil has become an environmental problem worldwide (Wang et al., 2020). Most of the toxic environmental pollutants released to the environment due to anthropogenic activities finally reach the soil system as it is the ultimate sink for them. Depending upon the physical and chemical nature of numerous inorganic and organic pollutants, the seriousness of problems concerning public and environmental health varies. The presence of these toxic compounds in the soil directly influences the physicochemical and biological equilibrium of the soil. Heavy metals can easily enter the food chain and undergo magnification from trophic level to trophic level and ultimately reach to the human being in elevated concentration. When the concentration of pollutants is above a defined legal standard value, water, air, and soil are described as contaminated (Horta et al., 2015).

The parental materials such as rocks are the primary source of heavy metals in soils from which they are originated. Of the total Earth's crust, 95% is made up of igneous rocks and about 5% of sedimentary rocks (Sarwar et al., 2016). In general, basaltic igneousrocks are rich in heavy metals such as copper (Cu), cadmium (Cd), nickel (Ni), and cobalt (Co), whereas shales contain large amounts of lead (Pb), copper (Cu), zinc (Zn), manganese (Mn), and cadmium (Cd). Heavy metals present in rocks can enter the soil environment through natural processes such as meteoric, biogenic, terrestrial, and volcanic processes like erosion, leaching, and surface winds. The disturbance of nature's slowly occurring geochemical cycle of heavy metals by anthropogenic processes such as ore melting, smelting, mining, overuse of fertilizers, pesticides, and sewage sludge results in the accumulation of one or more of heavy metals in the soil (Dixit et al., 2015). The main group is heavy metals that are not biodegradable and can easily accumulate in living organisms. Heavy metal(loid)s present in soils include nickel (Ni), lead (Pb), cadmium (Cd), arsenic (As), chromium (Cr), copper (Cu), cobalt (Co), zinc (Zn), manganese (Mn), aluminum (Al), and mercury (Hg). Among these metal(loid)s, As, Pb, Cd,

Integrative Strategies for Bioremediation of Environmental Contaminants, Volume Two. https://doi.org/10.1016/B978-0-443-14013-6.00005-6

and Hg are included in the top 20 Hazardous Substances of the Agency for Toxic Substances and Disease Registry and the United States Environmental Protection Agency (Khalid et al., 2017). Polluted soils significantly affect soil microbial communities by decreasing the biomass of soil microbes and the number of active bacteria colonies (Long et al., 2021).

The injudicious release of several toxic substances affects different soil properties adversely. As heavy metals are not biodegradable, they accumulate in the topsoil. Not only do they interfere with the physical and chemical properties of soil but also they appear in the biocenosis of soil. Toxicity of different heavy metals accumulated in the topsoil leads to disturbance in microbial diversity inhabiting. As soil health is primarily governed by the microbial community and their activity, it leads to imbalance in various soil chemical properties such as extreme pH, high electrical conductivity, micronutrient unavailability, low nitrogen and phosphorous availability, altered C/N ratio, low soil organic matter, and low fertility and toxicity. Heavy metal toxicity also indirectly affects different soil physical properties such as lower organic matter content, reduced water holding capacity, high bulk density, and low soil porosity. The deteriorated soil health owing to altered physical and chemical properties leads to stunted growth of both plants and microbes. Different physiological activities such as respiration, protein synthesis, cell membrane integration, and enzymatic activity of plants and microbes are affected (Ramamoorthy, 2015).

Globally, there are 5 million sites of soil pollution covering 500 million ha of land, in which the soils are contaminated by different heavy metals or metalloids, with the present soil concentrations higher than the geo-baseline or regulatory levels (Liu et al., 2018). In Europe, there are approximately 2.5 million sites that are potentially contaminated by heavy metals and organic contaminants (EEA, 2022). To promote human well-being and ensure healthy lives, the United Nations has set an ambitious sustainable development goal to reduce illnesses and deaths associated with soil contamination by 2030 (UN, 2015). Toxic metals and metalloids in soil have aroused much concern due to their toxic effects on human beings and ecosystems (O'Connor et al., 2020). Therefore, to achieve the goal of sustainable development, it is necessary to seek sustainable approaches for soil heavy metal remediation.

There are various approaches using which heavy metals can be removed or stabilized. Conventional remediation technologies, including portland cement-based solidification/stabilization (Shen et al., 2019), containment (Liu et al., 2018), soil washing (Zhang et al., 2010), thermal desorption (Park et al., 2015), and chemical oxidation/reduction (McCann et al., 2018), have been proven effective for the immobilization, removal, or the transformation/detoxification of heavy metals. However, they have been criticized due to environmental concerns such as high energy consumption, greenhouse gas emission, long-term metal leaching risks, and air pollution, economic concerns such as high capital cost, and social concerns such as low public acceptance. Therefore, green, cheap, and sustainable approaches need to be explored for large-scale applications.

2. Microbes: the invisible regulators of soil health

The term "soil health" was first introduced in the mid-1990s. Various scientists gave various definitions of soil health in different time. Doran and Safley (1997) proposed the most accepted definition. The definition states that "The continued capacity of soil to function as a vital living system, within ecosystem and land-use boundaries, to sustain biological productivity, promote the quality of air and water environments, and maintain plant, animal and human health." The above definition extensively includes a time component. Hence it reflects the importance of soil as continuous function over time owing to its dynamic nature and relation with plant (Sharma et al., 2016).

Soil is home to diverse rhizospheric microorganisms. The various microorganisms play a significant role in conditioning the soil. They not only provide nutrition to the microorganisms living in the soil but also increase buffering and filtering capacity of the soil system (Kumar et al., 2017). Hence rhizospheric microorganisms are considered vital for soil health regulation. They control nutrient cycling and regulate soil—plant system. They also play a crucial role in genesis and maintenance of soil structure. Hence description of soil health cannot be carried out without mention of soil microbial diversity. Soil microorganisms are an excellent indicator of soilhealth status. In terms of nutrient dynamics, root growth, flow of the nutrient through various compartments, and storage of nutrients are governed by soil biota under a healthy soil—plant system. Prior to utilization by the crop, even fertilization undergoes different biological transformations by the potential rhizospheric microbial species (Dubey et al., 2016).

2.1 Mineralization and humification

Organic matter present in the soil is in the most stable form. Microorganisms facilitate the availability of the organic matter through decomposition. Dead and decayed remains of plant and animals, and excretory material of living beings constitute the best source of organic matter in the soil. In order to be utilized by the plants, the constituents of the organic matter must be converted into simpler inorganic forms (minerals). The process of conversion of stable complex organic matter into simpler inorganic substances is called mineralization. The mineralization process is primarily facilitated by soil microorganism fungi and bacteria (Xu et al., 2015). The subsequent release of water, gases, and energy upon mineralization results in the formation of amorphous substance called humification. In plant communities, biological oxidation of sulfur, nitrogen, carbon, phosphorus, etc., is of utmost importance during mineralization. Hence the processes of mineralization and humification run parallelly in nature.

2.2 Microbial decomposition of organic matter and nutrient availability

Soil microorganisms produce extracellular hydrolytic enzymes, which entail enzymatic degradation of organic matter. Soil organic matter acts as a sink for many essential elements such as sulfur, nitrogen, and phosphorus. The important soil enzymes secreted by microorganisms are dehydrogenase, urease, amylase, arylsulfatase, cellulase, phosphatase, and chitinase. Several environmental factors such as temperature and microbial diversity affect the rate of decomposition of soil organic factor. Therefore, organic matter—amended soil improves many physical properties such as soil texture, C/N ratio, etc., which facilitate the growth of plants.

2.3 Nitrogen fixation

Nitrogen (N) is the most essential nutrient for plant growth. Although, it exists in molecular form and is not directly available to the plant use. It is available to plants through biological nitrogen fixation to ammonia. The process of nitrogen fixation is carried out by oxygen-sensitive, nitrogenase enzyme complex. Nitrogenase enzymes convert atmospheric nitrogen into ammonia using ATP as energy source. Different microbial species from various genera have the ability of nitrogen fixation such as Bacillus, Azospirillum, Pseudomonas, Enterobacter, Flavobacterium, Erwinia, and Rhizobium (Silva et al., 2016). Nitrogen-fixing microbes are categorized into (1) symbiotic nitrogen-fixing bacteria and (2) nonsymbiotic nitrogen-fixing bacteria. The symbiotic nitrogen-fixing microbes include members of rhizobia which form a symbiotic association with leguminous plants. The nonsymbiotic bacteria include Azospirillum, Azotobacter, *Gluconacetobacter diazotrophicus*, and cyanobacteria. Most of the atmospheric nitrogen is fixed by symbiotic association, while nonsymbiotic contributes only to little extent. Some of these microbes for root nodules with their respective host plant root product play a significant role in communication between plants and microbes. Microbes are more tolerant than host plants to drought and salinity. Nitrogenase enzymes responsible for nitrogen fixation are highly sensitive to salinity and their activity highly reduced under stress conditions. The microorganism residing in the environment can cause a dramatic change in plant growth and production through various regulators.

2.4 Phosphate solubilization

Phosphorus is an essential micronutrient and he most vital element for plant growth and development. Plants acquire phosphorus in the form of phosphate ions as most of the phosphorus present in the soil is in insoluble form and unavailable to the plants. Plant growth-promoting rhizobacteria (PGPR) have phosphate-solubilization ability. Phosphorus-solubilizing microbes (PSMs) govern the biogeochemical cycle in natural agriculture systems. The bioavailability of phosphate to plants depends on plants, microbes, and surrounding environment. The association of plant microbes could enhance the mobilization of phosphorus in soil and available to plants (Mehta et al., 2013). PSMs such as bacteria and fungi mobilize it by producing organic acid and phosphatase. Numbers of microbes have the capacity to solubilize phosphate.

3. Microbial system and the issue of heavy metal pollution in soil

3.1 Impact of heavy metals on soil microbiota

Metals can fill in as imperative segments in life forms at low concentrations. It regularly serves critical capacities in catalyst efficiency. In any case, over certain limit fixations, metals can end up noticeably lethal to numerous species. Lethal metals apply their poisonous quality in various ways including the relocation of fundamental metals from their typical restricting destinations on organic atoms, hindrance of enzymatic working, and interruption of nucleic corrosive structure. The growth and function of microorganisms as well as the community composition and diversity in soil ecosystems can be severely affected by pollution (Chen et al., 2014). Soil microorganisms, both free-living and symbiotic soil microbes in the rhizosphere of plants growing on metal-contaminated soils, can increase plant biomass production and enhance the phytoremediation process. However, heavy metals affect the growth, morphology, and metabolism of soil microorganisms, through functional disturbance, protein denaturation, or the destruction of the integrity of cell membranes (Deng et al., 2007). The elevated levels of heavy metals in soils have significant impacts on the population size and overall activity of the soil microbial communities. Several studies, depending on the isolation-based techniques used, have revealed that heavy metal contamination gave rise to shifts in microbial population (Zhang et al., 2010). Microorganisms are more sensitive to heavy metal stress than the other soil animals or plants growing on the same soil. Luckily, microorganisms can influence the reactivity and portability of metals.

3.2 Mineralization; material cycling

Heavy metal pollution in the soil, and the bioconcentration and transfer features of the soil—vegetation (root, stem, and leaf) system are not wellknown. Currently, progressing works aim to discover the characteristics of the transfer and bioconcentration of unwanted elements in the environment. Investigations on the status of heavy metal pollution within the ecosystem, trophodynamics, bioconcentration factor (BCF), and the translocation factor (TF) of trace metals in the food chain are imperative to implement habitat protection and appropriate conservation measures. To estimate the transferability of heavy metals from soil to crop, two indices, the BCF and the TF can be used.

$$\mathrm{BCF} = \mathrm{Croot}/\mathrm{Csoil}, \tag{3.1}$$

where Croot and Csoil are the concentration of each heavy metal in the crop root and the corresponding value in the soil, respectively, but also BCF can be expressed as the general ratio of concentration of toxic elements in the biomonitor (plant/animal) and concentration of toxic elements in water or soil. While the TF = Cgrain/Croot, where Cgrain is the amount of heavy metal in the crop grain and expresses the transfer within the plant. Biomagnification of heavy metals in organisms from different trophic levels (predator—prey relationship) can be quantified and evaluated by calculating the biomagnification factor (BF). BF = Cprey/Cpredator, where: Cprey is the concentration of toxic elements in the prey and Cpredator is the concentration of toxic elements in the predator.

3.3 Resistance/adaptative mechanisms developed in microbes to survive in HM-contaminated soil

Various microbes have tolerance capacity for heavy metals through various mechanisms such as compartmentalization, production of siderophores, and regulatory transport (Long et al., 2021). Recent advances in the field of molecular and genetic tool have contributed to a renewed understanding of the basis for heavy metal tolerance. Roux et al. (2012) examined the role that heavy metal—tolerant genes play showed that heavy metal tolerance can evolve rapidly and repeatedly in population exposed to heavy metals. Coubot et al. (2005) observed that major genes are often responsible for heavy metal tolerance. Therefore, exposure to heavy metals can lead to the evolution of new proteins through mutation. Studies on the genetics of heavy metals tolerance contribute greatly in remediation of polluted sites by the discovering of potential genes associated with remediation (Mamiroli and Maestri, 2008). Microbes may have genes that confer metal tolerance which might be transferred to relatively distantly related species.

4. Mechanisms involved in HM remediation by microbes

The remediation of the soil refers to the process of revitalizing the soil by the removal of contaminants or pollutants. Soil bioremediation is the process in which the restoration of polluted soil is carried out by living microorganisms. They are two methods by which microorganisms carryout bioremediation. They either release enzymes that degrade contaminants or develop various resistance mechanisms. When soil is contaminated with heavy metals, uptake of heavy metal is either energy-intensive or passive process. In case of active process, heavy metal uptake is carried out at the expense of energy through the process of bioaccumulation. Uptake of heavy metals is also done without any energy expenditure through the process of adsorption. The process can also be a combination of both accumulation and adsorption. After getting separated from the environment, the metal ions easily bind with different functional groups such as amino, hydroxyl, carboxylate, and phosphate. These metal ions are found within the cell wall of the microbes itself. The mechanisms of metal ions interaction with the microbial cell wall surface are van der Waals forces, covalent bonding, and electrostatic forces (Rajendran et al., 2003).

Those microorganisms which have inherent potential of transforming toxic elements into its less toxic or nontoxic form are considered potent metal accumulators.

There are several adaptive mechanisms by which microorganisms survive toxicity caused by metals.

- **Biotransformation**: Biotransformation refers to the process in which toxic metals are converted into nontoxic forms.
- **Extrusion system**: In extrusion system, chromosomal or plasmid-facilitated events push out metals are through the cells.
- **Production of exopolysaccharide (EPS)**: Microorganisms secrete EPSs to disrupt surrounding contaminants. EPSs are cell membrane developed outside the body of the organisms having efflux pumps. These efflux pumps are hydrophobic, thus help the organism adapt to the surrounding medium (Dixit et al., 2015; Wu et al., 2010).
- **Synthesis of metallothioneins**: Metallothioneins are cysteine-rich low molecular protein. These bind to metals to form different complex for the transformation of toxic metals into nontoxic form.
- **Use of oxidases and reductases**: Oxidases and reductases are the enzymes produced by the microorganisms which by different mechanisms such as oxidation, reduction, hydrolysis, etc., convert different pollutants to metabolic products.

There are different groups of microorganisms which are used in the bioremediation of heavy metals by exploring their potential.

4.1 Algae

A wide range of heavy metals are being removed by different conventional technologies such as lime precipitation or ion exchange in the past. The term "phycoremediation" is an umbrella term to denote the remediation process carried out by different types of algae and cyanobacteria including either removal or assimilation and degradation of heavy metals (Chabukdhara et al., 2017). The major drawback of using such technology is the ineffectiveness in removing the heavy metals at substantially concentration below 50 mg L^{-1} (Dwivedi, 2007). Moreover, they are expensive. To overcome these drawbacks, microalgae have been explored for their potential of removing heavy metals such as lead (Aksu and Kustal, 1992). There are several algal species which accumulate heavy metals such as

Anabaena inaequalis, Chlorella, Synechococcus sp., and *Stigeoclonium tenue.* The use of these organisms suffers from a major disadvantage that it needs specific operating conditions for practical application (Dwivedi, 2007).

Similar to bacteria, algae have various chemical moieties on their surface such as hydroxyl, carboxyl, phosphate, and amide, which act as metal-binding sites (Abbas et al., 2014; He and Chen, 2014). Phaeophyta macroalgae (brown algae) act as a sink for heavy metals. It has a large amount of carboxylic groups on its cell wall leading to high level of heavy metal accumulation due to electrostatic attraction between metals and cell wall. In a study, different species of brown algae were examined for their metal uptake activity. Among them, Padina sp. and Cystoseira sp. were reported to have effective metal uptake capabilities (Ali et al., 2013). Metals are taken up by these algae through their adsorption process. At first, the metal ions are adsorbed over their cell surface very quickly just in a few seconds or minutes; this process is called physical adsorption. Then, these ions are then transported slowly into their cytoplasm through a process called chemisorption. Polyphosphate bodies of algae enable freshwater unicellular algae to store other nutrients. Some researchers have established that metals such as Ti, Pb, Mg, Zn, Cd, Sr, Co, Hg, Ni, and Cu are sequestered in polyphosphate bodies in green algae. *Scenedesmus obliquus* is a typical example of green algae that is capable of accumulating increased Cd and Zn. These bodies complete two different functions in algae; first of all, they provide a "storage pool" for metals, and secondly, they act as a detoxification mechanism (Dwivedi, 2007). They act as hyper-accumulators and hyper-adsorbents with high discrimination for diverse elements. Besides, they produce high alkalinity which is essential for the precipitation of heavy metals during the treatment of acid mine drainage (Bwapwa et al., 2017).

4.2 Bacteria

The use of bacteria either whole or their extract to deal with pollutants has become a promising technology because of its low cost and environmentally friendly nature (Guerra et al., 2018). The remediation of contaminated soils which relies on the abilities of bacteria to degrade hydrocarbons component has been established to be efficient (Folwell et al., 2015). A study on marine-living bacteria reported various other strategies (apart from above listed) to resist high concentrations of Pb/Hg. These strategies include extracellular sequestration, alteration in cell morphology, altered permeability, or intracellular accumulation, precipitation of heavy metals, biosorption of heavy metals, or their volatilization, and demethylation (Naik and Dubey, 2017). Few sulfur-reducing bacteria (SRB) utilize heavy metals as their electron acceptor for respiration and precipitate them as their metal sulfides and hence help in bioremediation of heavy metals. These bacteria produce hydrogen sulfide by using sulfate as electron acceptor, which helps in oxidizing the organic matter.

The various mechanisms could be used to clean-up the environment, among which techniques like biostimulation and bioaugmentation are greatly employed. The technique can be performed as natural attenuation or as assisted microorganism clean-up, the addition of nutrient to fortify and stimulate the growth and metabolism of indigenous bacteria (biostimulation) and the introduction of suitable exogenous bacteria to enhance the clean-up of targeted contaminant (Kure et al., 2018). Adaptation is the first criterion of the selection of appropriate bacteria strain used for bioremediation. Thus, successful environmental decontamination using biological approaches requires bacteria strains that can degrade targeted pollutants (Katarina et al., 2015).

4.3 Fungi

Fungi are also being utilized as a contrivance for remediation of heavy metal–contaminated areas because of their ability to accumulate toxic metals. *Coprinopsis atramentaria* is studied for its bioaccumulation capacity of 76% of Cd^{2+}, when the concentration of Cd^{2+} was 1 mg L^{-1}, and 94.7% of Pb^{2+}, when the concentration of Pb^{2+} was 800 mg L^{-1}. Hence, it has been recognized as an able accumulator of heavy metal ions and a very important tool for mycoremediation (Lakkireddy and Kües, 2017).

Table 3.1 Heavy metal uptake efficiency of different microorganisms.

Heavy metals	Microbial strain	Total uptake (mg g^{-1})	References
Co, Cd, Cu, Zn	*Aspergillus niger*	6.71	Tsekova et al. (2014)
	Aspergillus awamori	7.18	
	Aspergillus ussamii	6.40	
	Rhizopus delemar	30.28	
	Penicillium brevicompactum	25.89	
	Saccharomycopsis lipolytica	0.62	
	Candida blankie	1.00	
	Saccharomyces cerevisiae	1.94	
Cr	*Bacillus circulans* *Bacillus megaterium*	34.5 and 32.0, respectively	Srinath et al. (2002)
Mn, Ni, Co, Hg, Cu, Pb	*Bacillus cereus*	35, 28, 27, 30, 17, and 28, respectively	Banerjee et al. (2015)
Mn, Ni, Co, Hg, Cu, Pb	*Bacillus subtilis*	34, 25, 29, 29, 13, and 30, respectively	
Cd, Cr, hg, Pb	*Bacillus subtilis 38 (B38)*	3.04, 1.83, 4.09, and 2.48, respectively	Wang et al. (2014)
Ni	*E. coli* AS21	0.27	Choudhury et al. (2016)

5. Applications of microbes in HM remediation in soil

The use of microorganisms has its own advantages over other conventional methods. These are economically feasible, practical, and moreover do not create any secondary pollution. The contaminant does not require any other treatment, and also the natural flora and fauna of the contaminated land can be restored to the original form. Below are few techniques being used for remediation using microbial agents (Fig. 3.1).

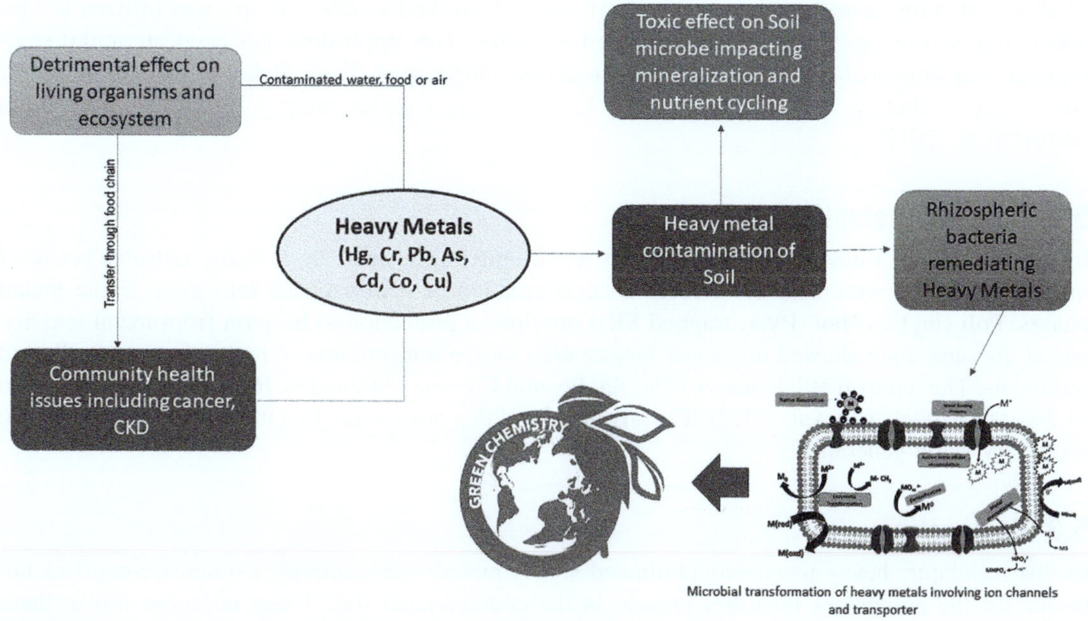

FIGURE 3.1

Schematic diagram of heavy metal remediation.

Adapted Verma, N., Bauddh, K., Kumar, S., Dwivedi, N., Singh, D.P., Barman, S.C., 2020. Accumulation of metals in weed species grown on the soil contaminated with industrial waste and their phytoremediation potential. Ecol. Eng. 61, 491–495.

5.1 Bioleaching

In this technique, microbes such as iron-oxidizing bacteria (*Acidithiobacillus ferrooxidans*) or sulfur-oxidizing bacteria are used for leaching purpose instead of chemical reagents. There are two methods by which the microbe aids in metal leaching as discussed below.

5.1.1 Direct surface attachment

In this technique, bacteria directly interact with heavy metal—contaminated sites by attaching to the metal salts and cause dissolution of metals as explained by the following equation:

$$MS_2 + H_2O + 3.5O_2 \rightarrow M^{2+} + 2SO_4{}^{2-} + 2H^+ \tag{3.2}$$

where MS_2 is an insoluble metal sulfide and M^{2+} is a free metal ion.

5.1.2 Indirect bacterial oxidation

Bacteria such as iron-oxidizing bacteria lead to the formation of Fe^{3+}. This Fe^{3+} then reacts with the metals and metal solubilization occurs.

$$14Fe^{2+} + 3.5O_2 + 14H^+ \rightarrow 14Fe^{3+} + 7H_2O \tag{3.3}$$

$$MS_2 + 8H_2O + 14Fe^{3+} \rightarrow M^{2+} + 14Fe^{2+} + 2SO_4{}^{2-} + 16H^+ \tag{3.4}$$

A recent study reveals, an active strain, Hyhel-1, identified as *Bacillus* sp., was utilized to bioleach copper from heavy metal-contaminated e-wastes. The strain does not need an acidic environment and works better at neutral pH and moderate temperature. Thus, Hyhel-1 strain offers a new cost-effective alternative to recover metals from e-waste without using solvents or acidification (Rozas et al., 2017).

5.2 Biological stabilization

In this technique, immobilized bacteria act as solubilizing agents. In a study, nutrient beads of immobilized SRB were used for effective transformation of heavy metals into more stable bound phases. Polyvinyl alcohol (PVA) trapped SRB provided a protection to bacteria from metal toxicity, and at the same time allowed increased surface area for the transmission of matter between cells and sediments. The removal efficiencies of Cu, Zn, Pb, and Cd were obtained as 76.3%, 95.6%, 100%, and 91.2%, respectively (Li et al., 2017). The advantages of this process are that it is recyclable and has no secondary waste generation.

5.3 Composting

In this technique, heavy metal–contaminated soil is mixed with compost. Compost comprises nutrients for the indigenous microbes present in the contaminated soil. These microbes utilize these nutrients for the remediation of contaminants present. Hence by this method, compost acts as both soil amendment agent and bioremediation technique. In a study of *Hibiscus cannabinus*, seeds were used to coremediate metals (Fe, Mn, Cu, Zn, and Cr). In this technique, compost along with plant technology remediated 72% Mn, 65% Fe, 60% Zn, and 42% Cu, respectively (Taiwo et al., 2016). Compost acts as a stabilization agent in the heavy metal–contaminated sites by forming the complexes or helps in the absorption or (co) precipitation of heavy metals (Chen et al., 2015).

6. Use of genetically modified microbes for HM remediations: recent advances and future prospects

Recently, advances in genetic engineering and the acceptability of genetically engineered microorganisms/biocatalysts for environmental remediation have proven to be more efficient than natural microorganism, especially for the elimination of recalcitrant compounds under natural conditions (Bhatt et al., 2020c). Bioremediation with microorganisms possesses the degradation potential to dissimilate chemicals of complex compounds and seeks sustainability in abating HM pollution from soil (Mosa et al., 2016; Bhatt et al., 2020a,b).

6.1 Biotechnological intervention in microbial bioremediation

Microorganisms are engineered by applying various genetic and metabolic engineering techniques to alter their genetic material to produce more efficient organisms, resulting in improved bioremediation. The engineered aspects can involve single-gene editing, pathway construction, and alteration of sequences of existing genes (both coding and controlling sequences) and focuses on manipulating the rate-limiting steps of metabolic pathways (Diep et al., 2018). Furthermore, these microbes employed in genetic engineering studies can also be utilized for the elimination of toxic HM pollution. The main purpose of genetic engineering in microbial bioremediation is to alter enzymes, metabolic pathways, gene expression system, etc. HMs are transported with the support of membrane transporters in microorganisms, that is, the proteins embedded in the lipid bilayer. These transporters mediate the movement of HM ions across the plasma membrane and the membranes of mitochondria, lysosomes, and vesicles (Chowdhury et al., 2018). On the other hand, enhanced bioremediation of the toxic metals is also achieved due to improved uptake of HM ions from the periplasm to the cytoplasm of gram-negative bacteria (Saier, 2016). The computational biology and multiomics (genomics, metagenomics, metabolomics, proteomics, and transcriptomics) approach has been implemented efficiently in gene mining, which aids in system biology studies of microbes at the genetic level in the context of HM bioremediation (Subashchandrabose et al., 2018). High-throughput and next-generation sequencing have identified the novel genes involved in biodegradation pathways of many heavy metal pollutants. In the series of modern technologies, gene-editing tools such as CRISPR-Cas have the potential to improve the bioremediation process by designing microbes with genes involved in the degradation of particularly obstinate compounds for biodegradation. CRISPR could be used to transfer a desired set of instructions in the microbe's genome in a candid mode, as it is a programmable, next-generation method for high-throughput genetic manipulation when compared with the conventional low throughput ZFNs and TALENs (Miglani, 2017). In general, genome editing alters the gene expression regulation at specified sites, and expedites new insights into the plant functional genomics (Hussain et al., 2018). Biotechnology approaches have produced many strains with the aid of multiomics. Typical examples include the following: *Mycobacterium*, *Rhodococcus wratislaviensis* (Gołebiewski et al., 2014), *Pseudomonas putida*, *Alcaligenes eutrophus* (Cycon and Pitrowska-seget, 2016), *Aspergillus niger*, *Candida pelliculosa* strain S-02 (Kumar et al., 2018), *Chlorella*, *Arthrobacter* (Gong et al., 2018), *Pseudomonas putida KT2440*, *Sphingobium* sp. Strain 1017-1 (Raman et al., 2018), *Stenotrophomonas maltophilia* (Cho et al., 2018), *Cyanobacterium synechocystis*, *Saccharomyces cerevisiae*, *Populus* sp. (Cai et al., 2019), *Sphingomonas* sp., and *Pseudomonas putida* strain KT2440 (Liu et al., 2019).

6.2 Nanotechnology for efficient microbial bioremediation

The nanoparticle inherently contains different functional groups such as $-COOH$, $-NH_2$, and $-OH$, and tailoring the desired functional groups by physical/chemical activation or surface modification has shown to yield enhanced removal of HMs. Bacterial strains are also able to synthesize the nanoparticle that can be beneficial for HMs bioremediation (Arshad et al., 2018). To generate a nanocomposite, microorganisms are immobilized or trapped within the nanomaterial, for example, gram-negative *Halomonas* sp. entrapped within polyvinylpyrrolidone-coated iron oxide nanoparticles was tested to remove Cd(II) and Pb(II) (Cao et al. 2020). Alternatively, the microorganism itself can act as a nanoparticle synthesizer, a method known as green synthesis. However, the separation or recovery of HMs from the nanomaterials is a very laborious procedure. Therefore, in recent years, magnetic nanomaterials have received more attention, wherein surface modification, encapsulation, and coating of iron/iron oxides have been emphasized for easy separation/recovery of the HMs (Jinyue et al., 2019).

6.2.1 Future research directions and the application of sustainable policies to reduce heavy metal pollution

Naturally, microorganisms possess many modes to endure HM pollutants in the environment; these processes and physiochemical behavior assist in enhancing their ability to uptake these pollutants and remediate the contaminated sites (Singh et al., 2016). Recently, yeast has been engineered to present plant-like characteristics and act as hyperaccumulators of different HMs from the water environment (Sun et al., 2020). Other microbes could also be engineered in this way to assist in HM remediation. Hence, in future research projects, more attention should also be given to algae as it is considered as a potent microorganism for HM sorption from the soil. More focus should also be on the advancement of new biotechnologies that involves synthetic biology tools so that nonmodel organisms can be used in place of laboratory or industrial strains for HM uptake. An example is the recovery/uptake of HMs from agricultural soils (Qian et al., 2020). The omics-based approaches are useful for the development of improved industrial strains that are resilient to the prevailing environmental conditions, that is, due to their improved genetic capabilities and HM tolerance. Apart from the recent microbial technologies, emphasis should be placed on plant breeding and biotechnology programs to improve root exudate characteristics in plants to enhance the HM eradication capabilities of associated microorganisms (Mosa et al., 2016). In addition to this, the application of nano- and nano (bio)-technologies as discussed in this chapter has tremendous potential to support the application of microbial technologies for dealing with HMs contamination.

7. Conclusion

HM pollution in soil, water, and agricultural land poses a great risk to all life forms and the natural environment. To protect the deterioration of environmental quality due to HM pollution, several measures have already been implemented in developed countries. More priority should be given to understand the physical, chemical, and biological behavior of microorganisms in the presence of HMs in soil, water, and gaseous environments in order to develop techniques that foster higher HM tolerance in microbes (Njoku et al., 2020). Recently, gene-editing technology is considered as a milestone in the field of HM pollution and its cleanup from contaminates sites (Hussain et al., 2018). With recent advancements in analytical and molecular biology tools, the manipulation of genomic sequences of these microbes would certainly facilitate the characterization and developments of genomic region that can be correlated with the activity, HM withstanding capacity/survival of microbes in the presence of high levels of HMs.

References

Abbas, S.H., Ismail, I.M., Mostafa, T.M., Sulaymon, A.H., 2014. Biosorption of heavy metals: a review. J. Chem. Sci. Technol. 3 (4), 74–19.

Aksu, Z., Sag, Y., Kutsal, T., 1992. The biosorption of copper(II) by *C. vulgaris* and *Z. ramigera*. Environ. Tech. 13, 579–586.

Ali, H., Khan, E., Sajad, M.A., 2013. Phytoremediation of heavy metals concepts and application. Chemosphere 91 (7), 869–881. https://doi.org/10.1016/j.chemo spher e.2013.01.075.

Arshad, F., Selvaraj, M., Zain, J., Banat, F., Haija, M.A., 2018. Polyethylenimine modified graphene oxide hydrogel composite as an efficient adsorbent for heavy metal ions. Separat. Purif. Technol. 209, 870–880. https://doi.org/10.1016/j.seppur.2018.06.035.

Bwapwa, J.K., Jaiyeola, A.T., Chetty, R., 2017. Bioremediation of acid mine drainage using algae strains: a review. South Afr. J. Chem. Eng. 24, 62–70. https://doi.org/10.1016/j.sajce.2017.06.005.

Banerjee, S., Gothalwal, R., Sahu, P.K., Sao, S., 2015. Microbial observation in bioaccumulation of heavy metals from the Ash dyke of thermal power plants of Chhattisgarh, India. Adv. Biosci. Biotechnol. 06 (02), 131–138. https://doi.org/10.4236/abb.2015.62013.

Bhatt, P., Bhatt, K., Huang, Y., Ziqiu, L., Chen, S., 2020a. Esterase is a powerful tool for the biodegradation of pyrethroid insecticides. Chemosphere 244, 125507. https://doi.org/10.1016/j.chemosphere.2019.125507.

Bhatt, P., Huang, Y., Zhang, W., Sharma, A., Chen, S., 2020b. Enhanced cypermethrin degradation kinetics and metabolic pathway in Bacillus thuringiensis strain SG4. Microorganisms 8, 223. https://doi.org/10.3390/micro organ isms8 02022 3.

Bhatt, P., Sethi, K., Gangola, S., Bhandari, G., Verma, A., Adnan, M., Singh, Y., Chaube, S., 2020c. Modeling and simulation of atrazine biodegradation in bacteria and its effect in other living systems. J. Biomol. Struct. Dyn. 12, 1–11. https://doi.org/10.1080/07391102.2020.18466 23.

Bhatt, K., Maheshwari, D.K., 2019. Decoding multifarious role of cow dung bacteria in mobilization of zinc fractions along with growth promotion of C. annuum L. Sci Rep 9 (1), 14232. https://doi.org/10.1038/s4159 8-019-50788 -8.

Cai, P., Gao, J., Zhou, Y., 2019. CRISPR-mediated genome editing in non-conventional yeasts for biotechnological applications. Microb. Cell Fact. 18, 63. https://doi.org/10.1186/s12934-019-1112-2.

Cao, M.A., Becerril, J.M., Barrutia, O., Artetxe, U., Garbisu, C., Hernández, A., 2020. Field assessment of the effectiveness of organic amendments for aided phytostabilization of a Pb-Zn contaminated mine soil. J. Geochem. Explor. 145, 181–189.

Chabukdhara, M., Gupta, S.K., Gogoi, M., 2017. Phycoremediation of heavy metals coupled with generation of bioenergy. In: Algal Biofuels: Recent Advances and Future Prospects. Springer, pp. 163–188. https://doi.org/10.1007/978-3-319-51010-1_9.

Chen, J., He, F., Zhang, X., Sun, X., Zheng, J., Zheng, J., 2014. Heavy metal pollution decreases microbial abundance, diversity and activity within particle-size fractions of a paddy soil. FEMS (Fed. Eur. Microbiol. Soc.) Microbiol. Ecol. 87 (1), 164–181.

Chen, M., Xu, P., Zeng, G., Yang, C., Huang, D., Zhang, J., 2015. Bioremediation of soils contaminated with polycyclic aromatic hydrocarbons, petroleum, pesticides, chlorophenols and heavy metals by composting: applications, microbes and future research needs. Biotechnol. Adv. 33 (6), 745–755.

Cho, S., Shin, J., Cho, B.K., 2018. Applications of CRISPR/Cas system to bacterial metabolic engineering. Int. J. Mol. Sci. 19, 1089. https://doi.org/10.3390/ijms1 90410 89.

Choudhury, R., Majumder, M., Roy, D.N., Basumallick, S., Misra, T.K., 2016. Phytotoxicity of Ag nanoparticles prepared by biogenic and chemical methods. Int. Nano Lett. 6, 153. https://doi.org/10.1007/s40089-016-0181-z.

Chowdhury, R., Ren, T., Shankla, M., Decker, K., Grisewood, M., Prabhakar, J., Baker, C., Golbeck, J.H., Aksimentiev, A., Kumar, M., Maranas, C.D., 2018. Pore designer for tuning solute selectivity in a robust and highly permeable outer membrane pore. Nat. Commun. 9, 3661. https://doi.org/10.1038/s4146 7-018-06097 -1.

Coubot, M.N.U., Korkmaz, H., Arikan, B., 2005. Plasmid mediated heavy metal resistance in *Enterobacter* spp. isolated from Sofulu landfill, in Adana, Turkey. Ann. Microbiol. 55 (3), 175–179.

Cycon, M., Piotrowska-Seget, Z., 2016. Pyrethroid-degrading microorganisms and their potential for the bioremediation of contaminated soils: a review. Front. Microbiol. 7, 1463. https://doi.org/10.3389/fmicb.2016.01463.

Deng, J., Liao, B., Ye, M., Deng, D., Lan, C., Shu, W., 2007. The effects of heavy metal pollution on genetic diversity in zinc/cadmium hyperaccumulator Sedum alfredii populations. Plant Soil 297 (1–2), 83–92.

Diep, P., Mahadevan, R., Yakunin, A.F., 2018. Heavy metal removal by bioaccumulation using genetically engineered microorganisms. Front. Bioeng. Biotechnol. 6, 157. https://doi.org/10.3389/fbioe.2018.00157.

Dixit, R., Wasiullah, M.D., Pandiyan, K., Singh, U.B., Sahu, A., Shukla, R., Singh, B.P., Rai, J.P., Sharma, P.K., Lake, H., Paul, D., 2015. Bioremediation of heavy metals from soil and aquatic environment: an overview of principles and criteria of fundamental processes. Sustainability 7, 2189–2212. https://doi.org/10.3390/su7022189.

Doran, J.W., Safley, M., 1997. Defining and assessing soil health and sustainable productivity. In: Pankhurst, C.E., Doube, B.M., Gupta, V.V.S.R. (Eds.), Biological Indicators of Soil Health. CAB International, Wallingford, pp. 1–28.

Dubey, G., Kollah, B., Gour, V.K., Shukla, A.K., Mohanty, S.R., 2016. Diversity of bacteria and archaea in the rhizosphere of bioenergy crop Jatropha curcas. 3Biotech 6, 257.

Dwivedi, B.U., 2007. Assisted phytoremediation of chromium spiked soils by sesbania sesban in association with *Bacillus xiamenensis* PM14: a biochemical analysis. Plant Physiol. Biochem. 146, 249–258. https://doi.org/10.1016/j.plaphy.2019.11.010.

European Parliament; European Council, 2002. Directive 2002/32/EC of the European Parliament and of the Council of 7 May 2002 on Undesirable Substances in Animal Feed; OJ L 140, 30.5.2002. European Parliament and European Council, Brussels, Belgium, pp. 10–22, 2002.

Gołebiewski, M., Deja-Sikora, E., Cichosz, M., Tretyn, A., Wróbel, B., 2014. 16S rDNA pyrosequencing analysis of bacterial community in heavy metals polluted soils. Microbial. Ecol. 67, 635–647. https://doi.org/10.1007/s0024 8-013-0344-7.

Gong, T., Xu, X., Dang, Y., Kong, A., Wu, Y., Liang, P., 2018. An engineered Pseudomonasputida can simultaneously degrade organophosphates, pyrethroids and carbamates. Sci. Total Env. 628, 1258–1265. https://doi.org/10.1016/j.scitotenv.2018.02.143.

Folwell, S., Barra, B., Caporaso, J.G., Seeger, M., 2015. From rare to dominant: a fine-tuned soil bacterial bloom during petroleum hydrocarbon bioremediation. Appl. Environ. Microbiol. 82, 888–896. https://doi.org/10.1128/AEM.02625-15.

Guerra, A.B., Oliveira, J.S., Silva-Portela, R.C., Araujo, W., Carlos, A.C., Vasconcelos, A.T.R., et al., 2018. Metagenome enrichment approach used for selection of oil-degrading bacteria consortia for drill cutting residue bioremediation. Environ. Pollut. 235, 869–880. https://doi.org/10.1016/j.envpol.2018.01.014.

He, J., Chen, J.P., 2014. A comprehensive review on biosorption of heavy metals by algal biomass: materials, performances, chemistry, and modeling simulation tools. Bioresour. Technol. 160, 67–78. https://doi.org/10.1016/j.biortech.2014.01.068.

Horta, A., Malone, B., Stockmann, U., Minasny, B., Bishop, T.F.A., McBratney, A.B., Pallasser, R., Pozza, L., 2015. Potential of integrated field spectroscopy and spatial analysis for enhanced assessment of soil contamination: a prospective review. Geoderma 241–242, 180–209.

Hussain, I., Aleti, G., Naidu, R., Puschenreiter, M., Mahmood, Q., Rahman, M.M., Wang, F., Shaheen, S., Syed, J.H., Reichenauer, T.G., 2018. Microbe and plant assisted-remediation of organic xenobiotics and its enhancement by genetically modified organisms and recombinant technology: a review. Sci. Total Environ. 1, 628–629. https://doi.org/10.1016/j.scito tenv.2018.02.037.

Jinyue, Y., Baohong, H., Jingkang, W., Beiqian, T., Jingtao, B., Na, W., Xin, L., Xin, H., 2019. Nanomaterials for the removal of heavy metals from wastewater. Nanomaterials 9, 424. https://doi.org/10.3390/nano9030424.

Katarina, X., Li, Y., Yan, X., 2015. Removal of heavy metals and arsenic from a co-contaminated soil by sieving combined with washing process. J. Environ. Sci. 1–9.

Khalid, S., Shahid, M., Niazi, N.K., et al., 2017. A comparison of technologies for remediation of heavy metal contaminated soils. J. Geochem. Explor. 182, 247–268.

Kumar, A., Maurya, B.R., Raghuwanshi, R., Meena, V.S., Islam, M.T., 2017. Co-inoculation with enterobacter and rhizobacteria on yield and nutrient uptake by wheat (*Triticum aestivum* L.) in the alluvial soil under indo-gangetic plain of India. J. Plant Growth Regul. https://doi.org/10.1007/s00344-016-9663-5.

Kure, G.P., Yadav, S.K., Thawale, P.R., Singh, S.K., Juwarkar, A.A., 2018. Growth of *Jatropha curcas* on heavy metal contaminated soil amended with industrial wastes and Azotobacter – a greenhouse study. Bioresour. Technol. 99, 2078–2082.

Kumar, S., Kaushik, G., Dar, M.A., Nimesh, S., Lopez-Chuken, U.J., Villarreal-Chiu, J.F., 2018. Microbial degradation of organophosphate pesticides: a review. Pedosphere 28, 190–208. https://doi.org/10.1016/S1002-0160(18)60017 -7.

Lakkireddy, K., Kües, U., 2017. Bulk isolation of basidiospores from wild mushrooms by electrostatic attraction with low risk of microbial contaminations. Amb. Express 7 (1), 28.

Li, X., Dai, L., Zhang, C., Zeng, G., Liu, Y., Zhou, C., Xu, W., Wu, Y., Tang, X., Liu, W., 2017. Enhanced biological stabilization of heavy metals in sediment using immobilized sulfate reducing bacteria beads with inner cohesive nutrient. J. Hazard Mater. 324, 340–347.

Liu, L., Li, W., Song, W., Guo, M., 2018. Remediation techniques for heavy metal-contaminated soils: principles and applicability. Sci. Total Environ. 633, 206–219. https://doi.org/10.1016/j.scitotenv.2018.03.161.

Liu, X., Hu, X., Cao, Y., Pang, W., Huang, J., Guo, P., Huang, L., 2019. Biodegradation of phenanthrene and heavy metal removal by acidtolerant Burkholderiafungorum FM-2. Front. Microbiol. 10, 408. https://doi.org/10.3389/fmicb .2019.00408.

Long, Z., Huang, Y., Zhang, W., Shi, Z., Yu, D., Chen, Y., et al., 2021. Effect of different industrial activities on soil heavy metal pollution, ecological risk, and health risk. Environ. Monit. Assess. 193, 1–12.

Mamiroli, N., Maestri, E., 2008. Health implication of trace elements in the environment and the food chain. In: Prasad, M.N.V. (Ed.), Trace Elements as Contaminants and Nutrients: Consequences in Ecosystems and Human Health. John Wiley & Sons, Inc. Publishing, pp. 23–53.

McCann, C.M., Peacock, C.L., Hudson-Edwards, K.A., Shrimpton, T., Gray, N.D., Johnson, K.L., 2018. In situ arsenic oxidation and sorption by a Fe-Mn binary oxide waste in soil. J. Hazard Mater. 342, 724–731. https://doi.org/10.1016/j. jhazmat.2017.08.066.

Mehta, P., Walia, A., Chauhan, A., Kulshrestha, S., Shirkot, C.K., 2013. Phosphate solubilisation and plant growth promoting potential by stress tolerant *Bacillus* sp. isolated from rhizosphere of apple orchards in trans Himalayan region of Himachal Pradesh. Ann. Appl. Biol. 163 (3), 430–443.

Miglani, G.S., 2017. Genome editing in crop improvement: present scenario and future prospects. J. Crop Improv. 31, 453–559. https://doi.org/10.1080/15427 528.2017.13331 92.

Mosa, K.A., Saadoun, I., Kumar, K., Helmy, M., Dhankher, O.P., 2016. Potential biotechnological strategies for the cleanup of the heavy metals and metalloids. Front. Plan. Sci. 7, 303. https://doi.org/10.3389/fpls.2016.00303.

Njoku, K.L., Asunmo, M.O., Ude, E.O., Adesuyi, A.A., Oyelami, A.O., 2020. The molecular study of microbial and functional diversity of resistant microbes in heavy metal contaminated soil. Environ. Technol. Innov. 17, 100606. https://doi.org/10.1016/j.eti.2020.100606.

Naik, M.M., Dubey, S., 2017. Lead-and mercury-resistant marine bacteria and their application in lead and mercury bioremediation. In: Marine Pollution and Microbial Remediation. Springer, pp. 29–40.

O'Connor, D., Hou, D., Ok, Y.S., Lanphear, B.P., 2020. The effects of iniquitous lead exposure on health. Nat. Sustain. 3, 77–79. https://doi.org/10.1038/s41893-020-0475z.

Park, C.M., Katz, L.E., Liljestrand, H.M., 2015. Mercury speciation during in situ thermal desorption in soil. J. Hazard Mater. 300, 624–632. https://doi.org/10.1016/j. jhazmat.2015.07.076.

Qian, X., Chen, L., Sui, Y., Chen, C., Zhang, W., Zhou, J., Dong, W., Jiang, M., Xin, F., Ochsenreither, K., 2020. Biotechnological potential and application of microbial consortia. Biotechnol. Adv. 40, 107500. https://doi.org/10.1016/j.biotechadv.2019.107500.

Rajendran, P., Muthukrishnan, J., Gunasekaran, P., 2003. Microbes in heavy metal remediation. Indian J. Exp. Biol. 41 (9), 935–944.

Ramamoorthy, D., 2015. Assessment of heavy metal pollution and its impacts on soil physical, chemical properties and β-glucosidase activities in agricultural lands, Puducherry region. Pol. J. Environ. Stud. Project report submitted to Department of Science, Technology and Environment, Government of Puducherry.

Raman, N.M., Asokan, S., Sundari, N.S., Ramasamy, S., 2018. Bioremediation of chromium (VI) by Stenotrophomonas maltophilia isolated from tannery effluent. Int. J. Environ. Sci. Technol. 15, 207–216. https://doi.org/10.1007/s1376 2-017-1378-z.

Roux, M., Chamber-Pérez, M.A., Palomares, A.J., Pajuelo, E., 2012. In situ" phytostabilisation of heavy metal polluted soils using *Lupinus luteus* inoculated with metal resistant plant-growth promoting rhizobacteria. J. Hazard. Mater. 177 (1–3), 323–330.

Rozas, E.E., Mendes, M.A., Nascimento, C.A., Espinosa, D.C., Oliveira, R., Oliveira, G., Custodio, M.R., 2017. Bioleaching of electronic waste using bacteria isolated from the marine sponge *Hymeniacidon heliophila* (Porifera). J. Hazard Mater. 329, 120–130.

Saier, M.H., 2016. Transport protein evolution deduced from analysis of sequence, topology and structure. Curr. Opin. Struct. Biol. 38, 9–17. https://doi.org/10.1016/j.sbi.2016.05.001.

Sarwar, N., Imran, M., Shaheen, M.R., Ishaque, W., Kamran, M.A., Matloob, A., Rehim, A., Hussain, S., 2016. Phytoremediation strategies for soils contaminated with heavy metals: modifications and future perspectives. Chemosphere 171, 710–721. https://doi.org/10.1016/j.chemosphere.2016.12.116.

Sharma, A., Shankhdhar, D., Shankhdhar, S.C., 2016. Potassium-solubilizing microorganisms: mechanism and their role in potassium solubilization and uptake. In: Meena, V.S., Maurya, B.R., Verma, J.P., Meena, R.S. (Eds.), Potassium Solubilizing Microorganisms for Sustainable Agriculture. Springer, India, pp. 203–219. https://doi.org/10.1007/978-81-322-2776-2_15.

Shen, Z.T., Jin, F., O'Connor, D., Hou, D.Y., 2019. Solidification/stabilization for soil remediation: an old technology with new vitality. Environ. Sci. Technol. 53, 11615–11617. https://doi.org/10.1021/acs.est.9b04990.

Singh, N.P., Sharma, J.K., Santal, A.R., 2016. Biotechnological approaches to remediate soil and water using plant–microbe interactions. In: Phytoremediation. Springer, Cham, pp. 131–152. https://doi.org/10.3389/fpls.2018.01476.

Silva, W.O., Stamford, N.P., Silva, E.V., Santos, C.E., Freitas, A.D.S., Silva, M.V., 2016. The impact of biofertilizers with diazotrophic bacteria and fungi chitosan on melon characteristics and nutrient uptake as an alternative for conventional fertilizers. Sci. Hortic. 209, 236–240.

Srinath, T., Verma, T., Ramteke, P., Garg, S., 2002. Chromium (VI) biosorption and bioaccumulation by chromate resistant bacteria. Chemosphere 48 (4), 427–435.

Subashchandrabose, S.R., Venkateswarlu, K., Krishnan, K., Naidu, R., Lockington, R., Megharaj, M., 2018. Rhodococcuswratislaviensis strain 9: an efficient p-nitrophenol degrader with a great potential for bioremediation. J. Hazard Mater. 347, 176–183. https://doi.org/10.1016/j.jhazm at.2017.12.063.

Sun, L., Ma, Y., Wang, H., Huang, W., Wang, X., Han, L., et al., 2020. Overexpression of PtABCC1 contributes to mercury tolerance and accumulation in arabidopsis and poplar. Biochem. Biophys. Res. Commun. 497, 997–1002. https://doi.org/10.1016/j.bbrc.2018.02.133.

Taiwo, A., Gbadebo, A., Oyedepo, J., Ojekunle, Z., Alo, O., Oyeniran, A., Onalaja, O., Ogunjimi, D., Taiwo, O., 2016. Bioremediation of industrially contaminated soil using compost and plant technology. J. Hazard Mater. 304, 166–172.

Tsekova, K., Kaimaktchiev, A., Tzekova, A., 2014. Bioaccumulation of heavy metals by microorganisms. Biotechnol. Biotechnol. Equip. 12 (2), 94–96. https://doi.org/10.1080/13102 818.1998.10818 998.

UN, 2015. Transforming Our World: The 2030 Agenda for Sustainable Development. USA^2, New York.

Verma, N., Bauddh, K., Kumar, S., Dwivedi, N., Singh, D.P., Barman, S.C., 2020. Accumulation of metals in weed species grown on the soil contaminated with industrial waste and their phytoremediation potential. Ecol. Eng. 61, 491–495.

Wang, T., Sun, H., Mao, H., Zhang, Y., Wang, C., Zhang, Z., Wang, B., Sun, L., 2014. The immobilization of heavy metals in soil by bioaugmentation of a UV-mutant *Bacillus subtilis* 38 assisted by Novo- Gro biostimulation and changes of soil microbial community. J. Hazard Mater. 278, 483–490. https://doi.org/10.1016/j.jhazmat.2014.06.028.

Wang, S.Y., Kalkhajeh, Y.K., Qin, Z.R., Jiao, W.T., 2020. Spatial distribution and assessment of the human health risks of heavy metals in a retired petrochemical industrial area, South China. Environ. Res. 188, 109661.

Wu, G., Kang, H., Zhang, X., Shao, H., Chu, L., Ruan, C., 2010. A critical review on the bio-removal of hazardous heavy metals from contaminated soils: issues, progress, eco-environmental concerns and opportunities. J. Hazard Mater. 174 (1−3), 1−8. https://doi.org/10.1016/j.jhazmat.2009.09.113.

Xu, Z., Yu, G., et al., 2015. The variations in soil microbial communities, enzyme activities and their relationship with soil organic matter decomposition along the northeren slope of Changbai mountain. Appl. Soil Ecol. 86, 19−29.

Zhang, W., Tong, L., Yuan, Y., Liu, Z., Huang, H., Tan, F., Qiu, R., 2010. Influence of soil washing with a chelator on subsequent chemical immobilization of heavy metals in a contaminated soil. J. Hazard Mater. 178, 578−587. https://doi.org/10.1016/j.jhazmat.2010.01.124.

Further reading

Transforming our world: the 2030 agenda for sustainable development. In: A New Era in Global Health, 2018. https://doi.org/10.1891/9780826190123.ap02.

CHAPTER

4

Application of biochar in wood-based industry as low-cost sustainable green technology

Ranjana Yadav[1] and Jitendra Kumar[2]

[1]*Indian Plywood Industries Research & Training Institute, Mohali, Punjab, India;* [2]*Department of Biosciences, UIBT, Chandigarh University, Mohali, Punjab, India*

1. Introduction

Renewable resource materials may now be employed as reinforcement in composites, perhaps helping to solve environmental issues. Biochar has been synthesized with agriculture waste by biomass pyrolysis (He et al., 2011), which helps to reduce the environment pollution (Jindo et al., 2014). It has the potential to minimize the usage of chemical fertilizers and pesticides in agriculture (Yaun et al., 2012). Biochar is also commonly utilized in the production of composites. The permeable design of biochars grants pressure in the biochar composite framework to be transmitted, enhancing mechanical properties (Behazin et al., 2017). Apart from the above advantages, biochars are also feasible due to their good stability, conductivity, and dispersion (Das et al., 2016).

Lee et al. (2013) synthesized the biochar by pyrolysis at 500°C with agricultural residues. They evaluated the biochar characterization, elemental composition, and pH. Biochar is made from rice husk pyrolysis and the unsaturated polyester biochar composites are prepared. Richard et al. (2016) synthesized the particle-loaded polyester composites and analyzed the different characteristics. Aamer Khan et al. investigated the physical, thermal, and mechanical properties of fillers modified epoxy composites. At various weight percentages, epoxy matrix was reinforced with biochar and carbon nanotubes. Biochar composites showed best results compared to other fillers. Although, the maximum electrical properties were achieved with carbon nanotube-filled composites (Khan et al., 2017). Imran Oral prepared the two different kinds of biochar-modified epoxy composites, which have been obtained by poplar and pine cone (PC) wastes. The mechanical properties of the poplar biochar-based composites 30% wt are compared to those of rubber composites (Oral, 2016). Steven C. Peterson has prepared the rubber matrix composites with different kinds of biochar (corn starch, flour, and stover). He found that mechanical performance of biochar composites was improved by adding corn starch (Peterson, 2012). Zhang et al. (2017) prepared HDPE (high density polyethylene) composites with rice husk biochar and wood flour. It was observed that composite materials initially tend to rise. Rice husk biochar-reinforced HDPE composite materials showed the strongest mechanical properties. Sundarakannan et al. (2019) investigated biochar-modified saturated polyester resin for composite fabrication with solution dispersion method. Different percentages of biochar (5%, 10%, and 15%) have been used. They have analyzed the different parameters like tensile

Integrative Strategies for Bioremediation of Environmental Contaminants, Volume Two. https://doi.org/10.1016/B978-0-443-14013-6.00009-3

strength, flexural strength, hardness, and impact strength. Kumar et al. (2021) synthesized the biochar with wood apple fruit shell waste (WAS) by ball milling and which has been used to remove harmful phenols as well as chlorophenols (4-CPh and 2,4-DCPh) from polluted aqueous solution. Absorption of phenol, 4-CPh, and 2,4-DCPh was found to be 102.71 mg g^{-1}, 172.24 mg g^{-1}, and 226.55 mg g^{-1} at 30 ± 1°C. From the result, it has been found that wood apple fruit shell-based biochar has effective and low-cost absorbent which is suitable for removing phenol and chlorophenol from polluted wastewater.

The effects of biochar on the mechanical and flame-retardant properties of wood plastic composites were investigated by Zhang et al. (2018). Composites with varying biochar content were made. They found that mechanical properties have been increased initially but declined as the biochar percentage grew. Magnesium and aluminum hydroxide enhanced their flame retardancy; however, magnesium hydroxide had a stronger impact than aluminum hydroxide.

The kind of biochar, polymer variety, biochar loading, and preparation temperature may alter the properties of biochar composites, implying that biochar properties are crucial in biochar composites (Li et al., 2020; Zhang et al., 2020a,b). Biochar composites have been studied extensively for their physical, mechanical, thermal, and conductivity characteristics (Ayadi et al., 2020), but manufacture and properties of these composites have rarely been investigated.

This chapter mainly focus on the synthesis and application of biochar and biochar-based composites.

2. Synthesis of biochars

Sun et al. (2014) prepared the biochars with different agro-residue (rice, cotton, wheat, and maize). They compared them to poplar (*Populus euramericana* cv.) wood char. After oven drying at 80°C, feedstocks were chopped into small pieces and sieved through 0.125 mm mesh. Van Soest (1963) described a method for determining the cellulose and lignin content of feedstocks. After being pyrolyzed in a N_2 environment at 600°C for 3 h, 5 g of each raw material was fed into a lab-scale tubular reactor in a muffle furnace. After the samples were cooled, they were brought to room temperature. To be employed, the chars were softly crushed and homogenized before passing through a 0.125 mm filter.

A study by Noumi et al. (2014) examined the optimization of charcoal production from eucalyptus wood using pyrolysis results. According to the study, they found that high temperature over 550°C, high pressure, and low heating rate gave the highest yield of charcoal. Using paper mulberry as feedstock, Jawaria et al. (2014) optimized the pyrolysis of biochar to obtain the highest yields at 350°C and 3 mm particle size. Ren et al., (2013) used response surface methodology to optimize bark pyrolysis to yield phenol-rich biooil. The optimal temperature and particle size were found to be 485°C and 0.35 mm, respectively. The Wretborn (2016) study determined that the optimum temperature for maximum charcoal production (32%) was 340°C through experimental optimization of pyrolysis of fire wood chips. Optimization of beech wood biochar production is still unresolved.

Idress et al. (2018) synthesized and optimized the method of production of biochar with forest wastes and residues on a large via pyrolysis. According to the results, biochar yield increased when temperatures decreased, but increased when nitrogen flow rate increased. Biochar's maximum proportion is 33.04% wt at temperature of 405.32°C and nitrogen flow rate of 2968.83 kg h^{-1} when both conditions are combined.

Miscanthus, wood chips, and dried distiller's grain were used to synthesize the biochar from three different North American feedstocks in a total of four pyrolysis reactor settings. Miscanthus biochar yields were nearly same in both the unmodified and continuous pyrolysis studies, with the mechanically fluidized reactor for slow pyrolysis producing the highest output. As the temperature climbed, the wood chips yielded more than the miscanthus, and the DDG yielded more at the same temperature range, albeit the yield declined more rapidly. A high-temperature treatment on DDG char resulted in an 80% yield (Cuthbertson, 2018).

Nizamuddin et al. (2019) developed a one-of-a-kind polymer composite containing rice husk hydrochar as an extender in a polylactide matrix, using microwave-assisted hydrothermal carbonization. A clean polylactide and a polylactide/hydrochar composite were synthesized, and mechanical, thermal, and rheological parameters have been examined.

Ayadi et al. (2020) used CarbonFX technology (Airex Energy Inc., Bécancour, QC, Canada) to create biochar. Pyrolysis was used to covert wood residues into torrefied wood (315°C) and biochar (400 and 455°C) in an oxygen-free atmosphere.

Akom et al. (2020) employed a locally built reactor to manufacture the biochar. 540 kg of biochar was made from 1460 kg of wood residues. The average charring ratio was 37%, with a feedstock weight of 122 kg. Biochar has been synthesized with 988 kg of firewood in 35 days.

3. Biochar composites and their properties

Zaverl et al. (2013) synthesized the wheat fodder biochar plastic composites in a pioneering study. The tensile and flexural modulus of the composite were found ~2.7 and ~2.3 GPa, with 10% biochar. Utilization of three sorts of biochar produced using plastic waste (PW), wood shavings (WSs) and PCs (prepared with pyrolysis at 450°C) was researched for the creation of epoxy composites (Ahmetli et al., 2013). PW-derived materials had the maximum tensile strength (0.59 MPa) at a 15% filler concentration, whereas WS and PC epoxy composites had the highest tensile strength at a 25% concentration (0.51 and 0.64 MPa, respectively). Furthermore, increased filler concentrations resulted in enhanced thermal stability (up to 30%). This shows that including biochar into the composite slowed thermal degradation and protected the epoxy from heat.

As of Mohanty et al. (2015), nylon 6-biochar composites were obtained by melt processing dried resin with biochar using a MiniCTW Micro-Conical Twin Screw Compounder (a twin-screw extruder integrated with an injection molding system) from HAAKETM. The nylon-6/biochar composites were processed at 240°C and 100 rpm, with a residence time of 120 s. As a result of the extrusion, the molten material was transferred into a preheated sample collector, injected into the mold at 230 par injection pressure, and held for 15 s before demolding.

As compared to corn-derived materials, Peterson (2013) synthesized the rubber composites along with biochar that contained lower ash contents (2.4%), as well as smaller particles. Biochar-based composites exhibited greater elongation and flexibility than carbon black-filled composites in this case. A 10% filler concentration of biochar and carbon black also improved the tensile, elongation, and toughness properties.

Biochar has been used partially and fully as an alternative of wood flour in composite, along with polypropylene (PP). The mechanical and physical properties of wood, biochar, and plastic composites were studied. 25% biochar composites have highest flexural strength and modulus of elasticity, 5% biochar composites have highest tensile strength, and 40% biochar composites showed the lowest water absorption. A variety of composite applications could benefit from the use of BC instead of traditional wood (DeVallance et al., 2015).

Biochar was made from wood scrap and combined with PP composites (Das et al., 2015). Experiments were carried out using biochar concentrations in the range of 6–30 wt% for the wood polymer composites (WPCs). Addition of 6 weight% biochar enhanced the composite's tensile/flexural strength and modulus compared to control samples, but 12 wt% and 18 wt% biochar composites showed more ductile and thermal stability. To create sustainable biocomposites, Bhattacharyya (2016) blended four different biomass wastes (rice husk, coffee husk, wool, and wood) with PP and pinewood waste and prepared the sustainable biocomposites. The biochar composites were examined by FTIR, XRD, DSC, TGA, cone calorimeter, and SEM.

Das et al. (2018) mixed four biomass wastes (rice husk, coffee husk, coarse wool, and landfill wood) with biochar and PP to synthesized biocomposites. Cone calorimeters were used to evaluate the combustion behavior of individual biomasses. The fire, thermal, mechanical, and morphological characteristics of biocomposites were examined. The wood constituent has an impact on the mechanical and fire properties of composites.

Das and Hedenqvist (2018) evaluated the wettability properties of biochar-reinforced wood and PP composites for outdoor use. Three composites with the highest water resistance were prepared by adding 12 wt% wood waste (*Pinus radiata*). To develop composites with acceptable water-repellent properties, biochar should be generated at low pyrolysis temperatures (resulting in a more hydrophobic biochar).

Giorcelli (2018) investigated biochar as a high-performance, low-cost filler for the production of polymer composites. Giorcelli et al. (2018) investigated biochar as a low-cost, environmentally friendly filler for polymer composites. They studied the tensile strength, young modulus, and resilience. They found that a very low percentage of biochar as filler increased the young modulus.

Dahal et al. (2019) studied thermal and mechanical properties of biochar as additive for glass fiber-reinforced composites. Different proportions of biochar and charcoal were used. In comparison with 5% biochar and without biochar composites, 10% biochar composites have better flame-retardant properties.

Natalio et al. (2020) developed lightweight biochar-based composites with gypsum and studied the electromagnetic shielding properties. Gypsum-biochar composites showed an improved electromagnetic shielding efficiency. In their study, they found that pyrolyzed biochar can be used as a future lightweight sustainable composite material.

Ayadi et al. (2020) synthesized the WPCs and biochar polymer composites (BPCs) and compared both biocomposites. Wood and biochar particles were sieved between 100 and 425 m (40 and 140 mesh). The physical and mechanical properties of WPC and BPC were examined at different pyro-gasification temperatures and fiber contents. BPCs showed the better dimensional stability than WPCs, but mechanical properties of BPCs were deteriorated due to inadequate adhesion between the polymer and the biochar.

Das et al. (2021) synthesized the polymer-based biochar composites and investigated the physical, mechanical, thermal, and electrical properties. They found that biochar has excellent additive for manufacturing composite materials. As a filler, biochar showed good strength and thermal characteristics.

Lepak-Kuc et al. (2021) synthesized the recycled material-based biochar composites. Conductivity, tensile strength, and material characterization were analyzed by thermal analysis, XRD, and SEM. Composites have good conductivity and flexibility. Biochar can be used as a substitute for carbon nanomaterial (CNT) in composite preparation because it is highly cost-effective.

4. Conclusion and outlook

Biochar is a plentiful renewable resource material made from the pyrolysis of biomass. Biochar can be used as a filler in composite panels. It can be considered as carbon sequestration also. Flexural, tensile, and strength properties were improved with an optimized ratio of biochar. Biochar composites showed the excellent properties. Biochar can be a substitute for CNTs in certain applications due to its lowest cost and environment friendly. Biochar composites have highest resistance toward water. Biochar composites can be used as bio-based light-weighted composites for future.

References

Ahmetli, G., Kocaman, S., Ozaytekin, I., Bozkurt, P., 2013. Epoxy composites based on inexpensive char filler obtained from plastic waste and natural resources. Polym. Compos. 34, 500–509.

Akom, M., Fanyin – Martin, A., Oti-Boateng, C., Otoo, E., Dawoe, E., 2020. Yield and cost of biochar produced by a locally fabricated reactor. J. Agric. & Envir. Sci. 9 (2), 1–5. https://doi.org/10.15640/jaes.v9n2a1.

Ayadi, R., Koubaa, A., Braghiroli, F., Migneault, S., Wang, H., Bradai, C., 2020. Effect of the pyro-gasification temperature of wood on the physical and mechanical properties of biochar-polymer biocomposites. Materials 13, 1327. https://doi.org/10.3390/ma13061327.

Behazin, E., Misra, M., Mohanty, A.K., 2017. Compatibilization of toughened polypropylene/biocarbon biocomposites: a full factorial design optimization of mechanical properties. Polym. Test. 61, 364–372.

Bhattacharyya, D., 2016. Effects of biomass types on flammability and mechanical properties of sustainable biochar-based composites. In: 3rd International Conference on Past and Present Research Systems of Green Chemistry September 19–21, 2016 Las Vegas, USA. Org. Chem. Curr. Res., vol. 5, 3Suppl. l.

Cuthbertson, D.M., 2018. The Production of Pyrolytic Biochar for Addition in Value-Added Composite Material, p. 5182. Electronic Thesis and Dissertation Repository. https://ir.lib.uwo.ca/etd/5182.

Das, O., Bhattacharyya, D., Hui, D., Lau, K.T., 2016. Mechanical and flammability characterisations of biochar/polypropylene bio composites. Compos. B Eng. 106, 120–128.

Das, O., Sarmah, A.K., Bhattacharya, D., 2015. A novel approach in organic waste utilization through biochar addition in wood/polypropylene composites. Waste Manag. https://doi.org/10.1016/J.WASMAN.2015.01.105. United States.

Das, O., Kim, N.K., Hedenqvist, M.S., Lin, R.J.T., Sarmah, A.K., Bhattacharyya, D., 2018. An attempt to find a suitable biomass for biochar-based polypropylene bio composites. Environ. Man 62, 403–413. https://doi.org/10.1007/s00267-018-1033-6.

Das, O., Hedenqvist, M.S., 2018. Wettability properties of biochar added wood/polypropylene composites. Academ. J. Polym. Sci. 1 (4), 66–69. AJOP.MS.ID.555570.

Dahal, R.K., Acharya, B., Saha, G., Bissessur, R., Dutta, A., Farooque, A., 2019. Biochar as a filler in glassfiber reinforced composites: experimental study of thermal and mechanical properties. Compos. B Eng. 175, 107169.

Das, C., Tamrakar, S., Kiziltas, A., Xie, X., 2021. Incorporation of biochar to improve mechanical, thermal and electrical properties of polymer composites. Polymer 13, 2663. https://doi.org/10.3390/polym13162663.

DeVallance, D.B., Oporto, G.S., Quigley, P., 2015. Investigation of hardwood biochar as a replacement for wood flour in wood–polypropylene composites. J. Elastom. & Plast. 48 (6), 510–522. https://doi.org/10.1177/0095244315589655.

Giorcelli, M., Kha, A., Pungo, N.M., Rosso, C., Tagliaferro, A., 2018. Biochar as a cheap and environmentally friendly filler able to improve polymer mechanical properties. Biomass Bioenergy 120, 2018. https://doi.org/10.1016/j.biombioe.2018.11.036.

Giorcell, M., 2018. Biochar as a high-performance low-cost filler for polymer composites. In: 3rd Edition of International Conference and Exhibitions on Polymer Chemistry, Vienna, Austria. https://doi.org/10.4172/2471-9935-C1-008.

He, X., Gong, Z., She, D., Zhang, B., Gao, H., 2011. Implications of production and agricultural utilization of biochar and its international dynamic. Trans. Chin. Soc. Agric. Eng. 27 (2), 1–7.

Idriss, B.M., Grema, A.S., Baba, D., Musa, M.A., Mohammed, H.I., 2018. Optimization of biochar production from beech wood pyrolysis. Int. J. Sci. Environ. Technol. 7 (3), 781–788.

Jawaria, A., Mahmood, T., Khalid, A., Siddique, T., Aziz, I., 2014. Optimization of pyrolysis yields of paper Mulberry (broussonetia). Int. J. Agric. Biol. 16 (5), 929–934.

Jindo, K., Mizumoto, H., Sawada, Y., Sanchez-Monederol, M.A., Sonoki, T., 2014. Physical and chemical characterization of biochars derived from different agricultural residues. Bio Geosci. 11, 6613–6621.

Khan, A., Savi, P., Quaranta, S., Rovere, M., Giorcelli, M., Tagliaferro, A., Jia, C., 2017. Low-cost carbon fillers to improve mechanical properties and conductivity of epoxy composites. Polymers 9 (12), 642.

Kumar, N.S., Shaikh, H.M., Asif, M., Al-Ghurabi, E.H., 2021. Engineered biochar from wood apple shell waste for high-efficient removal of toxic phenolic compounds in wastewater. Sci. Rep. 11, 2586.

Lee, Y., Park, J., Ryu, C., Gang, K.S., Yang, W., Park, Y.K., Hyun, S., 2013. Comparison of biochar properties from biomass residues produced by slow pyrolysis at 500°C. Bioresour. Technol. 148, 196–201.

Lepak-Kuc, S., Kicinski, M., Michalski, P.P., Pavlov, K., Giorcelli, M., Bartoli, M., Jakubowska, M., 2021. Innovative biochar-based composite fibres from recycled material. Materials 14, 5304. https://doi.org/10.3390/ma14185304.

Li, Z., Reimer, C., Wang, T., Mohanty, A.K., Misra, M., 2020. Thermal and mechanical properties of the biocomposites of miscanthus biocarbon and poly (3-Hydroxybutyrate-co-3-Hydroxyvalerate) (PHBV). Polymers 12, 1300.

Mohanty, A., Vivekanandhan, S., Anstey, A., Misra, M., 2015. Sustainable composites from renewable biochar and engineering plastic. In: 20th International Conference on Composite Materials Copenhagen.

Noumi, E.S., Blin, J., Rousset, P., de Cassia, O., Carneiro, A., 2014. Optimization of pyrolysis conditions for production of reductant charcoal. In: III Forum Nacional sobre Carvão Vegetal.

Natalio, F., Corrales, T.P., Feldman, Y., Lew, B., Graber, E.R., 2020. Sustainable lightweight biochar-based composites with electromagnetic shielding properties. ACS Omega 5, 32490–32497. https://doi.org/10.1021/acsomega.0c04639.

Nizamuddin, S., Jadhav, A., Qureshi, S.S., Baloch, H.A., Siddiqui, M.T.H., Mubarak, N.M., Grifn, G., Madapusi, S., Tanksale, A., Ahamed, M.I., 2019. Synthesis and characterization of polylactide/rice husk hydrochar composite. Sci. Rep. 9, 5445. https://doi.org/10.1038/s41598-019-41960-1.

Oral, I., 2016. Determination of elastic constants of epoxy resin/biochar composites by ultrasonic pulse echo overlap method. Polym. Compos. 37 (9), 2907–2915.

Peterson, S.C., 2012. Evaluating corn starch and corn stover biochar as renewable filler in carboxylated styrene–butadiene rubber composites. J. Elastomers Plastics 44 (1), 43–54.

Peterson, S.C., 2013. Utilization of low-ash biochar to partially replace carbon black in styrene–butadiene rubber composites. J. Elastomers Plastics 45, 487–497.

Ren, X., Gou, J., Wang, W., Li, Q., Chang, J., Li, B., 2013. Optimization of bark fast pyrolysis for the production of phenol-rich bio-oil. Bioresources 8 (4), 6481–6492.

Richard, S., Rajadurai, J.S., Manikandan, V., 2016. Influence of particle size and particle loading on mechanical and dielectric properties of biochar particulate-reinforced polymer nanocomposites. Inter. J. Polym. Analys. Charac. 21 (6), 462–477.

Sun, J., Lian, F., Liu, Z., Zhu, L., Song, Z., 2014. Biochars derived from various crop straws: characterization and Cd (II) removal potential. Ecotoxic. & Environm. Safety 106, 226–231. https://doi.org/10.1016/j.ecoenv.2014.04.042.

SundaraKannan, R., Arumugaprabu, V., Poomari Muthukumar, G., Vigneshwaran, S., Deepan, S.K., Muneesh, R., 2019. Biochar from sugarcane waste in polymer matrix composite. Inter. J. Innovative Techn. & Explor. Engg. (IJITEE) 9 (2S2), 350–352.

Van Soest, P.J., 1963. Use of detergents in the analysis of fibrous feeds. II. A rapid method for the determination of fibre and lignin. J. Assoc. Off. Agric.Chem. 46, 829–835.

Wretborn, T., 2016. Pyrolysis of Wood Chips: Influence of Pyrolysis Conditions On Charcoal Yield And Charcoal. Luleå University of Technology, Department of Engineering Sciences and Mathematics.

Yuan, Y., Tian, Y., Zhao, L., 2012. The study process of the biochar application. Renewable Energy Resour. 30 (9), 45–49.

Zhang, Q., Yi, W., Li, Z., Wang, L., Cai, H., 2018. Mechanical properties of rice husk biochar reinforced high density polyethylene composites. Polymer 10 (3), 286.

Zhang, Q., Cai, H., Yang, K., Yi, W., 2017. Effect of biochar on mechanical and flame-retardant properties of wood–Plastic composites. Results Phys. 7, 2391–2395. https://doi.org/10.1016/j.rinp.2017.04.025.

Zhang, Q., Xu, H., Lu, W., Zhang, D., Ren, X., Yu, W., Wu, J., Zhou, L., Han, X., Yi, W., 2020a. Properties evaluation of biochar/high-density polyethylene composites: emphasizing the porous structure of biochar by activation. Sci. Total Environ. 737, 139770.

Zhang, Q., Cai, H., Yi, W., Lei, H., Liu, H., Wang, W., Ruan, R., 2020b. Bio composites from organic solid wastes derived biochars: a review. Materials 13, 3923. https://doi.org/10.3390/ma13183923.

Zaverl, M.J., Misra, M., Mohanty, A.K., 2013. Using factorial statistical method for optimising co-injected biochar composites. In: Van Hoa, S., Hubert, P. (Eds.), Proceedings of the Proceedings of the 19th International Conference on Composite Materials, vol. 11. Concordia Centre for Composites, Montreal, QC, Canada, pp. 7802–7809.

CHAPTER

Biochar mechanisms of metal sorption

5

Kalpana Katiyar[1], Pooja Kumari[1], Nitya Singh[1] and Gagandeep Singh[2]
[1]*Department of Biotechnology, Dr. Ambedkar Institute of Technology for Handicapped, Kanpur, Uttar Pradesh, India;* [2]*Central Ayurveda Research Institute, Jhansi, Uttar Pradesh, India*

1. Introduction

Climate changes pose the most horrible challenge for living organisms in the current century. Greenhouse gases are the cornerstone of this. Among these, methane (CH_4), nitrous oxide (N_2O), and carbon dioxide are chief greenhouse gases. Both anthropogenic activities (deforestation, modern agricultural practices, metallurgy, mining, and fossils fuel burning) and biological activities (earthquakes, biomass decay, burning, volcanos, and oceans eruption) are accountable for the rise in atmospheric gases concentration.

Highly congested urban areas and industrial sites face the challenge of heavy metal—polluted soils. The principal heavy metals are mercury, copper, cadmium, arsenic, chromium, lead, and nickel. Soil polluted with heavy metals is a threat to ecosystem biodiversity. Some standard practiced sustainable remediation technologies to eradicate soil pollution and restore soil fertility are excavation, thermal treatment, phytoremediation, and thermal treatment. As the population's pressure continuously grows, more innovative and advanced remediation technologies to restore soil quality are urgently required.

Biochar used as the absorbent to restore the natural environment is recommended and practiced. There is extensive literature on the biochar potential for heavy metal removals from contaminated soils.

In 2011, the European Environment Agency (EEA) reported the soil contamination in approx. 284,000 sites in Europe devised a restoration project with an estimated six billion Euros/year. Heavy metals are the frequent reason for soil contamination, indicated by a national report. Moreover, heavy metals are omnipresent in the environment and adversely impact human health. The most common method used to remove metal contamination from soils is to dig out the affected dirt and dispose of it into landfills. Though these are not an option in a cost-effective and environmentally friendly sustainable remediation technique, an alternative option has been investigated. These include chemical precipitation, ion exchange, electrochemical treatment, and membrane technologies. Biosorption is a technique that is environment friendly and readily available in a large amount due to which it is common and cost-effective, for example, biochar.

Highly fertile soil can be found around the Amazon basin that fulfills the agricultural requirements for many centuries. It is a dark-colored soil found up to two meters in depth in the specific section of

Integrative Strategies for Bioremediation of Environmental Contaminants, Volume Two. https://doi.org/10.1016/B978-0-443-14013-6.00012-3

terra preta. This dark soil indicates large charcoal and organic matter concentration like plants and animal remains. The productivity of terra preta is because of good nutrient retention and a neutral pH, where the soil's nature is acidic (Talberg, 2009).

The pyrolysis of carbonaceous biomass such as crop residues, agricultural waste, and industrial biomass waste under oxygen-deprived and low temperature produces biochar as a by-product. Biochar is a stabilized biomass. The mixing of biochar to soil in proper ratio leads to improvement in crop yield, soil fertility, and the balanced nutrient cycle of the soil.

The low-temperature, oxygen-starved burning (also known as pyrolysis) of carbonaceous biomass such as crop residues, stall bedding, cull wood, and sawmill wastes produces biochar, a carbon-rich by-product (Lehmann, 2010). Biochar is a kind of stabilized biomass that may be kept in large amounts in the soil. Suppose these forms of carbon have the appropriate physical and chemical properties. In that case, they could improve crop yield by mixing with nutrients and soil mineral particles dynamically or reversibly (Sohi et al., 2010).

Biochar improves soil health and increases the potential of soils to sequester carbon (Talberg, 2009). Biochar plus inorganic fertilizer increased the yield of corn production in the most deteriorated soils compared with inorganic fertilizer alone yield. Since most carbon in biochar exists in aromatic form, which is very resistant in the environment, applying it to soil has been thought to have a high potential for enhancing long-term carbon sequestration (Zhang et al., 2013). Various eco-friendly solutions for sustainable waste management were routinely used. Presently, scientists are striving to develop cost-effective bioremediation methods for heavy metal–contaminated soils (Singh et al., 2020). Biochar properties are influenced by biomass species and preparation circumstances (such as pyrolysis temperature, residence time, and pyrolysis environment), which directly impact biochar's ability to remove heavy metals from soil (Wang et al., 2017).

Using activated charcoal, bioremediation of hydrocarbons and toxic metals is cost-effective and eco-friendly. The critical aspect is reducing the harmful effect of pentachlorophenol, pentachlorophenol, biphenyls, benzene, dioxins, linden, arsenic, chromium, and mercury compared to other value-adding technologies (Singh et al., 2020).

2. Biochar

The feedstock is similar to biochar and the soil potted for plantation (Bracmort, 2010). The International Biochar Initiative defines biochar as "a solid material derived from the thermochemical conversion of biomass in an oxygen-limited environment" (Qian et al., 2015). The charcoal produced from biomass (i.e., plant material and agricultural waste) is biochar (Talberg, 2009).

Biochar is an aging process when produced naturally. Artificially, pyrolysis of biomass produced biochar. Pyrolysis is a slow-burning process in the least amount or absence of oxygen, resulting in fine-grained charcoal. Biochar is used as a soil additive to improve nutrient retention and carbon storage. This is the reason why biochar is differentiated from charcoal. The science of biochar is still unknown as related, whether its history had been through thousands of years. The tactics in ancient slash and char were the reason for dark soil on the earth. The theories exist over biochar history. Slash and char are similar to slash and burn in clearing vegetation within a limited plot and lighting it up and allowing it to smolder rather than burning it down. The terra preta is formed eventually when these smoldering char combined with many other biomasses and are buried under a layer of dirt. Modern scientists have innovated techniques for manufacturing biochar compared to the early practices of slash-and-char (Talberg, 2009).

Biochar is utilized for soil remediation, amelioration, carbon sequestration, organic solid waste composting, water and wastewater purification, catalyst and activator, electrode materials, and electrode modification, all of which have been thoroughly addressed in the literature (Wang and Wang, 2019).

3. Characteristics of biochar

Biochar's ability to adsorb metals is influenced by its physicochemical qualities. Functional groups, surface area, mineral concentrations, pH, surface charge, and porosity must be accurately defined before exploring the potential of biochar for metal sorption (Li et al., 2017).

3.1 Surface area and porosity

The metal sorption capacity of biochar is determined by porosity and surface area factors. Micropores are formed in biochar when biomass is pyrolyzed. It is due to water loss during the dehydration process. The pore size of biochar is highly varied, ranging from nano (<0.9 nm) through micro (<2 nm) to macro (>50 nm). The pore size is an essential factor for metal sorption. Biochar with a small pore size cannot adsorb large sorbate, regardless of their polarity or charges: the surface area and porosity of biochar change in response to the pyrolysis temperature (Li et al., 2017). According to studies, higher pyrolysis temperature results in larger pore sizes and surface area. The effect of pyrolysis temperature on biochar surface area is shown in Fig. 5.1A. When pyrolysis temperature

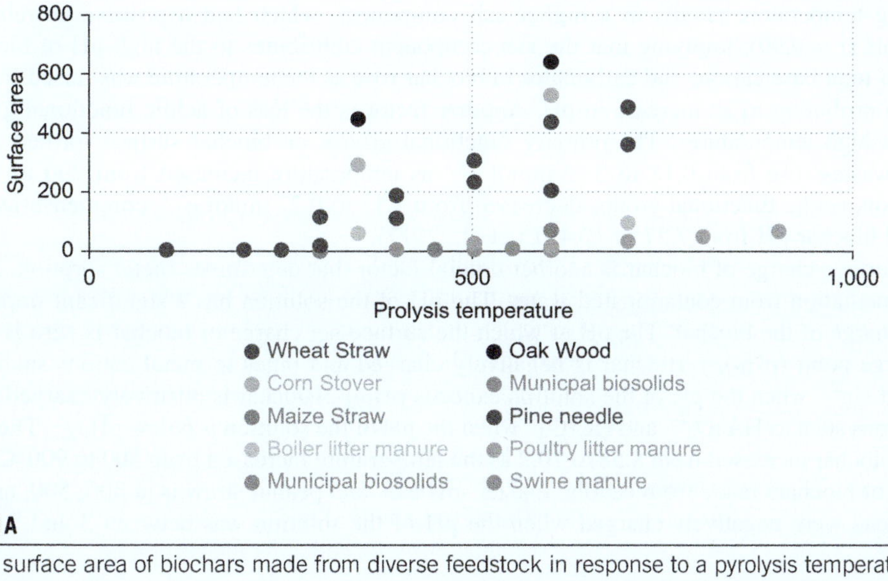

FIGURE 5.1A

Change in surface area of biochars made from diverse feedstock in response to a pyrolysis temperature.

increased from 500 to 900°C during the biosolids biochar formation, the porosity and surface area increased from 0.056 to 0.099 $cm^3\ g^{-1}$ and 25.4 to 67.6 $m^2\ g^{-1}$, respectively. However, biochar formed at high temperatures has a lower surface area and porosity in some circumstances. Biochar made from wheat straw has a smaller surface area at 700°C than 600°C (363 vs. 438 $m^2\ g^{-1}$). The porosity of biochar may be damaged or clogged by tar formed at high temperatures, resulting in a reduced surface area (Li et al., 2017). The composition of the biochar feedstock is also a determining factor for porosity and surface area along with pyrolysis temperature. Biochar formed from biosolids and manure has a significantly lesser surface area (5.4 $m^2\ g^{-1}$–94.2 $m^2\ g^{-1}$) than biochar created from different plant biomass. The effect of varying feedstock on biochar porosity is given in Table 5.1. In general, lignin-rich biomass (such as bamboo and coconut shell) produces macroporous-structured biochar, whereas cellulose-rich biomass (such as husks) produces mostly microporous-structured biochar (Li et al., 2017).

3.2 pH and surface charge

The pH of biochar also fluctuates with pyrolysis temperature and feedstock biomass, just like surface area and porosity. In general, the pH of biochar rises as the pyrolysis temperature increases, as shown in Fig. 5.1B. Biochar is generally alkaline, with occasional deviations depending on the feedstock biomass (Table 5.1). Biochar made from oak wood at 350 and 600°C was acidic (pH:4.84–4.91), whereas at 200°C, oak wood biochar also had a low pH (4.60), but at 400 and 600°C, the biochar was neutral to alkaline (pH:6.90–9.50). Biochars made from corn stover, wheat straw, and wastewater sludge were acidic at low temperatures (200–400°C), with pH ranging from 4.87 to 6.11 (Li et al., 2017).

Positive correlations among biochar pH and pyrolysis temperature have been observed for biochars made from oak wood, biosolids, wheat, corn, maize residues, manure, and conocarpus wastes. Increasing temperature results in a higher ash component, which had a positive correlation with biochar pH ($r = 0.99$), implying that the ash component contributes to the high pH of biochar. The content of total base cations and carbonates in biochar rose as the temperature was raised from 300 to 700°C, contributing to an increase in pH. Another factor is the loss of acidic functional groups like COOH at high temperatures. The primary functional groups on biochar surface formed from conocarpus wastes rose from 0.15 to 3.55 mmol g^{-1} as temperature increased from 200 to 800°C. In comparison, acidic functional groups decreased from 4.17 to 0.22 mmol g^{-1}, commensurate with the increased biochar pH from 7.37 to 12.4 (Li et al., 2017).

The surface charge of biochar is another crucial factor that determines metal sorption. For heavy metal remediation from contaminated water. The pH of the solution has a significant impact on the surface charge of the biochar. The pH at which the surface net charge of biochar is zero is known as zero charge point (pH_{PZC}). Biochar is negatively charged and binds to metal cations such as Cd^{2+}, Pb^{2+}, and Hg^{2+} when the pH of the solution exceeds pH_{PZC}. Biochar is positively charged and binds metal anions such as $HAsO_4^{2-}$ and $HCrO_4^-$ when the pH of the solution is below pH_{PZC}. The pH_{PZC} of biosolid biochar increased from 8.58 to 10.2 as the temperature increased from 500 to 900°C. The zeta potential of biochars made from canola, maize, soybean, and peanut straw is at 300, 500, and 700°C. All biochars were negatively charged when the pH of the solution was between 3 and 7 (Li et al., 2017).

Table 5.1 Physicochemical properties of biochars produced from various feedstock under different pyrolysis temperatures.

Biochar feedstock	Pyrolysis temperature (°C)	pH	Surface area (m^2g^{-1})	Atomic ratio			Mineral elements (%)						References
				N/C	H/C	O/C	P	Mg	K	Ca	AsIII (pH)	Pb (pH)	
Broiler litter manure	350	–	59.5										(Li et al., 2017; Uchimiya et al., 2010)
	700	–	94.2										
Swine manure	400	10.0	5.8				0.97	1.57	1.62	2.03	6.0	3.0	
	600	10.4	10.6				1.55	2.13	3.53	2.89			
Municipal solids	500	8.81	25.4	0.075	0.48	0.45	1.82	1.47	0.85	5.93			
	600	9.54	20.3	0.064	0.22	0.30	1.88	1.55	0.85	6.27			
	700	11.1	32.2	0.048	0.15	0.30	2.04	1.64	0.99	6.44			
	800	12.2	48.5	0.026	0.03	0.17	1.93	1.66	0.93	6.58			
	900	12.2	67.6	0.029	0.09	0.12	2.02	1.75	0.87	6.96			
Wheat straw	300		116		0.55	0.29					4.0	5.2	(Li et al., 2017; Chen et al., 2014)
	400		189										
	500		309		0.36	0.09							
	600		438										
	700		363		0.15	0.05							
Poultry litter manure	400	9.50	5.4				1.22	1.73	3.88	2.83			
	600	10.4	6.3				1.54	2.40	5.88	3.59			
Pine needle	100		0.65	0.012	1.44	0.62							
	200		6.22	0.013	1.91	0.48							
	250		9.52	0.012	1.08	0.40							
	300		19.9	0.014	0.75	0.28							
	400		112	0.013	0.45	0.17					3.5	5.0	
	500		236	0.012	0.33	0.14							
	600		207	0.010	0.26	0.10							
	700		491	0.011	0.18	0.10							
Oak wood	350	4.84	450	0.001	0.06	0.26					3.5	5.0	(Li et al., 2017; Nguyen et al., 2010)
	600	4.91	642	0.001	0.02	0.10							

Continued

Table 5.1 Physicochemical properties of biochars produced from various feedstock under different pyrolysis temperatures.—cont'd

Biochar feedstock	Pyrolysis temperature (°C)	pH	Surface area (m^2g^{-1})	Atomic ratio			Mineral elements (%)						References
				N/C	H/C	O/C	P	Mg	K	Ca	AsIII (pH)	Pb (pH)	
Maize straw	300	9.84	1.00	0.026	0.07	0.49							
	450	10.5	4.00	0.023	0.06	0.35							
	600	11.4	70.0	0.020	0.03	0.26							
Corn stover	350	5.88	293	0.013	0.07	0.37					5.0	6.0	
	600	6.71	527	0.012	0.03	0.21							
Wheat straw	200	6.11		0.018	1.42	0.58							
	400	10.8		0.012	0.63	0.20							
	600	11.0		0.011	0.26	0.13							
Municipal biosolids	400	8.46	5.49	0.120	1.01								
	450	9.74	7.21	0.120	0.87								
	500	9.75	7.73	0.110	0.68								
	550	10.5	8.45	0.110	0.58								
	600	11.7	5.99	0.090	0.43								
Conocarpus wastes	200	7.37		0.011	0.06	0.41	0.08	0.34	0.04	4.34			(Li et al., 2017; Al-Wabel et al., 2013)
	400	9.67		0.012	0.04	0.18	0.09	0.40	0.05	5.18			
	600	12.2		0.009	0.02	0.08	0.11	0.48	0.09	6.47			
	800	12.4		0.011	0.01	0.06	0.13	0.78	0.12	6.75			
Oak wood	200	4.60		0.014			0.03	0.04	0.13	0.39			
	400	6.90		0.021			0.06	0.15	0.38	1.18	3.5	5.0	
	600	9.50		0.029			0.06	0.18	0.44	1.39			
Wastewater sludge	300	532		0.111	1.05	0.24	2.50	0.35	0.10	3.47		5.0	
	400	487		0.102	0.76	0.17	2.80	0.43	0.11	4.17			
	500	727		0.090	0.52	0.02	3.30	0.46	0.18	4.62			
	700	12.0		0.050	0.30	0.00	3.60	0.54	0.20	5.35			

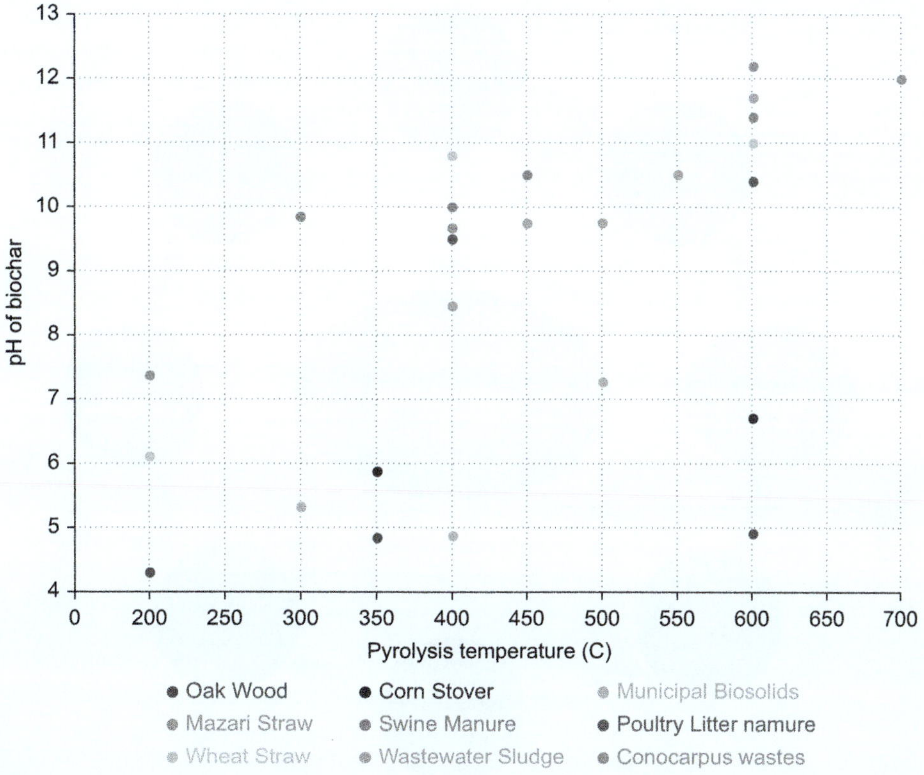

FIGURE 5.1B

pH of biochars made from diverse feedstock at temperatures ranging from 100 to 900°C during pyrolysis.

However, compared to biochars made at 300 and 500°C, those made at 700°C were less negatively charged, showing that increasing pyrolysis temperature increases pH_{PZC}. The number of negatively charged functional groups on biochars (e.g., −COO−, −COH, and −O.H.) decreases as the temperature rises, resulting in less-negative surface charges and greater pH_{PZC} (Li et al., 2017) (Fig. 5.2).

3.3 Functional groups

Metal sorption is aided by functional groups such as carboxylic, amino, and hydroxyl groups. The number of functional groups on the surface of biochar is controlled by two critical factors: pyrolysis temperature and biochar feedstock. However, unlike surface area, porosity, pH, and pH_{PZC}, the number of functional groups in biochar decreases as temperature rises, owing to the higher degree of carbonization (Li et al., 2017).

The atomic ratios of H/C, O/C, and N/C drop as temperature rises (Table 5.1), implying a decrease in the quantity of hydroxyl, carboxylic, and amino groups. The strength of peaks related to carboxylic

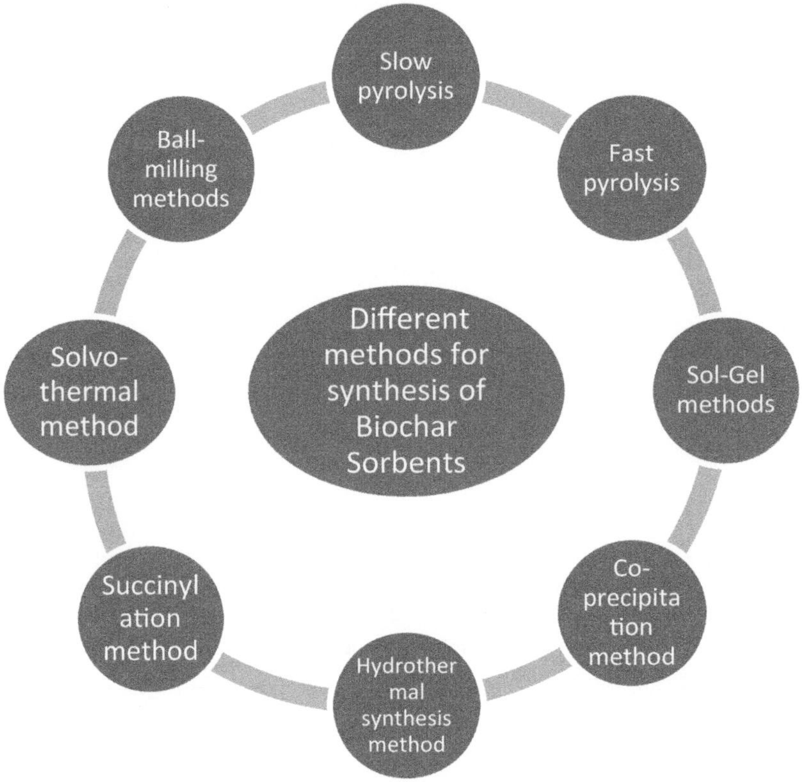

FIGURE 5.2

Schematic representation of different methods for synthesis of biochar sorbents.

(−COOH) and hydroxyl (−O.H.) groups reduced as the temperature climbed from 300 to 700°C for biochars made from canola, corn, soybean, and peanut straw, according to Fourier transform infrared (FTIR) photoacoustic spectroscopy. Manure biochars produced at 700°C showed a similar loss of FTIR spectrum characteristics as those made at 350°C. Most functional groups of lignocellulosic materials are lost during pyrolysis at increasing temperatures (Li et al., 2017).

3.4 Mineral composition

Mineral components also facilitate metal sorption from water in biochar, such as potassium (K), calcium (Ca), magnesium (Mg), and phosphorus (P). They can react with heavy metals and restrict their availability by exchanging or precipitating. The predominant mechanism governing Pb sorption by dairy-manure biochar was the precipitation of Pb phosphates such as Pb pyromorphite and Pb hydropyromorphite. Base cations such as Ca^{2+}, Mg^{2+}, Na^{+}, and K^{+} were released into the solution via cation exchange after the sorption of metal cations (Cd^{2+}, Cu^{2+}, Ni^{2+}, and Pb^{2+}) over broiler litter biochar (Li et al., 2017).

The amount of mineral components in biochar is controlled by the pyrolysis temperature and the feedstock (Table 5.1). For biochar made from biosolids, manure, conocarpus wastes, and oak wood, total K, Ca, Mg, and P concentrations rise as the pyrolysis temperature increases. Biochar yield is decreased at higher temperatures, resulting in mineral enrichment in biochar. Mineral component concentrations that are water-soluble, on the other hand, react differently than total concentrations. Water-soluble concentrations of K, Ca, Mg, and P generally increase when heated to 200°C during biochar synthesis but drop above that temperature. This is most likely due to enhanced crystallization, as demonstrated by the development of whitlockite [$(Ca, Mg)_3(PO_4)_2$] or integration into the silicon structure at 500°C, which is less soluble. Aside from temperature, feedstock has a significant impact on mineral component concentrations in biochar. In general, the P concentration of oak wood biochar (0.03%–0.06%) is lower than that of biosolid biochar (1.82%–3.60%). Biochars made from poultry litter and swine dung have greater K levels (1.6%–5.9%) than biochars made from other materials (Li et al., 2017).

4. Properties of biochar

The three organic molecules cellulose, hemicellulose, and lignin make up the majority of biomass. Because they perform differently during heat treatment, the biomass composition directly impacts product production and attributes (Weber and Quicker, 2018).

- Hemicellulose is a term that refers to a group of polysaccharides that have a branched chain structure. It is the most reactive of the three primary components, decomposing at temperatures between 220 and 315°C. The critical process in torrefaction is the breakdown of hemicellulose in the biomass. Torrefaction is a particularly delicate process that can be difficult to regulate because the most substantial changes to practically all parameters occur in the restricted temperature range of hemicellulose degradation (Weber and Quicker, 2018).
- Like hemicellulose, cellulose is a polysaccharide with an unbranched structure. It has higher thermal stability, disintegrating between 280 and 400°C. For decades, cellulose has been investigated as the most prevalent organic substance on the planet. The thermal decomposition, however, is still a mystery (Weber and Quicker, 2018).
- Lignin is a three-dimensional macromolecule made up of a variety of chemical linkages. As a result, unlike hemicellulose and cellulose, breakdown does not occur within a narrow temperature range. Due to the vast number of functional groups with varying thermal stabilities, lignin decomposes over a wide temperature range. Thermal deterioration begins at 200°C and can progress to temperatures as high as 900°C (depending on the residence period) (Weber and Quicker, 2018) (Table 5.2).

The proportion of cellulose, hemicellulose, and lignin determines the necessary treatment temperature and the mass yield for any given set of circumstances due to their varying thermal stabilities. Due to their varied origins, waste biomasses such as animal manure and sewage sludge do not contain significant amounts of these components and must be classified accordingly (Weber and Quicker, 2018).

Chain propagation of aliphatic carbon compounds can occur due to several activities. The reaction propagates with radicals collision and generation of a more stable carbon compound by random

Table 5.2 Chemical properties of biochar is explained in the given table.

		Chemical properties of biochar
1	Atomic ratios of C:H:O	❖ The van Krevelen diagram describes the evolution of C, O, and H during the carbonization process. ❖ Pyrolysis treatment temperature and decrease in the atomic ratio do not occur linearly. ❖ For woody and straw-like biomass, a rise in pyrolysis temperature results in low H/C and O/C ratios. ❖ For fossils fuel biomass, a rise in pyrolysis temperature results in a decrease in O/C ratio at a higher rate than the H/C ratio.
2	Elemental composition	❖ Biochar production leads to an increase in total carbon content and change in chemical composition in comparison to feedstock biomass. ❖ During pyrolysis, functional group detachment occurs and results in higher C content and lower H and O content.
3	Energy content	❖ Increase in pyrolysis temperature results in an increase in total energy content. ❖ From 250 to 350°C temperature, the most significant increase in energy content occurs. ❖ The residence time also affects the energy content.
4	Fixed carbon and volatile matter	❖ Fixed carbon refers to left out carbon after volatile matter release. ❖ 90%–95% fixed carbon biochar is used for the metallurgical process. ❖ When pyrolysis temperature increases up to 700°C, fixed carbon content increases from 10% to 90%, and volatile matter decreases from 90% to 10%.
5	Structural composition	❖ At the temperature range 250–350°C, partial decomposition of lignin and cellulose and complete disintegration of hemicellulose occurs. ❖ At 400°C, with 5 min residence time forms a mixture of hemicellulose, cellulose, and insoluble carbohydrate (holocellulose).
6	Functionality	❖ The carbonization process results in a low H/C ratio and more aromatic structures. ❖ Aromatic structures are responsible for high thermodynamic stability. ❖ The chief aromatic structures are randomly organized aromatic rings that exist in an amorphous phase, and condensed polyaromatic sheets exist in the crystalline phase.
7	pH value	❖ Biochar produced from hydrothermal carbonization and pyrolysis carbonization has different pH values. ❖ As the degree of carbonization increases, the pH value of biochar increases. ❖ Treatment temperature is the most determining factor for the pH value of biochar.
8	Cation exchange capacity (CEC)	The CEC represents the material's holding capacity of exchangeable cations (Ca^{+2}, Na^{+}, NH_4^{+}, Mg^{+2}, and K^{+}). The CEC of biochar determines surface charge, surface area, and surface structure.
9	Ash content and composition	❖ Biochar having high ash content is unsuitable for many industrial applications. ❖ The ash content of biochar is directly related to the ash content of the parent biomass. ❖ The ash contents are K, Ca, Na, and Mg referred to as alkali and alkaline earth metals (AAEM).

Table 5.2 Chemical properties of biochar is explained in the given table.—cont'd

	Chemical properties of biochar	
10	Reactivity	❖ The Boudouard reaction, the partial oxidation of carbon, and the water gas reaction are used to describe the reactivity of biochar. ❖ $C + CO_2 = 2CO$ ❖ $C + H_2O{=}CO + H_2$ ❖ $a.C + b.O_2 = c.CO + d.CO_2$ ❖ Reactivity represents the rate of these reaction. ❖ Reactivity depends on the concentration of the reactive gases and temperature. ❖ No standard method exists for the determination of biochar reactivity.
11	Self-heating and degradation during storage	❖ Due to high carbon content, biochar is associated with the risk of spontaneous combustion and self-burning. ❖ Exothermic reactions and thermal runway are responsible for self-ignition. ❖ The biochar factors like water content, particle size, ambient temperature, oxygen content, and storage conditions determine self-ignition and spontaneous combustion. ❖ Biochar is categorized as noninflammable material, and autoignition temperature remains lower than feedstock biomass.

depolymerization, the process comes to an end. The paths of free radical production and collision, on the other hand, are complicated. The pyrolysis of lignin is still not completely understood (Hassan et al., 2020).

5. Biochar production

Biochar generation by pyrolysis is considered a carbon-negative process because the biochar sequesters carbon while also improving the fertility of the soil on which the bioenergy feedstock grows. The biochar production system is powered by the energy it generates. Syngas, biooil, and biochar are the three significant outputs of a biochar production system (Bracmort, 2010).

Biochar is made by feeding biomass into a pyrolysis kiln, which burns with little or no oxygen. Crop residues, wood and wood waste, animal dung, and other organic resources could be used as feedstock biomass for biochar production. Two main products emerge from the kiln at the end of this process. The first is biochar, which typically accounts for around half of the carbon content of the biomass, and biofuel is another one. Syngas is a mixture of mostly hydrogen and carbon monoxide with a small amount of carbon dioxide used as a biofuel (Talberg, 2009).

Generally, pyrolysis occurs at temperatures below 700°C, although some characteristics, such as the pace of pyrolysis and the amount of oxygen present, can be changed. Faster pyrolysis yields more oils and liquids, while slower pyrolysis yields more syngas. The yield of biochar is improved by reducing the amount of oxygen present during pyrolysis (Talberg, 2009).

A second stage, gasification, can be performed after pyrolysis. The char is gasified to release more energy-rich syngases (usually hydrogen-based). A "gas cleanup" stage may also be used to remove particles, hydrocarbons, and soluble materials from the gas (Talberg, 2009).

The biofuel produced during the pyrolysis process can be utilized to generate the electricity required to run the kiln or other stages of the process. As a result, the system can operate without external power sources. Fig. 5.3 summarizes the pyrolysis process (Talberg, 2009).

The carbonization process in a traditional charcoal-producing method (charcoal pit) might take many weeks. Slow pyrolysis temperatures are typically approximately 500°C. However, they might vary depending on the desired product qualities. Treatment temperatures close to 1000°C that may be required for agricultural residues and other materials with low ash melting temperatures, which can be used for woody feedstock but create issues for agricultural residues and other materials with low ash melting temperatures. As a result, they are rarely heated above 700°C. Torrefaction is the term for pyrolysis between 200 and 300°C. Keeping and concentrating the majority of the energy in the solid is the main aim, as well as to improve significantly mechanical qualities of the biomass (such as grindibility), which can be limiting in some applications (Weber and Quicker, 2018).

In field trials, char and compost mixtures are frequently employed to create a soil amendment similar to the fertile terra preta soils of the Amazon region. Even though char is only a minor weight component of this soil amendment, it is a crucial component of the finished product (Meyer et al., 2011).

Although chemical pretreatment of biomass is a standard method in the manufacturing of activated carbons and has been shown to improve BET surface area, it is not employed in creating biochar. These impregnating chemicals were chosen to take advantage of the possibility of K and P leaking into the soil. In addition, the release of heavy metals from biochar was investigated using standard leaching assays to determine the viability of adding biochar to soils (Agrafioti et al., 2013).

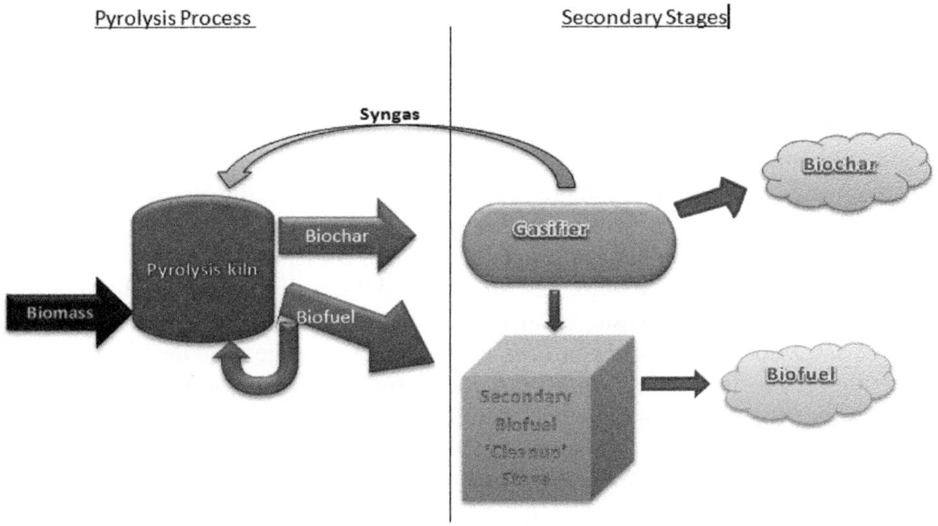

FIGURE 5.3

Simplified pyrolysis process flow diagram.

6. Pyrolysis

Pyrolysis engineers have traditionally sought to minimize char production as it has been viewed as a complex low-value waste fraction. Consequently, there has been much research effort on fast pyrolysis, which maximizes condensable oil and gasification (which is a process similar to fast pyrolysis but includes a limited supply of oxygen). These produce predominantly heavy oil or gas, yielding only small proportions of char. Developing integrated systems that make energy and char at high efficiency by slow pyrolysis is still, therefore, mainly on the research scale, with technology commercially deployed only at a handful of locations (Sohi et al., 2010).

Pyrolysis is defined as the chemical breakdown of a substance in the absence of oxygen at extremely high temperatures: the feedstock, pyrolysis temperature, and processing time influence biochar production amount and quality. "Rapid" pyrolysis (500°C) yields biochar in seconds, whereas "slow" pyrolysis yields far more biochar in hours (Bracmort, 2010).

Pyrolysis systems can be divided into three categories.

- Plants for central pyrolysis that processes all of a region's biomass.
- Individual or small groups of farmers can benefit from low-tech pyrolysis kilns (some later steps, such as gasification or gas cleanup, may be missing from these kilns).
- Syngas-powered pyrolysis trucks that may be driven around to treat biomass within a region. The biochar and biooil would be returned to the clients by truck (Talberg, 2009).

7. Advantages of biochar

- Biochar has the advantage of being able to be made from either tiny, essential mobile equipment or more extensive, fixed ones.
- Biochar has several potential environmental benefits, some of which are not quantifiable, whether used as a soil amendment or burned as an energy source (e.g., cooking and heating). Carbon sequestration, greenhouse gas emission reduction, and soil fertility are the three main benefits (Bracmort, 2010).

7.1 Carbon sequestration

Carbon sequestration captures and stores carbon so that it is not released into the atmosphere. Depending on the type of biomass feedstock being pyrolyzed, studies show that biochar sequesters roughly 50% of the carbon present within the biomass feedstock. The excess carbon is liberated during pyrolysis and can be collected for energy production. Large amounts of carbon can be sequestered in the soil for extended periods (hundreds to thousands of years), but specific estimates of carbon stored due to biochar application are difficult to come by. A 250-hectare farm, according to one scientist, could hold 1900 tons of CO_2 every year (Bracmort, 2010).

7.2 Greenhouse gas emission reduction

N_2O and CH_4 are the primary greenhouse gases linked with agriculture. Agricultural sources of N_2O emissions include cropland soils and grazing areas. CH_4 emissions from agriculture are caused mainly by livestock manure management and enteric fermentation. When applied to the earth, biochar can significantly reduce the release of N_2O, lowering greenhouse gas emissions. In one study, N_2O emissions were cut by 40%, despite the greenhouse gas being 298 times more potent than carbon dioxide in terms of global warming potential. Laboratory findings show that soil moisture and aeration influence N_2O emission reductions from biochar-treated soil. If biochar is put to the ground rather than burned for energy, greenhouse gas emissions might be reduced by 12%−84% (Bracmort, 2010).

7.3 Soil fertility

Biochar helps plants absorb nutrients and improves soil fertility. The use of biochar can help prevent the entry of toxic nutrients and pesticides into groundwater and soil erosion discharge into surface waters. Biochar may positively impact folks in underdeveloped countries if used for soil fertility. Crop yields may rise in impoverished tropical and subtropical areas with ample plant material feedstock, low-cost cooking fuel, and agricultural soil replenishment demands (Bracmort, 2010).

8. Disadvantages of biochar

- Given that biochar technology is still in its early phases of research, many people are concerned about its applicability in the United States. Feedstock availability, biochar management, and biochar system deployment are three factors to consider. The ability of the agricultural community to afford and run a system that is complimentary to present farming and forestry practices is critical to the successful introduction of biochar technology (Bracmort, 2010).

8.1 Feedstock availability

The availability of an ample feed supply for biochar production is an area that needs to be investigated further. Plant and crop residues, a core domain of the agricultural sector, have been the principal source of biochar feedstock. Because woody biomass is a cost-effective, readily available, and suitable feedstock, the forestry community may have a role to play. The benefits of using manure as a biomass feedstock are poorly understood. According to a group of Australian researchers, manure-based biochar offers advantages over plant-derived biochar because it is a by-product of another business and is regarded as a waste substance with little or no value in some areas. As a result, it can cut costs and address sustainability concerns associated with using purpose-grown biomass in the process (Bracmort, 2010).

8.2 Biochar handling

Biochar's use as a soil amendment seems ripe for further investigation. Specific concerns include when is the best time to apply biochar and how to ensure that it stays in place and does not endanger human health or impair air quality. If biochar is mistreated, particulate particles in the form of dust that is difficult for the human body to filter may be spread excessively. Because biochar is an explosive substance, there are possible public safety risks while handling it. Furthermore, the amount of land available for biochar application needs to be investigated further (Bracmort, 2010).

8.3 Biochar system deployment

Biochar systems are tailored to the feedstock to be decomposed as well as the energy requirements of a given operation. Expecting a "one-size-fits-all" conventional biochar system is unrealistic. Proponents claim that a succession of mass-produced biochar systems tailored to the demands of a particular part of the agriculture or forestry communities could be achievable (e.g., In the Southeast, there is a forestry community; in the Midwest, there is a maize grower community; and in the Mid-Atlantic, there is a poultry producer community.). The cost of biochar systems, their operation time, partnership with utility providers for the sale of biooil, and the availability of information concerning technology reliability would all play a role in their widespread implementation (Bracmort, 2010).

9. Biochar application

- The potential role of biochar in climate change mitigation has made it famous. Biochar is high in carbon, and depending on its final usage, the carbon may be retained, delaying or preventing carbon dioxide gas release into the atmosphere. Biochar's advantages extend beyond this, including the agricultural industry and various waste management applications. In addition, as previously stated, its manufacturing process produces biofuel, a renewable energy source that is both sustainable and renewable (Talberg, 2009).
- Biochar has been shown to be beneficial in the CH_4 fermentation process in several experiments. The absorbing characteristics of biochar were used in the following technique to block chemicals such as ammonia and limonene. The inclusion of biochar in the fermentation process helps faster organic matter breakdown and pH leveling, among other things (Czekała et al., 2019).
- Biochar is becoming more widely employed in the composting process, and its primary function is to reduce gas emissions and process temperature (Talberg, 2009).
- The structure-forming capabilities of biochar, which can alter the breakdown of organic matter and the degree of compost humification, are also essential in the composting process (Talberg, 2009).
- Biochar appears to be used to treat phosphorus in the aquatic environment, which is both economically and environmentally beneficial. This is related to the issue of water eutrophication and the scientific interest in biochar application in this field (Czekała et al., 2019).
- Biochar has been shown to improve biological nitrogen fixation (BNF) in legume crops when applied to soils (Shaaban et al., 2018).
- Applying acidic/neutral biochar to alkaline soils could lower pH, altering the solubility of soil nutrients like P and trace elements (El-Naggar et al., 2019).

10. Agricultural and waste management benefits

Because of its long-term persistence in soil, biochar can play a multifunctional role in agricultural and environmental sustainability development. There was a higher concentration of stable organic matter with around 70%–80% carbon levels and mineral stuff, which included nutrients (Singh et al., 2020).

When added to soil in sufficient amounts, pyrolyzed biomass of agricultural waste has a high surface area, porosity, variable charge, and functional groups (Fig. 5.4), which can improve soil water-

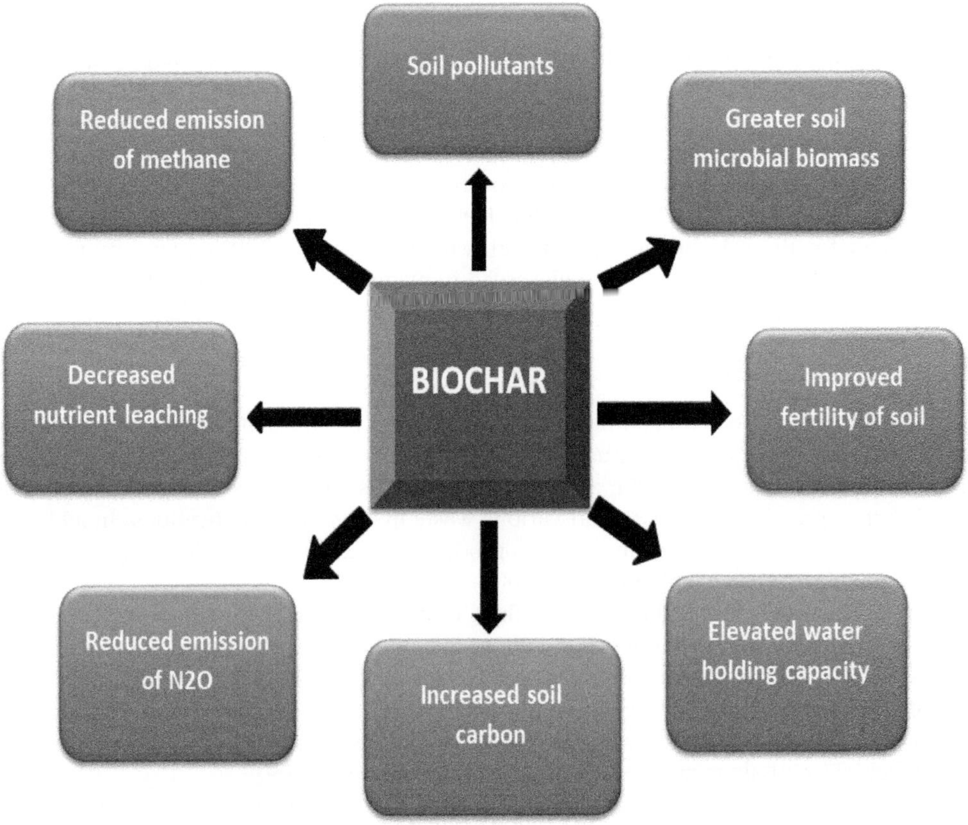

FIGURE 5.4

Biochar uses in agricultural and environmental advances (Singh et al., 2020).

holding capacity, pH, cation exchange capacity (CEC), base saturation, bioremediation of hydrocarbons and heavy metals, and finally crop resistance to disease. As the pyrolysis temperature and the types of feedstock change, the property varies depending on it (Singh et al., 2020).

Carbon sequestration, soil fertility improvement, pollutant immobilization, pollutant remediation, and waste management are among the most frequently touted benefits of biochar. Furthermore, adding biochar to soil can enhance bacterial abundance and reduce soil bulk density, which is vital for the bioremediation of soil contaminants. In agricultural, urban, and contaminated soils, biochar strongly absorbed salts and reduced the effect of salt on plants. Biochar has improved the accessibility of P and Zn and the total N concentrations in treated soil. Biochar can boost agricultural yields while avoiding the use of synthetic fertilizers. Due to its solid-binding pattern to soil particles, biochar can also help to reduce soil erosion and nutrient leaching (Singh et al., 2020).

Biochar can help the agricultural sector in two ways: soil enhancement and animal and crop waste disposal (Talberg, 2009).

Biochar production and use may be motivated by soil improvement and, as a result, greater productivity. Field trials have been going place worldwide since 1980, testing with the application of different forms of biochar on different soils. The type of biochar produced depends on the biomass used—in many cases, rice, wood, or bark—and product characteristics such as pyrolysis rate and kiln size. Acidic soils have been the focus of the majority of the investigations, which have all been conducted in tropical or semitropical climates. Experiments have also used different treatments, such as applying more or less biochar, as well as using and not using other fertilizers (Talberg, 2009).

The fact that biochar production employs organic waste is a second benefit to the agricultural sector (as well as a few others, such as the paper industry). If left to collect, animal and agriculture waste can contaminate ground and surface waters. Contamination can be avoided through waste management procedures, although they can be costly. If the economic expenses of waste treatment can be maintained below those of biochar, it is a viable option. The biochar production method turns waste into a resource by accepting organic material as an input. By working at a temperature exceeding 350°C, the pyrolysis process decreases the weight and volume of the feedstock while also removing possible pathogens that could be an issue if applied directly to soils. Green urban garbage and waste from various industrial processes like paper milling can be employed (Talberg, 2009).

District Chemical and Agricultural Stations with proper credentials conduct physical, physicochemical, and chemical examinations of fertilizers and soils. Using the results of these fertilizer assessments, one can determine the appropriate dosage of selected organic fertilizers to achieve the best results for the earth. Biochar has been the focus of various scientific and implementation research due to its multiple beneficial physical and chemical features. In a country where the population is still increasing, it is critical to cultivating every available space. A two-year field study employing biochar as an organic fertilizer was undertaken in the north of China under rainy conditions. Previously, fertilization in this area was characterized by high N and low P levels. The application of biochar on maize plantations resulted in a considerable increase in yields and reduced N_2O emissions (Czekała et al., 2019).

Biochar's adsorption characteristics are also used in water and wastewater treatment. This is supported by research on the capture of ions such as As^{3+}, Cd^{2+}, and Pb^{2+}, among other things. Biochar's ability to reduce N_2O emissions to the atmosphere is crucial from an environmental standpoint (Czekała et al., 2019).

11. Biochar and soil interaction

It is generally known that adding biochar to soil can bind soil particles and plant roots, resulting in increased phytoremediation and bioremediation due to their interactions. The high binding propensity to soil particles and long-term survival in soil, biochar may play an essential role in the situ bioremediation of contaminated soil. As a result, bioremediation can be accomplished by both biochar and the microbial population present inside the biochar (inside the porous zone) (Singh et al., 2020).

Even though roots are the first contact site between biochar particles and growing plants, few extensive studies of biochar–root interactions exist. Biochar has the potential to influence root growth and thus plant performance via two mechanisms: (i) as a direct nutrient source and (ii) via influencing nutrient availability (Prendergast-Miller et al., 2014). Biochar is a soil amendment that improves agricultural soils' physicochemical properties and biological activity. However, the impact of biochar on the components of the agricultural system is still unclear (Gorovtsov et al., 2020).

Biochar-aided bacterial-assisted bioremediation by attracting a larger community and surface area to absorb contaminants in polluted soil due to its vast surface area and porosity. Biochar has the trait of having a higher shallow surface. As a result, the size of the shallow surface is determined by the pyrolysis temperature and feedstock types. Furthermore, it has been noted that the size of surface area and pores is influenced by the volatile matter and tars present in the feedstock; as a result, high temperatures play an essential role in the removal of these volatiles from the feedstock. Although it can involve all of the mechanisms associated with bioremediation in polluted soil, the interaction between biochar and soil does not last long (Singh et al., 2020).

The types and rate of interactions in the soil were determined by four factors:

(a) Feedstock composition, especially mineral fraction;
(b) Pyrolysis settings;
(c) Biochar particle size and delivery mechanism; and
(d) Soil characteristics and local environmental circumstances (Singh et al., 2020).

Low-temperature biochar, with its less condensed "C" structure and higher nutrient content, is projected to have higher reactivity in soils and a better contribution to soil fertility than higher-temperature biochar (Singh et al., 2020). Biochar encourages organic carbon (O.C.) storage in soils. Organic carbon is unevenly distributed in soils, with varying structures, functions, and stability depending on particle size fraction (El-Naggar et al., 2018).

12. Sorption mechanisms of heavy metal

Because solution pH substantially impacts both metal speciation and surface charge of biochar, the pH for maximal metal sorption differs by metal. The complexation behavior of functional groups like carboxyl, hydroxyl, and amino is affected by changes in solution pH. Complexation, cation exchange, precipitation, electrostatic interactions, and chemical reduction are the five mechanisms suggested governing metal sorption by biochar from aqueous solutions (Fig. 5.5). However, each mechanism's role for each metal differs significantly depending on the target metals (Li et al., 2017).

Heavy metal adsorption by biochar is influenced by the dosage of biochar, pH of water, water temperature, presence of other cations in water, initial concentration and type of heavy metals, and the biochar aging process, in addition to the pore structure, surface CEC, and number and type of functional groups in biochar (Qiu et al., 2021).

Physical sorption, ion exchange, complexation, electrostatic interactions (chemisorption), and precipitation are some of the ways heavy metals are adsorbed onto biochar from the solution phase (Shakoor et al., 2020). As a complex carbon-rich substance, biochar's adsorption mechanism for heavy metals in aqueous solution is also quite diverse; its surface oxygen-containing functional groups and binding sites can easily make the biochar complexation or ion exchange with heavy metal ions; additionally, some insoluble inorganic salt (formed during the biochar preparation process) can coprecipitate with heavy metal ions on the surface (Chen et al., 2018).

Heavy metal removal mechanisms are critical because they control the efficacy of the process, provide for a greater understanding of the microaspects of the process, and allow for the development of novel sorbents through practical synthesis and modification processes. Physical adsorption, electrostatic interaction, complexation, ion exchange, and precipitation are all mechanisms used by biochar-based sorbents to remove heavy metals (Gupta et al., 2020).

Here are the main mechanisms of As, Cr, Cd, Pb, and Hg.

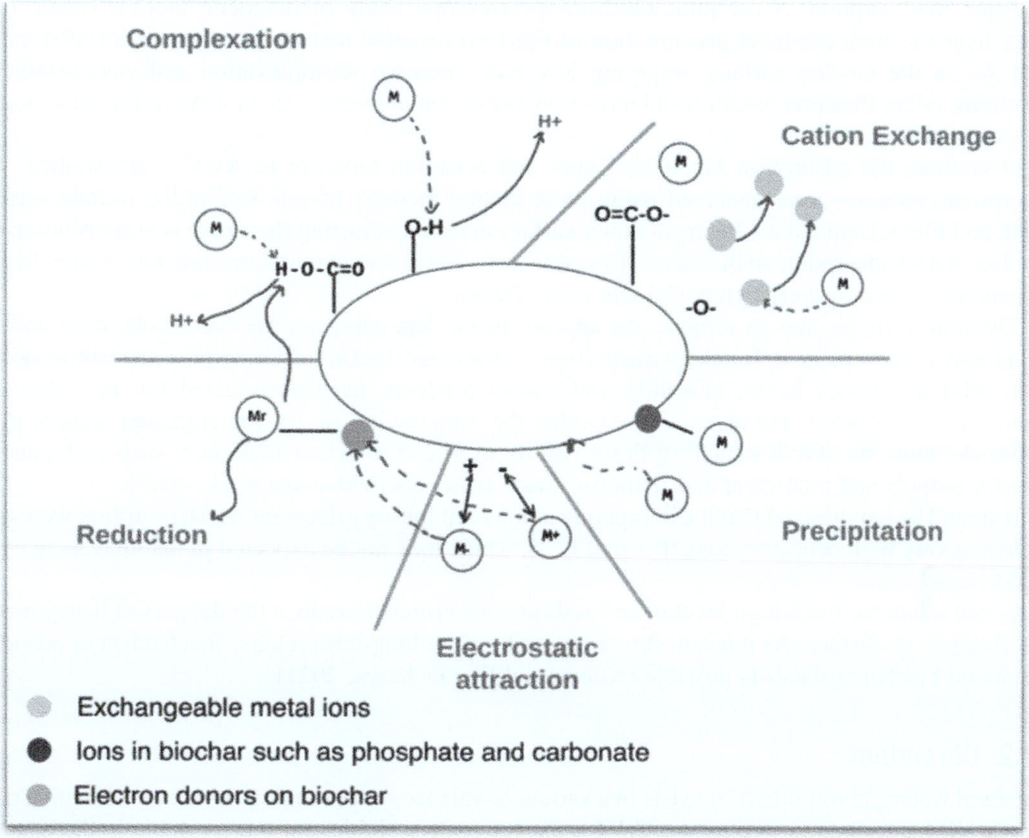

FIGURE 5.5

Heavy metal sorption mechanisms on biochar are depicted schematically.

12.1 Arsenic

Arsenic is a carcinogenic metalloid, and concentrations in natural waterways can vary by many orders of magnitude depending on the source and local geochemical environment. Arsenic can be found in various oxidation states in the environment (3, 0, +3, and +5). Arsenate (AsV) and arsenite (AsIII) are the most common inorganic arsenates (As+5), and arsenite (AsIII) in natural water, with (AsIII) being more poisonous and mobile than (AsV) (Li et al., 2017).

AsV is common in aerobic conditions and exists as $H_2AsO_4^-$ and $HAsO_4^{2-}$ at pH (3−11). AsIII is dominant in a reducing environment, forming H_3AsO^{3-} at pH 9.2 and H_2AsO^{3-} at pH > 9.2. As sorption by biochar is mediated by precipitation and reduction. No new peaks were found in X-ray diffraction (XRD) examinations of pinewood biochar before and after AsV sorption, indicating that no new minerals were formed and that the precipitation process was not relevant for AsV removal by biochar (Li et al., 2017).

After AsV sorption, X-ray photoelectron spectroscopy study of magnetic biochar (made from water hyacinth with chemical precipitation of Fe_2/Fe_3) revealed that AsV accounted for 89% of the total As on the biochar surface, implying low AsV decrease. Complexation and electrostatic interactions, rather than precipitation and reduction, are essential mechanisms for As sorption by biochar (Li et al., 2017).

Over time, the adsorption kinetics of pure and activated biochars to As(V) were studied. The adsorption processes were analyzed using three kinetic models: pseudo-first-order, pseudo-second-order, and Elovich equations. Enlarging their surface area and adjusting their pore size distribution can promote As(V) adsorption on biochars. However, surface area and pore volume are not the only factors influencing As removal efficiency (V) (Jin et al., 2014).

The ability of biochar to remove As appears to be depending on the feedstocks used and the technology used to make it. Biochars made from various feedstocks, including pine and oak wood and bark, solid waste, rice husks, biosolids, and animal products, have been tested for their ability to remove As from water. However, to maximize the remediation of As-contaminated waters, more feedstocks must be tested, including those readily available as waste materials, such as bioenergy industry outputs and municipal and domestic waste materials (Vithanage et al., 2017).

It should be emphasized that the competing anions' inhibitory effects on As(III) sorption were only evident at very high concentrations (0.1 mol L^{-1}), which may not be expected in nature (Zhang et al., 2015).

Because biochar has a high level of soil stability, it's critical to reduce the dangers of heavy metals adsorbing on its surface. As a result, future research on the long-term release mechanism of adsorbed arsenic on biochar surfaces is possible (Kumar and Bhattacharya, 2021).

12.2 Chromium

In natural waters, chromium (Cr) exists in a variety of valence states ranging from 2 to +6, with Cr(III) and Cr(VI) being the most common. CrVI is exceptionally soluble and transportable in the aqueous phase, and its carcinogenic, mutagenic, and teratogenic action in biological systems makes it a severe environmental concern (Li et al., 2017).

Cr(III) is a micronutrient deemed essential for humans because it is 300 times less harmful than Cr(VI). Although Cr(III) is less hazardous than Cr(VI), its disposal in natural waterways as a soluble species poses major health risks since it can be oxidized to Cr(VI) in the environment. The main Cr(VI) species under oxidizing circumstances are $HCrO_4^-$, CrO_4^{2-}, and $Cr_2O_7^{2-}$. $Cr_2O_7^{2-}$ and $HCrO_4^-$ dominate at pH 6.0 and >6.0, respectively, while CrO_4^{2-} dominates at pH > 6.0 (Li et al., 2017).

The sorption of Cr(III) by biochar colloids increases as the temperature rises from 100 to 400°C. Cr(III) sorption is pH-dependent, growing with increasing solution pH between 2.5 and 5.0. Higher concentrations of H+ hinder cation exchange between Cr(III) and minerals in biochar at lower solution pH. On the other hand, Cr(III) sorption by electrostatic attraction cannot be considered out (Li et al., 2017).

Biochars become increasingly negatively charged as the pH of the solution rises, allowing them to attract Cr more electrostatically (III). In contrast to Cr(III), two mechanisms for Cr(VI) sorption have been proposed: (a) electrostatic attraction between negatively charged Cr(VI) species and positively charged biochar; and (b) reduction of Cr(VI) to Cr(III) primarily by oxygen-containing functional groups such as carboxyl and hydroxyl groups, followed by Cr(III) complexation with functional groups on biochar (Li et al., 2017).

Cr(VI) removal is primarily due to Cr(VI) reduction in an acid media. Biomaterials' primary Cr(VI) elimination method is adsorption coupled with reduction. Cr(VI) is quickly reduced to Cr(III) under acid conditions in the presence of organic materials due to its high redox potential (+1.3 V) (Dong et al., 2011).

12.3 Cadmium

In an aqueous solution, divalent metal cations such as Cd and Pb have a high tendency to be hydrated, which is pH-dependent. Divalent metals have similar sorption mechanisms in aqueous solutions, including cation exchange, surface complexation, precipitation, and electrostatic interactions. The sorption of Cd and Pb by biochar has received the most excellent attention among divalent metals (Li et al., 2017).

At pH 8.0, cadmium has a low tendency to be hydrolyzed, but at pH 11.0, all cadmium exists as a hydroxyl complex. (Cd_2) is the most common species in natural aquatic solutions. Mechanisms for (Cd) sorption from aqueous solution by honey mesquite, cord grass, and loblolly pine biochars have been reported (Li et al., 2017).

The CEC of these plant biochars was divided into two categories: low and high. Cation exchange was the most common mechanism for Cd sorption in the high-capacity group. The behavior of K and Cd to replace Na-saturated biochar was studied using a flow calorimetry experiment. Based on flow calorimetry, the form and duration of heat signals for Cd exchange for Na were identical to those for K exchange for Na, showing that cation exchange is the primary process (Li et al., 2017).

Precipitation was proposed as the principal mechanism for Cd sorption due to the relatively large soluble concentrations of carbonate and phosphate in dairy manure biochar. Following Cd sorption, XRD investigations verified the production of Cd carbonate and phosphate minerals in water hyacinth biochar (Li et al., 2017).

By electrostatic interaction, ion exchange, surface complexation, and precipitation, mineral phases of biochars can contribute to Cd sorption (Qi et al., 2017).

12.4 Lead

Lead is a human carcinogen that can cause brain damage, kidney failure, hypertension, nervous system damage, and anemia (Ifthikar et al., 2017).

Pb is found mostly as Pb_2 and Pb(OH) at pH 5.5, $Pb(OH)_2$ at pH 5.5−12.5, and $Pb(OH)_4^{2-}$ at pH > 12.5 in natural aquatic systems. The amount of Pb adsorbed onto biochar was similar to the number of cations released, indicating that cation exchange is also the major mechanism for Pb sorption by oak wood and oak bark biochars. The three main mechanisms for Cd and Pb sorption by biochars are cation exchange, complexation, and precipitation. On the other hand, their sorption mechanisms are dependent on the properties of biochar, which are influenced by feedstock, pyrolysis temperature, and solution pH (Li et al., 2017).

Lead removal by biochar sorbents is due to either lead precipitation on the biochar surface or electrostatic interactions between lead species and negatively charged functional groups on the biochar surface. Other mechanisms, including complexation, chelation, and ion exchange, may govern the high affinity for lead and other metal ion species bound by biochar (Inyang et al., 2011).

12.5 Mercury

Mercury is found in three oxidation states: 0, +1, and +2, with divalent Hg most prevalent in the environment. Hg(II) exists in aqueous solution as Hg_2 at pH 3.0, and HgOH and $Hg(OH)_2$ at pH (3.0–7.0). Brazilian pepper charcoal's most common Hg sorption method was complexation with carboxylic and phenolic hydroxyl groups or a graphite-like domain (Li et al., 2017).

Traditional coagulation, chemical precipitation, reverse osmosis, chemical, and physical adsorption, ion exchange, and solvent extraction are popular physicochemical techniques for removing Hg from aqueous systems (Boutsika et al., 2014).

Biochar's lower Hg sorption capacity is likely due to diminished carboxylic and phenolic hydroxyl groups as the temperature rises. Hg is absorbed by different biochars through different complexation methods. In addition to complexation, the chemical reduction is responsible for Hg sorption. Hg sorption on activated C has been attributed to the presence of $Hg(OH)_2$, which exists as an uncharged soluble hydroxide salt, by several researchers. In contrast, others have claimed that if Hg precipitates with Cl during sorption, it is most likely Hg_2Cl_2. As a result, more research is needed to determine whether Hg^{2+} conversion to Hg^+ is a critical mechanism for controlling Hg sorption by biochar (Li et al., 2017).

In recent years, there has been a growing trend in the worldwide remediation area to promote more sustainable technology. Traditional Hg remediation methods, such as thermal treatment and soil washing, need to be improved to achieve better long-term results (O'Connor et al., 2018).

13. Modification of biochar

Biochar can sorb metals from water. However, its capacity is often lower than that of other common biosorbents like activated C. As a result, biochar has been changed in recent studies to improve its metal sorption capability. Surface area, porosity, pH_{PZC}, and functional groups, for example, have all been increased. Minerals, reductants, organic functional groups, and nanoparticles are all used to change biochars, as is activation with an alkali solution (Li et al., 2017).

Exogenous functional groups have been introduced to biochar because surface complexation between metals and functional groups such as carboxylic, amino, and hydroxyl groups plays a crucial role in metal sorption. For example, amino groups were added to rice husk and saw dust biochars through polyethyleneimine modification and nitration/reduction. Biochar is a porous material with a large surface area, influencing its metal sorption ability. As a result, increasing its surface area by adding nanoparticles improves its metal sorption capacity. Magnetic biochar/ZnS composites, for example, can be made by depositing ZnS nanocrystals onto magnetic biochar (Li et al., 2017).

Scholars investigated the effects of several modification methods on biochar to generate biochar with superior qualities. The activation of the initial biochar using physical and chemical processes to achieve the desired aim is modified. The properties of biochar are affected by the type of activator, soaking period, activation time, and activation temperature (Yang et al., 2019).

In addition to integrating nanoparticles, activation or modification of biochar with alkali solutions such as NaOH and KOH can increase the surface area of the material, resulting in increased metal sorption capacity (Li et al., 2017).

Biochar is a pyrogenic by-product of the pyrolysis process; changes to its surface morphology must be considered at the process level. In short, an effective strategy for modifying all pyrogenic products through a simple unit process for biomass pyrolysis is desirable (Kim et al., 2019).

A modification is made before or after the biochar is pyrolyzed (or both). While prepyrolysis adjustments involve treating the biochar feedstock before it is pyrolyzed, postpyrolysis modifications are more prevalent and entail treating biochar after it has been created (Sizmur et al., 2017).

Different elements are embedded into the biochar structure to create biochar-based composites before or postpyrolysis. This modification method has attracted the most study interest for making sorbents for remediation of polluted waters. The primary goal of these modifications is to improve the biochar's efficacy at removing contaminants from water by altering its physical or chemical qualities, such as surface area or surface functionality (Sizmur et al., 2017).

(1) Activation

Biochar is activated by forming a cluster of micro- and mesopores with steam or alkaline/acidic solutions, which dramatically increases the surface area for physical or chemical sorption of positively charged metal ions (Shakoor et al., 2020).

(a) Physical activation

Physical activation, also known as steam activation, is a commonly used modification procedure to increase the surface porosity of biochar for metal sorption and remove contaminants caused by incomplete combustion (Shakoor et al., 2020).

(b) Chemical activation

(i) Acids/oxidizing agent: In aqueous solutions, biochar contains negatively charged surface functional groups such as −O.H. and −COOH, which can form chemical interactions with positively charged cations. Adding acids to biochar results in additional oxygen-containing surface functional groups and an increase in biochar's ability to bind/attach cationic metals via the sorption process. The acid treatment of biochar increased the carboxylic groups on its surface and increased the porosity, increasing biochar surface area (Shakoor et al., 2020).

(ii) Alkaline solutions: Alkali activation usually entails treating charcoal with metal hydroxides like NaOH and KOH. Biochar activation with metal hydroxides improves the surface area, porosity, and oxygenated surface functional groups of biochar, resulting in increased heavy metal sorption. Biochar's surface functional groups provide positively charged exchange sites for cationic metals (such as Pb) sorption from contaminated water (Shakoor et al., 2020).

(iii) Impregnation of biochar: Impregnation/loading of biochar with clays, organic compounds, minerals, and carbonaceous materials such as carbon nanotubes or graphene oxide results in biochar-based composites. The impregnated biochar is used to improve the surface properties of the biochar (Shakoor et al., 2020). The adsorption of some heteroatoms or metal ions into the surface and pores of biochar is referred to as metal impregnation. On the one hand, the specific surface area is enhanced, while metal ions are coupled with the adsorbate to boost adsorption performance (Yang et al., 2019).

(iv) Minerals: Biochar has a vast surface area, a high pH, and a negative surface charge, making it a suitable sorbent for removing positively charged metal ions from the solution. Biochar with a high surface area may support the loading of minerals with different chemical properties with a positive charge, such as magnetite, hematite, goethite zero-valent Fe, calcium oxide, hydrous Mn oxide, and birnessite, making it an excellent sorbent for removing oxyanions like Cr from contaminated water (Shakoor et al., 2020).

(v) **Organic compounds:** Organic compounds containing functional groups can modify biochars. These functional groups have a high potential for forming solid interactions with both aqueous metal ions and the biochar surface. Creating stable complexes with amino groups can significantly improve metal sorption by adding amino acids to charcoal. Chemical reactions and the mixing of biochars with polymers (e.g., chitosan) containing an excess of amino groups can both be used to modify/treat the biochars (Shakoor et al., 2020).

14. Bioremediation and phytoremediation

14.1 Bioremediation

Bioremediation is a method that uses microorganisms to clean contaminated sources such as water, soil, and other environmental elements. Furthermore, several new reported bioremediation approaches are being applied, such as agricultural waste-mediated biochar, which can remediate soil contamination through its adsorption characteristics due to its distinctive surface and interior structure, and so on. The biological treatment of hazardous waste is comparable to wastewater, industrial, and solid waste. Industries have discharged massive amounts of sewage into canals, rivers, and other bodies without sufficient treatment of these solid and liquid wastes, which contain a variety of organic and inorganic pollutants that can harm living creatures (Singh et al., 2020).

Heavy metals are not biodegradable and can remain in contaminated soils for an extended period. Heavy metal removal from polluted soils is costly and time-consuming. To lower the bioavailability of heavy metals and minimize plant absorption, soil additives such as lime and compost are widely used to stabilize them in situ. Biochar can help to fix heavy metals in contaminated soils, enhance soil quality, and reduce fatty metal uptake by crops. As a result, applying biochar to soils affected by heavy metals could potentially give a new alternative for remediation (Zhang et al., 2013).

The use of biochar aids in the bioremediation of organic pollutants. In biochar-amended soils, the key mechanism is increased petroleum hydrocarbon-degrading microbial communities. Heavy metalloids can't be degraded or entirely removed from the environment, but they can be transformed from one form to another, from high to low concentration. Heavy metalloids can build up in the biota as well. As a result, there are two commonly used strategies for heavy metalloid bioremediation (Sun et al., 2018).

(1) Heavy metalloid absorption and accumulation in woody plants and bioenergy crops in contaminated farmlands, and heavy metalloid removal by harvesting the metalloids accumulated biomass (Sun et al., 2018);
(2) Heavy metalloid transformation into lower toxic products (e.g., complex state) and adsorption by microorganisms to reduce their toxicity and migration (Sun et al., 2018).

Heavy metals in the soil can be immobilized, allowing them to be held in the soil and released slower, resulting in minor environmental damage. Many studies have shown that lowering heavy metal concentrations in soil and water bodies reduces heavy metal uptake by plants (Tang et al., 2013). Biochar's physical qualities are crucial. The type of starting material and the parameters of the coal process determine the shape of the pores and their distribution and size in the biocarbon. Biochar can be activated (for example, using steam) in certain circumstances, allowing you to grow and expand the surface structure of the biochar. The design and size of this surface dictate a variety of biochar features and applications, such as a soil improver, because its presence aids in the adsorption of both pollutants and nutrients for plants (N, P, and K compounds) (Włodarczyk and Kwarciak-Kozłowska, 2021).

The particular surface, porosity, polarity, and aromaticity of biochar influence the biocarbon sorption process of organic contaminants in the aquatic environment. Biochar with a high specific character and microporosity is formed at temperatures exceeding 400°C (Włodarczyk and Kwarciak-Kozłowska, 2021).

14.2 Phytoremediation

Phytoremediation is a multidisciplinary approach that combines soil chemistry, microbiology, and plant physiology to immobilize contaminants in various mediums. Phytoextraction, phytostabilization, phytodegradation, phytovolatilization, and rhizoremediation are all examples of phytoremediation (Sun et al., 2018).

Phytoextraction and phytostabilization are covered in depth in this chapter. Phytoremediation is getting more attention than other heavy metalloid remedial approaches, such as excavating and dumping or chemical immobilization, because of its effectiveness and low cost. Additional advantages, such as erosion control and pollutant leaching, are critical for future land management and development (Sun et al., 2018).

Phytoremediation is a broad phrase that refers to treatments that integrate soil microbiology, chemistry, and plant physiology. Encapsulation or digging and dumping are now the most widely used methods for soil heavy metal remediation, although they do not address the problem of contamination. Because immobilization or extraction can be costly, phytoremediation might be seen as a more appealing option because it can be employed at a lower cost to restore or partially clean a site when compared to other options, with a price that is just 5% of that of other alternatives (Paz-Ferreiro et al., 2014).

Other benefits include its positive public perception as a remediation strategy and being more environmentally friendly than alternatives. The introduction of vegetation in the polluted area can help reduce erosion and contaminant leaching. The use of plants to remove toxins from the environment or change them into more minor hazardous forms is known as phytoremediation (Paz-Ferreiro et al., 2014).

Phytoremediation is a green approach that uses plants and associated microbes to remove toxins from contaminated locations. It was created in recent decades and is powered by solar energy. It is an environmentally beneficial, low-cost, safe, and long-term approach for heavy metal cleanup in soil. The critical barriers to promoting commercially feasible phytoremediation technology are biomass production, plant metal extraction efficiency, and economic costs (Zhou et al., 2020).

As a result, crops with high metal tolerance, biomass, and economic value, such as oil crops, are currently the research focus. An appropriate oil crop rotation strategy may allow for the remediation of contaminated soils with minimum impact on agricultural productivity while also providing economic benefits (Zhou et al., 2020).

15. Conclusion and future aspects

This chapter concludes that biochar is a carbon aged product produced through pyrolysis. The characteristics and properties of biochar are properly discussed above. The pros and cons of biochar are explained with sequestration, fertility, handling, system deployment, etc. Generally, the application of biochar is to improve or adsorb the harmful metals from soil and water for environmental purposes. The biochars are also used in organic fertilizers for soil enrichment.

Furthermore, the adsorption mechanism of heavy metals from biochar is a crucial element for soil nourishing treatment in organic fertilizers through remediation. Metals like arsenic (As), cadmium (Cd), chromium (Cr), lead (Pb), and mercury (Hg) have been explained how do these metal sorptions occur by using biochar and what is their natural form in which they exist. The bioremediation and phytoremediation for biochars are available. Since the heavy metal adsorption by biochar from the soil is still mostly unexplored, the adsorption of heavy metals from water is a vast hazard for a healthy environment. But yet, poisonous metals in soils are the most significant problem nowadays. Bioremediation can be considered for the sorption process of heavy metals from agricultural soil. Organic fertilizers can be produced instead of fertilizers that contain poisonous metalloids. This can help the environment overcome health-related issues from artificially created agricultural products. A group of scientists has been working on soil pollutants and the problems causing health hazards.

References

Agrafioti, E., Bouras, G., Kalderis, D., Diamadopoulos, E., 2013. Biochar production by sewage sludge pyrolysis. J. Anal. Appl. Pyrol. 101, 72–78.

Al-Wabel, M.I., Al-Omran, A., El-Naggar, A.H., Nadeem, M., Usman, A.R.A., 2013. Pyrolysis temperature induced changes in characteristics and chemical composition of biochar produced from conocarpus wastes. Bioresour. Technol. 131, 374–379. https://doi.org/10.1016/j.biortech.2012.12.165.

Boutsika, L.G., Karapanagioti, H.K., Manariotis, I.D., 2014. Aqueous mercury sorption by biochar from malt spent rootlets. Water Air Soil Pollut. 225 (1). https://doi.org/10.1007/s11270-013-1805-9.

Bracmort, K., 2010. Biochar: Examination of an Emerging Concept to Mitigate Climate Change. Congressional Research Service.

Chen, T., Zhang, Y., Wang, H., Lu, W., Zhou, Z., Zhang, Y., Ren, L., 2014. Influence of pyrolysis temperature on characteristics and heavy metal adsorptive performance of biochar derived from municipal sewage sludge. Bioresour. Technol. 164, 47–54. https://doi.org/10.1016/j.biortech.2014.04.048.

Chen, H., Xie, A., You, S., 2018. A review: advances on absorption of heavy metals in the waste water by biochar. IOP Conf. Ser. Mater. Sci. Eng. 301 (1), 12160 (IOP Publishing).

Czekała, W., Jezowska, A., Chełkowski, D., 2019. The use of biochar for the production of organic fertilizers. J. Ecol. Eng. 20 (1), 1–8. https://doi.org/10.12911/22998993/93869.

Dong, X., Ma, L.Q., Li, Y., 2011. Characteristics and mechanisms of hexavalent chromium removal by biochar from sugar beet tailing. J. Hazard Mater. 190 (1–3), 909–915.

El-Naggar, A., Lee, S.S., Rinklebe, J., Farooq, M., Song, H., Sarmah, A.K., et al., 2019. Biochar application to low fertility soils: a review of current status, and future prospects. Geoderma 337, 536–554.

El-Naggar, A., Awad, Y.M., Tang, X., Liu, C., Niazi, N.K., Jien, S., et al., 2018. Biochar influences soil carbon pools and facilitates interactions with soil: a field investigation. Land Degrad. Dev. 29 (7), 2162–2171.

Gorovtsov, A.V., Minkina, T.M., Mandzhieva, S.S., Perelomov, L.V., Soja, G., Zamulina, I.V., et al., 2020. The mechanisms of biochar interactions with microorganisms in soil. Environ. Geochem. Health 42 (8), 2495–2518. https://doi.org/10.1007/s10653-019-00412-5.

Gupta, S., Sireesha, S., Sreedhar, I., Patel, C.M., Anitha, K.L., 2020. Latest trends in heavy metal removal from wastewater by biochar based sorbents. J. Water Proc. Eng. 38, 101561.

Hassan, M., Liu, Y., Naidu, R., Parikh, S.J., Du, J., Qi, F., Willett, I.R., 2020. Influences of feedstock sources and pyrolysis temperature on the properties of biochar and functionality as adsorbents: a meta-analysis. Sci. Total Environ. 140714.

Ifthikar, J., Wang, J., Wang, Q., Wang, T., Wang, H., Khan, A., et al., 2017. Highly efficient lead distribution by magnetic sewage sludge biochar: sorption mechanisms and bench applications. Bioresour. Technol. 238, 399–406.

Inyang, M., Gao, B., Ding, W., Pullammanappallil, P., Zimmerman, A.R., Cao, X., 2011. Enhanced lead sorption by biochar derived from anaerobically digested sugarcane bagasse. Separ. Sci. Technol. 46 (12), 1950–1956.

Jin, H., Capareda, S., Chang, Z., Gao, J., Xu, Y., Zhang, J., 2014. Biochar pyrolytically produced from municipal solid wastes for aqueous As(V) removal: adsorption property and its improvement with KOH activation. Bioresour. Technol. 169, 622–629. https://doi.org/10.1016/j.biortech.2014.06.103.

Kim, Y., Oh, J.-I., Vithanage, M., Park, Y.-K., Lee, J., Kwon, E.E., 2019. Modification of biochar properties using CO_2. Chem. Eng. J. 372, 383–389.

Kumar, A., Bhattacharya, T., 2021. Removal of arsenic by wheat straw biochar from soil. Bull. Environ. Contam. Toxicol. 1–8.

Lehmann, J., 2010. To Give Biochar Seminar.

Li, H., Dong, X., da Silva, E.B., de Oliveira, L.M., Chen, Y., Ma, L.Q., 2017. Mechanisms of metal sorption by biochars: biochar characteristics and modifications. Chemosphere 178, 466–478.

Meyer, S., Glaser, B., Quicker, P., 2011. Technical, economical, and climate-related aspects of biochar production technologies: a literature review. Environ. Sci. Technol. 45 (22), 9473–9483.

Nguyen, B.T., Lehmann, J., Hockaday, W.C., Joseph, S., Masiello, C.A., 2010. Temperature sensitivity of black carbon decomposition and oxidation. Environ. Sci. Technol. 44 (9), 3324–3331. https://doi.org/10.1021/es903016y.

O'Connor, D., Peng, T., Li, G., Wang, S., Duan, L., Mulder, J., et al., 2018. Sulfur-modified rice husk biochar: a green method for the remediation of mercury contaminated soil. Sci. Total Environ. 621, 819–826.

Paz-Ferreiro, J., Lu, H., Fu, S., Méndez, A., Gascó, G., 2014. Use of phytoremediation and biochar to remediate heavy metal polluted soils: a review. Solid Earth 5 (1), 65–75. https://doi.org/10.5194/se-5-65-2014.

Prendergast-Miller, M.T., Duvall, M., Sohi, S.P., 2014. Biochar-root interactions are mediated by biochar nutrient content and impacts on soil nutrient availability. Eur. J. Soil Sci. 65 (1), 173–185. https://doi.org/10.1111/ejss.12079.

Qi, F., Yan, Y., Lamb, D., Naidu, R., Bolan, N.S., Liu, Y., et al., 2017. Thermal stability of biochar and its effects on cadmium sorption capacity. Bioresour. Technol. 246, 48–56.

Qian, K., Kumar, A., Zhang, H., Bellmer, D., Huhnke, R., 2015. Recent advances in utilization of biochar. Renew. Sustain. Energy Rev. 42, 1055–1064.

Qiu, B., Tao, X., Wang, H., Li, W., Ding, X., Chu, H., May 2021. Biochar as a low-cost adsorbent for aqueous heavy metal removal: a review. J. Anal. Appl. Pyrol. 155. https://doi.org/10.1016/j.jaap.2021.105081.

Shaaban, M., Van Zwieten, L., Bashir, S., Younas, A., Núñez-Delgado, A., Chhajro, M.A., et al., 2018. A concise review of biochar application to agricultural soils to improve soil conditions and fight pollution. J. Environ. Manag. 228, 429–440.

Shakoor, M.B., Ali, S., Rizwan, M., Abbas, F., Bibi, I., Riaz, M., et al., 2020. A review of biochar-based sorbents for separation of heavy metals from water. Int. J. Phytoremediation 22 (2), 111–126.

Singh, C., Tiwari, S., Singh, J.S., 2020. Biochar: a sustainable tool in soil pollutant bioremediation. In: Bioremediation of Industrial Waste for Environmental Safety. Springer, pp. 475–494.

Sizmur, T., Fresno, T., Akgül, G., Frost, H., Moreno-Jiménez, E., December 2017. Biochar modification to enhance sorption of inorganics from water. Bioresour. Technol. 246, 34–47. https://doi.org/10.1016/j.biortech.2017.07.082.

Sohi, S.P., Krull, E., Lopez-Capel, E., Bol, R., 2010. A review of biochar and its use and function in soil. Adv. Agron. 105, 47–82. https://doi.org/10.1016/S0065-2113(10)05002-9.

Sun, W., Zhang, S., Su, C., 2018. Impact of biochar on the bioremediation and phytoremediation of heavy metal(loid)s in soil. In: Advances in Bioremediation and Phytoremediation. https://doi.org/10.5772/intechopen.70349.

Talberg, A., 2009. The Basics of Biochar. Parliamentary Library, Australia.

Tang, J., Zhu, W., Kookana, R., Katayama, A., 2013. Characteristics of biochar and its application in remediation of contaminated soil. J. Biosci. Bioeng. 116 (6), 653–659.

Uchimiya, M., Lima, I.M., Thomas Klasson, K., Chang, S., Wartelle, L.H., Rodgers, J.E., 2010. Immobilization of heavy metal ions (CuII, CdII, NiII, and PbII) by broiler litter-derived biochars in water and soil. J. Agric. Food Chem. 58 (9), 5538–5544. https://doi.org/10.1021/jf9044217.

Vithanage, M., Herath, I., Joseph, S., Bundschuh, J., Bolan, N., Ok, Y.S., et al., 2017. Interaction of arsenic with biochar in soil and water: a critical review. Carbon 113, 219–230.

Wang, J., Wang, S., 2019. Preparation, modification and environmental application of biochar: a review. J. Clean. Prod. 227, 1002–1022.

Wang, H., Xia, W., Lu, P., 2017. Study on adsorption characteristics of biochar on heavy metals in soil. Kor. J. Chem. Eng. 34 (6), 1867–1873. https://doi.org/10.1007/s11814-017-0048-7.

Weber, K., Quicker, P., April 2018. Properties of biochar. Fuel 217, 240–261. https://doi.org/10.1016/j.fuel.2017.12.054.

Włodarczyk, R., Kwarciak-Kozłowska, A., 2021. 6 Biochar for Bioremediation of Toxic Metals.

Yang, X., Zhang, S., Ju, M., Liu, L., 2019. Preparation and modification of biochar materials and their application in soil remediation. Appl. Sci. 9 (7), 1365.

Zhang, X., Wang, H., He, L., Lu, K., Sarmah, A., Li, J., et al., 2013. Using biochar for remediation of soils contaminated with heavy metals and organic pollutants. Environ. Sci. Pollut. Control Ser. 20 (12), 8472–8483.

Zhang, W., Zheng, J., Zheng, P., Tsang, D.C.W., Qiu, R., 2015. Sludge-derived biochar for arsenic(III) immobilization: effects of solution chemistry on sorption behavior. J. Environ. Qual. 44 (4), 1119–1126. https://doi.org/10.2134/jeq2014.12.0536.

Zhou, J., Chen, L.H., Peng, L., Luo, S., Zeng, Q.R., 2020. Phytoremediation of heavy metals under an oil crop rotation and treatment of biochar from contaminated biomass for safe use. Chemosphere 247, 125856.

CHAPTER

Global restoration of contaminated agricultural soils through the utilization of potential microbes for the future generation

6

Kamal Prasad and Kavita Yadav
Centre for Advanced Agriculture Research, Absolute Bioscience, Delhi NCR, Gurugram, Haryana, India

1. Introduction

Contamination of soil owing to numerous anthropogenic activities is a serious concern all over the world. Increased release of a wide spectrum of xenobiotics has occurred from the intensification of agriculture and the expansion of enterprises. These toxins injure humans, animals, wildlife, crops, and native plants, causing ecological issues and a natural balance to be disrupted. Scientists from all over the world are working to find a solution using a variety of methods, including physical, chemical, and thermal procedures, as well as excavation and transfer of contaminated soil. Unfortunately, those procedures are costly and labor-intensive, do not always assure that pollutants are totally removed or eliminated, and frequently result in dramatic changes in the treated soil's physical, chemical, and biological features. Bioremediation techniques evolved as a result of the search for alternatives to established methods of cleaning polluted places. Bioremediation techniques have attracted a lot of attention, and a lot of research has been done and published on the use of diverse microorganisms for the decontamination of various types of pollutants in soil. However, further research is needed to obtain a thorough grasp of present strategies, as well as to modify them appropriately to get the most out of them while also exploring new possibilities based on day-to-day experiences.

2. Soil health

Soil health is often defined as "the capacity of a given soil to perform its functions as a living system capable of sustaining biological productivity, promoting environmental quality, and maintaining plant and animal health" (Doran and Zeiss, 2000). Soil contamination, along with other degradation processes, can have a negative impact on soil health (Gomez et al., 2012). However, soil is a hugely complex environmental matrix that serves a variety of, often competing, roles from both an ecocentric and anthropocentric perspective, necessitating the study of many distinct factors in order to accurately measure soil health. Most importantly, to appropriately assess soil health: physical, chemical, and biological properties with potential as indicators of soil functioning must always be included in the

Integrative Strategies for Bioremediation of Environmental Contaminants, Volume Two. https://doi.org/10.1016/B978-0-443-14013-6.00002-0

assessment (after all, physical, chemical, and biological processes in the soil ecosystem are not independent but interactive processes); (ii) chemical, (eco)toxicological, and ecological approaches must be incorporated into the evaluation; (iii) the intended use for the contaminated site must be carefully considered, as the very concept of contaminated sites must be taken into account; (iv) the system's inherent temporal and spatial variability (i.e., spatial heterogeneity, temporal dynamics), as well as the scale of both soil processes and the assessment itself, must be considered; and (v) a suitable (inevitably, often far from perfect) "healthy" reference soil should be identified for comparison and target setting purposes. Soil's physicochemical qualities such as pH, redox potential, organic matter content, texture, and others are important parameters that can significantly affect contaminant bioavailability and, hence, (eco)toxicity in soil. Unfortunately, under most environmental legislation, the overall concentration of contaminants is the most important component in contaminated soils (Environmental Risk Assessment [ERA]). However, such a factor (total soil contaminant concentration) is insufficient to appropriately analyze or quantify the possible negative impact of pollutants on soil functioning (Alvarenga et al., 2018). In reality, soil pollutants' mobility and bioavailability are both important factors in their uptake by organisms and, as a result, their (eco)toxicity (Vamerali et al., 2010; Megharaj et al., 2011). Contaminant bioavailability is a considerably more important element for adequate soil protection and risk assessment than total contaminant concentrations, since it represents the fraction that can be taken up by soil organisms and/or leached to other environmental compartments. Metal bioavailability is primarily influenced by physicochemical features of the soil, such as pH, redox potential, moisture content, organic matter content, clay content, anionic compound presence, and so on (Vangronsveld and Cunningham, 1998). Organic pollutants' bioavailability and mobility are largely determined by their solubility, hydrophobicity, and interaction with the mineral and organic fractions of the soil matrix via physicochemical processes such as sorption and complexation (Megharaj et al., 2011). As a result, when assessing soil health and, in particular, when choosing a soil remediation option and monitoring the success of the chosen remediation process, it is advised to always include the assessment of the bioavailable fraction of the pollutants. Despite this, there is no agreement on the optimal method for reliably estimating soil pollutant bioavailability. Chemical extractants, such as inorganic salts like $NaNO_3$, $(NH_4)_2SO_4$, and $CaCl_2$, are the most widely accepted approach for metallic pollutants (Vazquez et al., 2008). In any case, biological indicators are essential for a proper assessment of the impact of soil pollutants on soil health, in addition to total and bioavailable contaminant concentrations, because they directly represent the influence of contaminants on the soil biota (Alvarenga et al., 2018). Microorganisms have a significant role in many soil activities and the provision of ecosystem services, while also providing ecologically valuable information that incorporates numerous environmental elements (Jeffery et al., 2010). Similarly, standardized (eco) toxicological bioassays with model organisms such as *Eisenia fetida* (Irizar et al., 2015), *Vibrio fisheri* (Abbas et al., 2018), *Lactuca sativa* (Valerio et al., 2007), and *Cucumis sativus* have been created and proposed for soil (eco)toxicity research.

3. Source of pollutants in soil

Soil pollution is characterized as the accumulation of persistent poisonous substances, chemicals, salts, radioactive elements, or disease-causing agents in soils, all of which have negative impacts on plant development and animal health. Soil pollution can occur in a variety of ways, including seepage from a landfill (Adrian and Arnett, 2007), discharge of industrial waste into the soil (Alloway, 1995), percolation of contaminated water into the soil (Ang et al., 2005), rupture of underground storage tanks (April and Sims, 1995), and excessive application of pesticides, herbicides, or fertilizer (Ackerman, 2007). Table 6.1 lists the most well-known compounds that contribute to soil pollution. The environmentalists around the world are trying to overcome such huge load of pollutants in soil by several means.

4. Applications and technologies for biological soil remediation

Traditional physicochemical and chemical approaches for pollutant cleaning are frequently prohibitively costly. Furthermore, the amount of space accessible for disposal and storage is decreasing. Traditional clean-up procedures have a number of drawbacks, one of which is that, despite their high costs, they may not always assure that toxins are totally removed. As a result, the hunt for cost-effective and environmentally sound alternatives to traditional garbage disposal systems has exploded in the last two decades. Waste-related research, technological development, and implementation are now among the fastest-growing activities in the world. The most promising technologies are those that closely resemble time-tested, natural systems that have successfully restored ecosystems to their native states following undesired disruptions. Natural processes change, eliminate, remove, or otherwise stabilize chemicals from natural sources (animal, vegetable, or mineral) such that they do not accumulate to levels that endanger ecosystem balance or sustainability. Growing public awareness and concern about environmental pollution prompted government and business to collaborate on the development of safe and cost-effective waste management alternatives. Bioremediation has emerged as the most desirable strategy for cleaning up numerous environmental toxins among the technologies that have been explored. Bioremediation is a method for destroying or rendering harmless different pollutants through natural biological activity. As a result, it employs low-cost, low-technology approaches that are widely accepted by the public and may often be carried out on-site. Bioremediation is the process of degrading environmental pollutants into less hazardous forms using living creatures, typically bacteria. It degrades or detoxifies pollutants that are harmful to human health and/or the environment using naturally occurring bacteria, fungus, or plants. The microorganisms could be native to the contaminated location, or they could have been isolated elsewhere and transported to the contaminated site. Living organisms change contaminant substances through reactions that occur as part of their metabolic activities (Vidali, 2001). Microorganisms must enzymatically attack contaminants and transform them to harmless compounds for bioremediation to be effective. Because bioremediation is only effective when environmental conditions allow for microbial growth and

Table 6.1 Compounds that contribute to soil pollution.

Heavy metal/trace elements	Pesticides involved	Radionuclides	Munition waste	Others
Arsenic, cadmium, chromium, cobalt, copper, and mercury	Organophosphates, aldrin, and atrazine	Cesium, strontium, potassium, and thorium	Trinitrotoluene	Trichloroethylene and pentachlorophenol

activity, it is frequently used to manipulate environmental factors to speed up microbial growth and degradation. pH, temperature, and moisture all have an impact on microbial growth and activity (Prasad, 2020). Although microorganisms have been isolated in extreme settings, most of them grow best in a small range of temperatures, making it critical to obtain optimal conditions. If the pH of the soil is too acidic, lime can be used to neutralize it. The rate of many biological reactions is affected by temperature, and for every 10°C increase in temperature, the rate of many of them doubles. The cells, however, die at a particular temperature. In late spring, summer, and autumn, a plastic covering can be utilized to increase solar heating. All living species require access to water; therefore, irrigation is required to reach the ideal moisture level. Optimum environmental conditions for efficient performance of microorganisms are presented in Table 6.2.

5. Strategies for bioremediation

Organic contaminants are primarily converted to carbon dioxide, water, and biomass in bioremediation. Some contaminants can be bound to the humic substance fraction and hence immobilized. Degradation can take place in both aerobic and anaerobic environments. The aerobic method, which can be classed as ex situ or in situ, is mostly utilized for bioremediation. The amenability of the pollutant to biological transformation, the accessibility of the contaminant to microorganisms, and the ability to optimize biological activity are the three basic principles that guide the selection of appropriate technology among the wide range of bioremediation technologies developed to treat contaminants. The deterioration process is accelerated, and the degree of degradation is improved by using the right technologies and adjusting the circumstances, which lowers the treatment cost (Mohapatra, 2008). Ex-situ procedures are those that are used on soil and groundwater that have been removed from a place through excavation (soil) or pumping (groundwater) (water). The term "in situ" refers to treatments that are applied to soil and groundwater on-site with little disturbance. These approaches are the most popular since they are less expensive and cause fewer disruptions because they treat toxins in site rather than excavating and transporting them. However, the depth of soil that can be adequately treated limits in-situ treatment. Effective oxygen diffusion for optimal rates of

Table 6.2 Optimum environmental conditions for efficient performance of microorganisms.

S.no.	Environmental factors	Optimum conditions
1	pH	5.5–8.5
2	Temperature (°C)	15–45
3	Moisture	25%–28% water holding capacity
4	Soil type	Low clay or silt content
5	Oxygen	Aerobic, minimum air-filled pore space of 10%
6	Nutrient	N and P for microbial growth
7	Heavy metals	Total content 2000 ppm
8	Contaminants	Not too toxic

bioremediation in many soils extends from a few centimeters to about 30 cm into the soil; however, depths of 60 cm and greater have been successfully treated in other circumstances (Vidali, 2001).

5.1 Ex-situ methods

Land farming, also known as land treatment or land application, is an above-ground soil remediation process that uses biodegradation to lower organic pollutant concentrations. Aeration and/or the addition of minerals, nutrients, and moisture are commonly used to spread excavated contaminated soils in a thin layer on the ground surface and stimulate aerobic microbial activity inside the soils. Because contaminated soil is treated in thin layers up to 0.4 m deep, a considerable treatment area is required. Ploughing, harrowing, or milling at regular intervals helps to increase degradation by increasing oxygen supply and mixing. The treatment method is cost-effective, and it can be implemented if enough land is available (Mohapatra, 2008).

5.1.1 Biopiles

It's a cross between organic farming and composting. Excavated soils are mixed with soil additives, placed on a treatment area, and bioremediation with the use of forced aeration. Contaminants are broken down into carbon dioxide and water. A treatment bed, an aeration system, an irrigation/nutrient system, and a leachate collection system are all part of the basic biopile system. To minimize runoff, evaporation, and volatilization, as well as increase solar heating, soil mounds can be up to 20 feet tall and covered with plastic. Before entering the air stream, volatile organic compounds are treated if necessary (Shukla et al., 2010). Biopiles provide an ideal environment for aerobic and anaerobic bacteria to thrive.

5.1.2 Bioreactors

Contaminated soil is treated in either a solid or a slurry phase in this method. Solid phase reactors work on the basis of mechanical degradation of soil through attrition and vigorous mixing of the components in a confined container. Contaminants, bacteria, nutrients, water, and air are all brought into constant contact as a result of this. A slurry bioreactor is a containment vessel and apparatus used to create a three-phase (solid, liquid, and gas) mixing condition to increase the bioremediation rate of soil-bound and water-soluble pollutants as a water slurry of contaminated soil and biomass (usually indigenous microorganisms) capable of degrading target contaminants. Because the enclosed environment is more manageable and hence more controllable and predictable, the pace and amount of biodegradation in a bioreactor system are larger than in situ or in solid-phase systems. However, before being placed in a bioreactor, the contaminated soil must be pretreated (e.g., excavation) or the contaminant can be removed from the soil via soil washing or physical extraction (e.g., vacuum extraction) (Vidali, 2001).

5.1.3 Composting

It is used in bioremediation to degrade harmful organic molecules and may reduce the toxicity of metallic pollutants found in organic residues, garbage, and by-products. Organic wastes are decomposed by microorganisms in composting, which is comparable to what happens physiologically in soil. Composts have greater temperatures than soils, which leads to enhanced pollutant solubility and metabolic activity. Composts with a lot of substrates can cause organic pollutants to metabolize. Mechanical treatment of compostable materials, such as grinding, mixing, and sieving out nondegradable or undesired components like metals, plastics, glass, and stones, creates favorable circumstances for biological treatment. The nature of the organic pollutant, composting conditions and processes, microbial populations, and time all have an impact on the compost mechanism's efficacy (Barker and Bryson, 2002).

5.2 In-situ methods

5.2.1 Biosparging

It uses and promotes indigenous microbes in saturated soil to breakdown organic pollutants. Air is introduced into the saturated zone (below the water table) through boreholes to boost the activity of the soil's native microorganisms by increasing oxygen dissolution. Increased oxygen speeds up the aerobic biodegradation of pollutants in the soil or groundwater. Petroleum compounds that are adsorbed to the soil within the capillary fringe, below the water table, or dissolved in groundwater can be reduced through biosparging. Biosparging is often utilized at sites where mid-weight petroleum compounds, such as diesel fuel, are used; lighter petroleum molecules volatilize quickly and are removed quickly through sparging. The permeability of the soil is an important aspect in the technology's efficiency (Vidali, 2001; Mohapatra, 2008).

5.2.2 Bioventing

An in-situ remediation technique employs native microorganisms to biodegrade organic components adsorbing to unsaturated zone soils. It is based on soil vapor extraction with vacuum enhancement. Pressure variations in the subsurface generate an inflow of atmospheric air and, as a result, oxygen supply, which is required for aerobic pollutant decomposition. It works to clean up petroleum products such as gasoline, jet fuel, kerosene, and diesel fuel. If the contaminants to be treated are volatile, the extracted soil vapor must be treated by contaminant adsorption on activated carbon followed by biodegradation in a biofilter (Mohapatra, 2008).

5.2.3 Bioaugmentation

It includes the introduction of microbes, either native or exogenous, to polluted areas. The utilization of additional microbial cultures in a land treatment unit is limited by two factors: nonindigenous cultures rarely compete well enough with indigenous populations to develop and maintain viable population levels, and most soils with long-term exposure to biodegradable waste include indigenous microbes that are good degraders if the land treatment unit is adequately managed (Vidali, 2001).

6. Bioremediation of organic pollutants

In polluted areas, a wide range of organic contaminants are likely to be present, mandating the use of a wide range of microbes for efficient clean-up (Table 6.3). The first biological remediation agent was patented in 1974, and it was a strain of *Pseudomonas putida* (Prescott et al., 2002). Since then, a huge number of species from at least 11 distinct prokaryotic divisions have been added to the list (Glazer and Nikaido, 2007). Microorganisms degrade organic contaminants either in the presence of oxygen (respiration) or under anoxic circumstances (denitrification, methanogenesis, and sulfidogenesis). The bulk of contaminants in the environment degrade most quickly and completely when they are exposed to aerobic conditions. Oxidations, which are catalyzed by oxygenase and peroxidases, are important

Table 6.3 Potential of microorganism to degrade various organic pollutants.

Pollutants	Organism	References
Benzene, anthracene, and hydrocarbons	*Pseudomonas* spp.	Cybulski et al. (2003)
Halogenated hydrocarbons, linear alkyl benzene sulfonates, and polycyclic aromatics (PCBs)	*Alcaligenes* spp.	Lal and Khanna (1996)
Aromatics	*Flavobacterium* spp. *Azotobacter* spp.	Jogdand (1995)
Cycloparaffins and branched hydrocarbon benzene	*Mycobacterium* spp.	Sunggyu (1995)
Halogenated hydrocarbons	*Corynebacterium* spp.	Jogdand (1995)

enzymatic processes in aerobic biodegradation. Oxygenases are oxygen-dependent oxidoreductases that incorporate oxygen into the substrate. Degradative organisms require oxygen at two stages of their metabolism: the initial attack on the substrate and the last stages of the respiratory chain. Soluble carbon molecules are degraded sequentially to methane, carbon dioxide, ammonia, and hydrogen sulfide under strictly anaerobic settings by a synoptic interaction of fermentative and acetogenic bacteria with methanogens or sulfate reducers under strictly anaerobic conditions. In terms of kinetics and capabilities, anaerobic degradation has long been thought to be inferior to aerobic degradation. Anaerobic techniques have proven to be more efficient and less expensive than aerobic treatment when high loads of easily degraded organic pollutants are present (Chatterjee et al., 2008).

6.1 Pesticides

Atrazine and organophosphate are two of the most commonly used organic pesticides in agriculture. In conservation tillage systems, atrazine is the most commonly used herbicide to prevent soil erosion. It was first employed for weed control in agricultural production of crops such as maize, sorghum, and sugar cane in the 1950s. Atrazine is resistant to biodegradation, with a reported half-life of more than 170 days in soils harboring atrazine-degrading microorganisms and a solubility of roughly 30 mg L^{-1} (Protzman et al., 1999). Atrazine is commonly detected in surface and ground water samples due to its tenacity, posing a direct risk to humans through the intake of potable water. Like other triazine herbicides, atrazine works by attaching to the photosystem II protein plastoquinone-binding protein, which animals lack. Starvation and oxidative damage produced by a malfunction in the electron transport mechanism cause plant death. Several studies have called for its restriction in the United States due to its claimed endocrine disruptor effects, probable carcinogenic effect, and epidemiological link to reduced sperm levels in men (Ackerman, 2007). *Pseudomonas* sp. ADP has been the most closely investigated of the atrazine-degrading microbes. *Pseudomonas* sp. ADP transforms atrazine to cyanuric acid via the AtzA, B, and C enzymes in the biodegradation of atrazine. AtzA converts atrazine to hydroxyatrazine, which is subsequently hydrolytically deamidated to generate N-isopropylammelide. Finally, N-isopropylammelide is converted to cyanuric acid by AtzC, a hydrolytic deamidase similar to AtzB. Enzymes found in soil microorganisms then mineralize cyanuric acid, converting it to carbon dioxide and ammonia (Ang et al., 2005; Krutz et al., 2009). Over 100 OP pesticides are in use around the world, accounting for 38% of total pesticide usage (Singh, 2009). Insecticides and chemical warfare chemicals contain organophosphates, which are highly toxic neurotoxins. The organophosphate family includes paraoxon, parathion, chlorpyrifos disulfoton, ruelene, carbophenothion, and dimeton. The ability of this class of chemicals to suppress acetylcholinesterase and, as a result, block acetylcholinesterase from breaking down acetylcholine at the synaptic junction is primarily responsible for their neurotoxicological effects. These chemicals have also been linked to pathology and chromosomal damage in bladder cancer patients. The most important stage in the detoxification of organophosphates in soil is microbial degradation by hydrolysis of the P—O alkyl and P—O aryl bonds. The bacteria that degrade the most major organophosphates are listed in Table 6.4. Phosphotriesterases (PTEs) are a set of enzymes found in microbes, animals, and plants that are responsible for the breakdown of Organophosphorus (OP) compound. Lots of studies have been done on OP biodegradation to date, and as a result, our understanding of OP degradation has evolved significantly in recent years. It is being used for a variety of industrial applications as a result of a greater understanding of the issue (Singh, 2009).

Table 6.4 Microorganism known for metabolism of organophosphates in culture and in field conditions.

S.no.	Organophosphates	Bacteria	Fungi
1	Chlorpyrifos	*Pseudomonas* sp., *Bacillus* sp., *Kurthia* sp., *Streptococcus* sp.	*Aspergillus niger*, *Mucor* sp., *Fusarium* sp., *Claviceps* sp.
2	Diazinon	*Pseudomonas glycinea*, *Pseudomonas diminutum*, *Arthrobacter* sp.	*Aspergillus oryzae*, *Trichoderma* sp.
3	Malathion	*Bacillus cereus*, *Bacillus subtilis*, *Micrococcus* sp., *Rhizobium japonicum*	*Aspergillus* spp., *Penicillium* sp., *Trichoderma viride*
4	Dimethonate	*Pseudomonas putida*, *Rhizobium* sp., *Nocardia* sp.	*Claviceps* sp., *Mucor* sp., *Penicillium notatum*
5	Dichlorvos	*Bacillus coagulans*, *Pseudomonas diminutum*, *Pseudomonas fluorescens*, *Pseudomonas melophthora*	*Saprolegnia* sp., *Penicillium notatum*, *Aspergillus niger*

7. Bioremediation of inorganic pollutants

7.1 Heavy metals

Heavy metals are the most common inorganic contaminants, and they have contaminated a huge area of land as a result of mining, manufacturing, agricultural, and defense activities. Although metals are naturally present at various quantities in the earth's crust and many metals are required for cell function (e.g., copper, iron, manganese, nickel, and zinc), all metals are hazardous at larger amounts. Any metal (or metalloid) species may be deemed a "contaminant" if it exists in an unwelcomed location, or in a form or concentration that harms humans or the environment (McIntyre, 2003). Metal concentrations in soil typically range from less than one to as much as 70,000 mg kg^{-1}. Regardless of where the metals in the soil came from, high levels of several metals can cause soil deterioration, crop yield decline, and poor agricultural product quality (Long et al., 2002). Heavy metals are a long-term threat to both the environment and human health since they are not biodegradable and may infiltrate the food chain (Jarup, 2003). Arsenic (As), cadmium (Cd), chromium (Cr), copper (Cu), lead (Pb), mercury (Hg), nickel (Ni), selenium (Se), silver (Ag), and zinc (Zn) are among the metals/metalloids included. Aluminum (Al), cesium (Cs), cobalt (Co), manganese (Mn), molybdenum (Mo), strontium (Sr), and uranium (U) are some of the less frequent metallic species that can be considered pollutants (McIntyre, 2003). Pb, one of the most persistent metals, has a soil retention duration of 150–5000 years and has been shown to sustain high concentrations in soil for up to 150 years after sludge application (Nanda Kumar et al., 1995). Cd has an average biological half-life of roughly 18 years and a 10-year half-life in the human body (Knasmuller et al., 1998). Another difficulty with hazardous heavy metals is that they can be transported and deposited in the bodily tissues of animals or humans through the food chain, causing DNA damage and carcinogenic effects due to their mutagenic properties (Knasmuller et al., 1998). Some species of Cd, Cr, and Cu, for example, have been linked to health problems ranging from dermatitis to cancer (McLaughlin et al., 1999). Metal ions can fully limit the microbial population at high concentrations by inhibiting numerous metabolic activities such as protein denaturation, cell division inhibition, cell membrane disruption, and so on, or organisms can acquire resistance or tolerance to the heightened levels of metals. Although the speciation and bioavailability of metals may fluctuate with variations in environmental conditions, these metals cannot be destroyed biologically and are hence eternal (Shukla et al., 2010). In microorganisms, metal toxicity is caused by the displacement of critical elements from their native binding sites or by ligand interactions (Nies, 1999; Bruins et al., 2000). Hg^{2+}, Cd^{2+}, and Ag^{2+}, for example, tend to attach to SH groups, inhibiting the action of sensitive enzymes (Nies, 1999). Metals can also harm cell membranes, affect enzyme specificity, disrupt cellular activities, and destroy DNA structure at high levels (Bruins et al., 2000). Microorganisms have been compelled to evolve metal-ion homeostasis factors and metal-resistance determinants as a result of this (Nies, 1999; Bruins et al., 2000). Exclusion through permeability barrier, intra- and extracellular sequestration, active efflux pumps, enzymatic reduction, and reduction in the sensitivity of cellular targets to metal ions are six possible mechanisms for a metal resistance system (Ji and Silver, 1995; Nies, 1999; Bruins et al., 2000). Microorganisms (Table 6.5) can function in metal-contaminated environments because of one or more of these resistance mechanisms. Metal microbial transformations perform a variety of tasks in stressful environments and can be grouped into two major categories: inorganic redox conversions and conversions from inorganic to organic form and vice versa, most often methylation and demethylation. Microbes can obtain energy by oxidizing iron,

Table 6.5 Microorganism that utilizes heavy metals.

S.no.	Elements	Microorganism
1	Copper	*Pseudomonas aeruginosa, Chlorella vulgaris, Pleurotus ostreatus, Bacillus* sp.
2	Cobalt	*Zooglea* sp., *Phormedium valderium*
3	Cadmium	*Citrobacter* sp., *Aspergillus niger, Ganoderma applantus, Stereum hirsutum*
4	Zinc	*Bacillus* sp., *Aspergillus niger, Pleurotus ostreatus*
5	Silver	*Rhizopus arrhizus, Aspergillus niger,* Geobacter metallireducens
6	Mercury	*Volvariella volvacea, Chlorella vulgaris, Rhizopus arrhizus*
7	Chromium	*Desulfovibrio fructosovorans, Desulfovibrio vulgaris*
8	Nickle	*Zooglea* sp., *Chlorella vulgaris*

sulfur, manganese, and arsenic. Metal reduction, on the other hand, can take place by dissimilatory reduction, in which microbes use metals as terminal electron acceptors in anaerobic respiration. For example, oxyanions of As, Cr, Se, and U (Tebo and Obraztsova, 1998) can be used in microbial anaerobic respiration as terminal electron acceptors. In addition, microorganisms may possess reduction mechanisms that are not coupled to respiration, but instead are thought to impart metal resistance. For example, aerobic and anaerobic reduction of Cr (VI) to Cr (III), reduction of Se (VI) to elemental selenium (Lloyd et al., 2001), reduction of U(VI) to U(IV) (Chang et al., 2001), and reduction of Hg (II) to Hg (0) (Wagner-Dobler et al., 2000) are widespread detoxification mechanisms among microorganisms. Microbial methylation plays an important role in the biogeochemical cycle of metals because methylated compounds are often volatile. For example, mercury [Hg (II)] can be biomethylated by a number of different bacterial species (e.g., Pseudomonas sp., Escherichia sp., Bacillus sp., and Clostridium sp.) to gaseous methylmercury (Pongratz and Heumann, 1999), which is the most toxic and most readily accumulated form of mercury (Nikunen et al., 1990). Biomethylation of arsenic to gaseous arsines (Gao and Burau, 1997), selenium to volatile dimethyl selenide (Dungan and Frankenberger, 2000), and lead to dimethyl lead (Pongratz and Heumann, 1999) has also been observed in a variety of soil environments.

8. Future prospects and importance of bioremediation

Bioremediation is a technique for cleaning up polluted ecosystems such as soils, groundwater, and oceans. Bacteria, fungi, algae, and plant species can all be found in such systems. Toxic substances in their surroundings can be metabolized, immobilized, or absorbed by them. However, one of the key benefits of these systems is that they are less hazardous to the environment and produce little or no by-products. Furthermore, traditional physical and chemical treatments are ineffective and costly, causing more harm than help. As a result, by assessing previous bioremediation research, bioreactors or products that are more efficient and feasible could be designed. Furthermore, these systems may be able to eliminate all toxins from the environment. It also produces valuable chemicals as a by-product.

9. Conclusion

Sustainable biological soil remediation methods that are economically feasible are being developed to: efficiently remove contaminants from soil; reduce their bioavailability, mobility, (eco)toxicity, and potential risks to environmental and human health; and simultaneously restore soil health and ecosystem services. Traditional physicochemical approaches for restoring contaminated environments are being replaced by bioremediation. Due to the fact that it is a cost-effective, labor-intensive, safe, eco-friendly approach, considerable progress and advancements have occurred in this field during the last two decades. However, because bioremediation sometimes incorporates numerous treatment procedures and can endure for a long time (years or decades), and is frequently used in conjunction with other techniques, estimating its success can be challenging. More multidisciplinary research in connection to process optimization, validation, its impact on the environment, and the effectiveness and predictability of the approach should be carried out in this context to make it a widely acknowledged technique.

Acknowledgment

We are thankful to Director, Absolute Foods for providing research opportunities, lab support, and other necessary facilities.

References

Abbas, M., Adil, M., Ehtisham-ul-Haque, S., et al., 2018. Vibrio fischeri bioluminescence inhibition assay for ecotoxicity assessment: a review. Sci. Total Environ. 626, 1295–1309.

Ackerman, F., 2007. The economics of atrazine. Int. J. Occup. Environ. Health 13, 441–449.

Adrian, N.R., Arnett, C.M., 2007. Anaerobic biotransformation of explosives in aquifer slurries amended with ethanol and propylene glycol. Chemosphere 66, 1849–1856.

Alloway, B.J., 1995. Heavy Metals in Soils, second ed. Blackie Academic & Professional, London.

Alvarenga, P., Ferreira, C., Mourinha, C., et al., 2018. Chemical and ecotoxicological effects of the use of drinking-water treatment residuals for the remediation of soils degraded by mining activities. Ecotoxicol. Environ. Saf. 161, 281–289.

Ang, E.L., Zhao, H., Obbard, J.P., 2005. Recent advances in the bioremediation of persistent organic pollutants via biomolecular engineering. Enzym. Microb. Technol. 37, 487–496.

April, W., Sims, R.C., 1990. Evaluation of the use of prairie grasses for stimulating polycyclic aromatic hydrocarbon treatment in soil. Chemosphere 20, 253–265.

Barker, A.V., Bryson, G.M., 2002. Bioremediation of heavy metals and organic toxicants by composting. J. World Sci. 2, 407–420.

Bruins, M.R., Kapil, S., Oehme, F.W., 2000. Microbial resistance to metals in the environment. Ecotoxicol. Environ. Saf. 45, 198–207.

Chang, Y.J., Peacock, A.D., Long, P.E., Stephen, J.R., McKinley, J.P., MacNaughton, S.J., Hussain, A.K.M.A., Saxton, A.M., White, D., 2001. Diversity and characterization of sulfate-reducing bacteria in groundwater at a uranium mill tailings site. Appl. Environ. Microbiol. 67, 3149–3160.

Chatterjee, P., Chattopadhyay, S., Roy, S.K., 2008. Bioremediation: a tool for cleaning polluted environments. J. Appl. Biosci. 11, 594–601.

Cybulski, Z., Dzuirla, E., Kaczorek, E., Olszanowski, A., 2003. The influence of emulsifiers on hydrocarbon biodegradation by *Pseudomonadacea* and *Bacillacea* strains. Spill Sci. Technol. Bull. 8, 503–507.

Dungan, R.S., Frankenberger, W.T., 2000. Factors affecting the volatilization of dimethylselenide by *Enterobacter clocae* SLD 1a-1. Soil Biol. Biochem. 32, 1353–1358.

Doran, J.W., Zeiss, M.R., 2000. Soil health and sustainability: managing the biotic component of soil quality. Appl. Soil Ecol. 15, 3–11.

Gao, S., Burau, R.G., 1997. Environmental factors affecting rates of arsine evolution from and mineralization of arsenicals in soil. J. Environ. Qual. 26, 753–763.

Glazer, A.N., Nikaido, H., 2007. Microbial Biotechnology: Fundamentals of Applied Microbiology, second ed. Cambridge University Press, Cambridge, pp. 510–528.

Gomez-Sagasti, M.T., Alkorta, I., Becerril, J.M., et al., 2012. Microbial monitoring of the recovery of soil quality during heavy metal phytoremediation. Water, Air, Soil Pollution 223, 3249–3262.

Irizar, A., Rodríguez, M.P., Izquierdo, A., et al., 2015. Effects of soil organic matter content on cadmium toxicity in *Eisenia fetida*: implications for the use of biomarkers and standard toxicity tests. Arch. Environ. Contam. Toxicol. 68, 181–192.

Jarup, L., 2003. Hazards of heavy metal contamination. Br. Med. Bull. 68, 167–182.

Jeffery, S., Gardi, C., Jones, A., et al., 2010. European Atlas of Soil Biodiversity. Publication Office of the European Union, Luxembourg.

Ji, G., Silver, S., 1995. Bacterial resistance mechanisms for heavy metals of environmental concern. J. Indian Microb. 14, 61–75.

Jogdand, S.N., 1995. Environmental Biotechnology, first ed. Himalaya Publishing House, Bombay, pp. 104–120.

Knasmuller, S., Gottmann, E., Steinkellner, H., Fomin, A., Pickl, C., Paschke, A., God, R., Kundi, M., 1998. Detection of genotoxic effects of heavy metal contaminated soils with plant bioassay. Mutat. Res. 420, 37–48.

Krutz, L.Z., Burke, I.C., Reddy, K.N., Zablotowicz, R.M., Price, A.J., 2009. Enhanced atrazine degradation: evidence for reduced residual weed control and a method for identifying adapted soils and predicting herbicide persistence. Weed Sci. 57, 427–434.

Lal, B., Khanna, S., 1996. Degradation of crude oil by *Acinetobacter calcoaceticus* and *Alcaligenes odorans*. J. Appl. Bacteriol. 81, 355–362.

Lloyd, J.R., Mabbett, A.N., Williams, D.R., Macaskie, L.E., 2001. Metal reduction by sulphate-reducing bacteria: physiological diversity and metal specificity. Hydrometallurgy 59, 327–337.

Long, X.X., Yang, X.E., Ni, W.Z., 2002. Current status and perspective on phytoremediation of heavy metal polluted soils. J. Appl. Ecol. 13, 757–762.

McIntyre, T., 2003. Phytoremediation of heavy metals from soils. Adv. Biochem. Eng. Biotechnol. 78, 97–123.

McLaughlin, M.J., Parker, D.R., Clark, J.M., 1999. Metals and micronutrients—food safety issues. Field Crops Resour. 60, 143–163.

Megharaj, M., Ramakrishnan, B., Venkateswarlu, K., et al., 2011. Bioremediation approaches for organic pollutants: a critical perspective. Environ. Int. 37, 1362–1375.

Mohapatra, P.K., 2008. Textbook of Environmental Microbiology. I.K. International Publishing House Pvt. Ltd., New Delhi.

Nanda Kumar, P.B.A., Dushenkov, V., Motto, H., Raskin, I., 1995. Phytoextraction: the use of plants to remove heavy metals from soils. Environ. Sci. Technol. 29, 1232–1238.

Nies, D.H., 1999. Microbial heavy-metal resistance. Appl. Microbiol. Biotechnol. 51, 730–750.

Nikunen, E., Leinonen, R., Kultamaa, A., 1990. Environmental Properties of Chemicals. Ministry of the Environment, Environmental Protection Department, Research Report 91, pp. 685–689.

Pongratz, R., Heumann, K.G., 1999. Production of methylated mercury, lead and cadmium by marine bacteria as a significant natural source for atmospheric heavy metals in polar regions. Chemosphere 39, 89–102.

Prasad, K., 2020. Positive importance of arbuscular mycorrhizal fungi for global sustainable agriculture and environment management for green technology. Curr. Inves. Agric. & Curr. Res. 9, 1182–1184.

Prescott, L.M., Harley, J.P., Klein, D.A., 2002. Microbiology, fifth ed. McGraw-Hill, New York.

Protzman, R.S., Lee, P.H., Ong, S.K., Moorman, T.B., 1999. Treatment of formulated atrazine rinsate by *Agrobacterium radiobacter* strain J14A in a sequencing batch biofilm reactor. Water Resour. 33, 1399–1404.

Shukla, K.P., Singh, N.K., Sharma, S., 2010. Bioremediation: developments, current practices and perspectives. J. Genet. Eng. & Biotechnol. 1–19.

Singh, B.K., 2009. Organophosphorus-degrading bacteria: ecology and industrial applications. Nat. Rev. Microbiol. 7, 156–164.

Sunggyu, L., 1995. Bioremediation of polycyclic aromatic hydrocarbon-contaminated soil. J. Clean. Prod. 3, 255.

Tebo, B.M., Obraztsova, A.Y., 1998. Sulfate-reducing bacterium grows with Cr (VI), U(VI), Mn (IV), and Fe (III) as electron acceptors. FEMS (Fed. Eur. Microbiol. Soc.) Microbiol. Lett. 162, 193–198.

Valerio, M.E., García, J.F., Peinado, F.M., 2007. Determination of phytotoxicity of soluble elements in soils, based on a bioassay with lettuce (*Lactuca sativa* L.). Sci. Total Environ. 378, 63–66.

Vamerali, T., Bandiera, M., Mosca, G., 2010. Field crops for phytoremediation of metal-contaminated land. A review. Environ. Chem. Lett. 8, 1–17.

Vangronsveld, J., Cunningham, S.D., 1998. Metal Contaminated Soils: In Situ Inactivation and Phytorestoration. Springer-Verlag, Georgetown.

Vazquez, S., Moreno, E., Carpena, R.O., 2008. Bioavailability of metals and as from acidified multicontaminated soils: use of white lupin to validate several extraction methods. Environ. Geochem. Health 30, 193–198.

Vidali, M., 2001. Bioremediation. An overview. Pure Appl. Chem. 73, 1163–1172.

Wagner-Dobler, I., Lunsdorf, H., Lubbehusen, T., Von Canstein, H.F., Li, Y., 2000. Structure and speciescom positions of mercury-reducing biofilms. Appl. Environ. Microbiol. 66, 4559–4563.

CHAPTER

7

Biochar technology for sustainable environment: need of policy implementation for societal benefit

Vishakha Gupta and Riti Thapar Kapoor

Amity Institute of Biotechnology, Amity University, Noida, Uttar Pradesh, India

1. Introduction

Agricultural crop residue burning has emerged as an important challenge in the agricultural production system of India, which has resulted in rising air pollution incidents, greenhouse gas (like CH_4 and N_2O gases) emissions, and elevated levels of particulate matter and smog, thereby becoming a threat to biodiversity and human health (Wijitkosum and Jiwnok, 2019). The reasons for crop residue burning are labor scarcity, removal of weeds, increased stubble generation, and short time interval between seasonal harvesting crops (Venkatramanan et al., 2021). Around 500 million tonnes of crop residues are generated in India every year. Agricultural residue burning results in air pollution and leads to the loss of nutrients present in the crop residues. Huge proportions of carbon, nitrogen, phosphorous, potassium, and sulfur in the crop residues are lost in gaseous form and particulate matter (Jain et al., 2014). In India, the total amount of crop residue production and the burnt in the year 2017−18 was estimated to be 516 million tonnes and 116 million tonnes, respectively, which has generated around 176.1 million tonnes CO_2, 0.31 million tonnes CH_4, 10 million tonnes CO, and some amount of NH_3, N_2O, nonmethane volatile organic compounds (VOCs), and particulate matter (Porichha et al., 2021).

India is one of the leading agricultural countries; the conversion of agricultural wastes into biochar suggests a potential solution to managing crop residues (Venkatramanan et al., 2021; Wijitkosum and Jiwnok, 2019). Biochar is a solid organic residue produced by biomass pyrolysis at a temperature lower than 700°C under anoxia or limited oxygen supply (Chen et al., 2019). Biomass may include animal manures, food processing residues, municipal green waste, forestry cuttings, industrial and agricultural waste products, and biosolids (Ayaz et al., 2021).

Based on the heating rate and residence time employed for biochar production, pyrolysis processes are of two types: slow pyrolysis (conventional carbonization) and fast pyrolysis (Oni et al., 2019; Liang et al., 2021). The process in which biomass is subjected to a relatively long residence time (>2 h) for its conversion to biochar at a temperature between 300 and 700°C (Oni et al., 2019) is known as slow pyrolysis. On the other hand, when a relatively high temperature between 500 and 1000°C is applied at a low residence time for biochar production, the process is called fast pyrolysis (Kamali et al., 2021). The former process produces more syngas, while the latter generates more oils

Integrative Strategies for Bioremediation of Environmental Contaminants, Volume Two. https://doi.org/10.1016/B978-0-443-14013-6.00007-X

and liquids (Bolan et al., 2022; Oni et al., 2019; Tomczyk et al., 2020). Further, studies suggest that slow pyrolysis is the most effective process for biochar production and can result in an increased yield of up to 30%−35% from dry biomass weight (Jatav et al., 2021; Kamali et al., 2021; Liang et al., 2021; Oni et al., 2019; Panwar et al., 2019; Stewart et al., 2013).

Biochar has various applications, including soil health improvement, mitigation of greenhouse gas emissions, carbon sequestration, energy production, toxic metals remediation, and waste management, such as wastewater treatment, thereby providing resource for organic waste management (Panwar et al., 2019). Biochar usage can balance greenhouse gas emissions by storing carbon in a stable form and preventing the release of toxic gases into the atmosphere from biomass degradation (Xiang et al., 2020).

The inefficient and extensive use and disposal of dyes, phenols, antibiotics, and pesticides have drawn much attention because of their extreme toxicity and biodegradable environmental resistance. The rampant use of pesticides to control crop diseases and pests has led to soil pollution in agricultural lands, related ecosystems, and health risks to humans. Biochar, a cost-effective and sustainable substance, shows considerable potential to minimize the risk of soil contamination by immobilizing or degrading contaminants and can reduce pesticide uptake by plants (Khalid et al., 2020). Further, biochar also positively influences the soil's biological properties, including the soil microorganisms, by creating a habitat for their colonization due to its porous structure and serving as a mineral nutrient source. Biochar can also affect the soil enzymatic activity by influencing the synthesis of certain enzymes and their release in the soil by changing the pH, which can positively and negatively affect enzymes (Jatav et al., 2021).

2. Feedstock for biochar production

The classification of the feedstock material for an efficient and sustainable biochar production includes plant residue, sewage sludge, poultry litter, and other organic wastes and residues such as animal waste, industrial waste, urban refuse etc. (Panwar et al., 2019). Fig. 7.1 shows the production of biochar from agricultural residues. Depending on the type of feedstock used and pyrolysis conditions, the properties of biochar such as characteristics, physical properties, chemical composition, and nutrient contents significantly vary (Chen et al., 2019; Jatav et al., 2021; Oni et al., 2019). Plant biomass comprises cellulose, hemicellulose, and lignin with small amounts of sugar, protein, pectin, chlorophyll, etc. The major crops in the world, that is, maize, wheat, rice, and sugarcane, account for most lignocellulosic biomass (Bhuvaneshwari et al., 2019). Biochar raw material containing high lignocellulosic content yields higher biochar produce due to the transformation of lignin into char during the pyrolysis process (Jatav et al., 2021). Jatav et al. (2021) reported low carbon: nitrogen (C: N) ratio of feedstock material enhanced phosphorous availability and potassium content in biomass can influence the biochar pH value. Thus, it is essential to understand how feedstock properties impact biochar characteristics.

3. Properties of biochar

Structural information, composition, and physicochemical properties of biochar are essential for determining its applications.

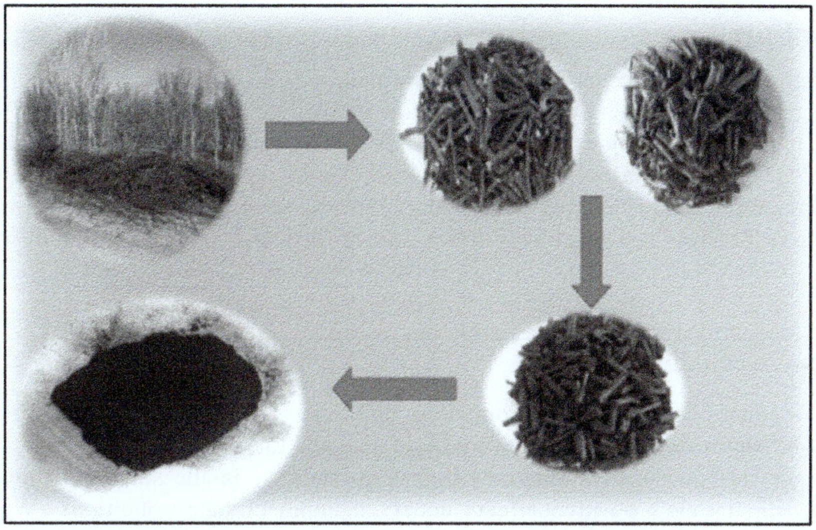

FIGURE 7.1

Production of biochar from agricultural residues.

Adopted from Vahedi, R., Rasouli Sadaghiani, M.H., Barin, M., Vetukuri, R.R., 2022. Effect of biochar and microbial inoculation on P, Fe, and Zn bioavailability in a calcareous soil. Processes. 10, 343.

4. Structural characteristics of biochar

Biochar skeletal structure comprises carbon and minerals of varied pore sizes (Oni et al., 2019). Biochar's porous structure consists of several functional groups and aromatic compounds (Oni et al., 2019). Its microporosity allows the sorption of dissolved organic matter and improves microbial activity, thereby expediting soil remediation (Chen et al., 2019; Oni et al., 2019).

5. Elemental composition of biochar

The composition and chemical texture of biochar are rich and may vary depending on the pyrolysis conditions, duration, and type of feedstock materials used in its production process (Oni et al., 2019). Biochar is highly stable and contains carbon, hydrogen, oxygen, nitrogen, sulfur, phosphorous, potassium, calcium, magnesium, sodium, silicon, etc., among which carbon content is the highest (more than 60%), followed by hydrogen and oxygen (Chen et al., 2019). Thermal degradation processes (including hydrothermal conversion, slow and fast pyrolysis, gasification, and torrefaction) are majorly distinguished by the availability of free water, atmospheric oxygen, feedstock residence time, gas environment, temperature, pressure, and heating rate (Lehmann and Joseph, 2009).

6. Physiochemical properties of biochar

The biochar physicochemical properties are governed by its pH, stability, water-holding capacity, porosity, contamination level, decomposition capability, and cation exchange capacity.

7. Effect of pH

Biochar is alkaline in nature because of inorganic minerals, type of raw materials used, and pyrolysis carbonization temperature (Chen et al., 2019). Usage of alkaline biochar improves the availability of phosphorous, nitrogen, and major cations such as Ca^{2+} (Oni et al., 2019) and increases the soil pH (Jatav et al., 2021). Biochar has an effective response to soils with low organic carbon content, poor soil physical properties, and acidic pH (Oni et al., 2019). However, in the case of soil with basic pH, frequent usage of biochar may negatively affect soil fertility and nutrient availability (Jatav et al., 2021).

8. Stability

Reports revealed that biochar of high aromaticity and degree of aromatic condensation has a high probability of having strong resistance toward biological and thermal degradation and therefore high stability (Leng et al., 2019). The biochar's long-term stability in the soil is essential for carbon sequestration, which shows positive effects under changing climate conditions (Wijitkosum and Jiwnok, 2019).

9. Water-holding capacity

Biochar exhibits high surface area, good soil porosity, and various aliphatic and functional groups, which contributes to an increase in the nutrient and water-holding capacity. The porous structure of biochar is capable of retaining water that provides a niche for the growth of microbial populations such as bacteria, fungi, or protozoa and thereby allowing them to recover more quickly under drought conditions (Ayaz et al., 2021; Jatav et al., 2021; Liang et al., 2014). Furthermore, the pore size in biochar depends on the production temperature. Biochar would have larger pores at high temperatures due to more water and biomass volatilization (Jatav et al., 2021).

10. Specific surface area

The specific surface area of biochar is directly related to the pyrolysis temperature. Low-temperature pyrolysis results in high volatile matter content, which fills the pore matrix of biochar and reduces its specific surface area. As the temperature increases, the biochar yield and volatile matter escape, resulting in microspores formation, simultaneously increasing the number of pores, leading to microporous structure and larger specific surface area. Due to the large specific surface area and porous structure, biochar has plenty of surface functional groups (Liang et al., 2021). However, when the temperature exceeds its threshold value, the specific surface area reaches its plateau and starts reducing (Bolan et al., 2022; Chen et al., 2019). Biochar has a specific surface area of approximately $1.5-500\ m^2\ g^{-1}$ (Chen et al., 2019).

Studies suggested that pore volume of biochar is directly related to its surface area (Ayaz et al., 2021; Bolan et al., 2022). Biochar has a high nutrient and water retention capacity due to its absorbing abilities and large surface area and porosity (Ayaz et al., 2021). High biochar surface area increases the ion exchange capacity with nutrients; thus, mixing biochar with soil can elevate soil's ion exchange capacity (Wijitkosum and Jiwnok, 2019).

11. Cation exchange capacity

Biochar can adsorb various essential ions for plant growth and soil fertility, such as calcium and ammonium ions. Biochar with cation exchange capacity can reduce nutrient leaching losses from soils (Chen et al., 2019). The cation exchange capacity is dependent on the biochar surface structure with functional groups that have surface charges. Literature suggested that biochar produced at higher pyrolyzed temperature (for instance, more than 500°C) has low cation exchange capacity, which attributes to biochar aromatization and disappearance of functional groups (carboxylic and phenolic groups) (Domingues et al., 2020). On the contrary, biochar shows high cation exchange capacity generated at a relatively lower temperature at which adequate functional groups are present in the structure to provide a negative charge that can attract cations to amend soil fertility (Oni et al., 2019).

12. Applications of biochar

Biochar application can replenish major soil nutrients in low fertility soils due to its unique surface charge density. It can also promote cation adsorption due to predominantly negative-charged surfaces (Shakoor et al., 2021), thereby improving crop production. It can help minimize any possible odor from organic nutrient sources by thoroughly mixing biochar with compost and manure (Lehmann and Joseph, 2009). Furthermore, biochar also has applications in providing alternative solutions for the remediation of other environmental problems (Fig. 7.2). One of the main applications of biochar is its use as a renewable fuel. The gradual depletion of conventional fossil fuel resources globally has challenged the introduction of new alternative energy sources such as biofuels. Biomass is the only renewable energy source that can be used without energy storage systems (Supriya et al., 2017).

Biochar can be used for wastewater treatment (Xiang et al., 2020). Biochar has shown efficacy in removing potentially toxic elements (PTEs) present in water (Muhammad et al., 2020). Such efficiency is affected by biochar's physicochemical and surface properties, pyrolytic conditions, type of feedstock, and pH (Bolan et al., 2022). Studies have reported biochar's ability for sorption/immobilization and removal of PTEs in wastewater samples is by various mechanisms such as ion exchange, electrostatic interaction, chemical precipitation, etc. These mechanisms are attributed to the properties of biochar (high porosity, high surface area, high buffering capacity, high ash content, alkalinity, and aromatic characters), soil type, metal type, the ageing process, and applied dosage (Bolan et al., 2022). Due to its beneficial physicochemical properties, biochar offers the possibility of removing hazardous pollutants in the air, such as VOCs that are carcinogenic in nature. Biochar with hydrophilic nature is used for the adsorption of polar VOCs such as ketones and aldehydes (Bolan et al., 2022).

13. Removal of organic contaminants by biochar

Removing organic pollutants from soil and wastewater such as fungicides, herbicides, pesticides, dyes, industrial chemicals, antibiotics/drugs, benzene, etc., is an essential application of biochar for soil remediation. Chemisorption (electrophilic interaction) and physisorption (hydrophobic interactions)

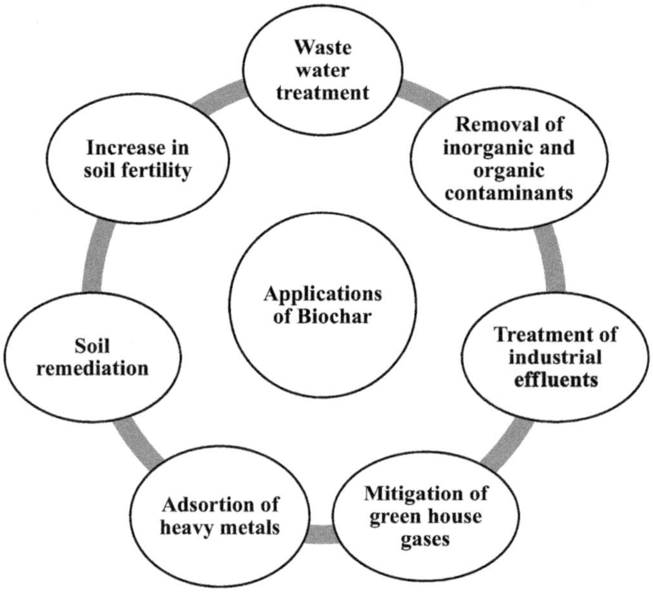

FIGURE 7.2

Application of biochar for sustainable environment.

are the primary methods for removing organic contaminants. Other mechanisms may include chemical transformation and partitioning (Oni et al., 2019). In addition, factors such as pyrolysis temperature, pH, targeted pollutant ratio to biochar, and type of biomass affect the elimination of organic contaminants by biochar-based material (Oni et al., 2019; Qiu et al., 2022).

Polychlorinated biphenyls (PCBs) are persistent organic pollutants that negatively affect the environment and human health. Biochar can remove PCBs in two ways: by chemical adsorption between PCBs and biochar, and by microspore-filling mechanism (Qiu et al., 2022).

Dyes are organic contaminants extensively used by the print, paper, and textile industries. They have good solubility and complex aromatic structures, making them highly toxic and less degradable. Biochar has been widely used to remove dye for water treatment (Liang et al., 2021; Kapoor et al., 2022). Fan et al. (2017) reported the production of biochar from municipal sludge and using it to remove methylene blue from wastewater by adsorption.

14. Removal of inorganic contaminants by biochar

Biochar is a cheap and effective tool for removing inorganic contaminants such as heavy metals like copper, chromium, lead, nickel, etc., from polluted soils and water. Heavy metals are nonbiodegradable, and they tend to accumulate in living organisms via food chain, leading to life-threatening disorders. The mechanisms by which biochar removes metal ions are adsorption, precipitation, ion exchange, etc. (Inyang et al., 2016; Liang et al., 2021). Xiang et al. (2020) reported that biochar has a high concentration of alkali salts and alkali earth minerals, which precipitates heavy metal contaminants.

15. Remediation of gases by biochar

Climate change has become a global concern and attributed to the greenhouse gases due to more anthropogenic activities (Ayaz et al., 2021). Thus, it is of utmost importance to take necessary actions to neutralize human-caused greenhouse gases emission. Agricultural waste burning is the second primary source of fire emissions following forest burning (Jain et al., 2014).

Soil is considered both a source and sink for greenhouse gases. Biochar offers various benefits for environmental remediation, such as enabling long-term storage of greenhouse gases by adding biochar as a soil amendment (Ayaz et al., 2021; Bolan et al., 2022). It has been reported that biochar application can balance approximately 12% of anthropogenic greenhouse gas emission annually (Jatav et al., 2021). Mixing biochar in the soil at a 13.5 t ha^{-1} rate might capture and store carbon in the ground for up to 200 years (Bolan et al., 2022; Matovic, 2011; Saletnik et al., 2019).

Literature revealed that incorporating biochar in soil results in the reduction of soil carbon dioxide content (by carbon sequestration), decrease in nitrous oxide emission, and enhances the methane uptake by soil, thereby enabling long-term storage and mitigation of greenhouse gas emissions (Chen et al., 2019; Lehmann and Joseph, 2009). Mechanisms include the change in abundance of soil microbes, improvisation on the adsorption ability of soil to ammonium or nitrate enhancing soil aeration, and reduction in denitrification rate.

Significant decline in N_2O and CH_4 emissions was observed in soils with neutral pH ($pH = 6.6-7.3$), whereas the most substantial mitigation effect was observed in acidic soils ($pH \leq 6.5$) for CO_2 compared to N_2O and CH_4 emissions (Shakoor et al., 2021). In rice crop fields, the effect of biochar has resulted in a significant reduction of soil CH_4 emissions while elevating soil net absorption of CH_4 (Ayaz et al., 2021; Chen et al., 2019). Furthermore, biochar can actively remove poisonous substances from gas. The biochar derived from sludge, pig manure, camphor, and bamboo efficiently removes hydrogen sulfide from biogas by adsorption. The adsorption is assisted by biochar pH, moisture content, surface area, and chemical bonding and showed efficacy >96% (Oni et al., 2019).

16. Biochar and human health risks

Biochar has shown significant potential for various applications. However, several adverse effects of biochar application have been reported. Biochar may contain contaminants itself, either coproduced during biochar production like polycyclic aromatic hydrocarbons (PAHs), metal cyanide and persistent-free radicals, or introduced by its feedstock such as PTEs (including metals and metalloids) (Han et al., 2021; Hilber et al., 2017).

During biochar production, specific issues may significantly risk human health and cause severe ecological effects (Wang et al., 2019). Studies have reported the foration of PAHs in biochar during the pyrolysis of feedstock in anoxia conditions, which depends on the feedstock and heat treatment temperature levels (de la Rosa et al., 2019; Wang et al., 2018, 2019; Zhang et al., 2018). PAHs were present in biochar produced by crop residues (Wang et al., 2019). The release of PAHs from biochar results in their sorption in soil and uptake by roots of crop plants grown in contaminated soils, which would get accumulated in large amounts in their edible plant parts, thereby maximizing the health risks (Jatav et al., 2021). The contaminants present in biochar may get collected and biomagnified through the food chain, thereby increasing the toxicity levels in living organisms. Wang et al. (2019) indicated that biochar application in soils might pose potential risks to human health. However, due to the significant variations in PAHs' concentrations and risks in soils and food crops, biochar should be assessed before adoption for agriculture.

Strategies such as optimization of pyrolysis parameters and posttreatment of biochar are employed to produce biochar with low PAH levels (Han et al., 2022). Microbial degradation and abiotic oxidation are natural processes that can decrease PAH concentrations in the environment (Han et al., 2022). The inhalation of small biochar particles during their production or usage can cause a severe respiratory disease known as pneumoconiosis (Jatav et al., 2021).

Chen et al. (2021) reported that their generated tobacco stem pyrolyzed biochar at different pyrolysis temperatures expedited the production of reactive oxygen species and may cause pulmonary toxicity in human lung epithelial cells. The presence of elevated levels of heavy metal concentration may increase human health risks as airborne pollutants on biochar particles can irritate eyes and mucous membranes (Jatav et al., 2021). More PTE concentrations in biochar were found when sewage sludge and pig manure were used as feedstock (Han et al., 2022; Hilber et al., 2017).

17. Economic importance of biochar

A biobased economy involves exploring and exploiting bioresources and creating a new bioproduct of economic importance. The bioresource is the biomass used to produce a bioproduct, that is, biochar. Biochar production is gaining attention due to its potential applications in the energy and environment sector. Biochar shows significant potential in addressing climate change and establishing a circular economy model (Hu et al., 2021). Biochar usage on soil fertility and crop nutrition improvement increased the agronomical benefits of biochar application (Jindo et al., 2020). The main reason for less utilization of biochar for land amendment is the uncertainty of its outcome and revenue implications (Vochozka et al., 2016). Biochar as a constituent of construction material, especially alternative to cement or sand, is one of the emerging applications (Hu et al., 2021). Biochar has a significant potential to be used as a filler material to bring significant alteration in composite manufacturing. Biochar's properties that would make composite manufacturing efficient include its carbon content, high surface area, surface functional groups, etc. These are also important in making durable wood-plastic composite (Bolan et al., 2022).

Economics is a crucial component in producing biochar and its application as a soil amendment. The economic feasibility of biochar production depends on the feedstock availability and processing, type and cost of pyrolysis, biochar yield, storage and utilization, labor, transportation, and application costs (Homagain et al., 2016). The economic cost of implementing biochar production and application is essential because it determines how easily the technology is executed and financed against other products (Srinivasarao et al., 2013). The products of biochar and biochar-blended soil enhancers are not yet economically competitive against other established products such as conventional fertilizers available in the market (Pourhashem et al., 2019).

Currently, there is no such major industrial market in India to obtain biochar prices and data for a comprehensive comparison. However, globally data are available on the mean price for pure biochar, and it has varied values. It ranges from as low as US \$0.09 kg^{-1} in Philippines to as high as US \$8.85 kg^{-1} in the United Kingdom. The price of biochar is highly variable depending on the origin of the biochar production sites and other input cost.

Roberts et al. (2010) reported that there are three components for the economic value of biochar: (1) presence of phosphorous and potassium nutrient contents in biochar, (2) improved fertilizer use efficiency, and (3) value of greenhouse gas reductions supported by the life cycle of the biochar. The standard analytical methods for various parameters can help in estimation of the cost of biochar production which includes biochar conversion efficiency, collection, processing and proximate analysis, and recovery of total carbon and nitrogen (Venkatesh et al., 2022).

18. Policy regulation on biochar

Biochar has demonstrated significant potential for environmental and agronomic benefits. However, its adoption for commercial purposes has not been widely employed by farmers as soil preservation or carbon sequestration practice, thereby raising concern about its wide implementation (Pourhashem et al., 2019). Any biochar regulation or certification that may develop must require complete scientific data and suggestions that would be crucial to the policy community in a comprehensive, robust, and independent of any conflict of interest (Verheijen et al., 2012). Sustainability schemes can be supported by biochar regulations that stipulate requirements for certification to access markets, enabling biochar producers and users to develop proper management system and get certification that would provide them access to the markets (Vereš et al., 2014). The policy influences the economic viability of biochar, and uncertainty over future policy may jeopardize investment in biochar production facilities (Pokharel and Comer, 2021). Presently, two globally accepted standards (International Biochar Initiative [IBI] and European Biochar Certificate [EBC]) have been developed to mitigate the health and environmental risk associated with biochar production and application especially in agriculture. However, these standards are insufficient to regulate biochar quality produced for other industries. Various countries have their own biochar regulations aligned with IBI and EBC biochar standards. Countries with no biochar norms comply with fertilizer standards or compost standards for biochar application (Hu et al., 2021).

Several European Union countries like Austria, Hungary, and Sweden approved the application for biochar certification under EBC-Agro certification (EBC, 2022). IBI has a scheme called "IBI Biochar Certification Program" to provide accreditation to biochar products by creating a standardized and recognized system that meets the IBI Biochar Standards. This program has a geographical restriction for its implementation only in the United States and Canada (IBI Certification Program Requirements, 2015).

The Australian New Zealand (ANZ) Biochar Initiative has been developing a roadmap for creating a strategic direction for biochar industry and market development. The ANZ Biochar Industry Group offers a policy to its members for supporting the adaption of legal regulations regarding biochar production and utilization (ANZBIG, www.anzbc.org.au). The US government launched several programs in the 20th century to spread public awareness of soil preservation practices and policies. Pourhashem et al. (2019) found 35 US policy programs associated with supporting biochar production. These programs were focused on climate change mitigation, agricultural waste management, and food and energy management. The programs were of three types: explicit program (designed for or mention biochar in their description), implicit program (which do not mention biochar but focus on biobased products), and potential program (which do not mention biochar, nor their definition of products and processes covers biochar production, but slight modifications in language or program can target biochar).

Currently, biochar application in India is limited and mainly practised in villages and small cities. Despite broad applicability, biochar production protocols have not yet been standardized in India (Lakaria et al., 2021). The systematic biochar production has been attempted only at a few locations in India as biochar plant was designed by ICAR-Central Institute of Agriculture Engineering, Bhopal to produce small quantities of biochar; batch-type continuous torrefaction unit and system of constant screw-type biochar production has been developed by Maharana Pratap University of Agriculture and Technology, Udaipur (Lakaria et al., 2021).

19. Conclusion and future perspectives

Sustainable management of biomass, particularly crop residue, is a great challenge especially in a developing country such as India. Burning crop residues at such a large scale ensues the generation of harmful greenhouse gases in the atmosphere, thereby deteriorating the air quality and causing human health risks. The Government of India took several campaigns and initiatives to impede crop residues burning such as encouragement for the use of crop residues in the energy sector as raw material and promoting research on the measures to handle agricultural waste without burning. Biochar has many applications, including soil amendment, renewable fuel, and wastewater treatment, which plays a vital role in addressing critical environmental issues. Biochar application is considered a potential carbon sequestration strategy in soil. It removes hazardous gases from the atmosphere and enables long-term storage of greenhouse gases in soil, thereby mitigating climate change. However, there are certain health risks associated with biochar usage due to the presence of heavy metals, PAHs, VOCs, etc. Therefore, guidelines by reputed platforms such as IBI have stipulated threshold values of such compounds, resulting in hindering their detrimental effects on human beings. The economic feasibility of biochar production depends on several factors from its availability as raw material to its processing. However, considering the current economic situation, farmers/users are not keen on adopting such practices because of lesser returns. Providing sufficient incentives for crop residue recycling, higher income for processing of crop residues, and farm machinery for on-farm recycling may urge farmers to adopt biochar applications. A well-designed policy can play a crucial role in achieving benefits from biochar. However, there is no such conventional policy for biochar production in India. Hence, there is need of policy guidelines by government for the production of biochar and its sustainable applications, which can promote the growth of biochar industry. Further investigations are required on the function of biochar and its modification with nanomaterials under changing climate conditions, which can explore its potential use in the coming years.

References

Australian New Zealand Biochar Industry Group (ANZBIG), www.anzbc.org.au.

Ayaz, M., Feizienė, D., Tilvikienė, V., Akhtar, K., Stulpinaitė, U., Iqbal, R., 2021. Biochar role in the sustainability of agriculture and environment. Sustainability 13, 1330.

Bhuvaneshwari, S., Hettiarachchi, H., Meegoda, J.N., 2019. Crop residue burning in India: policy challenges and potential solutions. Int. J. Environ. Res. Publ. Health 16 (5), 832.

Bolan, N., Hoang, S.A., Beiyuan, J., Gupta, S., Hou, D., Karakoti, A., Joseph, S., Jung, S., Kim, K.H., Kirkham, M.B., Kua, H.W., Kumar, M., Kwon, E.E., Ok, Y.S., Perera, V., Rinklebe, J., Shaheen, S.M.,

Sarkar, B., Sarmah, A.K., Singh, B.P., Singh, G., Tsang, D.C.W., Vikrant, K., Vithanage, M., Vinu, A., Wang, H., Wijesekara, H., Yan, Y., Younis, S.A., van Zwieten, L., 2022. Multifunctional applications of biochar beyond carbon storage. Int. Mater. Rev. 67, 150–200.

Chen, W., Meng, J., Han, X., Lan, Y., Zhang, W., 2019. Past, present and future of biochar. Biochar 1, 75–87.

Chen, Y.C., Chen, K.F., Lin, J.H., Huang, S.W., Chen, H.H., Andrew Lin, K.Y., Lin, C.H., 2021. The impact of pyrolysis temperature on physicochemical properties and pulmonary toxicity of tobacco stem micro-biochar. Chemosphere 263, 128349.

De la Rosa, J.M., Sánchez-Martín, Á.M., Campos, P., Miller, A.Z., 2019. Effect of pyrolysis conditions on the total contents of polycyclic aromatic hydrocarbons in biochars produced from organic residues: assessment of their hazard potential. Sci. Total Environ. 667, 578–585.

Domingues, R.R., Sanchez-Monedero, M.A., Spokas, K.A., Melo, L.C.A., Trugilho, P.F., Valenciano, M.N., Silva, C.A., 2020. Enhancing cation exchange capacity of weathered soils using biochar: feedstock, pyrolysis conditions and addition rate. Agronomy 10, 824.

EBC, 2012-2022. European Biochar Certificate - Guidelines for a Sustainable Production of Biochar. European Biochar Foundation (EBC), Arbaz, Switzerland. http://european-biochar.org. (Accessed 7 May 2022). Version 10.1 from 10th Jan 2022.

Fan, S., Wang, Y., Wang, Z., Tang, J., Tang, J., Li, X., 2017. Removal of methylene blue from aqueous solution by sewage sludge-derived biochar: adsorption kinetics, equilibrium, thermodynamics and mechanism. J. Environ. Chem. Eng. 5, 601–611.

Han, H., Buss, W., Zheng, Y., Song, P., Khalid Rafiq, M., Liu, P., Mašek, O., Li, X., 2022. Contaminants in biochar and suggested mitigation measures – a review. Chem. Eng. J. 429, 132287.

Han, L., Zhang, B., Chen, L., Feng, Y., Yang, Y., Sun, K., 2021. Impact of biochar amendment on soil aggregation varied with incubation duration and biochar pyrolysis temperature. Biochar 3, 339–347.

Hilber, I., Bastos, A.C., Loureiro, S., Soja, G., Marsz, A., Cornelissen, G., Bucheli, T.D., 2017. The different faces of biochar: contamination risk versus remediation tool. J. Environ. Eng. Landsc. Manag. 25, 86–104.

Homagain, K., Shahi, C., Luckai, N., Sharma, M., 2016. Life cycle cost and economic assessment of biochar-based bioenergy production and biochar land application in Northwestern Ontario, Canada. Forest Ecosyst. 3, 21.

Hu, Q., Jung, J., Chen, D., Leong, K., Song, S., Li, F., Mohan, B.C., Yao, Z., Prabhakar, A.K., Lin, X.H., Lim, E.Y., Zhang, L., Souradeep, G., Ok, Y.S., Kua, H.W., Li, S.F.Y., Tan, H.T.W., Dai, Y., Tong, Y.W., Peng, Y., Joseph, S., Wang, C.H., 2021. Biochar industry to circular economy. Sci. Total Environ. 757, 143820.

IBI-CER-02.1 IBI Biochar Certification Program Manual: Requirements and Procedures for Biochar Certification Version 2.1, 2015. (Accessed 7 May 2022).

Inyang, M.I., Gao, B., Yao, Y., Xue, Y., Zimmerman, A., Mosa, A., Pullammanappallil, P., Ok, Y.S., Cao, X., 2016. A review of biochar as a low-cost adsorbent for aqueous heavy metal removal. Crit. Rev. Environ. Sci. Technol. 46, 406–433.

Jain, N., Bhatia, A., Pathak, H., 2014. Emission of air pollutants from crop residue burning in India. Aerosol Air Qual. Res. 14, 422–430.

Jatav, H.S., Rajput, V.D., Minkina, T., Singh, S.K., Chejara, S., Gorovtsov, A., Barakhov, A., Bauer, T., Sushkova, S., Mandzieva, S., Burachevskaya, M., Kalinitchenko, V.P., 2021. Sustainable approach and safe use of biochar and its possible consequences. Sustainability 13 (18), 10362.

Jindo, K., Sánchez-Monedero, M.A., Mastrolonardo, G., Audette, Y., Higashikawa, F.S., Silva, C.A., Akashi, K., Mondini, C., 2020. Role of biochar in promoting circular economy in the agriculture sector. Part 2: a review of the biochar roles in growing media, composting and as soil amendment. Chem. Biol. Technol. Agric. 7, 16.

Kamali, M., Appels, L., Kwon, E.E., Aminabhavi, T.M., Dewil, R., 2021. Biochar in water and wastewater treatment - a sustainability assessment. Chem. Eng. J. 420, 129946.

Kapoor, R.T., Rafatullah, M., Siddiqui, M.R., Khan, M.A., Sillanpaa, M., 2022. Removal of reactive black 5 dye by banana peel biochar and evaluation of its phytotoxicity on tomato. Sustainability 14, 4176.

Khalid, S., Shahid, M., Murtaza, B., Bibi, I., Asif Naeem, M., Niazi, N.K., 2020. A critical review of different factors governing the fate of pesticides in soil under biochar application. Sci. Total Environ. 711, 134645.

Lakaria, B.L., Jha, P., Biswas, A.K., Patra, A.K., Chaudhari, S.K., 2021. Protocol for Biochar Use in Indian Agriculture. ICAR-Indian Institute of Soil Sciences, Bhopal, MP, pp. 1−13.

Lehmann, J., Joseph, S., 2009. Biochar for Environmental Management. Science and Technology. Routledge, ISBN 9780367779184.

Leng, L., Huang, H., Li, H., Li, J., Zhou, W., 2019. Biochar stability assessment methods: a review. Sci. Total Environ. 647, 210−222.

Liang, C., Zhu, X., Fu, S., Méndez, A., Gascó, G., Paz-Ferreiro, J., 2014. Biochar alters the resistance and resilience to drought in a tropical soil. Environ. Res. Lett. 9, 064013.

Liang, L., Xi, F., Tan, W., Meng, X., Hu, B., Wang, X., 2021. Review of organic and inorganic pollutants removal by biochar and biochar-based composites. Biochar 3, 255−281.

Matovic, D., 2011. Biochar as a viable carbon sequestration option: global and Canadian perspective. Energy 36, 2011−2016.

Muhammad, N., Nafees, M., Khan, M.H., Ge, L., Lisak, G., 2020. Effect of biochars on bioaccumulation and human health risks of potentially toxic elements in wheat (*Triticum aestivum* L.) cultivated on industrially contaminated soil. Environ. Pollut. 260, 113887.

Oni, B.A., Oziegbe, O., Olawole, O.O., 2019. Significance of biochar application to the environment and economy. Ann. Agric. Sci. (Cairo) 64, 222−236.

Panwar, N.L., Pawar, A., Salvi, B.L., 2019. Comprehensive review on production and utilization of biochar. SN Appl. Sci. 1, 168.

Porichha, G.K., Hu, Y., Rao, K.T.V., Xu, C.C., 2021. Crop residue management in India: stubble burning vs. other utilizations including bioenergy. Energies 14 (14), 4281.

Pokharel, R., Comer, B., 2021. Economics of Biochar. Michigan State University. https://www.canr.msu.edu/news/economics-of-biochar. (Accessed 7 May 2022).

Pourhashem, G., Hung, S.Y., Medlock, K.B., Masiello, C.A., 2019. Policy support for biochar: review and recommendations. GCB Bioenergy 11, 364−380.

Qiu, M., Liu, L., Ling, Q., Cai, Y., Yu, S., Wang, S., Fu, D., Hu, B., Wang, X., 2022. Biochar for the removal of contaminants from soil and water: a review. Biochar 4, 19.

Roberts, K.G., Gloy, B.A., Joseph, S., Scott, N.R., Lehmann, J., 2010. Life cycle assessment of biochar systems: estimating the energetic, economic and climate change potential. Environ. Sci. Technol. 44, 827−833.

Saletnik, B., Zagula, G., Bajcar, M., Tarapatskyy, M., Bobula, G., Puchalski, C., 2019. Biochar as a multifunctional component of the environment-a review. Appl. Sci. 9 (6), 1139.

Shakoor, A., Arif, M.S., Shahzad, S.M., Farooq, T.H., Ashraf, F., Altaf, M.M., Ahmed, W., Tufail, M.A., Ashraf, M., 2021. Does biochar accelerate the mitigation of greenhouse gaseous emissions from agricultural soil? - a global meta-analysis. Environ. Res. 202, 111789.

Srinivasarao, C., Gopinath, K.A., Venkatesh, G., Dubey, A.K., Wakudkar, H., Purakayastha, T.J., Pathak, H., Jha, P., Lakaria, B.L., Rajkhowa, D.J., Mandal, S., Jeyaraman, S., Sikka, A.K., 2013. Use of Biochar for Soil Health Enhancement and Greenhouse Gas Mitigation in India: Potential and Constraints National Initiative on Climate Resilient Agriculture Central Research Institute for Dryland Agriculture Hyderabad, p. 51.

Stewart, C.E., Zheng, J., Botte, J., Cotrufo, M.F., 2013. Co-generated fast pyrolysis biochar mitigates green-house gas emissions and increases carbon sequestration in temperate soils. GCB Bioenergy 5, 153−164.

Supriya, A., Samantray, R., Mishra, S.C., 2017. Biochar, a substitute for fossil fuels. In: IOP Conference Series: Materials Science and Engineering. Institute of Physics Publishing, p. 178.

Tomczyk, A., Sokołowska, Z., Boguta, P., 2020. Biochar physicochemical properties: pyrolysis temperature and feedstock kind effects. Rev. Environ. Sci. Biotechnol. 19, 191−215.

Vahedi, R., Rasouli Sadaghiani, M.H., Barin, M., Vetukuri, R.R., 2022. Effect of biochar and microbial inoculation on P, Fe, and Zn bioavailability in a calcareous soil. Processes 10, 343.

Venkatesh, G., Gopinath, K.A., Reddy, K.S., Reddy, B.S., Prabhakar, M., Srinivasarao, C., Visha Kumari, V., Singh, VK., 2022. Characterization of biochar derived from crop residues for soil amendment, carbon sequestration and energy use. Sustainability 14 (4), 2295.

Venkatramanan, V., Shah, S., Rai, A.K., Prasad, R., 2021. Nexus between crop residue burning, bioeconomy and sustainable development goals over north-western India. Front. Energy Res. 8, 614212.

Vereš, J., Kolonicný, J., Ochodek, T., 2014. Biochar status under international law and regulatory issues for the practical application. Chem. Eng. Trans. 37, 799–804.

Verheijen, F.G.A., Montanarella, L., Bastos, A.C., 2012. Sustainability, certification, and regulation of biochar. Pesq agropec bras Brasília 47 (5), 649–653.

Vochozka, M., Maroušková, A., Váchal, J., Straková, J., 2016. Biochar pricing hampers biochar farming. Clean Technol. Environ. Policy 18, 1225–1231.

Wang, J., Odinga, E.S., Zhang, W., Zhou, X., Yang, B., Waigi, M.G., Gao, Y., 2019. Polyaromatic hydrocarbons in biochars and human health risks of food crops grown in biochar-amended soils: a synthesis study. Environ. Int. 130, 104899.

Wang, J., Xia, K., Waigi, M.G., Gao, Y., Odinga, E.S., Ling, W., Liu, J., 2018. Application of biochar to soils may result in plant contamination and human cancer risk due to exposure of polycyclic aromatic hydrocarbons. Environ. Int. 121, 169–177.

Wijitkosum, S., Jiwnok, P., 2019. Elemental composition of biochar obtained from agriculturalwaste for soil amendment and carbon sequestration. Appl. Sci. 9 (19), 3980.

Xiang, W., Zhang, X., Chen, J., Zou, W., He, F., Hu, X., Tsang, D.C.W., Ok, Y.S., Gao, B., 2020. Biochar technology in wastewater treatment: a critical review. Chemosphere 252, 126539.

Zhang, G., Zhao, Z., Guo, X., Han, Z., He, Q., Zhang, F., Xu, H., 2018. Levels of persistent toxic substances in different biochars and their potential ecological risk assessment. Environ. Sci. Pollut. Control Ser. 25, 33207–33215.

CHAPTER

8

Potential of microbes for the remediation of heavy metal–contaminated soil

Kumari Manorma[1], Shweta Sharma[2], Abhishek Sharma[3] and P.K. Chauhan[4]

[1]*Dr. Yashwant Singh Parmar University of Horticulture and Forestry, Solan, Himachal Pradesh, India;* [2]*ICAR-Directorate of Mushroom Research, Solan, Himachal Pradesh, India;* [3]*CSIR-Imtech Chandigarh, Chandigarh, Punjab, India;* [4]*Shoolini University, Solan, Himachal Pradesh, India*

1. Introduction

Soil pollutants in the form of heavy metals (HMs) pose a serious threat to the surroundings and food stuff protection because of the rapid expansion of industry and agriculture, as well as the disturbance of natural ecosystems due to anthropogenic pressure associated to the growth of human populations. There are five million webpages of soil infected with heavy metals/metalloids with cutting-edge concentrations over regulatory levels across the sector (Li et al., 2019). Contamination by heavy metals causes numerous threats to the environment and humans, and has an impact on food chain protection, food quality, and the ability to use land for agricultural production.

Bioremediation and natural reduction are also noticeable as solutions for emerging contaminant issues; microbes are extremely useful in cleaning up a contaminated environment. In the bioremediation process, a variety of microorganisms, as well as aerobic and anaerobic bacteria and fungi, are involved. Bioremediation is concerned with the degradation, eradication, immobilization, or detoxification of various chemical wastes and physical dangerous chemicals from the environment through the use of microorganisms. The main premise is to degrade and transform pollution into less hazardous forms. The rate of deterioration is determined by two types of elements: biotic and abiotic factors. Different bioremediation techniques and strategies are now being used.

2. Heavy metals

Heavy metals are classified based on three distinct characteristics: density, atomic variety, and chemical properties. Heavy metals have a significant economic value in commercial application, but they have also become a major environmental problem around the world. Transition metals, metalloids, lanthanides, and actinides are among the heavy metals, which include transition metals, metalloids, and actinides. Some are required cofactors for specific enzymes, whereas others are not. The trace elements cobalt, copper, iron, manganese, molybdenum, selenium, and zinc make up the first institution, and their concentrations are tightly controlled by interactions with binding proteins because they are potential threats to cellular function (Theron et al., 2012). The second facility houses

Integrative Strategies for Bioremediation of Environmental Contaminants, Volume Two. https://doi.org/10.1016/B978-0-443-14013-6.00001-9

nonvital metals such as arsenic, cadmium, lead, mercury, plutonium, tungsten, and vanadium, which are toxic to cells and tissues due to their physicochemical features, such as ionic charge. The latter are of particular relevance because they have no critical function in living creatures, are extremely deadly at low exposure levels, and are regarded as the primary threat to life bureaucracy Atobatele and Olutona (2015). The most hazardous metals in the environment, according to the Environmental Protection Agency (EPA), are arsenic, cadmium, lead, and mercury. Furthermore, the US Agency for Toxic Substances and Disease Registry (ATSDR) lists more than 20 heavy metals with reported toxicity, but only four are of particular concern to human health: arsenic is the most common cause of acute heavy metal poisoning and thus ranks first on ATSDR's "top 20 list," lead is second, and cadmium is seventh (Flora et al., 2011).

The rapid growth of the human population, as well as the increased demand for industrial facilities to meet human requirements, has resulted in issues such as overuse of available resources and increased pollution in the land, air, and water environments. Heavy metal is important in the industrial sector, but it has also become a major environmental issue around the world (Igiri et al., 2018). Heavy metal-based environmental pollutants have become a serious threat to living organisms in the environment. At some point during the microbiological procedure, microorganisms use chemical pollutants as an energy source via their metabolic processes. Excessive amounts of inorganic nutrients in soil, on the other hand, limit microbial growth (Ahirwar et al., 2016). Microorganisms, in particular, are capable of degrading, detoxifying, and even accumulating hazardous organic and inorganic substances.

HM pollution is a serious environmental issue that affects crop output and food quality as a result of excessive use of agricultural inputs such as fertilizers, pesticides, and mulch, which has resulted in heavy metal contamination of soils. The majority of pesticides are natural chemicals, and a few are natural—inorganic compounds or pure minerals, with Hg, As, Cu, Zn, and other heavy metals present in some pesticides. Metals, unlike organic pollutants, do not degrade and hence remain in the environment for long periods of time; when present in high amounts, metals can interfere with plant metabolism (Ferraz et al., 2012). For the removal of heavy metal ions from soil, water bodies, and wastewater, advanced treatment technologies are required. Different microorganisms have been offered as efficient and cost-effective solutions for heavy metal removal from soil and water (Ahirwar et al., 2016). Recent improvements in bioremediation techniques have been made in the last two decades, with the ultimate goal of effectively repairing damaged areas in an environmentally acceptable and cost-effective manner.

Researchers have created and modeled exceptional bioremediation approaches, but due to the nature and/or form of pollution, no one bioremediation strategy can be used to heal contaminated settings as a "silver bullet." Autochthonous (indigenous) microorganisms present in polluted environments hold the key to solving most of the issues associated with polluting material biodegradation and bioremediation (Azubuike et al., 2016), provided that environmental conditions are suitable for their proliferation and metabolism.

2.1 Heavy metals contribute to pollution and diseases

In developing countries, heavy metals in the environment and commercial effluent are polluting ecosystems and posing a hazard to human health. All ecosystems have different concentrations of heavy metal components. Zn, Cu, Ni, Fe, and Mn, for example, are essential trace elements with a wide range of properties. High-level buildup of such metals, or their ingestion in greater quantities than the stated attention, can cause serious problems in the direction of living things, such as humans. In order for aquatic living species to thrive, heavy metal concentrations in water must be reduced. Furthermore, when plants, animals, and humans are exposed to high quantities of heavy metals, they can have significant fatal impacts. Metal-containing wastes are dumped into the environment either directly or indirectly, causing substantial contamination or posing a serious threat to human, soil, and sediment health (Siddiquee et al., 2011). Heavy metal concentrations have the ability to have extended biological half-lives, as well as to be resistant to degradable techniques and to show chemical toxicity in soluble fluids. As a result, even when present in low concentrations, heavy metals are considered a concern to humans and other species. Heavy metals can bind to many biomolecules, such as proteins and nucleic acids, and interfere with their functioning when they enter the human body through absorption.

2.2 Contaminations of heavy metals in environment

Heavy metals cannot be biologically destroyed ("no degradation," since changes in the nuclear form of the element occur), but may only be converted from one oxidation nation or natural complex to another. Furthermore, microorganisms are effective in heavy metal bioremediation. Adsorption, absorption, methylation, oxidation, and reduction are only a few of the ways that microorganisms have developed to protect themselves from heavy metal toxicity. Heavy metals are actively absorbed by microorganisms (bioaccumulation) and/or passively absorbed by microorganisms (adsorption). Because methylated substances are typically volatile, microbial methylation is critical in heavy metals bioremediation. Mercury, Hg (II), for example, can be biomethylated to gaseous methyl mercury by bacteria such as *Alcaligenes faecalis*, *Bacillus pumilus*, *Bacillus* sp., *Pseudomonas aeruginosa*, and *Brevibacterium iodinium* (Jaysankar et al., 2008).

Rapid industrialization and urbanization tendencies have had a direct impact on the environment. The resulting environmental degradation and contamination has become a major threat to all living things on the planet, including humans. Various human activities that result in the pollution of open water and aquatic ecosystems around the world have contaminated them with different heavy metals. A few water bodies, such as rivers, lakes, oceans, and groundwater, have experienced the most serious water contamination. Furthermore, an excessive amount of materials might alter the water's qualities and pollute it, rendering it unfit for its intended use. Point sources and nonpoint resources are the two types of contaminants that can be found in water. Single identified localized sources of water contaminants are referred to as point sources of contamination. It can occur when hazardous materials are released directly into bodies of water. Exxon Valdez, which ran aground in the Prince William Sound in 1989 and spilled 11 million gallons of crude oil into the Alaskan environment, is an example of water pollution (Siddiquee et al., 2011). Polluted runoff from agricultural regions flowing into river and ocean waters, as well as nonpoint supply contaminants, is influencing the water body through diffusion resources. Furthermore, nonpoint source pollution is derived and originates from a diverse range of sources, making it difficult to identify a precise approach for preventing pollution.

3. Toxic effects of heavy metals

3.1 Zinc (Zn)

Zinc is one of the principal metals that may be discovered in effluents discharged from industries. Most of the industries launched zinc as their waste material, which incorporates electroplating, manufacturing of batteries, galvanization, and metallurgical industries. Zn can play an essential function in plant growth. The growth slows down while Zn is poor in the environment. The functional group of Zn can help plants stabilize themselves by altering their structure, molecule, and membrane, as well as acting as a defense mechanism against disease. In general, zinc in metallic form is no longer harmful to the environment and has a low bioavailability. Other molecules, such as acids and oxygen, can, however, combine with zinc to generate a highly poisonous substance that can cause serious injury to biological systems Fosmire (1990).

3.2 Copper (Cu)

Copper is a metal that contributes to heavy metal pollution and infection much like any other. Copper (Cu) is usually found deep within the soil. It's an uncommon element that can be found in nature in its pure form as well as in ores like chalcopyrite (Igiri et al., 2018). Animal dung may have also contributed to the Cu contamination in the area. Cu is then employed as a parasite inhibitor supplement as well as in industrial factories, such as the paint producing facility, and is disposed of through digital and electric factories. A copper element's anthropogenic resources are primarily found in polluted locations such as landfills, mine sites, fossil fuel combustion, and home wastewaters. It could enter the ecosystem by volcanic ash, dust, and forest fires, or as dissolved chemicals in water. When taken in high proportions, copper contamination can harm the kidneys, liver, and even cause death. Despite the fact that copper is required for the production of hemoglobin by multiple enzymes, it has a negative impact on various organism statistics. Cu is essential for living creatures because it functions as an antioxidant and is found in collagen and elastin, as well as the electron transport chain.

3.3 Nickel (Ni)

Nickel, along with copper, is a major contaminant in the environment. It has the potential for being clastogenic, toxic, and carcinogenic. Nickel compounds' extraordinary solubilities have distinct carcinogenic potentials. According to Rani (2017), insoluble Ni_3S_2 or NiO is a powerful carcinogen, whereas soluble nickel salts are a vulnerable carcinogen. Nickel is a type of airborne pollutant that enters the body through the respiratory system in the form of nickel carbonyl $Ni(CO)_4$, an intermediate result of nickel refining. It contains a highly dangerous unstable liquid that, when inhaled, can result in death and cause a variety of fitness issues such as pulmonary edema, pneumonia, and respiratory failure.

3.4 Arsenic (As)

Arsenic (As) is a naturally occurring and abundant element in the Earth's crust with metallic and nonmetallic characteristics. It is chemically classified as a metalloid. In the soil, it can be found in both organic and inorganic forms, with the latter being the more dangerous. Mobility, toxicity, and environmental behavior are all influenced by the relative abundance of As^{5+} and As^{3+}, with inorganic As^{3+} species being more mobile and poisonous than inorganic As^{5+}species (Johnston et al., 2019). Because it forms complexes on solid surfaces, particularly hydroxides and oxides of iron (III) as well as calcite, As^{5+} is less cellular than As^{3+}. Arsenic is discharged from such surfaces into groundwater when iron is reduced, and it stays mobile following reduction to As^{3+}. Arsenite can also inactivate over 200 enzymes by attaching to thiol or sulfhydryl structures on proteins, whereas arsenate can replace phosphate, which is important in many metabolic reactions. Chronic exposure to this metalloid produces skin lesions such as melanosis (hyperpigmentation), keratosis, and leukomelanosis (hypopigmentation); lung, bladder, liver, and kidney malignancies; ischemic coronary heart disorders; and impaired cognitive, motor, and hormonal functions.

3.5 Cadmium (Cd)

Cadmium (Cd) is a nonessential metal that is also a major contaminant due to its high toxicity and solubility in water. It's a metal having a proclivity for forming strong dissolved complexes with inorganic and organic ligands, preventing sorption and precipitation. Furthermore, cadmium can disrupt enzyme functions and hinder DNA-mediated transformation in microorganisms; the largest anthropogenic sources of cadmium in soils are waste products from mine, industry, and agriculture (Kubier et al., 2019).

3.6 Lead (Pb)

Lead (Pb) is a toxic nonessential heavy metal that is widely distributed and has a wide range of negative effects on living organisms at the morphological, physiological, and biochemical levels due to its persistent presence in water and soil, generally immobile and buildup in the upper eight inches of the ground. Despite the fact that it is a naturally occurring element, anthropogenic activities such as mining, fossil fuel combustion, and manufacturing contribute to its increased concentrations in soils (Tchounwou et al., 2012). Poisoning with lead (Pb) is most commonly caused by ingesting it in food or drinking water. It is readily absorbed into the bloodstream, causing damage to a variety of organs. Various research have revealed that lead has harmful effects on the nervous system, including irritability, agitation, headaches, confusion, ataxia, drowsiness, convulsions, and coma, as well as on renal capacities, body improvement, and the lymphatic system.

3.7 Mercury (Hg)

Mercury (Hg) can be found in the environment in the form of both organic and inorganic (Hg, Hg_2^{2+}, Hg^{2+}) compounds. Mercury accumulates in many parts of the human body, causing damage to the brain, thyroid, breast, heart, muscular tissues, liver, kidneys, skin, and pancreas. The nervous system is the most damaged, according to Andreoli and Sprovieri (2017). Inorganic mercury can stop enzymes from working properly inside the body and disrupt cell metabolism. The organic form has a negative impact on brain function and could reach the food supply through the food chain.

3.8 Chromium (Cr)

Chromium (Cr) is found in several oxidation states (−2 to +6) in the environment, but the most solid forms are trivalent (III) and hexavalent (X) (VI). Because of its high solubility and mobility, chromium (VI) is more readily absorbed by the human body through inhalation, ingestion, and skin contact than chromium (III). Chromium (VI) enters the cell through enhanced diffusion through a nonspecific anion channel, whereas chromium (III) enters through passive diffusion or phagocytosis. The liver, kidney, spleen, and bone are the main organs that are affected. The hexavalent structure can easily enter red blood cells. Ulcers, dermatitis, rupture of the nasal septum, and lung cancer are the most common chromium side effects in humans. Chromium changes the form of microbial communities in soil and inhibits their growth (Mishra et al., 2016).

4. Heavy metals and metalloids: their origins, characteristics, and properties

Metals and metalloids with a density greater than 5000 kg m^{-3} and an atomic mass greater than 20 are referred to as "heavy metals" in the environment. According to this classification, heavy metals/metalloids are defined as 51 elements from the periodic table. Ali and Khan (2018). Their chemical characteristics, as well as those of the soil, contribute to their mobility and bioavailability in soil. The interactions between soil components and metals/metalloids are influenced by pH, adsorbent surface characteristics, the presence of cations, and anions.

Chromium, manganese, nickel, copper, zinc, cadmium, lead, and arsenic are the most prevalent heavy metals and metalloids found in the environment. The very poisonous elements are chromium, copper, zinc, cadmium, lead, mercury, and arsenic (Ali et al., 2019). They are usually discovered as ores as a result of the weathering of underlying bedrock (sulfides of Pb, Co, Fe, As, Pb, Zn, Ag, and Ni and oxides of Se, Al, Mn, and Sb). Arsenic, mercury, lead, and cadmium sulfides, as well as copper sulfides (chalcopyrite, $CuFeS_2$) and iron sulfides, are widely found in soils (pyrite, FeS_2). The environmental problem is exacerbated by anthropological activities that result in an increase in heavy metal and metalloid concentrations, particularly in the refining and mining of ores, pesticide packages, fertilizer manufacturing, and solid wastes. As illustrated in Fig. 8.1, the anthropogenic assets of metals and metalloids can be separated into five groups.

Heavy metals and metalloids are used in a variety of industries, resulting in increased market demand and global production. Copper, selenium, zinc, iron, vanadium, and manganese, in trace levels, are essential for a variety of organic processes, including respiratory systems, complex

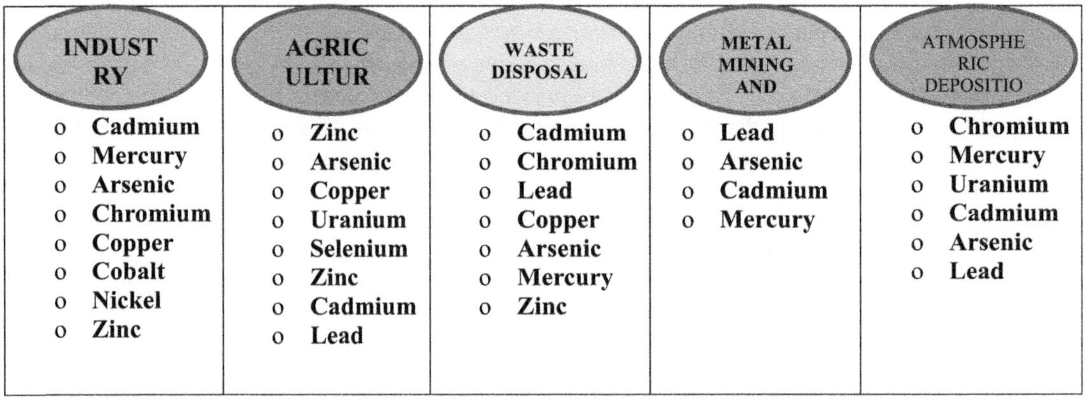

FIGURE 8.1

Sources of contaminating metals and metalloids that are anthropogenic.

compound biosynthesis, the nervous system, enzyme regulation, and function. Iron, zinc, tin, lead, copper, and tungsten all play important roles in digital electronics, especially in semiconductor recognition, according to Koller and Saleh (2018). Table 8.1 summarizes the features of the most hazardous heavy metals and metalloids.

Table 8.1 Characteristics of heavy metals and metalloids.

Elements	Applications	Chemical and physical properties	World production ton year^{-1}
Zinc (Zn)	Paints, rubber, cosmetics, pharmaceuticals, plastics, inks, soaps, batteries, textiles, and electrical equipment	Density: 7140 kg m^{-3} Atomic mass: 65.38 Heat of fusion: 7.32 kJ mol^{-1}	13,500,000 (year: 2019)
Copper (Cu)	Electric and electronics, construction, pesticides, and industrial machinery and transport equipment	Density: 8960 kg m^{-3} Atomic mass: 63.55 Heat of fusion: 13.26 kJ mol^{-1}	20,000,000 (year: 2017)
Mercury (Hg)	Pharmaceuticals, fungicides, industrial process waters, seed dressings, dental preparations, and fluorescent and ultraviolet lamps	Density: 13,530 kg. m^{-3} Atomic mass: 200.59 Heat of fusion: 2.29 kJ mol^{-1}	4000 (year: 2019)
Lead (Pb)	Building construction, cable coating ammunition, electrical accumulators, and batteries	Density: 11,340 kg. m^{-3} Atomic mass: 207.2 Heat of fusion: 4.77 kJ mol^{-1}	11,600,000 (year: 2018)
Chromium (Cr)	Industrial applications, tanning agents, alloys, paint pigments, photography, and catalysts	Density: 7190 kg. m^{-3} Atomic mass: 51.99 Heat of fusion: 21.00 kJ mol^{-1}	15,000,000 (year: 2017)
Arsenic (As)	Pesticides, pharmaceuticals, alloys, and semiconductors	Density: 5730 kg. m^{-3} Atomic mass: 74.92 Heat of fusion: 24.44 kJ mol^{-1}	33,000 (year: 2019)
Cadmium (Cd)	Silver cadmium batteries, paint pigments, electroplating, television, photography, machinery, and baking enamels	Density: 8650 kg. m^{-3} Atomic mass: 112.41 Heat of fusion: 6.21 kJ mol^{-1}	24,670 (year: 2019)

5. Heavy metals are removed using traditional methods

Traditional heavy metal removal methods such as adsorption, electrodialysis, precipitation, and ion exchange have limitations. Chemical precipitation entails incorporating anions to precipitate metals as suspended debris, which can subsequently be removed. The method isn't always exact, and it can't get rid of heavy metals in small amounts. Heavy metals can be eliminated down to parts per billion using the ion alternative technique. It is, however, a nonspecific, pH-sensitive, and expensive technique. Membranes are used in the reverse osmosis process. These traditional methods have problems such as sluggish and inefficient elimination, the generation of polluted sludge that requires cautious disposal, the high cost and power required in the processes, and membrane blockage Firdousi (2017). There is a need for a low-cost, high-efficiency method of removing heavy metals that is also environmentally friendly. As an alternative to traditional procedures, there has been a growing interest in the use of biological agents for heavy metal removal.

6. Bioremediation

Bioremediation is a technique in which biological organisms are employed to remove or neutralize a contaminant from the environment through a metabolic process. The "biological" species include microscopic organisms such as fungi, algae, and microorganisms, as well as "remediation"—the process of resolving the problem. Microorganisms thrive in a wide variety of settings across the biosphere. They grow in soil, water, plants, animals, the deep sea, and the frozen ice. Microorganisms are the best candidates for acting as our environmental stewards because of their sheer numbers and desire for a wide range of substances.

Bioremediation is a technique for removing/changing hazardous contaminants such as heavy metals into less hazardous substances; and/or removing toxic elements from contaminated environments; or degrading organic substances and remaining mineralization of organic substances into carbon dioxide, water, nitrogen gas, and so on, using dead or alive biomass. Bioremediation can be applied to soil and water media using both in-situ and ex-situ techniques. Fig. 8.2 depicts a brief description of the specific bioremediation techniques for various contaminants.

The in-situ method does not require excavation or removal and does not disrupt the soil structure. It can use either intrinsic (intrinsic microbial flora stimulation) or engineered (intrinsic microorganism introduction) bioremediation Hazen (2010). A few factors to consider for this method include the type and character of contaminants, the degree of contamination, the type of soil or site geochemistry, and the geographical location. Contaminants may also become adsorbed on soil particles and thus become unavailable for bioremediation. Other challenging situations during bioremediation include microbial opposition and dying after inoculation, as well as the temperature and moisture situation of the media. High temperatures increase contaminant solubility and, as a result, their mobilization. In-situ treatment typically entails pumping oxygen/nutrients into the soil (bioventing/biostimulation). The texture also plays an important role in some aspects of bioventing and biostimulation. When compared to fine-textured soils, it is easier to pump and distribute oxygen and vitamins in coarse-textured soils. Fine soils, such as clay, retain moisture in their various smaller pores with a high surface area, preventing oxygen from dispersing uniformly throughout the contaminated soil. However, this technique may not be suitable for all types of soils, as natural conditions (such as temperature) may be limiting. Numerous

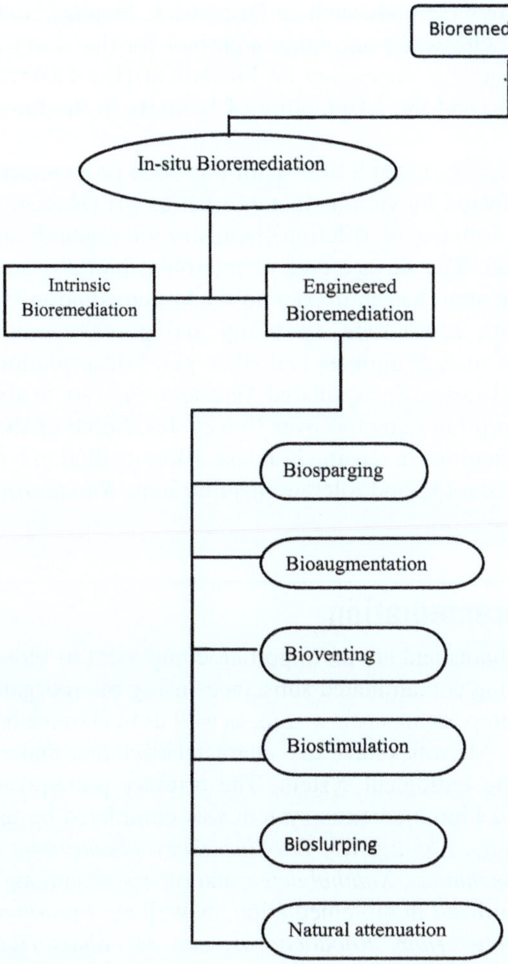

FIGURE 8.2

Types of bioremediation employed environmental clean-up.

studies have shown that microbial cells have greater adsorption ability than clay particles. Native species isolated from contaminated sites have been shown to have exceptional resistance and biosorption competence to heavy metals. Two indigenous traces, AK1 and AK9, of the genus *Pseudomonas*, were isolated from As-contaminated water in the Ganga basin. The strains have been reported to be resistant to As and other heavy metals such as silver (Ag), cadmium (Cd), cobalt (Co), chromium (Cr), copper (Cu), mercury (Hg), nickel (Ni), and lead (Pb) (Satyapal et al., 2018).

The ex-situ technique entails transporting contaminated soil and water from the contaminated site to another location for further processing. It can be classified as a stable-segment approach (for land treatment), a slurry-phase approach, or a pile strategy (for a mixed medium containing stable and liquid phases in fermenters) Tak (2020). It employs methods such as fermenters, biopiles, and land farming. In a fermenter, infected soil is blended with water and other additives for the slurry-phase method. However, the performance of the fermenter is dependent on biosorbent (live/useless), the primary conditions required for microbial growth, and the adaptability of biomass to the fermenter configuration.

The removal of mercury from synthetic wastewater using a bioreactor has been pronounced.

The wastewater bioremediation process is affected by various factors such as pH (Hassan et al., 2020). The pH influences their bioavailability by influencing solution chemistry via methods such as complexation, hydrolysis, redox, and precipitation. The surface area of microbial biomass and pre-treatment procedures (which increases the surface area) has an impact on the bioremediation system. To increase a suitable industrial biosorbent with appropriate electricity and porosity, microbial biomass may need to be immobilized in matrices such as alginate and silica gel. Encapsulation provides physicochemical stability as well as heat resistance. Encapsulated *Agrobacterium* sp. in alginate with Fe nano-debris demonstrated excellent adsorption potential over five cycles. Some of the processes' limitations include low selectivity and difficulties in reusing biomass. Bioremediation has also been mediated by microbial biofilms that are resistant to and tolerate metallic ions. *Rhodotorula* sp. has a removal efficiency of up to 95.39%.

7. Microorganisms exploited in bioremediation

Microorganisms play an important role in food chains and are an important component of biological stability. Bioremediation is the process of removing contaminated substances using microorganisms, fungi, algae, and yeast. Microbes can develop in temperatures below zero, as well as in extreme heat, in the presence of hazardous substances or garbage. Microbes have two characteristics that make them suited for the remediation process: flexibility and biological system. The primary prerequisite for microbial interest is carbon. In unusual settings, a bioremediation system was completed by using a microbial consortium. *Achromobacter, Arthrobacter, Alcaligenes, Bacillus, Corynebacterium, Pseudomonas, Flavobacterium, Mycobacterium, Nitrosomonas, Xanthobacter*, and others are among these microbes. There are microbial agencies that are utilized in bioremediation, as well as: *Pseudomonas, Acinetobacter, Sphingomonas, Nocardia, Flavobacterium, Rhodococcus*, and *Mycobacterium* are examples of aerobic bacteria that have the ability to breakdown complex chemicals. Pesticides, hydrocarbons, alkanes, and polyaromatic chemicals have all been documented to be degraded by these bacteria. Many of these bacteria use pollutants as a source of carbon and energy. Anaerobic bacteria are not as commonly used as aerobic bacteria. The employment of aerobic bacteria for bioremediation of chlorinated fragrant compounds, polychlorinated biphenyls, and dechlorination of the solvents trichloroethylene and chloroform, which degrade and convert harmful substances to less poisonous forms, is gaining popularity.

Microorganisms are also important in the biogeochemical cycle of metals and the cleanup of heavy metal-contaminated habitats. Microorganisms can survive in their presence despite their harmful effect by using a variety of methods to reduce or tolerate their toxicity. There are five basic mechanisms of heavy metal resistance in bacteria, according to Bruins et al. (2000) and Choudhury and Srivastava (2001): (1) Extracellular barriers: the cell wall, plasma membrane, or capsule can all prevent metal ions from entering the cell. (2) Active metal ion transport (efflux): proteins belonging to three families:

P-kind ATPases, CDF (cation diffusion facilitator), and RND (resistance nodulation cell division) form a mechanism to export toxic metals from the cytoplasm. (3) Extracellular sequestration: involves the buildup of metal ions in the periplasm, the outer membrane, or the complexation of metal ions as insoluble compounds via cellular additives. (4) Intracellular sequestration: this type of metal resistance is primarily based on the accumulation of metals in nonbioavailable forms within the cytoplasm to avoid exposure to vital cellular components. Two examples of this type of metal resistance are metallothionein production in *Synechococcus* sp. and cysteine-rich proteins in *Pseudomonas* sp. and (5) metal ions are reoxidized: metal oxidation, such as Cu(I) to Cu(II) with the help of CueO, and As(V) to As(III) with the help of As(III) then effluxing A(III) out of cells are both important detoxifying mechanisms Ianeva (2009).

Biochemical and molecular mechanisms are two types of mechanisms utilized by bacteria (Ma et al., 2016). Metal detoxification, mobilization, immobilization, transformation, transport, and dispersion are examples of biochemical pathways that microbes use to interact with extracellular soluble metallic ions and play a role in microbe tolerance. The genetic determinants of heavy metal resistance are known as molecular mechanisms, and they can be found on the chromosome or in extrachromosomal genetic elements. In addition to biochemical and molecular causes, bacteria's heavy metal tolerance can be measured in terms of the minimum inhibitory concentration (MIC), which is defined as the lowest concentration of a metal that stops microorganisms from growing.

Heavy metals in soils must be remedied if the ecosystem and living beings are to be preserved. Physical and chemical solutions, as well as engineering restoration, are examples of traditional remediation procedures. They are quick, but they are also expensive and ineffective since they produce secondary pollutants and have negative impacts on soil qualities. Biological techniques, on the other hand, are being considered as an effective option for heavy metal remediation. These include bioremediation, which utilizes microorganisms to remove contaminants from soils, and phytoremediation, which uses plants to remove heavy metals. Microbial bioremediation is a cost-effective, efficient, and ecologically friendly strategy for reducing the expense of heavy metal infection cleaning. Biosorption (which includes precipitation, chemical adsorption, ion alternate, surface precipitation, the formation of complexes with natural ligands, and redox reactions), biomineralization (which includes bioleaching that involves the mobilization of heavy metal ions from insoluble ores through dissolution or complexation), and biooxidation are the mechanisms behind microbial heavy metal remediation (Jin et al., 2018). Phytoremediation, on the other hand, is a wholly gradual and seasonally effective approach that relies heavily on the climate, water, and soil conditions. Phytoremediation is made up of a variety of processes that are typical of plant—soil—atmosphere interactions; four phytoremediation processes are identified for heavy metal-infected soil: phytoextraction, phytostabilization, phytovolatilization, and rhizofiltration.

Plant—microbe interactions play a major role in the adaptation to heavy metal-polluted settings, and as a result, they should be thoroughly researched in order to improve microbe-assisted phytoremediation approaches. Plant-related microorganisms, particularly plant growth-promoting rhizobacteria (PGPR), play an important role in the remediation of polluted soils and the enhancement of plant growth through a variety of mechanisms that include biological nitrogen fixation, phosphate solubilization, phytohormone production (indole-3-acetic acid [IAA], cytokinins, and gibberellins), siderophore production, and the production of 1-aminoacyl (El-Mehy et al., 2019). In heavy metal-infected soils, PGPR stands for substantial metal resistant bacteria and phytoremediation boosting agents, whose application might result in a noticeable increase in metal mobilization or immobilization in soils and plant biomass.

8. Biosorbent materials

Choosing an efficient, highly selective, and low-cost biosorbent is a critical condition. The biosorbent should be easily accessible or exhibit rapid development. The efficiency of the biosorbent is determined by the experimental requirements and the pretreatment of the bioagent. There have been numerous studies on wastewater treatment using various biosorbents Demirbas (2008). Biological agents of various types had been used for cleanup in both ex- and in-situ settings. These include agricultural wastes such as wheat/rice straw, tea/espresso/yeast waste, cotton waste, and so on. Microorganisms (bacteria, fungi, yeast, or algae) derived from their natural environments can be excellent biosorbents. Heavy metals can be absorbed by these biosorbents at very low concentrations. The removal of heavy metals is aided by functional groups such as amide, amine, carbonyl, carboxyl, and many more. Microorganisms have unique enzyme profiles that are essential for strong metallic resistance. However, the biosorption system is influenced by steric and conformational factors, as well as the quantity and availability of reactive sites. Microorganisms such as fungi transform heavy metals into less hazardous chemicals and use them to proliferate; for example, *Pleurotus* sp., *Klebsiella oxytoca*, and others have metal-binding capacity. In lead-contaminated soil, *Cephalosporium aphidicola* has been shown to be effective. Table 8.2 displays some of the most often used biosorbents.

9. Mechanisms of bioremediation

In heavily diseased locations, microorganisms adapt to and withstand heavy metals. Extracellular polymeric compounds found on the biomass cell wall can bind to heavy metals via mechanisms such as proton alternate or metal microprecipitation. Because of the existence of carboxyl, amino, phosphoryl, and sulfo structures as possible ion alternative sites and metal sinks, biomass surfaces have a high cost. Bioremediation occurs through a variety of mechanisms including as redox system, adsorption, complexation, ion-trade, precipitation, and electrostatic appeal.

Microorganisms can also cause metallic mobilization/immobilization through redox reactions, influencing bioremediation strategies. Heavy metals, such as Fe, As, Cr, and Hg, go through oxidation and reduction cycles. Bioremediation is made possible by converting an element from its insoluble and stationary form in sediments to its mobile and soluble form. While harmful metal ions are redistributed

Table 8.2 Biosorbents include a wide range of species.

Biosorbent	Type/species
Bacteria	*Pseudomonas* sp., *Bacillus* sp., *Vibrio* sp., *Desulfovibrio* sp., and *Geobacter* sp.
Fungi	*Aspergillus* sp., *Penicillium* sp., *Rhizopus* sp., *Pleurotus* sp., and *Saccharomyces* sp.
Algae	*Cladophora* sp., *Spirogyra* sp., *Fucus* sp., and *Chlamydomonas* sp.
Agricultural residues	Wheat straw, rice bran, etc.
Industrial wastes	Fermentation/food/beverages wastes
Others	Cellulose and chitosan-based substances

and propelled from their stable segment in sediments into the solution phase, mobilization can have negative consequences Fomina and Gadd (2014). This improves their bioavailability and allows heavy metals to enter microbial metabolic systems. Hg(II) is reduced by bacteria to a more basic and volatile form of Hg(0). Microbial reduction can also improve the solubility of ions such as Fe(III) and As(V) by reducing them to Fe(II) and As(III), respectively, and may aid in soil leaching. Bacteria from one-of-a-kind natural aquifers have been identified as having the potential to modify As, according to research. Pokhrel and Viraraghavan used *Aspergillus niger* to eliminate As(V) and As(III). Heavy metal bio-methylation is an important process in soil and water that can change the toxicity, volatility, and mobility of heavy metals. It also serves as a vital technique of cleansing by removing unstable methylation species from cells. Dimethyl mercury and alkyl arsines, the methylated derivatives of Hg and As, respectively, are dangerous because they evaporate and are lost from the soil. The methyl donor is the natural organic component of soil. Any other indirect route of metal mobilization, on the other hand, involves the microbial degradation of organic count number, which increases the discharge of these ions. Heavy metals and dissolved organic materials have been shown to be released by *Schizophyllum commune*. Microbe excretion of metabolites such as carboxylic acids and amino acids is a key method for chelating metallic ions.

Microbes immobilize metals and operate as metal sinks by using unique mechanisms (ex- or in-situ) such as biosorption, bioaccumulation, bioconversion, and/or inter/intracellular precipitation (as oxalates of Zn, Cu, Co, Cd, and Ni) that work in unique ways. A detail can be easily removed from its aqueous phase in groundwater or wastewater by immobilization. As(III) is immobilized and maintained in sediments due to bacterial oxidation to As(V). To convert Cr(VI) to Cr(III) and immobilize it in hydroxide/oxide forms, *Methanothermobacter thermautotrophicus* was used. For *Bacillus cereus* and *Shewanella* sp., bacterial reduction and immobilization for Cr(VI) have also been reported. Consider how cellular components such as the cell wall and plasma membrane operate as barriers to metal ion entry into cells (Igiri et al., 2018).

10. Biosorption and bioaccumulation

Biosorption and bioaccumulation are appealing alternatives to typical heavy metal remediation procedures. Heavy metal absorption utilizing live biomass (metabolism structured/active uptake) is referred to as bioaccumulation, and it is defined as the uptake of pollutants using living biomass/cells (Fig. 8.3). Due to highly harmful metals that can accumulate in cells and disrupt metabolic activity, resulting in cell death, and using living biomass for cleanup may not be a viable option. Lifeless biomass (biosorption), on the other hand, is unaffected by toxicity, requires no growth/dietary medium, and is adaptable to environmental circumstances. Heavy metals are adsorbed on the surface in a passive mode (without regard for strength consumption) until equilibrium is reached. Biosorption, in compared to active uptake/bioaccumulation, is enormous since it is metabolism-independent, but it is highly reliant on the biomass/biosorbent type and pollutants involved. Microbial biomasses of fungi, algae, or yeast have been used for bioremediation for in-situ treatments because of these benefits. Heavy metal bioremediation in the form of metallic nanoparticles with the help of microorganisms, as well as the use of genetically altered microorganisms as part of the bioremediation process, has also been mentioned (Paliwal et al., 2012).

The concentration of metallic ions within microbial cells is known as intracellular sequestration. Because of surface contacts, heavy metal ions become complexed and then transported into the cellular

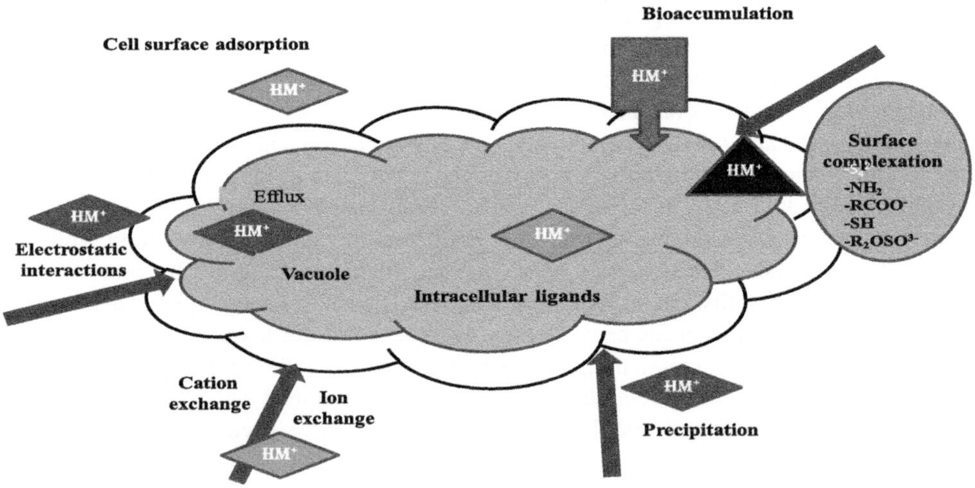

FIGURE 8.3

Mechanisms of heavy metal uptake by microorganisms.

system. Metal ion concentration in the periplasm or their complexation as insoluble precipitates are examples of extracellular sequestration. Cadmium precipitation in *Pseudomonas aeruginosa* and *Klebsiella planticola* has been reported (Sharma et al., 2000).

11. Bioremediation by bacteria

Bacteria can be found in almost any environment. Bacteria come in a variety of morphologies, including rods (*Bacillus*), cocci (*Streptococcus*), filamentous (*Actinomyces*), and spirals (Spirillum) (*Vibrio cholera*). Biosorption with bacteria is a cost-effective and efficient way to remove contaminants from wastewater, including nonbiodegradable elements such as heavy metals. Bacterial biomass can consist of both living and nonliving cells. Mustapha and Halimoon (2015) have developed and evolved mechanisms for metal ions resistance and remediation in order to survive. Bioremediation of heavy metal ions with bacterial agents has been studied extensively. Metals such as Cu, Zn, Pb, Cd, and Cr are quickly eliminated by bacterial biomass. Heavy metal ions and bacterial species have an impact on biosorption effectiveness (thanks to their extraordinary cellular systems in terms of peptidoglycans like N-acetylmuramic acid and poly-N-acetylglucosamine). The most important physical interface between metal ions and bacterial biomass is the bacterial cell wall. The negative charge of anionic functional groups (such as amine, hydroxyl, carboxyl, sulfate, and phosphate) found in gram-positive bacteria (peptidoglycan, teichoic acids, and teichuronic acids) and gram-negative bacteria (peptidoglycan, lipopolysaccharides, and phospholipids) confers metal-binding capability on or within the cell wall. Extracellular heavy metal removal occurs through the use of dead biomass cells. These interactions are caused by functional groups in the cell wall, such as carboxyl, phosphonate, amine, and hydroxyl.

Complexation allows the carboxyl groups to bind Cd on the surface. The amino groups have demonstrated effective Cr removal via chelation and electrostatic interactions. Bacterial species must be exposed to pollutants for enzymatic induction before being used for bioremediation. There is a minimal contaminant concentration required to induce enzymatic expression, which is required for the technique Adenipekun and Lawal (2012). For bioremediation, bacteria such as *Pseudomonas, Desulfovibrio, Bacillus,* and *Geobacter* were utilized.

12. Bioremediation with algae (phycoremediation)

Algae of various types can be found in vast numbers in marine habitats. Algae are autotrophic organisms that require little nutrition and produce a large amount of biomass. Brown algae were found to have a higher biosorption capacity than the other two algal categories, Phaeophyta, Rhodophyta, and Chlorophyta (brown, green, and red, respectively) (phycoremediation). The kind and structure of the algal biomass, as well as the charge and chemical constitution of the heavy metal ion, all influence metal ion biosorption. For in-situ cleanup, several algae, in live or dead forms, have been utilized singly or in combination, in batches or columns. Algal proteins contain capacity metallic sites including amine, hydroxyl, carboxyl, sulfate, and phosphate, which function through sophisticated production ways during heavy metal cleanup (Abbas et al., 2014). Ion exchange allows heavy metal ions to replace calcium, magnesium, and sodium ions in the cellular wall. Fungi are well-known for their pervasiveness in the natural world, and they are widely used in industrial applications. Fungi adapt to their surroundings (in terms of morphology, ecology, and metabolism) and are responsible for natural processes such as decomposition and nutrient cycling. It has been claimed that they can withstand and survive stress circumstances such as moisture, nutrients, pH, and so on. Abdi and Kazemi (2015) define mycoremediation as the use of fungus (alive or dead) to remove pollutants from specified environmental segments. Mycoremediation is a cost-effective technique that does not produce hazardous waste. As a result of the thorough mineralization of contaminants in nature, it poses a complete solution. The identification and use of an appropriate fungus species for the goal heavy metal or other contaminants is critical to the success of mycoremediation. *The presence of fungi, the chemical behavior of the components, and the presence or absence of the fungi following sequestration all influence the fate availability of heavy metals and other pollutants in the media.*

According to reports, *Saccharomyces cerevisiae* may sequester up to 79% of Pb and Cd from polluted soil. Fungal cellular partitions (containing chitin, proteins, glucans, lipids, pigments, and polysaccharides) and functional groups such as hydroxyl, carboxyl, amino, sulfate, or phosphate are involved in biosorption, which is mediated by interactions such as adsorption, ion change, and complexation. Ismail and Moustafa (2016) found that *Aspergillus* sp. could remove Cr from tannery wastewater; it removed 65% of the Cr from the wastewater compared to 85% from the synthetic medium.

Wood-decaying species (white- and brown-rot fungus), mushrooms, and other fungi make up the phylum basidiomycetes. Mushrooms have long been an important part of the human diet due to their nutritional and therapeutic characteristics. Because of their capacity to absorb heavy metals, they could be utilized for mycoremediation in addition to being eaten as food. Metal uptake in mushrooms decreases as contact time, mycelia age, and fructification increase. Some wild mushrooms that are safe to consume can accumulate heavy metals above their background level. Heavy metals are dispersed unevenly throughout the mushroom fruiting body. White-rot fungi such as *Pleurotus ostreatus* and *Termitomyces clypeatus* have been found to breakdown chronic pollutants, according to Singh (2006).

13. Factors affecting microbial remediation of heavy metal

Bioremediation is the process of using bacteria, fungus, and plants to degrade, remove, change, immobilize, or detoxify a variety of chemicals and physical pollutants from the environment. Microorganisms operate as biocatalysts, facilitating the development of biochemical reactions that destroy the preferred pollutant, thanks to their enzymatic pathways. Microorganisms can only operate against pollution if they have access to a variety of chemical molecules that help them manufacture energy and vitamins to build additional cells. The chemical composition and concentration of contaminants, as well as the physicochemical characteristics of the surroundings and their availability to microorganisms, all influence bioremediation efficiency El Fantroussi and Agathos (2005). Because bacteria and contaminants do not come into contact with one other, the pace of deterioration is influenced. Furthermore, microbes and pollution do not always spread evenly in the environment. Because of several elements, regulating and optimizing bioremediation processes is a complex system. The lifestyles of a microbial population capable of degrading pollutants, the supply of toxins to the microbial population, and environmental factors are all treated here (form of soil, temperature, pH, the presence of oxygen or different electron acceptors, and nutrients).

14. Biological variables

Biotic variables influence natural chemical breakdown through competition between microbes for limited carbon resources, unfavorable interactions between microorganisms, and microorganism predation by protozoa and bacteriophages. The rate of contamination degradation is frequently determined by the contaminant concentration and the amount of "catalyst" present.

The quantity of "catalyst" in this context refers to the number of organisms capable of metabolizing the contaminant, as well as the quantity of enzymes(s) produced by each cell. The pace of contaminant breakdown can be accelerated or slowed by the production of certain enzymes by cells. Furthermore, the degree to which individual enzymes involved in contaminant metabolism must participate, as well as their "affinity" for the contamination and accessibility to the contaminant, is largely essential. Mutation, horizontal gene switch, enzyme interest, interaction (competition, succession, and predation), its own growth until considerable biomass is reached, population range, and composition are all mentioned Madhavi and Mohini (2012).

15. Environmental determinants

The metabolic characteristics of the bacteria and the physicochemical qualities of the centered pollutants determine whether or not they will interact at some point throughout the process. The interplay between the two is unmistakable; however, the environmental conditions at the interaction point are crucial. pH, temperature, moisture, soil structure, solubility in water, vitamins, site characteristics, redox capability, and oxygen content, lack of educated human resources in this field, and physicochemical bioavailability of contaminants all influence microorganism growth and activity (contaminant concentration, type, solubility, chemical structure, and toxicity). The kinetics of degradation is determined by the factors described above. Biodegradation can occur in a wide range of pH conditions; however, in most aquatic and terrestrial structures, a pH of 6.5–8.5 is usually ideal for biodegradation.

Moisture affects contaminant metabolism by influencing the type and quantity of soluble materials available, as well as the osmotic stress and pH of terrestrial and aquatic systems. The following is a list of the most common environmental elements.

15.1 Nutrient accessibility

The addition of nutrients affects the critical nutritional balance for microbial growth and replication, as well as the rate and effectiveness of biodegradation. By adjusting the bacterial C: N: P ratio, nutrient balancing, particularly the delivery of essential nutrients such as N and P, can improve biodegradation efficiency. Microorganisms require a variety of nutrients, including carbon, nitrogen, and phosphorus, to survive and maintain their microbial activity. The degree of hydrocarbon breakdown is also limited at low concentrations. In cold conditions, adding the right amount of nutrients is a good way to increase the metabolic interest of microorganisms and, as a result, the pace of biodegradation (Couto et al., 2014). The availability of nutrients limits biodegradation in aquatic environments. Oil-consuming bacteria, like all other creatures, require nutrients for optimal growth and development. These nutrients are present in the natural environment, although in little concentrations.

15.2 Temperature

Temperature is the most important physical element in determining the survival of microorganisms and the composition of hydrocarbons. Oil degradation via natural ways is very slow in freezing conditions like the Arctic, putting the microbes under extra stress to smooth out the spilt petroleum. Most oleophilic bacteria are metabolically inactive because to the subzero temperature of the water in this region, which causes the shipping channels inside the microbial cells to shut down or possibly freeze the entire cytoplasm, rendering them metabolically dormant Macaulay (2014). Biological enzymes involved in the degradation pathway have a preferred temperature and do not have the same metabolic turnover at all temperatures. Furthermore, the degradation process for a specific component necessitates a specific temperature. Temperature affects microbial physiological features; hence, it can speed up or slow down bioremediation techniques. Temperature increases the rate of microbial activity, which reaches its maximum at the optimum temperature. It has begun to diminish abruptly when the temperature rises or falls, finally stopping after reaching a certain point.

15.3 Oxygen concentration

Different species require oxygen, whereas others do not, allowing for a faster rate of biodegradation. Because oxygen is a gaseous necessity for most living organisms, biological degradation occurs in both aerobic and anaerobic environments. In most circumstances, the presence of oxygen can improve hydrocarbon metabolism Macaulay (2014).

15.4 Moisture level

Microorganisms require a sufficient amount of water to grow. The moisture content of the soil has a negative impact on biodegradation agents.

15.5 pH

The pH of a chemical, which refers to its acidity, basicity, and alkalinity, has an impact on microbial metabolic activity as well as the rate at which it is eliminated. The ability for microbial development could be determined by measuring pH in soil, according to Asira (2013).

Lower or higher pH values verified poor outcomes; metabolic processes are extremely sensitive to even minor pH alterations.

15.6 Characterization and selection of the site

Prior to providing a bioremediation solution, sufficient remedial investigation activities must be completed to accurately define the value and degree of contamination. At a bare minimum, these efforts should include the following: determining the horizontal and vertical quantity of contamination, listing the characteristics and locations to be sampled and the reasons for their selection, and describing the sample acquisition and analysis procedures to be used.

15.7 Ions of metal

Metals are essential for bacteria and fungi in small amounts, but in large amounts, they inhibit cell metabolism. The pace of degradation is affected by metal compounds in both direct and indirect ways.

15.8 Toxic substances

When the hazardous nature of a few pollutants is present in large concentrations, it can cause toxic effects on microorganisms and slow down decontamination. Toxicants, their concentrations, and the microorganisms exposed all influence the degree and mechanisms of toxicity. Focused existence forms are killed by some organic and inorganic substances. Madhavi and Mohini (2012).

16. Applications of bioremediation

Bioremediation should be assessed as appropriate approaches that can be used on all states of matter in the environment:

- **i.** Solids (soils, sediment and sludge)
- **ii.** Liquids (groundwater, surface water, and waste water from industry)
- **iii.** Gases (commercial air emissions)
- **iv.** Subsurface conditions (saturated and vadose zones).

(1) Intrinsic (natural) bioremediation, (2) biostimulation (environmental modifications, such as nutrient utility and aeration), and (3) bioaugmentation are the three general bioremediation processes (addition of microbes).

The natural soil microflora is widely used as part of the organic network for bioremediation.

Better plants, on the other hand, can be controlled to improve toxicant removal (phytoremediation), which is mostly used to clean up metal-infested soils.

17. Bioremediation has the following advantages

- It is a natural system that takes a short time to complete, and it is an appropriate waste treatment procedure for diseased materials such as soil. Microbes have the ability to proliferate and decompose the contamination while present in that particular site. The biodegradative population appears to be declining as the pollutant is degraded. The treatment leftovers are mostly risk-free products made mostly of water, carbon dioxide, and cellular biomass.
- It requires relatively little effort and may be done on-site on a regular basis without causing major disruptions to ordinary operations. This also eliminates the need to transfer large amounts of waste off-site, as well as the potential dangers to human health and the environment that could arise during transit.
- It is carried out in a cost-effective manner since it loses far less than other standard approaches (technology) for hazardous waste cleanup. Treatment options for oil-infested locations are important.
- It also allows for complete pollution destruction, since many harmful compounds can be transformed to harmless products, and it eliminates the risk of future legal liability associated to the remediation and disposal of contaminated material.
- It does not include any potentially hazardous chemical components. Nutrients, namely fertilizers, were supplied to promote active and rapid microbial development. Typically seen in lawns and gardens. The dangerous chemical compounds are entirely destroyed as a result of bioremediation, which converts harmful chemicals into water and innocuous gases.
- Because of their natural purpose in the ecosystem, they are simple, require less effort in depth, and are inexpensively priced.
- Eco-friendly and long-lasting.
- Contaminants are eliminated rather than really transported to different media in the environment.
- Nonintrusive, potentially allowing site usage to continue.
- Relatively simple to implement.
- Effective method of cleaning the natural environment of a variety of contaminants while also being environmentally friendly.

18. Bioremediation has the following drawbacks

- It is limited to biodegradable substances. Not all substances can be completely degraded in a short period of time.
- Some people are concerned that biodegradation products will be more persistent or dangerous than the parent molecule.
- Biological techniques are frequently very specific. The availability of metabolically viable microbial populations, adequate environmental growth circumstances, and proper ranges of nutrients and pollutants are all important site factors required for completion.
- Extrapolating from lab and pilot-scale investigations to full-scale discipline operations is difficult.
- Bioremediation technology must be expanded and engineered to be acceptable for locations with complex combinations of contaminants that are not evenly disseminated throughout the environment. Contaminants can exist in the form of solids, liquids, or gases.

- It usually takes longer than other treatment options, such as soil excavation and removal or incineration. The right performance standards for bioremediation remain a source of regulatory confusion. Because there is no universally agreed definition of "clean," assessing the overall performance of bioremediation is difficult.

Bioremediation is only possible with biodegradable chemicals. This approach is susceptible to rapid and complete deterioration. Biodegradation products can be more persistent or toxic than the original chemical in the environment.

1. Uniqueness
   Biological strategies are quite specific. The presence of metabolically competent microbial populations, proper environmental growth circumstances, and adequate stages of nutrients and pollutants are all required site factors.
2. Increase the difficulty of the challenge
   Scaling up bioremediation methods from batch and pilot scale research to large-scale subject operations is difficult.
3. Technological progress
   More study is needed to widen current engineer bioremediation solutions that can be used at locations with composite mixtures of toxins that are not evenly distributed across the environment. It can take the shape of solids, liquids, or gases.
4. Procedure for keeping track of time
   When compared to alternative remediation options, such as excavation and soil removal from a contaminated site, bioremediation takes longer.
5. Regulatory ambiguity
   We can't say for sure that the cleanup is 100% complete because there isn't a standard definition of clean. As a result, evaluating bioremediation's effectiveness is challenging, and there is no perfect endpoint for bioremediation treatments.

19. Prospects for bioremediation

Bioremediation procedures come in a variety of forms, all of which have been proven to be efficient in recovering polluted environments. Microorganisms play a critical role in bioremediation; thus, their diversity, abundance, and community structure in polluted environments provide insight into the risk of any bioremediation strategy that introduces additional environmental components that impede microbial activity. Advanced molecular approaches, including as genomics, proteomics, metabolomics, and transcriptomics, have aided in the discovery of microbial identity, functions, metabolic, and catabolic pathways, as well as the development of microbial-based holistic strategies. Pollutant bioavailability, nutrient availability, and the presence or lack of bacteria with degradative capabilities may all delay the completion of bioremediation.

Biostimulation and bioaugmentation techniques boost microbial activity in polluted locations, which is essential for bioremediation. By adding nutrients to a polluted sample, biostimulation increases microbial activity. Microorganisms are abundant in a variety of environmental conditions; however, it is far more significant that pollutant degrading microbes are naturally found in polluted contaminated sites; their proliferation and metabolic activities may also be influenced by pollutant type and concentration; later, we can use agro-commercial wastes, which contain nitrogen, phosphorus, and potassium, as a nutrient source for the most polluted sites. It has been claimed that microbial consortiums decompose pollution more effectively than pure isolates (Silva-Castro et al., 2012).

This activity is a result of the metabolic diversity of individual isolates, which is derived from their isolation supply, adaption technique, pollutant composition, and synergistic effects, which may result in complete and rapid pollution degradation when such isolates are mixed together.

Furthermore, as compared to a nonadjusted configuration, both bioaugmentation and biostimulation were successful in removing pollutant, including polyaromatic hydrocarbons (PAHs), from a closely polluted pattern (control).

Bioaugmentation has been shown to accelerate the degradation of several chemicals, despite the fact that it has been identified as an effective strategy. Specific microorganisms can be presented as "introduced microbes" to enhance current populations if proper biodegrading microorganisms are not present in soil or if microbial populations have decreased due to contaminant toxicity. However, the possibility that the inoculated microorganisms will not continue to exist inside the new environment makes bioaugmentation a very risky technique. Bioaugmentation is the term for this method. The use of natural or genetically designed microorganisms with specific metabolic profiles to remediate sewage, contaminated water, or soil is referred to as bioremediation. The use of carrier materials such as alginate, agar, agarose, gelatin, gellan gum, and polyurethane alleviates some of the problems associated with bioaugmentation.

Biosurfactants are chemical analogues with environmentally favorable and biodegradable properties. However, using biosurfactants to polluted sites is uneconomical due to their high creation value and lack scalability. Biosurfactant makers can benefit from agro-industrial waste mixtures as nutrient assets throughout the fermentation process. The use of a variety of bioremediation techniques will aid in the growth remediation process (Cassidy et al., 2015). Enhancing bioremediation capability by the systematic application of genetically modified microorganisms (GEMs) is a promising strategy. This is owing to the capability of creating a designer biocatalyst goal pollutant as well as recalcitrant substances by integrating novel and efficient metabolic pathways, expanding the substrate range of existing pathways, and improving catabolic activity stability.

However, in an environmental application, concurrent gene transfer and proliferation of GEM is a promising strategy. Bacterial containment systems, in which any GEM escaping a polluted environment is reconstructed. Furthermore, utilizing a biological approach, a by-product pathway of genetically engineering microorganisms with a specific contaminated substance should improve bioremediation efficacy. Nanomaterials minimize pollutant toxicity to microorganisms by increasing surface area and lowering activation energy, which speeds up and lowers the cost of bioremediation (Rizwan et al., 2014).

20. Microorganisms that have been genetically modified (GEMs)

Genetically engineered microorganisms have had their genetic material transformed through the use of genetic engineering procedures that are spurred by natural rather than artificial genetic exchange between microorganisms. Recombinant DNA technology is the name given to this type of creative work done in a scientific method. By generating genetically modified organisms, genetic engineering has advanced the usage and removal of dangerous undesirable wastes under laboratory circumstances (Jain et al., 2011). Living things that have been created using recombinant DNA techniques or through spontaneous genetic material exchange. An important component of bioremediation in the modern era is inserting the right gene for the production of a particular enzyme capable of degrading a variety of pollutants.

GEMs have demonstrated their potential for bioremediation programmers in soil, groundwater, and activated sludge environments, demonstrating better degradative abilities for a wide range of chemical pollutants. A variety of prospects for improving degradative overall performance utilizing genetic engineering technologies have recently emerged. For example, rate-limiting steps in known metabolic pathways can be genetically modified to provide increased degradation rates, or entirely new metabolic pathways might be added into bacterial genomes to degrade previously resistant substances. There are four activities/procedures to be carried out in GEMs:

(1) Altering the specificity and affinity of enzymes.
(2) The creation and regulation of pathways.
(3) Development, monitoring, and control of bioprocesses.
(4) Chemical detection, toxicity reduction, and cease factor evaluation bioaffinity bioreporter sensor programs.

Essential genes in microorganisms are found on a single chromosome; however, genes encoding enzymes essential for the catabolism of a variety of these unusual substrates can be found on plasmids. The catabolism has been linked to plasmids. As a result, GEMs can be employed correctly for biodegradation purposes, and the results establish/indicate a research frontier with far-reaching ramifications in the future.

Advantage of GEMs in bioremediation: The most important function of GEMs in bioremediation is to speed up the revival of polluted waste sites, increase substrate degradation, present a high catalytic or usage ability with a small amount of cell mass, and generate safe and purified environmental conditions by decontaminating or neutralizing any harmful materials.

Disadvantages of GEMs in bioremediation: The most significant downsides are not addressed by traditional processes; in certain cases, cells die as a result of work connected to their release into the environment. Finally, added overseas modified strain to the system results in unreacted and causes immeasurable damaging impact on the natural structural and useful microorganism's community composition and incidence; finally, it was confirmed that delay of growth and substrate degradation, seasonal version, and different abiotic element fluctuation have direct and indirect effect and relationship on microbial interest; finally, added overseas modified strain to the system ends in unreacted and causes immeasurable damaging impact on the natural structural and useful microorganism's community composition and incidence.

21. Recommendations

The unrestricted discharge of commercial effluents into agricultural fields or bodies of water enhances the likelihood of them entering the food chain via plants and aquatic animals, followed by bioaccumulation. Various in-situ and ex-situ bioremediation techniques suitable for various environmental circumstances were explored and recommended (Adams et al., 2015). Designing, improving, and implementing those techniques necessitates a careful selection of biological agents. Extensive research is being done on the use of specific microorganism strains for bioremediation. Microorganisms carry out redox reactions, and as a result, metallic mobilization/immobilization has an impact on bioremediation procedures. *Bacillus* sp. strains that oxidize manganese (II) produce oxidized Mn, which indirectly oxidizes Cr(III) into the cellular and accessible form of Cr(VI).

Heavy metal bioremediation is more effective when a variety of microbial traces are used simultaneously rather than a single species. Advances in genetic engineering and optimization approaches suggest that those technologies have a bright future. For various pollutants, genetically engineered microbes may have a superior bioremediation capability. Furthermore, the ability of agricultural and industrial waste biomass as bioremediations on a lab/industry scale is currently being investigated; for example, sugarcane bagasse, coconut shell waste, rice husk, and beer waste yeast are all being investigated Babel and Kurniawan (2004).

After several physical and chemical alterations, the biosorption capability of various biosorbents is improved, and studies are required to allow commercial usage of those biosorbents across industries. The bioremediation methodology necessitates a comprehensive and all-encompassing approach for systematic, viable, and long-term approaches that may be easily modified for every situation. Furthermore, there may be a pressing need for coordination at all levels, including research groups, the general public, governmental organizations, and corporations.

22. Conclusion

Biodegradation has the potential to be a very profitable and appealing choice for remediating, cleaning, managing, and developing improved methods for cleaning up polluted environments through microbial activity. Heavy metal contaminants can come from both man-made and natural sources. Heavy metals should be kept out of the environment since they are nonbiodegradable and harmful. Industrial wastewater discharged into the environment, including soil and rivers, necessitates immediate intervention by government agencies, regular monitoring, and rehabilitation using appropriate procedures. Traditional treatment methods have limitations and must be replaced by more efficient, cost-effective, and environmentally friendly methods, such as bioremediation with biological agents. The rate of breakdown of undesired waste chemicals is determined by competition with inorganic agents such as fungus, bacteria, and algae, as well as external abiotic factors (aeration, moisture, pH, and temperature), and limited bioavailability. Bioremediation is dependent on a number of parameters, including but not limited to cost, site features, pollutant type, and concentration. Adopting a suitable biosorbent in terms of efficiency and cost is a critical issue. Microorganisms produce redox reactions and, as a result, have an impact on bioremediation processes through metal mobilization/immobilization. Metal−microbe interaction affects microbiological procedures such as growth, colonization, and the production of microbial biofilms for remediation. More research is needed to translate those lab-scale options into industrial applications (e.g., microbial films), taking into account elements such as appropriate and adapted microorganisms/suitable biomass waste and/or combinations of different types of microbial biomass with conventional technologies and appropriate situations in ex- or in-situ environments.

References

Abbas, S.H., Ismail, I.M., Mostafa, T.M., Sulaymon, A.H., 2014. Biosorption of heavy metals: a review. J. Chem. Sci. Technol. 3 (4), 74−102.

Abdi, O., Kazemi, M., 2015. A review study of biosorption of heavy metals and comparison between different biosorbents. J. Mater. Environ. Sci. 6 (5), 1386−1399.

Adams, G.O., Fufeyin, P.T., Okoro, S.E., Ehinomen, I., 2015. Bioremediation, biostimulation and bioaugmention: a review. Int J. Environ. Bioremed. Biodegr. 3 (1), 28−39.

Adenipekun, C.O., Lawal, R., 2012. Uses of mushrooms in bioremediation: a review. Biotechnol. Mol. Biol. Rev. 7 (3), 62–80.

Ahirwar, N.K., Gupta, G., Singh, R., Singh, V., 2016. Isolation, identification and characterization of heavy metal resistant bacteria from industrial affected soil in central India. Intern. J. Pure Appl. Biosci. 4 (6), 88–93.

Ali, H., Khan, E., 2018. What are heavy metals? Long-standing controversy over the scientific use of the term 'heavy metals'—Proposal of a comprehensive definition. ToxicolEnviron Chem. 100, 6–19.

Ali, H., Khan, E., Ilahi, I., 2019. Environmental chemistry and ecotoxicology of hazardous heavy metals: environmental persistence, toxicity and bioaccumulation. J. Chem. 1, 1–15.

Andreoli, V., Sprovieri, F., 2017. Genetic aspects of susceptibility to mercury toxicity: an overview. Int. J. Environ. Res. Publ. Health 14, 93.

Asira, E.E., 2013. Factors that determine bioremediation of organic compounds in the Soil. Acad. J. Interdisc. Stud. 2, 125–128.

Atobatele, O.E., Olutona, G.O., 2015. Distribution of three non-essential trace metals (cadmium, mercury and lead) in the organs of fish from Aiba Reservoir, Iwo, Nigeria. Toxicol. Rep. 2, 896–903. https://doi.org/10.1016/j.toxrep.2015.06.003.

Azubuike, C.C., Chikere, C.B., Okpokwasili, G.C., 2016. Bioremediation techniques—classification based on site of application: principles, advantages, limitations and prospects. World J. Microbiol. Biotechnol. 32 (11), 180.

Babel, S., Kurniawan, T.A., 2004. Cr(VI) removal from synthetic wastewater using coconut shell charcoal and commercial activated carbon modified with oxidizing agents and/or chitosan. Chemosphere 54 (7), 951–967.

Bruins, M.R., Kapil, S., Oehme, F.W., 2000. Microbial resistance to metals in the environment. Ecotoxicol. Environ. Saf. 45 (3), 198–207. https://doi.org/10.1006/eesa. 1999.1860.

Cassidy, D.P., Srivastava, V.J., Dombrowski, F.J., Lingle, J.W., 2015. Combining in situ chemical oxidation, stabilization, and anaerobic bioremediation in a single application to reduce contaminant mass and leachability in soil. J. Hazard Mater. 297, 347–355. https://doi.org/10.1016/j.jhazmat.2015.05.030.

Choudhury, R., Srivastava, S., 2001. Zinc resistance mechanisms in bacteria. Curr. Sci. 81 (7), 768–775.

Couto, N., Fritt-Rasmussen, J., Jensen, P.E., Hojrup, M., Rodrigo, A.P., et al., 2014. Suitability of oil bioremediation in an artic soil using surplus heating from an incineration facility. Environ. Sci. Pollut. Res. 21, 6221–6227.

Demirbas, A., 2008. Heavy metal adsorption onto agro-based waste materials: a review. J. Hazard Mater. 157 (2–3), 220–229.

El Fantroussi, S., Agathos, S.N., 2005. Is bioaugmentation a feasible strategy for pollutant removal and site remediation? Curr. Opin. Microbiol. 8 (3), 268–275.

El-Mehy, R., Abou-Aly, H., Youssef, A., Tewfike, T., Elakshar, E., 2019. Efficiency of heavy metals-tolerant plant growth promoting bacteria for alleviating heavy metals toxicity on sorghum. Environ. Exp. Bot. 162.

Ferraz, P., Fidalgo, F., Almeida, A., Teixeira, J., 2012. Phytostabilization of nickel by the zinc and cadmium hyperaccumulator *Solanum nigrum* L. are metallothioneins involved? Plant Physiol. Biochem. 57, 254–260.

Firdousi, S.A., 2017. Bioaccumulation and bio-absorptions of heavy metals by the mushroom from the soil. J. Med. Chem. Drug Discov. 2 (3), 25–33.

Flora, S.J.S., Pachauri, V., Saxena, G., 2011. Arsenic, Cadmium and Lead. Reproductive and Developmental Toxicology. Academic Press, London.

Fomina, M., Gadd, G.M., 2014. Biosorption: current perspectives on concept, definition and application. Bioresour. Technol. 160, 3–14.

Fosmire, G.J., 1990. Zinc toxicity. Am. J. Clin. Nutr. 51 (2), 225–227.

Hassan, A., Periathamby, A., Ahmed, A., et al., 2020. Effective bioremediation of heavy metal–contaminated landfill soil through bioaugmentation using consortia of fungi. J Soils Sediments 20, 66–80.

Hazen, T.C., 2010. In situ: groundwater bioremediation. In: Handbook of Hydrocarbon and Lipid Microbiology. Springer, Berlin, Germany, pp. 2583–2594.

Ianeva, O.D., 2009. Mechanisms of bacteria resistance to heavy metals. Mikrobiol Z 71 (6), 54–65.

Igiri, B.E., Okoduwa, S.I., Idoko, G.O., Akabuogu, E.P., Adeyi, A.O., Ejiogu, I.K., 2018. Toxicity and bioremediation of heavy metals contaminated ecosystem from tannery wastewater: a review. J. Toxicol. 1–16.

Ismail, I.M., Moustafa, T.M., 2016. Biosorption of heavy metals. In: Heavy Metals: Sources, Toxicity and Remediation Techniques. Nova Science Publishers, Hauppauge, NY.

Jain, P.K., Gupta, V.K., Bajpai, V., Lowry, M., Jaroli, D.P., 2011. GMO's: perspective of bioremediation. In: Jain, P.K., Gupta, V.K., Bajpai, V. (Eds.), Recent Advances in Environmental Biotechnology. LAP Lambert Academic Publishing AG and Co. KG, Germany, pp. 6–23.

Jaysankar, D., Ramaiah, N., Vardanyan, L., 2008. Detoxification of toxic heavy metals by marine bacteria highly resistant to mercury. Mar. Biotechnol. 10, 471–477.

Jin, Y., Luan, Y., Ning, Y., Wang, L., 2018. Effects and mechanisms of microbial remediation of heavy metals in soil: a critical review. Appl. Sci. 8 (8), 1336.

Johnston, S.G., Karimian, N., Burton, E.D., 2019. Fire promotes arsenic mobilization and rapid arsenic (III) formation in soil via thermal alteration of arsenic-bearing iron oxides. Front. Earth Sci. 7, 139.

Koller, M., Saleh, H.M., 2018. Introductory chapter: introducing heavy metals. In: Saleh, H.M., Aglan, R. (Eds.), Heavy Metals. Intech Open, London, UK.

Kubier, A., Wilkin, R.T., Pichler, T., 2019. Cadmium in soils and groundwater: a review. Appl. Geochem. 108.

Li, C., Zhou, K., Qin, W., Tian, C., Qi, M., Yan, X., et al., 2019. A review on heavy metals contamination in soil: effects, sources, and remediation techniques. Soil Sediment Contam. Int. J. 28 (4), 380–394.

Ma, Y., Oliveira, R.S., Freitas, H., Zhang, C., 2016. Biochemical and molecular mechanisms of plant-microbe-metal interactions: relevance for phytoremediation. Front. Plant Sci. 7, 918.

Macaulay, B.M., 2014. Understanding the behavior of oil-degrading microorganisms to enhance the microbial remediation of spilled petroleum. Appl. Ecol. Environ. Res. 13, 247–262.

Madhavi, G.N., Mohini, D.D., 2012. Review paper on – parameters affecting bioremediation. Inter. J. Life Sci. & Phar. Res. 2, 77–80.

Mishra, S., Bharagava, R.N., 2016. Toxic and genotoxic effects of hexavalent chromium in environment and its bioremediation strategies. J. Environ. Sci. Heal Part C 34, 1–32.

Mustapha, M.U., Halimoon, N., 2015. Microorganisms and biosorption of heavy metals in the environment: a review paper. J. Microb. Biochem. Technol. 7 (5), 253–256.

Paliwal, V., Puranik, S., Purohit, H.J., 2012. Integrated perspective for effective bioremediation. Appl. Biochem. Biotechnol. 166 (4), 903–924.

Rani, I., 2017. Studies on Bioabsorption Ability of Algae and Fungi Inhabiting Coast of Karachi (Doctoral Dissertation). University of Karachi.

Rizwan, M., Singh, M., Mitra, C.K., Morve, R.K., 2014. Ecofriendly application of nanomaterials: nanobioremediation. J. Nanopart. 1–7.

Satyapal, G.K., Mishra, S.K., Srivastava, A., Ranjan, R.K., Prakash, K., Haque, R., Kumar, N., 2018. Possible bioremediation of arsenic toxicity by isolating indigenous bacteria from the middle Gangetic plain of Bihar, India. Biotechnol. Rep. 17, 117–125.

Sharma, P.K., Balkwill, D.L., Frenkel, A., Vairavamurthy, M.A., 2000. A new Klebsiella planticola strain (Cd-1) grows anaerobically at high cadmium concentrations and precipitates cadmium sulfide. Appl. Environ. 66 (7), 3083–3087.

Siddiquee, S., Yusof, N.A., Salleh, A.B., Tan, G.S., Bakar, F.A., Yap, C.K., Yusuf, U.K., Ismail, A., Tan, S.G., Naher, L., Ho, C.L., Yusuf, U.K., 2011. Assessment of surface water quality in the Malaysian coastal waters by using multivariate analyses. Sains Malays. 40 (10), 1053–1064.

Silva-Castro, G.A., Uad, I., Gónzalez-López, J., et al., 2012. Application of selected microbial consortia combined with inorganic and oleophilic fertilizers to recuperate oil-polluted soil using land farming technology. Clean Techn. Environ. Policy 14, 719–726.

Singh, H., 2006. Mycoremediation: Fungal Bioremediation. Wiley, Hoboken, NJ, p. 592.

Tak, I.R., 2020. Role of bio-tools in restoration of freshwater ecosystems. In: Qadri, H., Bhat, R.A., Mehmood, M.A., Dar, G.H. (Eds.), Fresh Water Pollution Dynamics and Remediation. Springer, Singapore, pp. 123–142.

Tchounwou, P.B., Yedjou, C.G., Patlolla, A.K., Sutton, D.J., 2012. Heavy metal toxicity and the environment. Experientia Suppl. 101, 133–164.

Theron, A.J., Tintinger, G.R., Anderson, R., 2012. Harmful interactions of non-essential heavy metals with cells of the innate immune system. J. Clin. Toxicol. S3:005.

CHAPTER

9

Eco-friendly and cost-effective metal sequestration mechanism by the use of biosensor microorganisms

Uzma Afaq and Syed Khalida Izhar

Department of Biosciences, Integral University, Lucknow, Uttar Pradesh, India

1. Introduction

Due to their long-persistence and nonbiodegradable nature, heavy metals are considered as one of the most lethal inorganic contaminants (Kanwar et al., 2020). Metals and metalloids which possess 5 g/cm^3 densities are categorized as heavy metals. In contrast, gave many examples such as Cd, Co, Cr, Cu, Hg, Ni, Pb, Zn, etc., as heavy metals that should have $>4 \pm 1$ g/cm^3 elemental density. Fossil fuel burning, atmospheric impurities, pesticides, and nitrogen and phosphorus fertilizers contribute majorly in the accumulation of heavy metals in agricultural soils. Mining and smelting operations may introduce Cd, Pb, As, and Hg metals. Ni, Cu, Co, Cr, As, Cd, Zn, and Hg metals may be added by industries. Cr, Hg, As, Cd, Pb, Cu, and U may be added through atmospheric dispersion and deposition. Excessive use of agrochemicals may result in the addition of Cd, Cu, As, U, Zn, Pb, and Se. Solid/liquid waste disposal done in improper way may result in the addition of Cu, Cr, Cd, As, Pb, Zn, and Hg. Heavy metals are carcinogenic and can cause serious health problems in living things if their concentrations exceed the World Health Organization (WHO) limit. Their presence in the soil environment can degrade the quality of food, groundwater, vegetation growth, and so on. Besides, they enter living systems via the food chain and cause fatal diseases (Ghrefat and Yusuf, 2006). Heavy metal toxicity has an inhibitory effect on enzymatic activity, plant growth, stomata function, nutrient element accumulation, photosynthesis activity, and root system. They have the ability to build up in various parts of the human body. Even at low concentrations, humans and other organisms are adversely affected. Heavy metals bind with various biomolecules in the human body, disrupting the body's natural functioning (Amin et al., 2013). They have an impact on the mitochondria, cell membrane, endoplasmic reticulum, lysosomes, nuclei, and enzymes that are involved

Integrative Strategies for Bioremediation of Environmental Contaminants, Volume Two. https://doi.org/10.1016/B978-0-443-14013-6.00013-5

in detoxification and damage repair. Heavy metals are also called trace elements as they are in trace concentrations (ppb to less than 10 ppm) in a variety of environmental matrices. Bioavailability is influenced by physical factors such as adsorption, temperature, phase association, and sequestration (Tchounwou et al., 2012).

Plethora of conventional methods, for example, chemical precipitation, ion exchange, electrodialysis, nano-filtration, reverse osmosis, ultrafiltration, coagulation, floatation, flocculation, etc., have been used so far for removing heavy metals. Since heavy metals already harm various living organisms, adding more chemicals to the environment to remove them would be ineffective. These methods are also time consuming and expensive. The Environmental Protection Agency (EPA) defined parameterization as a process in which microbes such as algae, bacteria, yeasts, and fungi are used to degrade, break down, or transform highly toxic contaminants into less toxic or nontoxic forms, thereby assisting the environment in eliminating such hazardous contaminants. It is a natural process that has a high surface-to-volume ratio, is easily accessible, has rapid kinetics of adsorption and desorption, and is inexpensive (Holan and Volesky, 1995).

2. Heavy metals

Elements with a high atomic weight and density at least five times that of water and occur naturally are known as heavy metals. Their use in a variety of fields, including domestic, industrial, medical, agricultural, and technological, has resulted in potential effects on the environment and human health. Their toxicity is determined by the route of exposure, dose, age, chemical species, genetics, gender, and nutritional status of those exposed. Due to their high toxicity, lead, cadmium, arsenic, chromium, and mercury are among the priority metals for public health significance. Since even at low levels of exposure they lead to multiple organ damage, these metals are also known as systematic toxicants. The US EPA and the International Agency for Research on Cancer (IARC) (known or probable) have classified heavy metals as human carcinogens (Tchounwou et al., 2012). The removal of heavy metals from industrial wastewater is extremely crucial. Cadmium, copper, zinc, mercury, lead, nickel, and chromium are all frequently found in industrial wastewater. The diagram below depicts the main sources of heavy metal pollutants and their effects on human life (Wang and Chen, 2009).

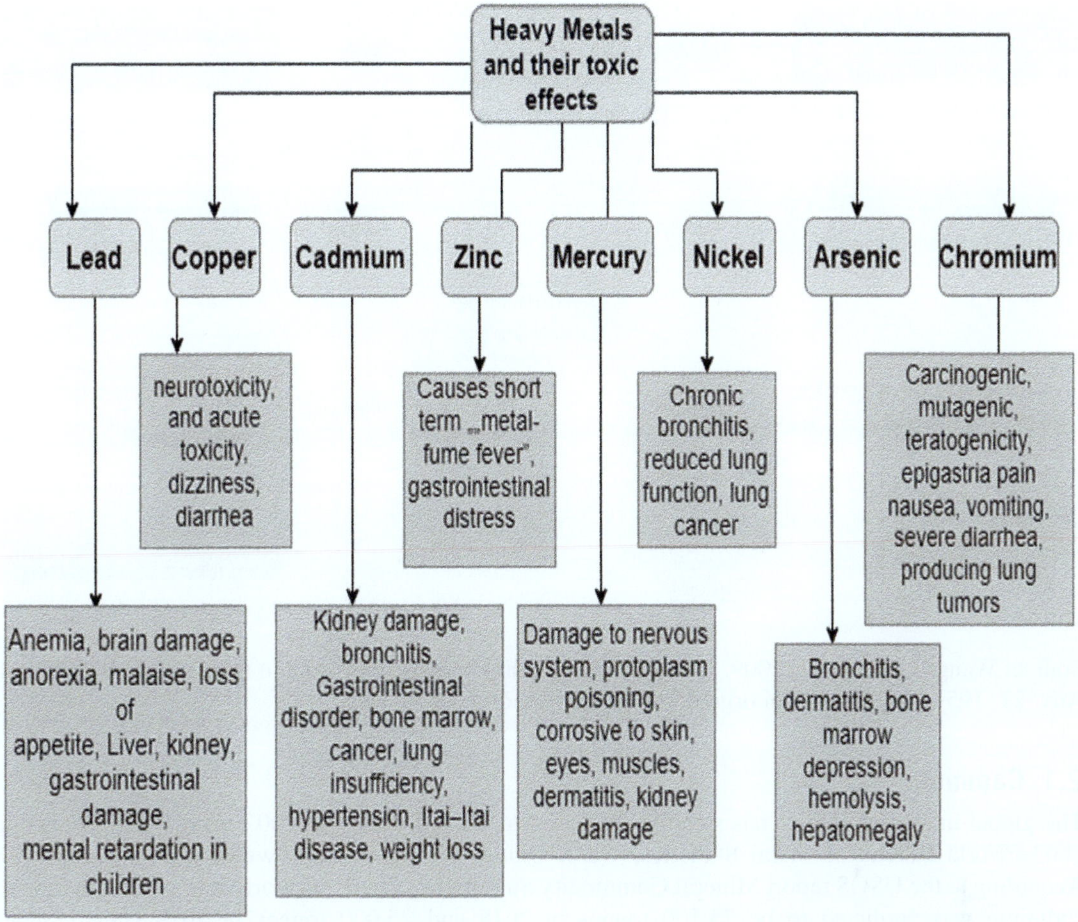
Heavy Metals and their toxic effects
Lead
Copper
Cadmium
Zinc
Mercury
Nickel
Arsenic
Chromium
neurotoxicity, and acute toxicity, dizziness, diarrhea
Causes short term „metal-fume fever", gastrointestinal distress
Chronic bronchitis, reduced lung function, lung cancer
Carcinogenic, mutagenic, teratogenicity, epigastria pain nausea, vomiting, severe diarrhea, producing lung tumors
Anemia, brain damage, anorexia, malaise, loss of appetite, Liver, kidney, gastrointestinal damage, mental retardation in children
Kidney damage, bronchitis, Gastrointestinal disorder, bone marrow, cancer, lung insufficiency, hypertensicn, Itai–Itai disease, weight loss
Damage to nervous system, protoplasm poisoning, corrosive to skin, eyes, muscles, dermatitis, kidney damage
Bronchitis, dermatitis, bone marrow depression, hemolysis, hepatomegaly

Textile, dyeing,paint and pigments, steel fabrication
Chromi-um
Zinc
Mining, refineries,brass manufacturing, plumping

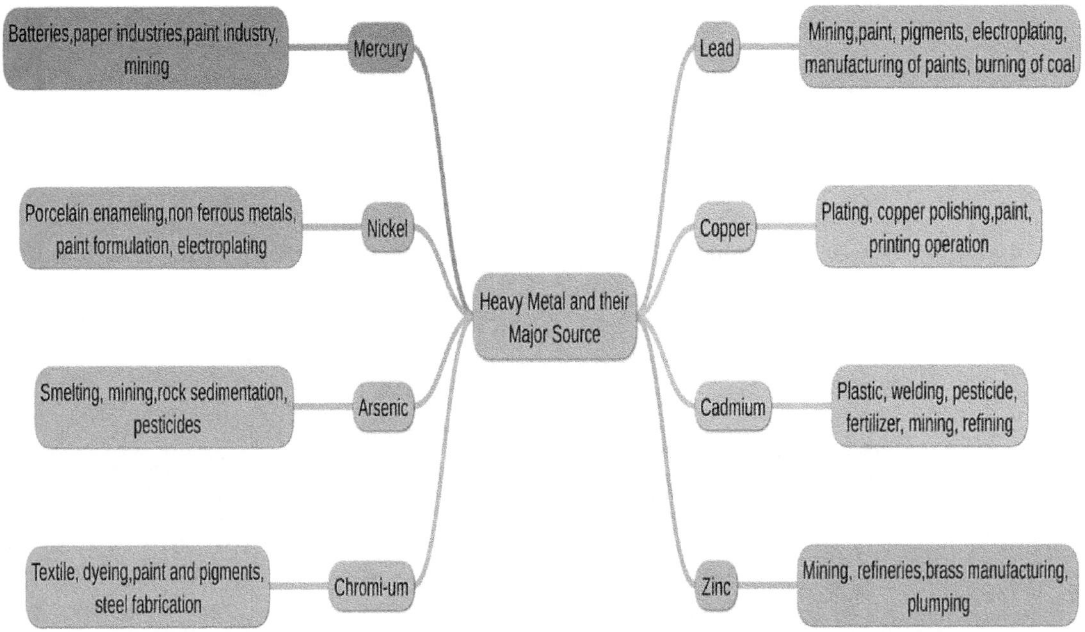

Source: Wang, J., Chen, C., 2009. Biosorbents for heavy metals removal and their future. Biotechnol. Adv. 27, 195–226. https://doi.org/10.1016/j.biotechadv.2008.11.002.

2.1 Cadmium

The global industrial Cd use has steadily increased to 20,400 tonnes in 2007 from 18,400 tonnes in 2003 (World Bureau of Metal Statistics, Ware, United Kingdom; http://www.world-bureau.com/). According to the USGS report Mineral Commodity Summaries, 2020, the world refinery production of cadmium was predicted to be 25,100 tonnes in 2018 and 25,000 tonnes in 2019 United States Geological Survey. As a chemical element, Cd cannot be decomposed, and its concentration in the environment rises steadily, owing largely to human activity. Recovery and safe removal of Cd from ecosystems is unachievable because of its pervasive prevalence in various forms in the environment. The use of phosphate fertilizers for agricultural purposes is a primary source of Cd in soils. Cd accumulates in animals and plants (root vegetables and shoots, such as wheat, rice, cocoa, peanuts, or tobacco) along the food chain (offal, crustaceans, and molluscs). As a result, Cd is becoming a more significant environmental contaminant. Cd is ranked as No. 7 (out of 275) on its priority list of hazardous compounds for the past 15 years by the Comprehensive Environmental Response, Compensation and Liability Act (CERCLA). The IARC in 1993, based on data from human occupational exposure, classified Cd and Cd-containing compounds as group-1 human carcinogens. Although the toxicity and carcinogenicity of Cd at high doses have long been recognized, accumulating epidemiological evidence suggests significant toxicity following long-term Cd exposure that is not limited to cancer and affects more people than previously thought. Chronic low-dose Cd exposure appears to be associated with an expanding spectrum of health hazards. Cd, according to current dogma, directly damages the liver and kidneys and indirectly damages bone via the kidneys. However, new research suggests that Cd may also cause osteoporosis by directly affecting bone tissues. Diabetic populations appear to be more vulnerable to Cd-induced renal toxicity. Surprisingly, environmental Cd exposure increases mortality in a continuous manner with no discernible threshold effect. In humans, "itai-itai," a disease characterized by excruciating bone pain is the most severe form of Cd toxicity. However, it has been proposed that overall nutritional status (rather than just food Cd content) is a more important factor in determining Cd exposure (Genchi et al., 2020).

2.2 Copper

Cu is an alloy or metal that is used in construction, machinery, military weapons, and transportation. It is also a component of other alloys used in imitation jewelery and white gold. Cu is also used in dental products, intrauterine devices, and cosmetics. While exposure to Cu via these items is not harmful, exposure to high concentrations of Cu is damaging. Since the liver is the first site of Cu deposition after it enters the blood, chronic Cu toxicity primarily affects it. Cu toxicity causes renal tubules damage, hemolysis, liver cirrhosis and affects brain and other organs leading to coma, hepatic necrosis, vascular collapse, and death. Chronic Cu toxicity has also been reported in dialysis patients receiving dialysis via Cu tubing, in workers using Cu-containing pesticides and in infants maintained for longer periods on intravenous parenteral nutrition. Consumption of Cu-contaminated water or foods results in the development of acute gastrointestinal symptoms but not with increased mortality from liver disease. Cu poisoning may result in anorexia, weakness, and lethargy in the early stages as well as epithelial lining erosion of the gastrointestinal tract, acute tubular necrosis in the kidney, and hepatocellular necrosis in the liver. The lethal dose of Cu estimated in an untreated adult is about 10–20 g. Various medical conditions, involving obstructive bile excretion such as primary biliary cirrhosis, obstructive hepatobiliary disease, extrahepatic biliary atresia, choledochal cysts, neonatal hepatitis, and a-1-antitrypsin deficiency, predispose the patient to increased Cu concentrations (Dixit et al., 2015; Gaetke and Chow, 2003).

2.3 Lead

For the past 6000 years, lead metal has been associated with human activities. People in ancient civilizations used lead to manufacture trays, kitchen utensils, and other decorative articles. Lead is toxic to humans, with the most deleterious effects on the hemopoietic, nervous, urinary tract, and the reproductive systems (Papanikolaou et al., 2005). Lead is known as the most important toxic heavy element in the environment. Its characteristics like ductility, malleability, softness, resistance to corrosion, and poor conductibility seem to make it difficult to give up its use. Due to its continuous use and nonbiodegradable nature, it accumulates in the environment with increasing hazards (Wani et al., 2015). Lead toxicity causes gastrointestinal, hematological, and neurological dysfunction in children and adults. The major sources of lead for children and infants are soil and dust, chips of leaded paint, food, and water. Ingestion of lead-contaminated soil and dust by hand-to-mouth activity is the major source for most children, but ingestion of leaded-paint chips by individuals suffering from pica is still an important source of exposure (Lockitch, 1993). The most deleterious effects of lead are on kidney function, erythropoiesis, and the central nervous system (Papanikolaou et al., 2005). Lead poisoning occurs mainly by ingestion of water or food contaminated with lead. However, accidental ingestion of contaminated dust, soil, or lead-based paint may also result in poisoning. Lead is believed to be quickly absorbed in the blood stream and has adverse effects on certain organ systems like the cardiovascular system, central nervous system, immune system, and the kidneys (Wani et al., 2015).

Hematological effects: In lead poisoning, erythrocyte protoporphyrin begins to rise and erythrocyte heme decreases at a blood lead concentration of 0.96 μ mol L^{-1}.

Effect on the kidneys: Chronic excessive lead exposure may cause interstitial nephritis and renal tubular injury leading to chronic renal failure.

Neurological effects: The precise effect of lead on the brain is not clear. Lead interacts with calcium-dependent systems in several different ways. Lead may utilize calcium transport systems to enter cells, block voltage-sensitive calcium channels on the plasma membrane of presynaptic nerve endings, or increase spontaneous neurotransmitter release from synaptosomes. Acute lead encephalopathy occurs with failure of blood—brain barrier function. High lead concentrations appear to disrupt second-messenger signals and cellular calcium metabolism. Normally tight junctions between brain endothelial cells widen and allow plasma to flow into the brain and cerebral edema develops.

Cardiovascular effects: Even low levels of lead in the blood have been shown to raise blood pressure in studies, but the results have not been consistent. Over a range of blood lead levels between 0.24 and 3.86 mol L^{-1} and 0.096—0.72 mol L^{-1}, and positive correlations between blood lead concentrations and diastolic blood pressure were discovered.

Effect of lead on growth, stature, and endocrine system: Lead has a minor but noticeable effect on growth. The link between lead poisoning and lower socioeconomic status, substandard housing, poor nutrition, and pica has been blamed for stunted growth in children exposed to high levels of lead. At the average age of 59 months, the growth effect attributed to blood lead was a 1.5% reduction in height compared to the height expected if blood lead was zero (Lockitch, 1993).

2.4 Zinc

Zinc plays a central role in the immune system, and zinc-deficient people experience increased susceptibility to a variety of pathogens. The immunologic mechanisms whereby zinc modulates increased susceptibility to infection have been studied for several decades (Shankar and Prasad, 2018). Most of the nutrition research on zinc has focused on its essential roles in the body. Many studies have examined the consequences of a deficient state on development, growth, and health and the prevalence of deficiency in various population groups. Relatively little attention has been directed toward toxic properties of zinc other than clear industrial hazard, such as with metal fume fever due to the inhalation of zinc oxide fumes or as a result of severe pollution of a localized environment (Li et al., 2006). The first type of well-studied toxic reactions to zinc in human beings was the "metal fume fever," which was induced by intense inhalations of industrial fumes which contains zinc oxide. The most prominent respiratory effects of metal fume fever include chills, fever, gastroenteritis, substernal chest pain, and cough. The impairment of pulmonary function may be accompanied by reduced lung volume and decreased diffusion capacity for carbon monoxide, and there may be rise in bronchiolar leukocytes (Nriagu, 2011). Skin contact with concentrated solutions or zinc powders can result in severe corrosive effects including blistering, ulceration, and permanent scarring. Zinc salts tend to be corrosive and if ingested may result in severe injury to the mouth, stomach, and throat. Initial symptoms include burning of the mouth, pharynx, and vomiting accompanied by erosive pharyngitis, gastritis, and esophagitis. Zinc salts are strong eye irritants, cause pain and erythema, which may be complicated by corneal ulcerations, edema and burns, iritis, hyperemia, hemorrhaging, bullous keratopathy, glaucoma, and cataract formation. Reported symptoms in people exposed to high levels of zinc include premature atrial beats, hypertension secondary to intravascular volume, hypovolemic shock (pulse over 120 beats per minute), and hypertension. Ingestion of high levels of zinc has resulted in lethargy, light-headedness, staggering, difficulty in writing clearly, anxiety, depression, somnolence, and comatose (Nriagu, 2011).

2.5 Mercury

Mercury is a rare element, ranking 16th from the bottom of the list of elements in abundance in the earth. It is ubiquitous, being present in trace amounts throughout the lithosphere, the hydrosphere, the atmosphere, and the biosphere, as well as in igneous rocks of all classes (Patra and Sharma, 2000). Mercury is a heavy metal of known toxicity, noted for inducing public health disasters in Minamata Bay, Japan (Bernhoft, 2012). In 1950, Minamata Bay tragedy caught the world unawares. It has been since recognized that the multiple pathways of mercury contamination through air, water, food, pharmaceuticals, cosmetic products, etc., pose serious concern because it persists in the environment and accumulates in the food web (Zahir et al., 2005). The Minamata disaster in Japan involved the conversion of effluents containing elemental mercury to bioavailable methyl mercury, which entered the aquatic food chain. Mercury poisoning has again become a problem of current interest as a result of environmental pollution on a global scale (Patra and Sharma, 2000). Mercury exists in three basic forms: elemental, inorganic, and organic. Dental amalgams are the most common source for elemental mercury vapor, which is a stable monoatomic gas. Inorganic mercury, which is a divalent compound, is the toxic species found in human tissue after conversion from the other forms. Organic mercury in the form of methyl and ethyl mercury is primarily from fish, sea mammals, and thimerosal vaccines (Houston, 2011). Mercury is the most dangerous of all the heavy metals. It will modify the distribution and retention of other heavy metals. Mercury has no known physiologic role in human metabolism, and the human body has no mechanisms to actively excrete mercury. Mercury thus accumulates during life so that the average 165 lb person has a total body burden of about 13 mg of mercury (Houston, 2011).

About 80% of metallic mercury vapor outgassed from amalgams is absorbed through inhalation. On entry to the body, mercury vapor has great affinity for sulfhydryl groups and bonds with sulfur-containing amino acids in the body. Mercury vapor is transported to the brain, either adherent to red cell membranes or dissolved in serum. Metallic mercury passes easily through the blood brain barrier and through the placenta, where it lodges in the fetal brain. Metallic mercury is, however, rapidly oxidized to mercuric on entry in the blood stream, although not so quickly as to prevent considerable uptake by the central nervous system while still in the metallic form. In addition to the brain, metallic mercury is also deposited in the thyroid, muscles, myocardium, breasts, adrenals, kidneys, liver, skin, sweat glands, lungs, enterocytes, pancreas, salivary glands, prostate, and testes and may be associated with dysfunction of those organs (Bernhoft, 2012). Mercury also induces mitochondrial dysfunction with reduction in ATP, increased lipid peroxidation, and depletion of glutathione (Houston, 2011).

2.6 Nickel

The 28th element, nickel, belongs to the iron group of transitional metals that can assume various valencies from −1 to +4 (Savolainen, 1996). Nickel is a well-known heavy metal found at very low levels in the environment. Nickel is available in all meteorites and soil types and also erupts from volcanic emissions. Nickel is principally bound with oxygen or sulfur in the environment and forms oxides or sulfides in earth crust. The vast industrial use of nickel during its recycling, production, and disposal has led to widespread environmental pollution (Das et al., 2018). Nickel is a very versatile component in such corrosion-resistant metal alloys as stainless steel, in coins to replace silver, in endoprostheses, and others. The modern commercial uses of nickel include nickel—cadmium batteries, electronic circuits, and catalysis of chemical processes. Nickel can be found in ceramics, many household products, and in jewelery. The production of nickel exceeds one million metric tons worldwide (Savolainen, 1996). Nickel concentrations vary widely in different environmental niches, with low nanomolar (i.e., B0.1 ng g^1 [ppb]) to low micromolar (i.e., B0.1 mg g^1 [ppm]) levels in many aquatic systems, 10—40 ppm in most soils, and several 100 ppm in serpentine soils. Nickel is a common industrial pollutant with concentrations in wastewater around industrial parks sometimes reaching the low millimolar range. Nickel toxicity to humans has received intensive attention due to nickel's links to cancer. Until recently, nickel toxicosis in microorganisms was believed to be a problem limited to cells exposed to industrial pollution or found in naturally occurring nickel-contaminated soils. This assumption was overturned, however, when a nickel defense system (RcnA) was characterized in the mesophile *Escherichia coli*. Furthermore, putative homologs of RcnA were detected in archaea and throughout bacteria. This discovery suggests that excessive nickel is a routine physiological concern of microorganisms (MacOmber and Hausinger, 2011). Nickel ions can stimulate lipid peroxidation as do other transitional iron-group metal ions. Lipid peroxidation is a nonspecific proximate mechanism of toxicity for many chemicals and even for hyperoxia and hypoxia (Savolainen, 1996). Acute nickel toxicity is characterized by such central nervous system effects as giddiness, lassitude, and headache, which are associated with nausea, vomiting, diarrhea, cough and shortness of breath, and subnormal body temperature (Savolainen, 1996). The environmental sources of lower levels of nickel include tobacco, dental or orthopedic implants, stainless steel kitchen utensils, and inexpensive jewelery. Tobacco smoking is another source of nonoccupational exposures to nickel. It has been observed that each cigarette contains 1.1—3.1 μg of nickel and that about 10%—20% of the nickel inhaled is present in the gaseous phase (Das et al., 2018). It has been reported that nickel contact caused allergic dermatitis and immunologic urticarial; hence, nickel can be marked as both immune sensitive as well as an allergen (Das et al., 2018).

2.7 Arsenic

The elements exist in the environment in various oxidation states and as various species, such as(V), As(III), As(0), and As(-III). Arsenic is a heavy metal whose name is derived from the Greek word arsenikon, which means "powerful." In an oxidized environment, As mostly appears as oxyanions. Arsenic cannot be easily destroyed and can only be converted into different forms or transformed into insoluble compounds when combined with other elements like iron. Many impurities, such as iron, lead, and selenium, may be mixed in with arsenic wastes, making removal uneconomical (Choong et al., 2007). Humans are exposed to arsenic through either the oral route, which includes contaminated food and water, or the inhalation route, which includes agricultural pesticides and mining activities. Arsenic contamination is a major public health concern in many countries, including Bangladesh, India, Mexico, China, Taiwan, Chile, Argentina, and the United States. According to a fact sheet published by the WHO in 1999, arsenic contamination is a major public health concern that necessitates immediate action. According to WHO guidelines, arsenic has a safety limit of 10 $\mu g\ L^{-1}$ and a maximum permissible limit of 50 $\mu g\ L^{-1}$ in drinking water (Singh et al., 2011). Over 200 million people are at risk worldwide, and exposure to the Arsenic toxicity is heavily influenced by its metabolism. Reduction to a trivalent state is followed by oxidative methylation to a pentavalent state, that is, arsenicals. Metalloid arsenic is a daily occurrence due to its environmental pervasiveness. Arsenic, which exists in a variety of chemical forms and oxidation states, has both acute and chronic negative health effects, including cancer. The precise mechanism of arsenic action is unknown, but several hypotheses have been proposed. Inorganic arsenic in the pentavalent state may replace phosphate in a number of biochemical reactions. In the trivalent state, inorganic and organic (methylated) arsenic can bind to critical thiols in proteins, inhibiting their activity (Hughes, 2002). Arsenic is a naturally occurring element that can be found in the environment in both organic and inorganic forms (Hughes, 2002). Inorganic arsenic is far more toxic than organic arsenic. Peripheral neuropathy is the most common neurological effect of long-term arsenic toxicity, and toxic hepatitis with increased levels of liver enzymes is perhaps the most common gastrointestinal effect (Jomova et al., 2011).

It is toxic to the majority of organ systems, with the kidney being the most vulnerable. The severity of arsenic poisoning is determined by several factors, including dose, individual susceptibility to arsenic, and the age of the affected individuals. While chronic arsenic exposure affects the vascular system, resulting in hypertension and cardiovascular disease, acute arsenic toxicity can result in cardiomyopathy and hypotension.

2.8 Chromium

It is a transition element in the periodic table's group VI-B (Shanker et al., 2005). Chromium is a common contaminant in surface and ground water and occurs naturally in high concentrations in ultramafic rocks. The French chemist Vauquelin discovered chromium (Cr) in the Siberian red lead ore (crocoite) in 1798. The widespread presence of chromium in surface and ground water reflects its importance as an industrial metal in a wide range of processes. Chromium, a lustrous silver–gray metal, is rarely found in the Earth's crust as an element (or metal), but rather as compounds or ions in water. Chromium is mined as a primary product from chromite ($FeCr_2O_4$), which is the most important commercial source of chromium. Cr is found everywhere in nature and in varying concentrations in the environment. The primary applications for chromium are 18% refractory 67% metallurgical, and 15% chemical (Saha et al., 2011). Chromium (Cr) is a major industrial waste produced by many industries such as textiles, tanneries, electroplating, and metallurgical, which causes health problems in humans and animals and has an impact on marine life. Cr(III) is naturally present in the environment, whereas Cr(VI) is entirely produced by human activity. Cr(VI) is used as an anticorrosive agent in various cooling systems, combustion (for example, cigarette smoke), and power plant ash (Shekhawat et al., 2015). The majority of people are exposed to Cr(VI) through industrial emissions. The primary routes of chromium exposure at hazardous levels are dermal and inhalation, though there have been cases of accidents or attempted suicide where significant oral exposure has occurred (Gad, 1989).

Respiratory exposure: Workers in the chromate and Cr(VI) producing industries are irritated as a result of the exposure to the respiratory tract. It has been reported that it causes nasal ulceration and perforation (Shekhawat et al., 2015).

2.8.1 Oral

Oral chromium compound exposure is generally regarded as a life-threatening hazard and a medical emergency. This emergency necessitates treating (i) burns and corrosion along the oral cavity and upper end of the trachea, as well as (ii) acute renal toxicity from absorbed hexavalent chrome compounds (Gad, 1989).

Chromium compounds are also extremely toxic to plants and have a negative impact on their growth and development (Shanker et al., 2005). Cr's first interaction with a plant occurs during the uptake process. Because Cr is a toxic, nonessential element for plants, they lack specific mechanisms for its uptake. As a result, this heavy metal is taken up by carriers similar to those used for the uptake of essential metals for plant metabolism. Cr's toxic effects are primarily determined by metal speciation, which governs its uptake, translocation, and accumulation. The Cr(VI) transport pathway is an active mechanism that involves carriers of essential anions such as sulfate. Fe, S, and P have also been shown to compete with Cr for carrier binding (Shanker et al., 2005).

2.8.2 Allergy to chromium

Due to their solubility and reactivity, Cr(VI) compounds cause sensitization and allergic reactions. Contact dermatitis has been reported as a result of the tetravalent form of CrO_7.

- **Skin allergy:** Cr(VI) is a sensitizing agent. Workers in Cr salt industries were exposed to Cr and were reported to have contact dermatitis, which has become a serious problem in the industry. A large proportion of the population who comes into contact with Cr compounds develops skin patches, which becomes a serious problem.
- **Respiratory system:** Asthma has been reported in people who have come into contact with or have been exposed to welding fumes from stainless steel. Cr(VI) compounds are held responsible. The maximum amount of Cr(VI) that can be inhaled through the air is 0.02 g/m^3 (Shekhawat et al., 2015).

2.8.3 Dermal

Dermal effects are largely determined by the compound's or mixture's acidity and oxidizing potential. Irritation or corrosive reactions are quite likely with Cr(VI); such effects with Cr(III) are less likely and less severe. While chromium compounds are unlikely to be absorbed sufficiently through intact skin to cause systemic toxicity (kidney damage), absorption can occur if the skin's integument is significantly disrupted (as in the well-known "chrome burn" process), and acute kidney damage may result as a secondary effect (Gad, 1989).

3. Conventional methods for heavy metal removal from industrial effluents

Many processes, such as adsorption, precipitation, coagulation, ion exchange, cementation, electrodialysis, electrowinning, electrocoagulation, and reverse osmosis, have been developed to reduce heavy metal pollution (Ahluwalia and Goyal, 2007).

1. Precipitation is the method for removing toxic heavy metals from water at parts per million levels as some metal salts are insoluble in water and precipitate when the correct anion is added. Despite its low cost, the process's efficiency is hampered by low pH and the presence of other salts (ions). The process involves the addition of other chemicals, which results in the production of a high-water content sludge, the disposal of which is expensive (Gray, 1999). At low concentrations,

precipitation with lime, bisulfide, or ion exchange lacks specificity and is not effective in removing metal ions.

2. Electrowinning is widely used in heap leaching and acid mine drainage in mining and metallurgical industrial operations. It is also used in metal transformation, electronics, and electrical industries for metal removal and recovery. Metals found in effluents such as Ag, Co, Au, Cr, Ni, Cd, Pb, Sn, and Zn can be recovered through electrodeposition using insoluble anodes (Gray, 1999).
3. Ion exchange is another method that has been used successfully in the industry to remove heavy metals from effluents. Though it is more expensive than the other methods, it can achieve ppb levels of clean up while handling a relatively large volume. An ion exchanger is a solid that can exchange cations or anions from its surroundings. Synthetic organic ion exchange resins are commonly used matrices for ion exchange. The drawback of this method is that it cannot handle concentrated metal solutions because the matrix becomes easily fouled by organics and other solids in the wastewater. Furthermore, ion exchange is nonselective and highly sensitive to solution pH (Ahluwalia and Goyal, 2007).
4. Electrocoagulation is an electrochemical method for removing metals from solution that employs an electrical current. The electrocoagulation system can also remove suspended solids, dissolved metals, tannins, and dyes. The contaminants in wastewater are kept in solution by electrical charges. These ions and other charged particles become destabilized and precipitate in a stable form when they are neutralized with ions of opposite electrical charges provided by an electrocoagulation system (Ahluwalia and Goyal, 2007).
5. Reverse osmosis and electrodialysis use semipermeable membranes to recover metal ions from dilute wastewater. In electrodialysis, selective membranes (alternating cation and anion membranes) are fitted between the electrodes in electrolytic cells, and the associated ion migrates under continuous electrical current, allowing metal recovery (Ahluwalia and Goyal, 2007).
6. Cementation is a type of another precipitation method that involves an electrochemical mechanism in which a metal with a higher oxidation potential enters solution, for example, oxidation of metallic iron, Fe(0) to ferrous iron(II) to replace a metal with a lower oxidation potential. Copper is most commonly separated by cementation, along with noble metals such as Ag, Au, and Pb, as well as, Cd, Ga, Pb, Sb, and Sn (Ahluwalia and Goyal, 2007).

4. Understanding biosorption

Biosorption, the process of using a microorganism to recover and remove heavy metals from aqueous solutions, has been known for a few decades but has recently evolved as a low-cost promising technology. Heavy metal uptake occurs as a result of physicochemical interactions of metal ions with cellular compounds of biological species in this process (Holan and Volesky, 1995).

Biosorption is a metabolically independent and passive process that removes heavy metals through a passive binding process with nonliving microorganisms such as algae, filamentous fungi, bacteria, and yeasts, as well as other biomass types capable of efficiently collecting heavy metals (Migahed et al., 2017). Biosorption is the ability of certain types of inactive, nonliving microbial biomass to bind and concentrate heavy metals from even very dilute aqueous solutions. This property is exhibited by biomass, which acts solely as a chemical substance, an ion exchanger of biological origin. This phenomenon was discovered to be caused by the cell wall structure of certain fungi, algae, and bacteria (Volesky, 1990).

The biosorption process demonstrated many desirable characteristics, including rapid kinetics of desorption and adsorption, selective removal of metals over a wide pH and temperature range, and low capital and operating costs. Biosorption could separate heavy metals from various waste materials, including wastewater effluent, based on the metal binding capacities of various biological materials (Migahed et al., 2017). Biosorbent could be easily manufactured with low-cost growth media or obtained as an industrial by-product.

5. Advantages of biosorption process (Holan and Volesky, 1995)

Overall, the potential advantages of the biosorption process over conventional heavy metal removal methods include:

- the use of naturally abundant renewable biomaterials that can be produced cheaply;
- the ability to treat large volumes of wastewater due to rapid kinetics;
- high selectivity in terms of removal and recovery of specific heavy metals;
- the ability to handle multiple heavy metals and mixed wastes;
- high affinity, reducing residual metals to less than 1 ppb.
- There is less need for expensive reagents, which typically cause disposal and space issues.
- Operation under a wide range of physiochemical conditions, such as temperature, pH, and the presence of other ions (such as Ca(II) and Mg(II)).
- Low capital investment and low operational costs.
- Significantly increased recovery of bound heavy metals from biomass.
- Significantly reduced the amount of hazardous waste generated.

6. Disadvantages (Ahluwalia and Goyal, 2007)

- Early saturation can be a problem, which means that when metal interactive sites are occupied, metal desorption is required prior to further use, regardless of metal value.
- Because cells do not metabolize, the potential for biological process improvement (e.g., through genetic engineering of cells) is limited. Because the adsorptive agent is produced during pregrowth, there is no biological control over the biosorbent's characteristics. This is especially true if waste biomass from a fermentation unit is used.
- There is no possibility of biologically changing the metal valency state. For instance, less soluble forms or even the degradation of organometallic complexes.

7. Factors affecting on biosorption of metals (Holan and Volesky, 1995)

Biosorption is affected by a variety of factors. Some of these factors are associated with biomass and metal, while others are associated with environmental conditions. The following are the major factors that influence the biosorption process:

(1) Temperature: Unlike the bioaccumulation process, biosorption efficiency remains unaffected within the temperature range of 20–35°C; however, high temperatures, such as 50°C, may increase biosorption in some cases, but these high temperatures may cause permanent damage to microbial living cells, resulting in decreased metal uptake. Adsorption reactions are generally exothermic, with decreasing temperature increasing the extent of adsorption.

(2) The surface area to volume ratio: Individual cells or particles, as well as the available surface area of immobilized biofilms, may benefit from it. Furthermore, biomass concentration may have an effect on biosorption efficiency, with increasing biomass concentration resulting in a decrease in sorption per unit weight.

(3) Acidity: The most important parameter in biosorption processes appears to be pH. Biosorption is similar to an ion-exchange process, and biomass can be thought of as natural ion-exchange materials with weakly acidic and basic groups. As a result, the pH of the solution influences the nature of biomass binding sites and metal solubility; it affects metal solution chemistry, the activity of functional groups in biomass, and metallic ion competition. Metal biosorption has been repeatedly demonstrated to be pH dependent in almost all systems studied, including bacteria, cyanobacteria, algae, and fungi. Due to competition between cations and protons for binding sites, biosorption of metals such as Cu, Cd, Ni, Co, and Zn is frequently reduced at low pH levels. In general, when the pH of the metal solutions is reduced from 6.0 to 2.5, the heavy metal uptake of most biomass types decreases significantly. Metal ions are removed from solutions with minimal or negligible removal at pH less than 2. When the pH rises from 3.0 to 5.0, metal uptake rises. The optimal pH value is critical for maximum metal sorption, and this capacity decreases as the pH value increases.

(4) Characteristics of biomass: The nature of the biomass or derived product, including the nature of its application, such as freely suspended cells, immobilized preparations, living biofilms, and so on, may be considered one of the important factors. Physical treatments such as boiling, drying, autoclaving, and mechanical disruption all have an effect on binding properties, whereas chemical treatments such as alkali treatment frequently improve biosorption capacity, which is especially noticeable in some fungal systems due to DE acetylation of chitin to form chitosan–glycan complexes with higher metal affinities.

(5) Biomass concentration: The specific uptake is affected by the concentration of biomass in solution. At a given equilibrium concentration, low cell densities adsorb more metal ions than high cell densities. As a result, electrostatic interaction between cells is important in metal uptake. Because an increase in biosorbent concentration leads to interference between the binding sites, the specific uptake of metals increases at lower biomass concentrations. Metal ions' access to binding sites is hampered by high biomass concentration.

(6) Initial metal ion concentration: The initial concentration acts as a powerful driving force in overcoming all metal mass transfer resistance between the aqueous and solid phases. The initial concentration of metals will increase the amount of metal adsorbed by the biomass. At low initial metal concentrations, the maximum percentage of metal removal can be obtained. Thus, at a given concentration of biomass, metal uptake increases as the initial concentration increases.

(7) Metal affinity to the biosorbent: The permeability and surface charges of the biomass are affected by physical/chemical pretreatment, and metal binding groups are made available for binding. It can be manipulated by pretreating the biomass with alkalis, acids, detergents, and heat, which may increase metal uptake.

8. Application of biosorption

Over the last 30 years, a large amount of research on metal biosorption has been published in order to elucidate the principles of this effective metal-concentration phenomenon. Biosorption is regarded as a potentially cost-effective biotechnology for treating high volume, low-concentration complex wastewaters containing heavy metals. Some efficient natural biosorbents that require little modification in their preparation have been identified. The biosorption has not yet been used, but it appears that biosorption will have little competition in many types of large-scale environmental metal removal applications (Volesky and Naja, 2005; Wang and Chen, 2009).

There have been few studies to determine the biosorbent's compatibility with real industrial effluents. However, several attempts to scale up or commercialize the biosorption process based on conventional sorption operations have so far been unsuccessful.

9. The future of biosorption

Biosorption application is facing a significant challenge, and some researchers have proposed several solutions. According to Volesky and Naja (2005), the failure of process commercialization is primarily due to nontechnical pitfalls involved in the commercialization of technological innovations as some solid capitalization is needed for innovative process ventures; a partnership approach may be preferable. The selection of partners, on the other hand, appears to be critical. There are two trends in biosorption development for metal removal in the future. One emerging trend is the use of hybrid technology for pollutant removal (Tsezos, 2001), especially using living cells. Another trend is to create good commercial biosorbents, similar to ion exchange resins, and to aggressively pursue the market (Volesky, 2007). The previously stated bioprocesses, as well as other nonbiotechnology-based processes such as chemical precipitation, flotation, electrochemical processes, and membrane technology, may be useful for large-scale wastewater treatment, including the simultaneous removal of organic substances and heavy metal ions in solution. Due to the difficulties associated with biosorption application, people are considering using hybrid technology, which consists of multiple processes to treat real effluents. Several biotechnology-based processes, including biosorption, bioreduction, and bioprecipitation, were proposed. As a result, the use of living cells rather than dead cells for biosorption has resurfaced (Malik, 2004). Some examples have been provided (Brady et al., 1994) . All these processes even may possibly be realized in a single reactor; hence, the corresponding novel reactors should be designed (Tsezos, 2001). Malik (2004) recommended that growing microbes be used as a viable alternative to pure biosorption removal of metal contaminants from complex industrial effluents Tsezos (2001). A hybrid technology, either intra-biotechnological or inter-technological, using a combination of various processes, including biosorption, was proposed. Because biosorption can effectively sequester dissolved metals out of dilute complex solutions with high efficiency and rapid intrinsic kinetics, it is a desirable component in the design of flow sheets. With these properties, biosorption is an excellent candidate for the treatment of high volume, low concentration complex waste waters (Tsezos, 2001). Biosorption appears to be suitable for secondary or polishing applications for metal removal from dilute waste streams, which would be competitive with ion-exchange resin based on a final cost—benefit analysis, and the greatest use for biosorption may be in modular systems for small businesses, for example, for specific treatment (Edyvean et al., 1997). Another trend necessitates the enhancement of biomaterial immobilization, as well as the optimization of biosorption process parameters and physicochemical conditions, including reuse and recycling. The mechanisms underlying biosorption or metal–microbe interactions should be thoroughly investigated. Market factor for successful biosorption application should be considered. From Volesky's viewpoint (Volesky, 2007), when knowledge of biosorption is adequate, applications of certain types of biosorption are on the horizon, inviting "new technology" enterprise ventures and presenting new and quite different challenges. Volesky (2007) thought that metals are only the "tip of the iceberg". He emphasized a completely different type of biosorption, focusing on the purification and recovery of high-value proteins, steroids, pharmaceuticals, and drugs such as digoxin or vinblastin, rather than on environmentally friendly low-cost biosorption. For example, antibodies can be used as a biosorbent to lock and thus extract, recover, and purify the desired target molecule from a mixture. The sources and type of biosorbent have a significant impact on the overall cost of the biosorbent material. If the biomass must be cultured specifically for this purpose, manufacturers will include maintenance and production costs, as well as a commercial fee, in the total cost. Because of their low growth requirements, certain biomass types, such as photoautotrophic algae (e.g., Chlorella and Oscillatoria spp.), can be successfully grown for large-scale commercial use (water, sunlight, and CO_2). Although marine algae such as *Sargassum fluitans* and *Ascophyllum nodosum* have shown biosorptive potential, the costs of harvesting the biomass may be prohibitive to its use. Many different types of industrial waste biomass have been studied for their biosorptive potential. These include yeasts such as *S. cerevisiae* from the food and beverage industry and *Candida albicans*, a clinical isolate; molds such as *Rhizopus arrhizus* from the food industry, *Penicillium chrysogenum* from antibiotic manufacturers, and *Aspergillus niger* from citric acid and industrial enzyme producers; bacteria such as Bacillus spp. used in amino acid and antibiotic fermentations, and *Streptomyces noursei* from the pharmaceutical industry. Since they already pose disposal issues, these potential biosorbents can usually be obtained relatively free of charge from the respective producers. The only costs should be those of drying and transport, if necessary. These low-cost biosorbents will make the process highly economical and competitive, particularly for environmental applications such as detoxifying effluents from, for example, metal-plating and metal-finishing operations; mining and ore processing operations; metal processing, battery and accumulator manufacturing operations; thermal power generation (particularly coal-fired plants); nuclear power generation, etc. The advancement of the biosorption process necessitates additional research in the areas of modeling, regeneration, and immobilization of biosorbents, and treatment of real industrial wastewater. Working together, researchers from various scientific backgrounds, ranging from engineering to biochemistry, will make significant contributions to elucidating biosorption mechanisms. Interdisciplinary efforts are required and present a significant challenge. The optimization of specific biosorption process applications must be done in collaboration with industrial users/clients and necessitates specific process engineering expertise as well as a significant investment in development capital. Several studies have shown that biosorption is a viable alternative to conventional systems for removing heavy metal ions from aqueous solutions (Wang and Chen, 2009).

10. Types of biomass

Under various conditions, agricultural products such as coconut husks, rice straw, waste coffee powder, dried plant leaves, wool, cotton seed hulls, waste tea, and cork biomass have all been tested for their metal-binding capacity. Sewage treatment plant sludge and microbial cells such as bacteria, fungi, algae, yeast, and peat moss; *S. cerevisiae* waste biomass from fermentation and the food industry, as well as other polysaccharide materials, and so on—these biosorbents primarily fall into the following categories as native biomass. To improve the competitiveness of industrial processing operations, it is critical to develop and implement cost-effective metal removal/recovery processes. Because of these disadvantages, as well as the need for more cost-effective and efficient methods of recovering metals from wastewater, alternative separation technologies have been developed. A recent trend has been the implementation of passive treatment schemes. These improve water quality through naturally occurring geochemical and biological processes that require minimal operation and maintenance. Biological removal, which includes the use of plants (living or dead), microorganisms (fungi, algae, and bacteria), and biopolymers, could be an effective method for removing heavy metals from wastewater. Some waste mycelia, in particular, are abundant for heavy metal removal. Microorganisms such as bacteria, fungi, yeast, and algae are excellent sources of biosorbent in their natural environments. Because of their metal-sequestering properties, these biosorbents can be used to reduce heavy metal ion concentrations in solution from ppm to ppb levels. It has the ability to efficiently and quickly sequester dissolved metal ions from dilute complex solutions, making it an excellent candidate for the treatment of high volume and low concentration complex wastewaters.

11. Factors affecting biomass choice

The availability and low cost of biomass are the most important factors to consider when selecting it for large-scale industrial applications. Taking these factors into account, native biomass can be derived from: (i) free industrial waste; (ii) organisms found in large quantities in nature; and (iii) organisms that can be grown quickly or specially cultivated or propagated for biosorption purposes. Metals, for example, can be removed by living microorganisms as well as dead biological material. Certain microorganisms' chemical composition determines their strong biosorbent behavior toward metallic ions. This biosorbent is composed of dead, metabolically inactive cells. Some biosorbents have broad activity, binding, and collecting the majority of heavy metals with no specific activity, whereas others are specific for specific metals. Biosorption can be done with either living or dead biomass. While the use of dead biomass or derived products may simplify life by reducing complexity, the impact of metabolic processes on sorption is frequently underestimated. In fact, feasibility studies for large-scale applications using nonliving biomass are more applicable than bioaccumulative processes using living microorganisms. Because the latter requires a nutrient supply as well as a complex bioreactor system. Furthermore, maintaining a healthy microbial population is difficult due to the toxicity of the pollutants being extracted and other unsuitable environmental factors such as temperature and pH of the solution being treated. The recovery of valuable metals in living cells is also limited because they may be intracellularly bound. As a result of these factors, interest in the use of nonliving biomass as biosorbents has grown. Dead biomass, as previously stated, has an advantage over living microorganisms. Many properties of living microorganisms, on the other hand, remain unexploited in an industrial context and are all worth further investigation because they may be useful for specific applications. Recent biosorption research has concentrated on waste materials that are byproducts or waste from large-scale industrial operations. Some examples include waste mycelia from fermentation processes, olive solid residues, activated sludge from sewage treatment plants, biosolids, and aquatic macrophysics.

12. Economic aspects of biosorption technology

The most important economic aspects of biosorption technology are as follows: (i) require that the biomass used be both natural and inexpensive (low cost and large availability). (ii) Heavy metal selectivity elimination must take place under a wide range of pH, temperature, and adsorption and desorption kinetics. (iii) Microorganisms with a high surface-to-volume ratio must be discovered, as must microorganisms with a superior ability to detoxify heavy metals. Several important parameters that influence the performance of living biosorbents must be considered, including: (i) the physiological state of the organism, (ii) the age of the cells and the availability of micronutrients during their growth, as well as (iii) the environmental conditions that exist during the biosorption process.

13. Types of biological biosorbents

In general, the most important microbial biosorbents can be divided into the following categories:

(1) Bacterium: Bacteria are the most abundant and versatile microorganisms, accounting for a significant portion of the total living terrestrial biomass of approximately 1018 g. Bacteria have a straightforward morphology. Bacteria are most commonly found in three basic shapes: spherical or ovoid (coccus), rod (bacillus, with a cylindrical shape), and spiral (spirillum), though there is a wide range of shapes due to genetic and ecological differences. Bacteria vary in both size and shape. There is a great deal of variation among different bacterial species when it comes to: (i) The number of surface binding sites; (ii) the strength of binding for different ions; and (iii) the binding mechanisms. Bacterial polysaccharide slime layers readily provide amino, carboxyl, phosphate, and sulfate groups for metals biosorption.
Bacterial biomass is typically generated as a waste byproduct of industrial operations or can be propagated on a large scale. Bacterial uptake capacities typically range from 0.23 to 0.90 mmol g^{-1}. Bacteria are used as biosorbents due to their small size, ubiquity, ability to grow under controlled conditions, and resistance to a wide variety of varying environmental conditions. Bacteria species such as Bacillus, Pseudomonas, Streptomyces, Escherichia, Micrococcus, etc., has been tested for uptake metals or organics.

(2) Algae: Algae are considered one of the most promising types of biosorbents which are interest in search for and development as a new biosorbent materials because they have high sorption capacity and are readily available in large quantities in seas and oceans. Algae have low nutrient requirements; being autotropic, they produce a large biomass, and unlike other biomass and microbes, such as bacteria and fungi. They generally do not produce toxic substances. Three groups of algae are found:

- **(1)** Microalgal (green algae or fresh water algae);
- **(2)** Macro-algal (brown algae or marine algae;
- **(3)** In comparison to red and green algae, brown algae have a higher uptake capacity.
  As a result, the metal uptake capacities of nonliving biomass of microalgal (green algae) and macroalgal species (brown algae) range from 0.066 to 1.20 mmol g^{-1} and 0.65 to 1.21 mmol g^{-1}, respectively. When compared to other types of biomass, such as fungi (15%) and bacteria (10%), the use of algae as a biosorbent material has been less (85%). Metal ion binding on algal surfaces is influenced by a number of factors, including:

(1) Ionic charge of metal ion;
(2) Algal species; and
(3) The metal ion solution's chemical composition. Carboxyl, amine, imidazole, phosphate, sulfate, sulfhydryl, hydroxyl, and chemical functional groups are found in algal cell proteins and sugars as potential metal cation-binding sites. Heavy metal ions can be sequestered by microalgae using the same biosorption mechanisms as other microbial biomass, as well as the formation of phytochelatins, which they produce in response to toxic heavy metal stress. There are two main mechanisms that are involved in microalgae biosorption:
   (i) Heavy metal ions displace ions like Na, Mg, and Ca in an ion exchange process.
   (ii) Complexation of metal ions with functional groups such as carboxyl, amino, thiol, hydroxy, phosphate, and hydroxy-carboxyl, which can interact with heavy metal ions in a coordinated manner. Biosorption is based on a number of metal-binding processes that occur with components of the algae cell wall in general. The algae cell walls can biosorb metals in a reversible manner, similar to an ion-exchange resin. As a result, the biosorption mechanism is thought to be influenced by the composition of the algal cell wall. Mannan, xylan, alginic acid, chitin, and other polysaccharides can be used to make algal cell walls. Acid-binding sites such as amino, amine, hydroxyl, imidazole, phosphate, and sulfate groups can be provided by these components, in combination with the proteins present. The biosorption mechanism has been described as not involving van der Waals forces at the cellulose network of the cell walls, so the metal biosorption process involves both ionic charge and covalent bonding. Proteins and polysaccharides are thought to be the main components responsible for biosorption. Ionic charge bonding with carboxyl and sulfate groups associated with these components could be expected, as well as covalent bonding with amino and carboxyl groups. The cell wall of the microalgal *Scenedesmus obliquus* behaved like a weak acidic cation exchanger, containing various cell wall ligands with different exchange capacities, according to studies.

(3) Fungi: Fungi are a type of waste biomass from industrial fermentation that makes excellent metal sorbets. As a result, fungi, including yeasts, have gotten a lot of attention. Fungi are effective and cost-effective at sequestering heavy toxic metals from dilute aqueous solutions through biosorption because:
   (i) It has the benefit of containing a high proportion of cell wall materials;
   (ii) It has exceptional metal binding properties;
   (iii) The antibiotic and food industries both produce large amounts of it;
   (iv) It creates an environmentally friendly environment. Molds, yeasts, and mushrooms are three types of fungi that have significant practical value. Filamentous fungi and yeasts have been used to bind metallic elements in many wastewater treatments. Fungi can be found in all natural settings and play an important role in industrial processes. Fungi come in a variety of shapes and sizes, from unicellular yeasts to polymorphic and filamentous fungi, and many have complex macroscopic fruiting bodies.

Both living and dead fungal cells have a remarkable ability to absorb toxic and precious metals in the field of biosorption. Metal ions are removed by living and dead cells in two different ways: the first mode of metal uptake, referred to as biosorption or passive uptake, is independent of cell metabolic activity and involves surface binding of metal ions to cell walls and extracellular material. Intracellular uptake, active uptake, or bioaccumulation is the second mode of metal uptake into the cell across the cell membrane, which is dependent on cell metabolism. For yeast, active mode can help with metal removal, but it may not help with metal uptake at high metal concentrations, especially for filamentous fungi. Intracellular uptake has a complicated and poorly understood mechanism. In comparison to living cells, the biosorptive capacity of dead fungal cells has been extensively studied. Dead cells' biosorptive capacity may be higher, equal to, or lower than that of living cells. Dead biomass has some advantages over living cells in industrial applications. The dead biomass can be obtained as a waste product from established fermentation processes from industrial sources. Intracellular uptake of metal ions occurs through the metabolism of living cells, whereas cell surface sorption allows toxic metal ions to interact with functional groups on the cell surface such as carboxylate, hydroxyl, sulfate, phosphate, and amino groups. Ion exchange, complexation, and physical adsorption all play a role in these interactions. Fungal cell walls are complex macromolecular structures that contain chitins, glucans, mannans, and proteins, as well as other polysaccharides, lipids, and pigments such as melanin. The cell wall has been linked to the sequestration of heavy metal ions by fungal biomass. The main (up to 90%) constituents of the fungal cell wall are a variety of polysaccharides. Phosphate groups, carboxyl groups on uranic acids and proteins, and nitrogen-containing ligands on protein, as well as on chitin or chitosan, all affect metal uptake capacity in the fungal cell wall. Bioaccumulation can accumulate higher concentrations of heavy metals in yeasts than biosorption. For many filamentous fungi, however, general biosorption is responsible for the majority of heavy metal uptake. Fungi have been found to have biosorption capacities ranging from 0.012 to 1.979 mmol g^{-1}. Because mushrooms grow in natural habitats, pollutants of heavy metals found in the soil or natural substrates are removed by the fruiting bodies of mushrooms, resulting in metal accumulation in the mycelia and sporocarps (Holan and Volesky, 1995).

14. Biosorption mechanisms

Due to microorganism's complexity, there are a variety of ways for the metal to be captured by the cell. As a result, biosorption mechanisms are diverse and, in some cases, still poorly understood. The following criteria can be used to categorize them.

According to the location where the metal removed from the solution is found, biosorption may be classified as:

1. extracellular accumulation/precipitation;
2. cell surface sorption/precipitation;
3. intracellular accumulation.

Biosorption mechanisms can be classified based on their reliance on the cells' metabolism:

1. metabolism dependent;
2. nonmetabolism dependent.

Metal transport across the cell membrane results in intracellular accumulation, which is influenced by the cell's metabolism. This type of biosorption can only occur in living cells. It's frequently linked to a microorganism's active defense system, which reacts in the presence of a toxic metal. Biosorption is obviously not instantaneous in this case, as it requires time for the microorganism to react. We have cell surface sorption, which is not dependent on metabolism, in the case of physicochemical interaction between metal and functional groups of the cell surface, based on physical adsorption, ion exchange, and complexation. This nonmetabolism-dependent physicochemical phenomenon of metal biosorption is relatively fast and reversible. Metal-binding functional groups such as carboxylate, hydroxyl, sulfate, phosphate, and amino groups are abundant in microbial biomass cell walls, which are primarily composed of polysaccharides, proteins, and lipids. When such a mechanism is present, which is fortunately the most common, biomass has all of the chemical properties of an ion exchange resin or activated carbon, implying numerous advantages in the industrial application of biosorption. In the case of precipitation, there is no single classification. In fact, metal precipitation can occur in solution as well as on the cell surface. Furthermore, if the microorganism produces compounds that favor the precipitation process in the presence of toxic metals, it may be dependent on the cells' metabolism. On the other hand, precipitation may not be dependent on the cells' metabolism, occurring after a chemical interaction between the metal and the cell surface.

14.1 Physical adsorption

The phenomena associated with the presence of van der Waals' forces are included in this category. Radionuclides present in an aquatic (sea) environment are accumulated by marine microorganisms through direct adsorption from the water, according to Polikarpov in 1966. Tsezos and Volesky confirmed that the biosorption of thorium and uranium by *R. arrhizus* fungal biomass is also based on physical adsorption in the cell-wall chitin structure. Biosorption of uranium, cadmium, zinc, copper, and cobalt by dead biomass of algae, fungi, and yeasts is thought to occur through electrostatic interactions between ions in solution and cell walls, according to Kuyucak and Volesky. Copper biosorption by the bacteria *Zoogloea ramigera* and the algae *Chlorella vulgaris*, chromium biosorption by the fungi *Ganoderma lucidum* and *A. niger*, and cadmium biosorption by marine algae have all been linked to electrostatic interactions. Copper, nickel, zinc, cadmium, and lead biosorption by *R. arrhizus* are also due to physical adsorption.

14.2 Ion exchange

Polysaccharides are the basic building blocks of microorganism cell walls. The ion exchange properties of natural polysaccharides have been thoroughly investigated, and it is well known that bivalent metal ions exchange with the polysaccharide's counterions. Alginates are natural salts of K+, Na+, Ca2+, and/or Mg2+ found in marine algae. These metallic ions can exchange with counter-ions like Co2+, Cu2+, Cd2+, and Zn2+, resulting in metal uptake through biosorption. Ion exchange between metal ions and phosphodiester residue counterions, both present in the cell wall and cytoplasmic fractions, is thought to be involved in uranium and lead biosorption by *Streptomyces longwoodensis*, according to Friis and Myers-Keith. Copper biosorption by the fungi *Ganoderma lucidum* and *A. niger* has also been linked to ion exchange.

14.3 Complexation

After interaction between the metal and active groups, the metal may be removed from solution by forming a complex on the cell surface. Metal ions can bind to unidentified (single) ligands or chelate with other metal ions. The biosorption of thorium and uranium by *R. arrhizus* is based on complex formations as well as physical adsorption, as previously mentioned. Metals coordinate with the nitrogen of the chitin cell wall network. *C. vulgaris* and *Z. ramigera* biosorb copper through both adsorption and coordination bonds between metals and amino and carboxyl groups of cell wall polysaccharides, according to Aksu et al. *Pseudomonas syringae* accumulation of calcium, magnesium, cadmium, zinc, copper, and mercury was discovered to be due solely to complexation.

14.4 Transport across the cell membrane

As previously stated, this phenomenon is linked to cell metabolism. Unfortunately, due to the toxicity of some elements, biosorption in the presence of high metal concentrations cannot be investigated. In fact, there is very little information about this type of mechanism. The same mechanism that transports metabolically important ions like potassium, magnesium, and sodium across microbial cell membranes could be used to transport heavy metals. The presence of heavy metal ions with the same charge and ionic radius can cause the metal transport system to become confused. This type of mechanism frequently occurs concurrently with biosorption events that are unrelated to metabolic activity. Biosorption by living microorganisms has two basic steps, according to many examples in the literature. The first is metabolism-independent cell wall binding, and the second is metabolism-dependent intracellular uptake, in which metal ions are transported across the cell membrane and into the cell.

14.5 Precipitation

As previously stated, precipitation can be dependent or independent of cellular metabolism. In the former case, metal removal from solution is frequently linked to a microorganism's active defense system. When they react with a toxic metal, they produce compounds that aid in the precipitation process. According to Scott and Palmer, detoxification systems that precipitate cadmium on the cell surface determine cadmium elimination from solution by some Arthrobacter and Pseudomonas species. Precipitation that is not caused by cellular metabolism could be the result of a chemical reaction between the metal and the cell surface. The formation of the complex uranium-chitin, as mentioned above, is followed by complex hydrolysis and the precipitation of the hydrolysis product (uranyl hydroxide) in the cell wall, which is the final step in uranium biosorption by *R. arrhizus*. According to Holan et al. an additional mechanism, such as metal entrapment in the form of insoluble microdeposits, can significantly contribute to cadmium biosorption by marine algae biomass. Biosorption mechanisms are not only diverse, but they can also occur simultaneously, as evidenced by the literature cited (Veglio' and Beolchini, 1997).

15. Conclusion

For the removal of toxic metals from industrial effluents, biosorption has been shown to be a viable alternative to conventional systems. The development of the biosorption process necessitates additional research in the areas of modeling, biosorbent material regeneration, and testing immobilized raw biomasses with real industrial effluents (Veglio' and Beolchini, 1997).

Transport across the cell membrane, complexation, ion exchange, precipitation, and physical adsorption are all involved in the biosorption process. In the case of nonliving biomass, the only mechanism is metabolism-independent metal binding to cell walls and external surfaces. Adsorption processes such as ionic, chemical, and physical adsorption are used in metabolism-independent uptake. Cell walls contain the majority of functional groups that are responsible for metal binding to biomass. Metal adsorption is accomplished by a variety of functional sites in the biosorbent, including carboxyl, imidazole, sulfhydryl, amino, phosphate, sulfate, thioether, phenol, carbonyl, amide, and hydroxyl moieties. As a result, biomass has been identified as a natural source with a variety of functional sites and the ability to adsorb metal ions (Javanbakht et al., 2014).

References

Ahluwalia, S.S., Goyal, D., 2007. Microbial and plant derived biomass for removal of heavy metals from wastewater. Bioresour. Technol. 98, 2243–2257. https://doi.org/10.1016/j.biortech.2005.12.006.

Amin, N.U., Hussain, A., Alamzeb, S., Begum, S., 2013. Accumulation of heavy metals in edible parts of vegetables irrigated with waste water and their daily intake to adults and children, District Mardan, Pakistan. Food Chem. 136, 1515–1523. https://doi.org/10.1016/j.foodchem.2012.09.058.

Bernhoft, R.A., 2012. Mercury toxicity and treatment: A review of the literature. J. Environ. Public Health 2012. https://doi.org/10.1155/2012/460508.

Brady, D., Rose, P.D., Duncan, J.R., 1994. The use of hollow fiber cross flow microfiltration in bioaccumulation and continuous removal of heavy metals from solution by *Saccharomyces cerevisiae*. Biotechnol. Bioeng. 44, 1362–1366.

Choong, T.S.Y., Chuah, T.G., Robiah, Y., Gregory Koay, F.L., Azni, I., 2007. Arsenic toxicity, health hazards and removal techniques from water: An overview. Desalination 217, 139–166. https://doi.org/10.1016/j.desal.2007.01.015.

Das, K.K., Reddy, R.C., Bagoji, I.B., Das, S., Bagali, S., Mullur, L., Khodnapur, J.P., Biradar, M.S., 2018. Primary Concept of Nickel Toxicity – An Overview. https://doi.org/10.1515/jbcpp-2017-0171 1–12.

Dixit, R., Wasiullah, Malaviya, D., Pandiyan, K., Singh, U.B., Sahu, A., Shukla, R., Singh, B.P., Rai, J.P., Sharma, P.K., Lade, H., Paul, D., 2015. Bioremediation of heavy metals from soil and aquatic environment: An overview of principles and criteria of fundamental processes. Sustain. Times 7, 2189–2212. https://doi.org/10.3390/su7022189.

Edyvean, R.G.J., Williams, C.J., Wilson, M.W., Aderhold, D., 1997. Biosorption using unusual biomasses. In: Wase, J., Forster, C. (Eds.), Biosorbents for Metal Ions. CRC Press, London, UK, pp. 165–182.

Gad, S.C., 1989. Acute and chronic systemic chromium toxicity. Sci. Total Environ. 86, 149–157. https://doi.org/10.1016/0048-9697(89)90201-5.

Gaetke, L.M., Chow, C.K., 2003. Copper toxicity, oxidative stress, and antioxidant nutrients. Toxicology 189, 147–163. https://doi.org/10.1016/S0300-483X(03)00159-8.

Genchi, G., S, S., Graziantono, L., Carocci, A., Catalano, A., 2020. The effects of toxicity. Int. J. Environ. Res. Publ. Health 17, 1–24.

Ghrefat, H., Yusuf, N., 2006. Assessing Mn, Fe, Cu, Zn, and Cd pollution in bottom sediments of Wadi Al-Arab Dam, Jordan. Chemosphere 65, 2114–2121. https://doi.org/10.1016/j.chemosphere.2006.06.043.

Gray, N.F., 1999. Water Technology. John Wiley & Sons, New York, pp. 473–474.

Holan, Z.R., Volesky, B., 1995. Biosorption of heavy metals: Review. Biotechnol. Prog. 11, 235–250.

Houston, M.C., 2011. Role of mercury toxicity in hypertension, cardiovascular disease, and stroke. J. Clin. Hypertens. 13, 621–627. https://doi.org/10.1111/j.1751-7176.2011.00489.x.

Hughes, M.F., 2002. Arsenic toxicity and potential mechanisms of action. Toxicol. Lett. 133, 1–16. https://doi.org/10.1016/S0378-4274(02)00084-X.

Javanbakht, V., Alavi, S.A., Zilouei, H., 2014. Mechanisms of heavy metal removal using microorganisms as biosorbent. Water Sci. Technol. 69, 1775–1787. https://doi.org/10.2166/wst.2013.718.

Jomova, K., Jenisova, Z., Feszterova, M., Baros, S., Liska, J., Hudecova, D., Rhodes, C.J., Valko, M., 2011. Arsenic: Toxicity, oxidative stress and human disease. J. Appl. Toxicol. 31, 95–107. https://doi.org/10.1002/jat.1649.

Kanwar, V.S., Sharma, A., Srivastav, A.L., Rani, L., 2020. Phytoremediation of toxic metals present in soil and water environment: A critical review. Environ. Sci. Pollut. Res. 27, 44835–44860. https://doi.org/10.1007/s11356-020-10713-3.

Li, Y., Zhu, Y., Shen, W., Zhang, A., Zhang, J., Ren, H., Yuan, G., Gu, L., 2006. Study on risk factors of sporadic hepatitis E virus cases in some districts of Shanghai. Zhonghua Liuxingbingxue Zazhi 27, 298–301.

Lockitch, G., 1993. Perspectives on lead toxicity. Clin. Biochem. 26, 371–381. https://doi.org/10.1016/0009-9120(93)90113-K.

MacOmber, L., Hausinger, R.P., 2011. Mechanisms of nickel toxicity in microorganisms. Metallomics 3, 1153–1162. https://doi.org/10.1039/c1mt00063b.

Malik, A., 2004. Metal bioremediation through growing cells. Environ. Int. 30, 261–278.

Migahed, F., Abdelrazak, A., Fawzy, G., 2017. Batch and continuous removal of heavy metals from industrial effluents using microbial consortia. Int. J. Environ. Sci. Technol. 14, 1169–1180. https://doi.org/10.1007/s13762-016-1229-3.

Nriagu, J., 2011. Zinc toxicity in humans. Encycl. Environ. Heal. 801–807. https://doi.org/10.1016/B978-0-444-52272-6.00675-9.

Papanikolaou, N.C., Hatzidaki, E.G., Belivanis, S., Tzanakakis, G.N., Tsatsakis, A.M., 2005. Lead toxicity update. A brief review. Med. Sci. Monit. 11.

Patra, M., Sharma, A., 2000. Mercury toxicity in plants. Bot. Rev. 66, 379–422. https://doi.org/10.1007/BF02868923.

Saha, R., Nandi, R., Saha, B., 2011. Sources and toxicity of hexavalent chromium. J. Coord. Chem. 64, 1782–1806. https://doi.org/10.1080/00958972.2011.583646.

Savolainen, H., 1996. Biochemical and clinical aspects of nickel toxicity. Rev. Environ. Health 11, 167–173. https://doi.org/10.1515/REVEH.1996.11.4.167.

Shankar, A.H., Prasad, A.S., 2018. Shankar1998 68.

Shanker, A.K., Cervantes, C., Loza-Tavera, H., Avudainayagam, S., 2005. Chromium toxicity in plants. Environ. Int. 31, 739–753. https://doi.org/10.1016/j.envint.2005.02.003.

Shekhawat, K., Chatterjee, S., Joshi, B., 2015. Chromium toxicity and its health hazards. Int. J. Adv. Res. 3, 167.

Singh, A.P., Goel, R.K., Kaur, T., 2011. Mechanisms pertaining to arsenic toxicity. Toxicol. Int. 18, 87–93. https://doi.org/10.4103/0971-6580.84258.

Tchounwou, P.B., Yedjou, C.G., Patlolla, A.K., Sutton, D.J., 2012. Molecular, clinical and environmental toxicology volume 3: Environmental toxicology. Mol. Clin. Environ. Toxicol. 101, 133–164. https://doi.org/10.1007/978-3-7643-8340-4.

Tsezos, M., 2001. Biosorption of metals. The experience accumulated and the outlook for technology development. Hydrometallurgy 59, 241–243.

Veglio', F., Beolchini, F., 1997. Removal of metals by biosorption: A review. Hydrometallurgy 44, 301–316. https://doi.org/10.1016/s0304-386x(96)00059-x.

Volesky, B., 1990. Biosorption of Heavy Metals. CRC Press, Boca Raton, FL.

Volesky, B., 2007. Biosorption and me. Water Res. 41, 4017–4029.

Volesky, B., Naja, G., 2005. Biosorption: Application Strategies. 16th Internet. Biotechnol. Symp. Compress Co., Cape Town, South Africa.

Wang, J., Chen, C., 2009. Biosorbents for heavy metals removal and their future. Biotechnol. Adv. 27, 195–226. https://doi.org/10.1016/j.biotechadv.2008.11.002.

Wani, A.L., Ara, A., Usmani, J.A., 2015. Lead toxicity: A review. Interdiscip. Toxicol. 8, 55–64. https://doi.org/10.1515/intox-2015-0009.

Zahir, F., Rizwi, S.J., Haq, S.K., Khan, R.H., 2005. Low dose mercury toxicity and human health. Environ. Toxicol. Pharmacol. 20, 351–360. https://doi.org/10.1016/j.etap.2005.03.007.

CHAPTER 10

Application of biochar for soil remediation

Jagdish Gabhane[1], **Sneha T. Bankar**[2] **and Vivek P. Bhange**[3]
[1]*Chintamani College of Art and Science, Gondwana University, Gondpipri, Maharashtra, India;* [2]*Department of Chemistry, Rashtrasant Tukadoji Maharaj Nagpur University, Nagpur, Maharashtra, India;* [3]*Department of Biotechnology, Priyadarshini College of Engineering, Nagpur, Maharashtra, India*

1. Introduction

The tropical region receives the world's utmost rain and sunlight which are the most important factors for plant growth and support much of the world's agriculture. Nearly 40% of the world's population is dependent on the tropical region for food; however, soil organic matter (SOM) loss is also predominant in the tropical region (Glaser et al., 2002). Also, intensive agricultural practices due to incredible intensification in the global human population result in the deterioration of soil, reduced fertility, and loss in soil ecosystems (Rashid et al., 2016). The deterioration is physiochemical, biological, structural, or a combination of all. The reason behind this is the use of high concentrations of fertilizers, pesticides, exhaustive cropping, and use of weedicides (Dregne, 2002). Increased costs of chemical fertilizers, pesticides, and weedicides with reduced crop productivity tremendously increased the process of food products which took the world toward food catastrophes in a few years. To solve this issue, soil remediation is an essential tool.

Several agronomists and researchers have suggested several remediation strategies, such as thermal soil remediation, encapsulation, air sparging, and bioremediation. Some researchers documented that thermal soil remediation is faster and more reliable to remove toxicants or pathogens (Sierra et al., 2016); nevertheless, as it removes or burns most of the organic matter, it also has limitations to worldwide application. Soil encapsulation is a method of covering an infected area with the help of lime, fly ash, or cement. Such encapsulated soil is not useful for agriculture. In the air sparging method, a large volume of pressurized air is injected into groundwater or contaminated soil, removing volatile organic compounds (Sierra et al., 2016). This method is also not useful for soil remediation for agricultural purposes.

Bioremediation is the most reliable method of remediation of infected soil, which includes the use of microbial consortia (Bodor et al., 2021; Villaverde et al., 2017), organic matter amendment (Ros et al., 2010), activated carbon amendments (Berglunda et al., 2004), compost amendment (Cerda et al., 2017), biochar amendments (Gabhane et al., 2020; Lehmann et al., 2011), etc. Microorganisms are considered degraders of soil pollutants, which are highly pollutant specific. Both indigenous and exogenous microorganisms are used for the bioremediation of polluted soil. Several microbes are used

Integrative Strategies for Bioremediation of Environmental Contaminants, Volume Two. https://doi.org/10.1016/B978-0-443-14013-6.00011-1

for bioremediation, such as *Pseudomonas putida*, *Dechloromonas aromatica*, *Deinococcus radiodurans*, *Methylibium petroleiphilum*, *Alcanivorax borkumensis*, *Phanerochaete chrysosporium*, *etc*. (Bodor et al., 2021; Ros et al., 2010).

Bodor et al. (2021) studied the effects of organic matter from *Micrococcus luteus* on soil polluted with lubricant oils. Villaverde et al. (2017) studied the effects of microbial consortia isolated from different agricultural soils. Several similar studies reported positive stimulatory effects of the application of microorganisms to remediate contaminated or degraded soil. In microbial remediation, limiting nutrients should be added to polluted soil to boost the required soil microbiota, which accelerates the natural biodegradation process (Ossai et al., 2020). The influence of charcoal or activated carbon is an ancient method of soil remediation where farms are burned after crop harvesting. However, effects on soils remain poorly understood until the study of Berglunda et al. (2004). The conclusion of their study was charcoal or activated carbon plays an important role in N cycling. Compost amendment to deteriorated soil is a favorite bioremediation technique as its high nutrient contains high biodegradation efficiency, a simple and eco-friendly technique (Cerda et al., 2017). Composting helps convert highly complex organic waste into humus-like compost, increasing soil's efficiency (Ossai et al., 2020).

Soil remediation by biochar is an excellent method of bioremediation as it has several advantages over other techniques. Biochar is a carbon-rich solid by-product produced through high-temperature pyrolysis or degasification of organic residues produced under low oxygen or anaerobic condition (Gabhane et al., 2020). Compared to activated carbon biochar has many micropores, which increases internal surface area (Chan et al., 2007). Also, it has more adsorptive and absorptive capacity than charcoal (Amonette et al., 2012). When compost, organic manure, or fertilizers are mixed with biochar, they get absorbed on and in the pores and remain available longer (Ippolito et al., 2016). The efficiency of compost in soil fertility significantly increases by the amendment of biochar. When soil microbes are amended with biochar, their rapid colonization and multiplication occur, which again influences soil remediation. Recently enzyme immobilization has also been tried by several researchers and showed stimulatory and long-lasting effects of enzymes on soil remediation and improving soil fertility (Ippolito et al., 2016; Gabhane et al., 2020).

Fungi and soil bacteria have a significant role in the fixation of atmospheric nitrogen in the soil. *Arbuscular mycorrhizal* fungi and biological nitrogen-fixing *rhizobacteria* annually contribute 5%–20% to the total N of grassland and savannah (Van Der Heijden et al., 2008). The biochar pores provide space for the colonization of soil microbes, fungi, and actinomycetes, which synthesize extracellular enzymes such as cellulase, glucosidase, glycosidase, amylase, pectinase, lignin peroxidase, L-leucine aminopeptidase, urease, β-xylosidase, etc. (Wang et al., 2015). These enzymes further degrade the organic matters of soil and ultimately made soil fertile. Soil microenvironment is exceedingly important for crop productivity as it is a property of soil structure. The application of biochar amends the soil microenvironment to support the ecological life of soil (Liao et al., 2016). Others extensively studied the effects of biochar amendment in soil and documented significant improvement in the physical structure of the soil, particularly amplified porosity, surface area, water adsorption, holding capacity, and oxygen uptake. Increased soil porosity could help maintain moisture and aeration, limiting factors for microbial life that stimulate enzyme activities and nitrifier activities (Mukherjee et al., 2013). Additionally, biochar serves as a source of micronutrients such as boron, molybdenum, K, P, Ca, etc., which are essential elements for the nodulation of rhizobia (Rondon et al., 2007). Recently several researchers studied the role of biochar as an immunity booster of soil. It shows a reduction in disease severity of different pathogens and also shows induction in system-wide defense response in the host plant (Meller et al., 2012; Elad et al., 2010).

Therefore, the present chapter highlights the application of biochar as soil remediation with a major focus on its significant role in improving soil health parameters such as soil microstructure, porosity, and water-holding capacity. We also covered the single and combined application of biochar and compost and their effects on microbial diversity. Our focus extends to the adsorption and absorption of extracellular enzymes of soil and their stimulatory or inhibitory behavior toward biochar amendment. Overall, we reviewed all positive and negative impacts of biochar application as a soil health conditioner.

2. Significance of feedstock

Biochar production from organic biomass or feedstock is of utmost important steps in the bioremediation of soil. Feedstock availability, its chemical composition, is the basic component for producing good quality biochar with cost economy; otherwise, the technique should become costly. Thus, in the present section, we discuss types, availability of feedstocks, and the significance of their compositions. For the biochar production, different varieties of feedstocks are being used such as agricultural residues, urban waste, paper waste, woody biomass, aquatic biomass, animal and human excreta, industrial waste, food and kitchen waste, dairy and paper mill waste, poultry waste, etc. (Gabhane et al., 2020). For producing good quality of biochar, low cost and abundant available municipal solid waste and agricultural residues are more useful. Both fractions comprise nearly 90% of total organic waste generation. The quality of biochar is based on high fiber content, and thus agricultural waste is considered the most promising feedstock for biochar production (Tripathi et al., 2016).

2.1 Quantitative importance of feedstock availability

As per a recent investigation by the Indian Ministry of New and Renewable Energy (MNRE), India generates 500 million tons of crop residues per year of which 92 Mt is burned each year (NPMCR, 2019), nearly 410 Mt pose burden of its management. As per the study of Gabhane et al. (2016), the utilization of agro-waste is 21.61%, cattle feeding 4.57%, farm wastage 21.99%, burning in the field 4.49% composting, and remaining 47.34% is net available waste (NAW) as shown in Fig. 10.1. Thus,

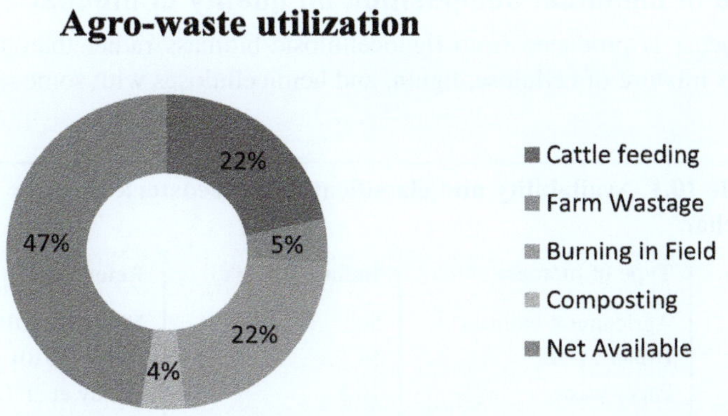

FIGURE 10.1

Agro-waste utilization and net availability for biochar production.

as per present data, approximately 240 million tonnes is NAW. In India, the practice of burning agricultural waste in the field is quite old and routine. The most probable reason is it is an easy and economical process to prepare a farm for the next crop. But such burning of agro-waste contributes to global warming and humiliating soil fertility, and increases the level of particulate matter in the air. Thus, biochar from agro-waste or conversion to other energy products is the best strategy for their management. The second most important feedstock for biochar production is urban waste. The rate of urban waste generation is so high that it will reach 3.40 billion tonnes in 2050 in the world (Kaza et al., 2018), which was 2.01 billion tonnes in 2016.

A third most abundant feedstock is paper waste, which composes a significant share of municipal and industrial wastes and accounts for nearly 11 million metric tonnes (Nandy et al., 2015). Though the paper waste is recyclable, repeatedly recycling loses its quality, and further, it is unutilized for paper making. Waste paper is high cellulose content and should be used for cellulosic ethanol, but its contamination with dye and ink makes it defective. So, it is the best material for biochar production. India alone covers 69.8 million ha of forest area (FAO, 2010), which produces 3000 metric tons of paper annually, consuming nearly 10,000 metric tons of wood. As per quality concerns, wood is an excellent feedstock as it contains a high concentration of lignin, resin, and cellulose. Lai et al. (2013) used hardwood and Douglas fir wood chips to produce biochar, and documented quality and yield are better than other feedstock.

The dry floor of rivers, lakes, and ponds is the richest source of aquatic waste which can easily be used for biochar production. Aquatic biomass includes algae, giant kelp, other seaweeds, and marine microflora with phytoplankton and zooplankton are a major resource for biochar production; however, its collection and processing are very costly. The popular algae used for energy purposes *are Laminaria japonica* (brown alga) and *Gelidium amansii* (red alga). These are vital feedstock for biochar production as it contains good quality cellulose (Nong et al., 2020). The exact quantity of aquatic feedstock could not be calculated. Still, it is a promising feedstock for biochar production as it is often fast-growing and able to grow in freshwater, seawater, or damped oils. Bird et al. (2011) documented algal biochar is more nutritionally rich than other feedstocks, biochar, and the concentration of macronutrients (N, P, K, Ca, and Mg) and micronutrients (Mo) is higher. Table 10.1 shows the availability and classification of feedstock used for biochar.

2.2 Significance of chemical composition on quality of biochar

Good quality of biochar is produced from lignocellulosic biomass rather than starchy biomass as it comprises a complex mixture of cellulose, lignin, and hemicelluloses with some resins, pectin, and ash

Table 10.1 Availability and classification of feedstock used for biochar.

S.No.	Type of biomass	India (MMT/Y)	References
1.	Agricultural biomass	500	NPMCR (2019)
2.	Urban waste	54	Singh (2020)
3.	Paper waste	10.5	Nandy et al. (2015)
4.	Forest/woody biomass	0.01	FAO (2010)

(Gabhane et al., 2016), which decomposes at high temperature with a diverse mechanism with a series of chemical reactions to convert them into char having nano-pore structure. During this conversion process, nonaromatic cellulose converts into condensed aromatic char with a higher surface area. Similarly, lignin, which is initially aromatic material, further increases its aromaticity; however, the surface area is lower than cellulose-derived biochar (Lawal et al., 2021). Cellulose is a promising carbon former to prepare adsorbents of various contaminants; thus, the more the cellulose content, the more the adsorption capacity (Yang et al., 2021). Detailed study of cellulose content and adsorption capacity of biochar was carried out by Lawal et al. (2021) by preparing biochar from commercial cellulose (100% cellulose), an oil palm frond (39.5% cellulose), and palm kernel shell (20.5% cellulose) at 630°C and studying their adsorption properties by using iodine and organic contaminants. They reported an increase in external surface area and average pore size with cellulose content with the largest adsorption capacities. In the studies of Deng et al. (2016), while comparing the textural properties of biochar from the different components of wood *Pinus radiate*, pyrolyzed at 400–600°C, reported cellulose consistently produced biochar with elevated surface area and total pore volume with a wider average pore size than biochar from lignin. He further confirms his observation by studying the micropores percentage of cellulose-derived and lignin-derived biochar. He reported 87% of microspores were observed in lignin-derived biochar, whereas 80% microspores were observed in cellulose-derived biochar. The information is highly useful for the preparation of biochar with the desired application. That means if biochar application is of adsorption of contaminant or toxicant, one should prefer high cellulose content feedstock. In contrast, if the application is adsorption/absorption of nutrients or enzymes, one should prefer high lignin content feedstock. Thus, with these studies, conclusion can be absolute by lengthening the scope into relating biomass components to biochar adsorption capacity.

3. Methods of biochar preparation

The classic difference between charcoal and biochar production initially is the result of vegetation fire or intentionally induced fire by humans. It later is anaerobic burning in a specific reactor under controlled conditions. In the present section, different biochar production methods and their effects on the consequential product are discussed. The biochar production methods are mainly classified into traditional and modern approaches based on their advancements and modernization. We emphasized our study on the mechanism of production method, effects on their compositional and structural changes, and their application. Table 10.2 covered the methods with various biochar production conditions.

3.1 Slow pyrolysis

When biomass is heated under low oxygen supply or complete anaerobic conditions, the process is called pyrolysis, which is either slow or fast depending on heating temperature and heating rate. According to Czernik and Bridgwater (2004), pyrolysis point of cellulose is 350°C, and lignin starts to melt above 350°C based on thermogravimetric analysis. Thus, the effective temperature range for pyrolysis was found to be 300–700°C. In slow pyrolysis, biomass is heated in the range of 300–600°C for more than 2–3 h. Generally, the heating rate is kept at 5–7°C/min and when it reaches the required temperature, it stands for 2–3 h to finish complete pyrolysis. The heating rate and heating duration are variable and are

Table 10.2 Methods with various biochar production conditions.

S. No.	Method of biochar production	Temperature (^{0}C)	Duration (minute)	Heating rate (per minute)	Product yield	References
1.	Slow pyrolysis	300–600	60–120	5–7°C	• 40%–50% biochar • 25%–30% biooil • 20–25 syngas	Lai et al. (2013)
2.	Fast pyrolysis	Above 500	10–20	300°C	• 60% biooil • 20% biochar • 20% syngas	Dai et al. (2017)
3.	Flash pyrolysis	Above 1000	≤1	1000°C	• Biooil is main product • 60%–70% biooil • 10%–15% biochar • 20% syngas	Dai et al. (2017)
4.	Magnetic	450–600	10–20	–	• Biochar is main product	Zhou et al. (2017)
5.	Microwave	200–300	10–20	200°C	• Biooil is main product	Arafat Hossain et al., 2017
6.	Hydrothermal carbonization	220–240	Several hours	Slow	• Biochar is main product	Gao et al. (2013)

dependent on types and amounts of biomass (Gabhane et al., 2020). Biochar is the major product of slow pyrolysis, and biooil and syngas are minor products (Lai et al., 2013). The yield of biochar in slow pyrolysis is 40%–50%, whereas biooil is 20%–25% and syngas is less than 20% (Lai et al., 2013). As per the study of Sahoo et al. (2021), biomass composition plays an important role in the quality and yield of biochar, while biochar production from bamboo and pigeon pea stalk at 400, 500, and 600°C. They reported a high lignin fraction of bamboo (24.53%) biomass yield more biochar (32.20%–27.00%) than pigeon pea stalk (18.30%) biomass (29.80%–21.70%). Also, the biochar with the highest mass fraction of carbon with more surface area was found at a pyrolysis temperature of 600°C. Sahoo et al. (2021) reported alkalinity of all biochar increased with pyrolysis temperature, which is due to the removal of acidic and polar functional groups and resulted in the more hydrophobic product. Slow pyrolysis biochar enriched with sylvite, calcite, and silicates of Mg, Mn, and Ca with high ash content.

3.2 Fast pyrolysis

In the fast pyrolysis, biomass is heated at 500°C for 10–20 min where biooil is the main product (50%–60%) and biochar (20%) as the by-product (Dai et al., 2017). Fast pyrolysis is a high-efficient thermochemical process. Compared with slow pyrolysis, the process is more advantageous in the view of eco-compatibility because of its short retention time and high product recovery. However, it is not the main method for biochar production, seeing as; the premium product is biooil. Biochar quality and production are highly influenced by different pyrolysis conditions and various activation methods. He et al. (2021) studied coimpacts temperature and duration on biochar properties, also the effects of different pretreatments such as steam, CO_2, acid, etc., and reported a temperature of 650 °C was effective in forming carbonized structure in biochar, while 750°C was critical for the porous structure development. Also, a longer duration increased pore volume without changing the yield. The biochar obtained by fast pyrolysis has several adsorptive applications. Karunanayake et al. (2017) studied the application of biochar to remove 4-nitroaniline, salicylic acid (SA), benzoic acid, and phthalic acid from water and concluded that the biochar from fast pyrolysis has more advantages for adsorptive actions.

Another variation is flash pyrolysis where biomass decomposes above 1000°C within a very short period, mostly less than a minute, and the heating rate is above 1000°C per second. Thus, complete process operates within a minute. Generally, the process used for biooil production is not recommended for biochar production. Also, the process needs some specialized reactors such as fluidized bed reactor and twin-screw mixing reactor. The process is not economically compatible as it requires a very high temperature and heating rate.

3.3 Microwave heating

The mode of heating greatly influenced the quality and production of biochar. Direct heating through a hot plate is a convective mode of heat transfer. On the other hand, the microwave is molecular collision due to dielectric polarization; thus, biochar production through microwave heating is an excellent superior technique over slow and fast pyrolysis. Here, energy is introduced into the reactor distantly devoid of contact between the energy source and the reaction mixture. This method shows more advantages over slow and fast pyrolysis as it reduced the temperature requirement by 200°C while achieving similar results (Arafat Hossain et al., 2017). Arafat Hossain et al. (2017) studied microwave processing for biochar and hydrogen production and reported 450°C temperature, 400 W microwave power, and 4–6 min resident time as the best conditions for microwave pyrolysis for biochar production; however, for hydrogen production, 700°C temperature and 400 W power and 4–6 min retention time are the best. Biochar produced by this method has more calorific value than conventional biochar processing.

3.4 Magnetic biochar

This is a different biochar producing method with specific applications for adsorption and recovery of adsorbent material. Biochar is prepared from the addition of ferrous or ferric ion (iron) on their surface with the help of a specific binding agent (gelatine) at a high temperature (above 450°C). The biomass heating could be carried in the microwave or simple furnace (Zhou et al., 2017). As per the study of Zhou et al. (2017), magnetic biochar is precisely applicable for removing various pollutants via adsorption. In this study, he documented the removal of arsenic from wastewater. Some other researchers documented the removal of other groundwater pollutants such as lead, zinc, and copper. The adsorption capacity with magnetic biochar increases up to 3–4 folds when compared with conventional biochar. Several researchers reported similar observations. Its high specific adsorptive capacity could be a better choice for nutrient sinks during the application of soil remediation.

3.5 Vacuum pyrolysis

Vacuum pyrolysis is a type of slow pyrolysis with the heating condition and heating rate being similar. The main difference between vacuum and slow pyrolysis is former operates under vacuum or very low pressure. The end products of both methods are significantly different, which is due to the effective removal of vapor during vacuum pyrolysis. Compared to slow pyrolysis, a product from vacuum pyrolysis is qualitatively good with a high yield due to the prevention of devolatilized inorganic components. Vacuum pyrolysis is used to produce high-quality biochar, which shows high porosity and, thus, is useful in the adsorption of minerals and nitrogen while applying as a soil amendment (Uras et al., 2012).

3.6 Torrefaction

Thermal degradation of biomass under low temperatures (230 and 300°C) is termed torrefaction. Torrefaction is not a biochar production method; however, it is a pretreatment step to achieve good quality biochar. Torrefaction improves the physical, chemical, and biological properties of biomass subjected to biochar production. In torrefaction, hemicelluloses are decomposable components as their melting point is below 300°C, which after torrefaction become more stable and carbon-rich solid products (Almeida et al., 2010). Like other biochar producing methods, torrefaction is the thermochemical process that operates under the absence of oxygen under atmospheric pressure which is carried out in the fixed bed reactor.

3.7 Hydrothermal carbonization

All the methods discussed above need dry biomass or the absence of moisture; however, hydrothermal carbonization (HTC) is a process where biochar is produced from high moisture-containing feedstock such as sludge, animal waste, and compost. HTC operates under a low temperature (220–240°C) for several hours under high pressure (2–10 MPa) in rotary drums, kilns, and stoves. The physicochemical properties of HTC biochar differ from other methods in size, shape surfaces, and functional groups, and most of the initial carbon remains as soluble forms. The advantage of HTC produced biochar is the retention of nutrients such as N and P, which is useful in soil fertility; thus, it is useful in soil remediation (Gao et al., 2013).

4. Biochar as soil remediation

4.1 Soil health stimulator

4.1.1 Effect of biochar on soil structure

Soil structure can be defined as the spatial arrangement of the different components or properties of soil. Soil structure contains living organisms' physical habitat and controls many important physical, chemical, and biological soil functions and associated ecosystem services. The incorporation of biochar may improve the soil's physical structure, especially increased porosity, surface area, water adsorption and holding capacity, oxygen uptake, etc.

There are several theories for the mechanism of soil aggregation, such as the hierarchical theory of aggregation and the concentric theory. Hierarchical theory suggests that macroaggregates are formed by joining microaggregates, and the bonds between microaggregates are weaker than those within microaggregates. The concentric theory of aggregation proposes that the external layers are concentrically built on the external surface of the aggregate, with younger carbon in outer layers of aggregates than in aggregate interiors. All mechanisms can be accountable for the formation and stabilization of soil structure following biochar application. Applied biochar can be part of the soil aggregates or combined with mineral particles in the soil. The soil's physical structure can be improved by the assimilation of biochar, like increased macroaggregates, porosity, surface area, oxygen uptake, higher SOM content, enhanced microbial activities, etc. (Liu et al., 2014; Lei and Zhang 2013). The colonization of the soil bacteria and fungi, useful in absorbing nutrients from the soil, increases as the surface area and pore structure increase (Chan et al., 2007). Moisture and aeration could be maintained by the increased porosity of the soil, which is crucial to microbial life and thus stimulates nitrifier activities. There is stimulation of nitrification due to enhanced porosity of soil by absorbing nitrifier inhibitors (e.g., phenolics). But the process of biochar to provide habitat for nitrifiers to colonize takes several months. Biochar amendment in the soil alters its property that helps nitrogen-fixing bacteria to make habitat inside the biochar pore.

Soil aggregate stability is more affected by hydrophobic components of organic matter than hydrophilic components. Biochar is a highly aromatic carbon structure. It can improve aggregation by helping to bind native SOM by making aggregates more resistant to physical disturbance. By changing the ionic composition of the soil solution, biochar can improve soil aggregates. In biochar-induced enhancements, in soil aggregation, various mechanisms may be involved. Several types of research show that biochar can impact soil aggregation by variations of soil pH and improve aromaticity of soil organic carbon pool. Flocculation of clay particles can be improved by higher soil pH, resulting in the formation of water-stable aggregates. Biochar amendment increases soil pH, which can benefit soil aggregation (Wang et al., 2017). Soil microbial communities may be affected by biochar amendment by supplying feedstock to produce cementing agents for soil aggregates, extracellular polymeric substances. Variations in the biochar reactivity may give different effects (i.e., amount of reactive functional groups). Applied doses of biochar and size particles of biochar significantly affect soil aggregation. Biochar reactivity strongly depends on feedstock, its production conditions, and contact between biochar and soil particles. According to Wang et al. (2017), the addition of biochar has increased C storage in macroaggregates and improved aggregate stability in Yolo soil compared with Vina soil. According to Lei and Zhang (2013), there was a decrease in soil bulk density; similarly, Laird et al. (2010) evaluated the impacts of biochar produced with mixed hardwoods; there was increased specific surface area as the biochar concentration increased. Effects of biochar amendment on different soil type as a soil stabilizer are shown in Table 10.3.

4.1.2 Effects of biochar on water-holding capacity

To improve water use efficiency in agricultural production, increasing the water-holding capacity of soil is the key. It may help provide enough energy and food for an increasing global population. In recent researches, biochar amendment was found out to be very helpful to increase soil water-holding capacity. This indicates that soils added with biochar could hold more water from rainfall, which should increase crop production in nonirrigated dry land regions. In irrigated regions, it could reduce the amount of irrigation water needed to grow crops. To determine the water-holding capacity by mass, the following equation is used (ASTM 2010):

$$Water\ holding\ capacity(\%) = \frac{(mass_{wet} - mass_{dry})}{mass_{dry}} \times 100\%$$

Nonamended sandy loam soil has a water-holding capacity of 16%, while pure biochar can hold over 2.7 times its mass of water. Biochar has high total porosity, and it can both hold water in small pores and thus increase water-holding capacity and let the water flow through the larger pores after heavy rain from topsoil to deeper soil layers (Asai et al., 2009). Biochar produced at low temperature contains more aliphatic compounds in the biochar pores that increase the hydrophobicity. At the same time, high-temperature pyrolysis allows a much smaller number of aliphatic compounds to remain in the pores. Soil mixed with biochar pyrolyzed at 300^0C showed water repellence. The same soil mixed with biochar pyrolyzed at 500°C was wettable. The treatments with biochar produced at higher pyrolysis temperatures resulted in higher water content values at the same suction during the incubation time.

Total porosity and pore size distribution (PSD) are affected by the increased surface area and microporosity of the particles in the amended substances. Mesopores are significantly increased at the expense of macropores in coarse-textured soil because the addition of finer substrates may clog the existing coarser pores and make them finer. Biochar material has a substantial volume of micropores, enhancing the amount of finer pores; this combinedly improves moisture retention capacity in the amended soil and boosts the plant's available moisture content. Amendment of biochar significantly influenced the water retention curves (Lei and Zhang, 2013). The large inner surface area of biochar is responsible for the direct effect of biochar application on water-holding capacity, and the indirect effect is linked to the soil aggregation or structure enhanced by biochar. At lower suctions, water holding depends on the content of larger pores, which is strongly influenced by the soil structure. In contrast, water retention at high suctions is influenced more by soil texture and surface area.

According to Lei and Zhang, 2013, there is a decrease in the soil bulk density and an increase in the water content at low suctions. Since soil structure and adsorption properties are affected by SOM and composition, soil water-holding capacity may also be affected by changes in SOM. The treatment of biochar remarkably improved the SOM content, which might increase the water-holding capacity. Other researchers also suggested that biochar's high inner surface area and porous structure could improve soil pore structure and water storage and thus increase soil water-holding capacity (Verheijen et al., 2009).

4.2 Biochar as nutrient sink

Root growth and plant performance can also be affected by biochar through two mechanisms: (i) as a direct nutrient source and (ii) through impacts on nutrient availability. Biochar can supply micronutrients, for example, boron, molybdenum, K, P, Ca, etc., which are essential elements for the nodulation of Rhizobia. Plant roots respond to soil amendment because biochar provides direct nutrients through the addition of soluble P and biochar-N retention. The increased N retention in the rhizosphere and bulk soils is the indirect effect of biochar on soil nutrients. The bigger rhizosphere proved these direct and indirect interactions in biochar-amended soils; the rhizosphere contained more biochar particles; and biochar-P availability affected the rhizosphere heath growth. Combined, biochar and compost increase nutrients' availability and improve crop productivity (Fischer and Glaser, 2012). Compared to chemical fertilizers, there is a 25% improvement in crop productivity and noted the increased soil organic carbon from 0.93% (by fertilizer) to 1.25% (biochar amended), soil water content from 18% to over 23% by biochar amended and with a significant increase in nodulation number (NN), leaf chlorophyll content, leaf nutrient concentration, etc. (Agegnehu et al., 2015). When applied to the soil, a chemical fertilizer, manure, and compost get rapidly depleted by leaching or degrading to another form. Leaching major plant nutrients such as nitrate-nitrogen (NO3 − N), P, and K potentially leads to environmental pollution. Hence the biochar amendment in combination with fertilizers or compost could be more beneficial. Fischer and Glaser (2012) reported a similar observation where they stated increased peanut yield and an increase in available nitrogen/phosphorus, soil pH, and cation exchange capacity. The positive impact of biochar on soil fertility and the productivity of a wide range of crops are supported by several research types.

Nutrients in soil can be affected by biochar in several ways: (1) as a nutrient sink, thus affecting the flexibility and bioavailability of nutrients; (2) as a source of nutrients for plants and soil microorganisms; and (3) as a soil conditioner, thus altering soil properties that affect the reactions and cycling of nutrients in the soil. Biochar can provide nutrients such as nitrogen (N), phosphorus (P), potassium (K), and other trace elements naturally present in the original feedstock used for biochar production. During the weathering of biochar in soil, most nutrients are released, and they become available for plant uptake. The nature of the feedstock materials and the pyrolysis conditions decides the nutrient content of biochar. Manure- and biosolid feedstock-based biochar contains higher levels of N and P than those derived from wood- and straw-based feedstock materials (Zhao et al., 2018). While through gaseous emission, N content decreases with an increase in pyrolysis temperature due to an increase in ash content the P and K contents. As a nutrient sink, biochar can hold nutrients, thereby dropping their losses through leaching and gaseous emission. Biochar porosity, cation, and anion exchange capacity (surface charge) affect its nutrient retention capacity (Yu et al., 2018). The loss of N, P, and K through leaching and N via nitrous oxide emission is reduced by biochar amendment. However, the pH of the biochar affects the loss of N through ammonia emission; biochar with a slightly acidic or near-neutral pH reduces ammonia volatilization from the soil. Biochar amendment influences several soil properties, including pH, bulk density, cation exchange capacity, water retention, and biological activity. Reactions of nutrients on soil particles and microbial transformation of nutrients are affected by changes in the above soil properties. By increasing the soil nutrient contents and the mobility of nutrients, biochar enhances soil fertility and crop productivity. It improves microbial activity, aeration, and water retention, buffers soil reactions, decreases the bulk density, and maintains soil aggregate structure (Zhang et al., 2020). Biochar reduces nitrification in soil resulting in decreased nitrate leaching.

4.3 Effects of biochar on microbial diversity

The soil in the continuous agricultural operation, forest soil, and soil having vegetations possess diverse useful biota which after death impart organic carbon, nitrogen, and other useful nutrients. Due to the availability of nutrients and water, several fungal and bacterial species grow and multiply, and after their death, they also incorporate other nutrients and growth factor. Such microbes, which decompose the organic fraction of plant residues and help enrich nutrients, are called positive stimulators of soil. Not all soil biotas are worked as decomposers; some with this action show adverse effects on plant growth. They are called negative stimulators or pathogens (Rashid et al., 2016). Some early studies reported biochar is not a good source of energy for microorganisms; however, several studies show biochar amendment stimulates microbial population and activates dormant soil microorganisms, ultimately increasing microbial respiration. In addition, biochar offers a habitat for microorganism and reduce the bioavailability of various soil toxins through adsorption.

In the present section, we emphasized how biochar shows synchronous effects with positive stimulator microbe and how they inhibit pathogens' growth. As per Rashid et al. (2016) study, soil enrichment with organic matter improves soil fertility and organic carbon quality, producing beneficial effects on plant growth. Several researchers documented biochar is a sympathetic organic amendment to improve soil fertility, nutrients, and soil aggregate stability (Zheng et al., 2018). It is due to large and highly negatively charged surface area, enriched recalcitrant carbon, rich porosity, abundant surface oxygen functional groups, and alkaline properties (Lehmann et al., 2011). As per Lehmann et al. (2011), the porous surface of biochar serves as a sanctuary for microbes from predation; also, it improves soil physicochemical properties, which ultimately enhance microbial activity and functions.

Moreover, biochar's rich recalcitrant carbon structure affords competitive living spaces for specific bacterial species such as *Acidobacteria* and *Actinobacteria*, which can degrade complex carbon sources. While studying the stimulatory effects of biochar addition in soil, Ren et al. (2018) documents rapid colonization of fungal species *Zygomycota* and *Basidiomycota* shown on biochar pores due to the higher competitive power of hyphae-colonizing biochar pores. A similar observation was also documented by Zhang et al. (2021) while studying the impact of biochar amendment on the population of *Bacteroidetes* and *Proteobacteria*. As per the study of Rashid et al. (2016), the application of bacterial and fungal inocula on soil induces the fertility of degraded land by a different process. Thus in combination with biochar, the effect of fungi and bacteria increased tremendously (Table 10.3). Some studies documented the negative impact of biochar on the abundance of fungal and bacterial species, such as *Actinobacteria, Chloroflexi, Planctomycetes, Verrucomicrobia, and Nitrospirae* (Zhang et al., 2021). However, others (Wang et al., 2021) documented a significantly increased abundance of bacteria and fungi by 102% and 178%, respectively. This might be due to the increased availability of total organic carbon, total nitrogen, and other nutrients.

Table 10.3 Effects of biochar amendment on different soil type as a soil stabilizer.

S. No.	Amendment (biochar)	Soil type	Effects on soil	References
1.	Walnut shell biochar	Yolo silt loam	• Greater increase in pH • Improved aggregate stability • Increased C storage in macroaggregates	Wang et al. (2017)
		Vina soil	• Significant increase in pH	
2.	Algae aquaculture in whitefish biochar	Yolo soil	• Increased pH • Improved aggregate stability • Increased C storage in macroaggregate	
		Vina soil	• Significant increase in pH • Improved aggregate stability	
3.	Wheat straw biochar (350−550°C)	Red soils	• Biochar amendment with the amount of 40 t ha^{-1} significantly increased the proportion of acroaggregates in the 0−15 cm soil layer	Liu et al. (2014)
4.	Wheat straw and olive tree pruning(370°C)	Slightly acidic loamy sand soil	Decrease in soil bulk density from 1.56 to 1.49 g cm^3, with wheat straw biochar at the highest application rate (i.e., 2.5%)	
5.	Biochar with dairy manure and woodchip	The soil was a loamy soil with 40% sand, 35% silt, and 25% clay	Promotes the formation and stabilization of macroaggregates at the early incubation period	Lei and Zhang (2013)
6.	Grass clippings, cotton trash, and plant prunings (450°C)	Alfisol soil	Reduction in tensile strength and higher field capacity water content	Chan et al. (2007)

Continued

Table 10.3 Effects of biochar amendment on different soil type as a soil stabilizer.—cont'd

S. No.	Amendment (biochar)	Soil type	Effects on soil	References
7.	Biochar derived from mixed hardwoods	Surface (0–15 cm) soil (clarion, fine-loamy, mixed, superactive, mesic Typic Hapludolls)	The specific surface area increased from 130 to 153 $m^2\ g^{-1}$ as the biochar concentration increased from 0 to 20 $g\ kg^{-1}$	Laird et al. (2010)
8.	Corn stover (*Zea mays*) (CS)	Typic Fragiaqualf/ Tokomaru soil (Alfisol)	Total soil pore volumes significantly increased soil water-holding capacity	Laird et al. (2010)

However, archaea show a slight increase after biochar amendments. As per the study of Wang et al., (2021), fungi are the most dominating species and show significant positive stimulation compared to bacterial species, which significantly increase the fungal to bacterial ratio. He further studied the stimulatory effects of biochar amendments on the relative abundance of *Acidobacteria*, which favors soil organic carbon accumulation. Also, there is a significant increase in *Mortierella* and *Westerdykella*, which are more helpful to plant growth and TOC degradation. Biochar amendment negatively impacts the abundance of *Athelia* and *Penicillium*, which are phytopathogens localized in soil and harming plant growth. As per Chen et al. (2013), biochar amendment significantly increased bacterial abundance from 28% to 64%, whereas fungal abundance increased from 35% to 45%. Overall in all the reports, both the species show positive influence by the amendment of biochar. As per the researchers, the main reason behind this is that biochar provides space, nutrients, and food in total nitrogen and total organic carbon. Also, biochar helps to keep away pathogenic competitive bacteria and fungus. Due to a significant increase in the abundance of decomposer fungi, the nutrients were increased by the decomposition of protein, starch, lipids, and other complex organic forms. Though there are lots of positive effects of biochar amendment on microbial abundance and diversity, some other studies have shown the negative effects of biochar on soil quality and fertility, which includes short-term reductions in soil mineral N availability and decreased performance of crops on calcareous soils.

4.4 Effects of biochar on soil enzymes

Soil microorganism is the source of extracellular enzymes responsible for organic matter decomposition and is involved in the cycling of nutrients. There are several enzymes present in the ecosystem of soil such as cellulose, amylase, lipase, protease, acid phosphatase, alkali phosphatase, dehydrogenase, urease, b-D-cellobiosidase, b-glucosidase, b-xylosidase, a-glucosidase, N-acetyl-b-glucosaminidase, leucine aminopeptidase, phosphomonoesterase, sulfatase, oxidoreductases (peroxidase and phenol oxidase), etc. Most of these enzymes are essentially required to decompose complex organic material to a simple form for either energy purposes or metabolic substrates (Burns et al., 2013). Cellulase, hemicellulase, ligninase are essentially required to degrade plant biomass present in the agricultural and forest floor soil as cellulose, hemicelluloses, and lignin are the main components of the plant. These enzymes are referred to as C-cycling enzymes as they metabolize organic carbon for energy purposes.

Some enzymes are nitrogen mobilizers especially proteases. Protein consists of nitrogen in the form of NH_2, which gets released in the soil after the degradation of the protein. Acid and alkali phosphatases are phosphate mobilizers. Inorganic phosphate is present in the cell's nucleic acid, also in the form of ATP, GTP, TTP, and CTP, an energy currency for the cell. After decomposition of the cell, these high-energy phosphate groups get converted into available phosphate by the action of phosphatases. With phosphatases, phosphomonoesterase also works in phosphate cycling. Dehydrogenase is an enzyme mostly useful in the metabolic activities of cells. It removes protons and pairs of an electron from NADH and donates them to a respiratory chain component. The enzyme is not a decomposer, but it is useful in microbial metabolism. Urease is an enzyme that hydrolyzes urea (carbamide) into ammonia and carbon dioxide produced by several bacterial species. Several recent studies show positive impacts of biochar amendment on enzymes activity. Presently most researchers focus their attention on understanding the effects of biochar on the activity of these enzymes. Some studies have described that biochar amendment to soil frequently increases the soil enzyme activities related to N and P cycling; however, some studies documented a reduction in soil enzyme activities involved in C cycling.

Conversely, some studies show inconsistent results, which suggest variable effects of biochar amendment on different soils, enzymes, and assay types Wang et al. (2015). A similar observation was also reported by Song et al. (2019) while studying the effectivity of biochar application on soil, N_2O emission in agricultural soil. They conclude that biochar application decreased soil N_2O emission in the Moso bamboo plantation by decreasing labile N concentrations, N-cycling enzyme activities, and nitrification/denitrification rates. Dominchin et al. (2021) reported biochar addition can influence microbial community structure and enzyme activity in different ecosystems but only for the short term. As per his report, combined effects of biochar with nitrogen fertilizer (e.g., urea) are more significant on enzyme action. Especially unease responds positively when biochar is added with urea. Biochar amendment to the soil increases the activity of glucosidase, glucuronidase, and phosphomonoesterase, being on average 65%, 43%, and 85% significantly higher in biochar 3% than in the control treatment, respectively. It is concluded that there is a strong correlation between soil macronutrients and enzymes such as glucosidase and urease. Also, total phospholipid fatty acid (PLFA) content was not affected by biochar application; however, some PLFA markers suggested an increase in the gram-negative and fungal population. Further, Dominchin et al. (2021) made some conclusions:

1. Effects of biochar amendment on generalist enzymes such as FDA (fluorescein diacetate) and DHA (dehydrogenase) are poor or contradictory.
2. The effects of biochar amendment on specific enzymes such as enzymes involved in C and P cycling are significant.
3. Biochar amendment significantly influences the activities of glucosidase, glucuronidase, and phosphomonoesterase.
4. Urease shows different effects, and it is not related to biochar addition; however, urea addition to soil increases urease activity.
5. Most gram-negative bacteria show positive effects of biochar amendment, whereas gram-positive bacteria did not show any response with biochar addition.

6. Fungal/total PLFA is highest in biochar amended soil.
7. Biochar amendment significantly increases fungal biomass, which in turn increases decomposition activities and C- cycling.

Wang et al. (2015) studied the effects of biochar on the eleven soil enzymes including four C-cycling enzymes (b-D-cellobiosidase, b-glucosidase, b-xylosidase, and a-glucosidase), 1C and N cycling enzyme (N-acetyl-b-glucosaminidase), 2 N cycling enzymes (leucine aminopeptidase and urease), 1 P cycling enzyme (phosphomonoesterase), 1 S cycling enzyme (sulfatase), and 2 oxidoreductases (peroxidase and phenol oxidase). They conclude that soluble organic carbon and total nitrogen increased with increased biochar addition, ultimately increasing soil microbial community. The enzymes associated with C and S cycling show significant activities at 0.5% biochar addition, whereas increasing further biochar concentration reduces their activities. The enzymes involved in nitrogen utilization, for example, L-leucine aminopeptidase and urease, show a significant positive response with biochar amendment as it improves total nitrogen and exchangeable Ca. Similar studies of other researchers also documented stimulatory effects of biochar amendment on N-acetyl-β-glucosaminidase, urease, and β-glucosidase. A group (Huang et al., 2017) studied stimulatory effects of biochar on invertase, urease, and alkaline phosphatise by adding 1%–2% of biochar; they documented further increase in biochar concentration a negative response. β-1,4-glucosidase, invertase, alkaline phosphatase, and some other enzymes show negative responses with high concentration (above 2%) of biochar addition, which is probably due to (i) strong adsorption property of biochar adversely binds the enzyme, which limits its catalytic activity, (ii) it might be detrimental effects on microbial growth and enzyme production, and (iii) high pH might have influenced the metabolic activities due to biochar addition (Gabhane et al., 2020). The effects of biochar amendment on soil enzymes are covered in Table 10.4.

Table 10.4 The effects of biochar amendment on soil enzymes.

S.N	Name of enzyme	Enzyme action	Effects of biochar addition	References
1.	L-leucine aminopeptidase	Involved in N cycling	Increased with the increase of biochar addition rate	Wang et al. (2015)
2.	Urease	Involved in N cycling	Effect is inconsistent and increase after addition of urea	
3.	β-xylosidase	Not involved in S and C cycling	Stimulatory effects of biochar addition	
4.	glucosaminidase	–	Activity increase by biochar amendments	
5.	N-acetyl-β-glucosaminidase	–	Activity increase by biochar amendments	
6.	phosphomonoesterase	Involved in P cycling	Activity increase by biochar amendments	
7.	Sulfatase	1 S cycling enzyme	Increased with the increase of biochar addition rate	Dominchin et al. (2021)

Table 10.4 The effects of biochar amendment on soil enzymes.—cont'd

S.N	Name of enzyme	Enzyme action	Effects of biochar addition	References
8.	Peroxidase and phenol oxidase	Oxidoreductases	Increased with the increase of biochar addition rate	
9.	Alkaline phosphatase	P cycling	Increase activity by increasing biochar concentration	Huang et al. (2017)
10.	Invertase	Sugar inversion	Negative response of biochar addition	
11.	α-1,4-glucosidase	Starch degradation and C cycling	Positive response of biochar addition	
12.	β-Dcellobiohydrolase	Cellulose digestion and C cycling	Positive response of biochar addition	
13.	β-1,4-glucosidase	Cellulose digestion and C cycling	Negative response of biochar addition	
14.	Phosphatase	P cycling	Negative response of biochar addition	

4.5 Effects of biochar amendment on immunity of plants

Soil physicochemical and biological properties, crop productivity and performance, plant health, or rhizosphere microbiology all these properties for biochar have been studied. However, to define biochar as a positive or negative operator on the crop-soil/substrate system, all these individual factors need to be incorporated to approximate the overall impact and realize the fundamental mechanism behind its realistic action. Thus, as discussed in earlier sections, biochar shows a variety of excellent properties which can work in different fields, including agricultural, industrial, and medicinal. In this section, the focus is on the utility of biochar as an immunity booster for plants either by increasing resistance against a pathogen or by suppressing the pathogenicity of the pathogen. Considering the extensive consumer pressure to lessen the use of chemical pesticides in agriculture and the mounting interest in utilizing biochar in soil for soil health booster, its spacious study as a disease suppressant or plant immunizer is essential.

A range of different types of biochar was reported to act to suppress the pathogenicity of fungus infection. Elad et al. (2010) carried detailed study on the pepper and tomato plants to show less susceptibility by the amendment of biochar. *Botrytis cinerea* and *Oidiopsis sicula* are two common pathogenic fungi routinely found in infected pepper and tomato plants that show less susceptibility by amendment of biochar in soil. Additionally, pepper plants were protected against damages from the broad mite pest (*Polyphagotarsonemus latus* [Banks]). One group (Meller Harel et al., 2012) studied the increased resistance of strawberry plant against foliar fungal pathogen *B. cinerea* (casual agent of gray mold), *Podosphaera apahanis, Braun and Takam* (responsible for powdery mildew in strawberry), and *Colletotrichum acutatum* (Simmonds) (pathogens responsible for anthracnose). All these fungi have a different mode of infection, such as *B. cinerea* (necrotrophic), *C. acutatum* (hemibiotrophic), and *P. apahanis* (biotropic). All these fungi show suppression of their pathogenic effects in biochar-amended soil.

A new group (De Tender et al., 2016) studied the effect of biochar amendment on two plants, that is, lettuce and strawberry grew in field soil and white peat, respectively. Both the plants were routinely studied for physiochemical properties of the plant growth for several months. They observed that the effect of biochar amendment in lettuce plants is negligible concerning crop growth, crop health, and rhizosphere microbiology. However, the strawberry plant shows a significant positive response of biochar addition (3%), such as a higher fresh and dry plant weight, a lower susceptibility for the fungal pathogen *B. cinerea* on both leaves and fruits, and changes in the rhizosphere microbiology which is analyzed by PLFA profiling and 16S rDNA amplicon sequencing. Further, disease suppression effects of biochar amendment are specific to plant species and pathogen-specific. Most of the researchers documented a higher proportion of biochar is not useful to boost the immunity of plants against the pathogen. Elad et al. (2010) stated 1%–3% of biochar amendments were generally effective against necrotrophic and biotrophic diseases. In the case of strawberry plants, 1% biochar does not show significant effects against gray mold when the young plant grows in the presence of biochar. Also, it is ineffective when plants grow in peat and are exposed to pathogen *anthracnose*. However, a different result is observed when planting system is coconut fiber: peat, where 1% biochar shows significant plant resistance against *anthracnose*. However, 1% biochar is ineffective against powdery mildew infection despite any peat or planting system. As per Meller Harel et al. (2012), 3% biochar is effective against all types of foliar pathogen.

The researcher has proposed several hypotheses to explain the exact mechanism behind disease suppression by biochar amendment. As per the study of Vallad and Goodman (2004), two pathways are responsible for the disease suppression capacity or boosting plant immunity. First is systemic acquired resistance (SAR), which usually starts with a hypersensitive reaction (HR) leading to local necrosis, involves synthesis of pathogenesis-related (PR) proteins and is mediated by the phytohormone SA. Second is induced systemic resistance (ISR), commonly triggered by plant growth-promoting rhizobacteria (PGPR) and fungi (PGPF), depending on the phytohormones ethylene (ET), jasmonic acid (JA), and methyl jasmonate (MeJA).

The molecular basis of plant disease suppression in biochar amendment has been documented by Meller Harel et al. (2012). As per their studies, biochar amendment induces the expression of five defense-related genes in the noninfected strawberry plant. The induced expression of these genes indicates the direct effect of biochar to enhance strawberry plant resistance. Gene FaPR1 belongs to the PR1 family, a classical marker for the SAR pathway. Faolp2, from the PR-5 family, is one of two strawberry genes known to encode thaumatin-like or osmotin-like proteins (OLPs). Acidic members of the PR-5 family, such as Faolp2, have been postulated to have antifungal properties. Fra a3 is one of three known strawberry genes encoding for fruit allergens and belonging to the PR10 family. PR10 proteins are small intracellular proteins that can exhibit antimicrobial activity in vitro against bacteria, fungi, and viruses, regulated by pathogen infection, various phytohormones, and abiotic stresses (Gomez-Gomez et al., 2011).

5. Summary

World population explosion creates an extensive burden on the demand for food which results in intensive agricultural practices. Repeatedly cropping reduced the fertility of the soil, resulting in low productivity and increased the price of food. Thus, to improve soil fertility and crop productivity, organic amendment such as biochar is very useful. Though the biochar amendment method is highly popular, utilization without its compressive study is like self-destruction, as once we start its field, application reversal cannot happen. Thus in the present chapter, we focused on all such parameters, which are highly essential. Various feedstocks and their availability with their compositional significance is discussed. This is a concern with the cost of biochar as it is the primary requirement of the farmer. This discussion is highly important that one should know the quality and economy of biochar amendment. The impact of biochar on the soil structure and water-holding capacity, and other soil physicochemical properties are also focused. After the application of biochar in soil, it should be synergistically working with soil parameters. The incorporation of biochar improves the physical structure of the soil. Especially it increases porosity, surface area, water adsorption and holding capacity, oxygen uptake, etc. Several mechanisms and theories for soil stabilization also support the application of biochar that could help in the formation and stabilization of soil structure. The applied biochar can be part of the soil aggregates or combined with mineral particles in the soil. Similarly, the physical structure of the soil can be improved by the assimilation of biochar. Thus, biochar amendment is useful for stabilizing the soil and improving physicochemical soil properties. Water is an essential requirement for better crop yield, which can be held in soil in between the pores of the soil. But the basic requirement of pores is they should be small in size. Otherwise, water should get drained out. Small size and a large number of biochar pores hold a good amount of water and are made available for a long period. Also, a large number of pores of biochar become a storehouse for nutrients. The leaching of nutrients is very low in biochar-amended soil. Due to the availability of nutrients and water, several fungal and bacterial species grow and multiply, and after their death, they also incorporate different nutrients and growth factor. Such microbes, which decompose the organic fraction of plant residues and help enrich nutrients, are called positive stimulators of soil. Also, these microorganisms secrete extracellular enzymes, which are useful in decomposing the organic fraction of soil. Cellulose, amylase, ligninase, hemicellulase, urease, phosphatase, protease, and other several enzymes are the true carbon, nitrogen, and phosphate mobilizes. These entire enzymes show good stimulatory action in biochar-amended soil. In the last, the effects of biochar amendment on the immunity of plants are discussed. The utility of biochar as an immunity booster for plants is either by increasing resistance against a pathogen or suppressing the pathogen's pathogenicity. The 3%–5% biochar amendment is useful for suppressing several fungal pathogens.

References

Agegnehu, G., Bass, A.M., Nelson, P.N., Muirhead, B., Wright, G., Bird, M.I., 2015. Biochar and biochar-compost as soil amendments: efects on peanut yield, soil properties and greenhouse gas emissions in tropical North Queensland, Australia. Agric. Ecosyst. Environ. 213, 72–85. https://doi.org/10.1016/J.AGEE.2015.07.027.

Almeida, G., Brito, J.O., Perré, P., 2010. Alterations in energy properties of eucalyptus wood and bark subjected to torrefaction: the potential of mass loss as a synthetic indicator. Bioresour. Technol. 101, 9778–9784. https://doi.org/10.1016/j.biortech.2010.07.026.

Amonette, J.E., Joseph, S., Joseph, S., 2012. Characteristics of Biochar: Microchemical Properties, pp. 65–84.

Arafat Hossain, M., Ganesan, P., Jewaratnam, J., Chinna, K., 2017. Optimization of process parameters for microwave pyrolysis of oil palm fiber (OPF) for hydrogen and biochar production. Energy Convers. Manag. 133, 349–362. https://doi.org/10.1016/j. enconman.2016.10.046.

Asai, H., Samson, B.K., Stephan, H.M., Songyikhangsuthor, K., Homma, K., Kiyono, Y., Inoue, Y., Shiraiwa, T., Horie, T., 2009. Biochar amendment techniques for upland rice production in Northern Laos 1. Soil physical properties, leaf SPAD and grain yield. Field Crop. Res. 111, 81–84.

ASTM, 2010. Laboratory Determination of Water (Moisture) Content of Soil and Rock by Mass. D 2216 - 10ASTM International: West Conshohocken, PA.

Berglunda, L.M., DeLuca, T.H., Zackrisson, O., 2004. Activated carbon amendments to soil alters nitrification rates in Scots pine forests. Soil Biol. Biochem. 36, 2067–2073.

Bird, M.I., Wurster, C.M., de Paula Silva, P.H., Bass, A.M., de Nys, R., 2011. Algal biochar—production and properties. Bioresour. Technol. 102, 1886–1891. https://doi.org/10.1016/J.BIORT ECH.2010.07.106.

Bodor, A., Bounedjoum, N., Feigl, G., Duzs, A., Laczi, K., Szilagyi, A., Rakhely, G., Katalin, P.K., 2021. Exploitation of extracellular organic matter from Micrococcus luteus to enhance ex situ bioremediation of soils polluted with used lubricants. J. Hazard Mater. 417, 125996.

Burns, R.G., DeForest, J.L., Marxsen, J., Sinsabaugh, R.L., Stromberger, M.E., Wallenstein, M.D., Weintraub, M.N., Zoppini, A., 2013. Soil enzymes in a changing environment: current knowledge and future directions. Soil Biol. Biochem. 58, 216–234. https://doi.org/10.1016/J.SOILBIO.2012.11.009.

Cerda, A., Artola, A., Font, X., Barrena, R., Gea, T., Sánchez, A., 2017. Composting of food wastes: status and challenges. Bioresour. Technol. 248, 57–67.

Chan, K.Y., Van Zwieten, L., Meszaros, I., Downie, A., Joseph, S., 2007. Agronomic values of greenwaste biochar as a soil amendment. Soil Res. 45, 629. https://doi.org/10.1071/SR07109.

Chen, J.H., Liu, X.Y., Zheng, J.W., Zhang, B., Lu, H.F., Chi, Z.Z., Pan, G.X., Li, L.Q., Zheng, J.F., Zhang, X.H., Wang, J.F., Yu, X.Y., 2013. Biochar soil amendment increased bacterial but decreased fungal gene abundance with shifts in community structure in a slightly acid rice paddy from Southwest China. Appl. Soil Ecol. 71, 33–34.

Czernik, S., Bridgwater, A.V., 2004. Overview of applications of biomass fast pyrolysis oil. Energy Fuel. 18 (2), 590–592. https://doi.org/10.1021/ef034067u.

Dai, L., Fan, L., Liu, Y., Ruan, R., Wang, Y., Zhou, Y., Zhao, Y., Yu, Z., 2017. Production of bio-oil and biochar from soapstock via microwave-assisted co-catalytic fast pyrolysis. Bioresour. Technol. 225, 1–8. https://doi.org/10.1016/j.biortech.2016.11.017.

De Tender, C.A., Debode, J., Vandecasteele, B., D'Hose, T., Cremelie, P., Haegeman, A., Ruttink, T., Dawyndt, P., Maes, M., 2016. Biological, physicochemical and plant health responses in lettuce and strawberry in soil or peat amended with biochar. Appl. Soil Ecol. 107, 1–12. https://doi.org/10.1016/J.APSOIL.2016.05.001.

Deng, J., Xiong, T., Wang, H., Zheng, A., Wang, Y., 2016. Effects of cellulose, hemicellulose, and lignin on the structure and morphology of porous carbons. ACS Sustain. Chem. Eng. 4, 3750–3756.

Dominchin, M.F., Verdenelli, R.A., Berger, M.G., Aoki, A., Meriles, J.M., 2021. Impact of N-fertilization and peanut shell biochar on soil microbial community structure and enzyme activities in a Typic Haplustoll under different management practices. Eur. J. Soil Biol. 104, 103298.

Dregne, H.E., 2002. Land degradation in the drylands. Arid Land Res. Manag. 16 (2), 99–132.

Elad, Y., David, D.R., Harel, Y.M., Borenshtein, M., Kalifa, H., Ben, H., Silber, A., Graber, E.R., 2010. Induction of systemic resistance in plants by biochar, a soil-applied carbon sequestering agent. Phytopathology 100, 913–921. https://doi.org/10.1094/PHYTO -100-9-0913.

Fischer, D., Glaser, B., 2012. Synergisms between compost and biochar for sustainable soil amelioration. In: Management of Organic Waste. InTech.

Food and Agriculture Organization of the United Nations, 2010. Global Forest Resources Assessment 2010: Main Report. Food and Agriculture Organization of the United Nations 51. Asia-Pacific Forestry Sector Outlook Study II india Forestry Outlook Study by the Ministry of Environment and Forests Government of india Food and Agriculture Organization of the United Nations Regional Office for Asia and the Pacific. 2009.

Gabhane, J.W., Bhange, V.P., Patil, D., Bankar, S.T., Kumar, S., 2020. Recent trends in biochar production methods and its application as a soil health conditioner: a review. SN Appl. Sci. 2, 1307. https://doi.org/10.1007/s42452-020-3121-5.

Gabhane, J., Tripathi, A., Athar, S., William, S.P.M.P., Vaidya, A.N., Wate, S.R., 2016. Assessment of bioenergy potential of agricultural wastes: a case study cum template. J Biofuels Bioenergy 2, 122. https://doi.org/10.5958/2454-8618.2016.00011.0.

Gao, Y., Wang, X., Wang, J., Li, X., Cheng, J., Yang, H., Chen, H., 2013. Effect of residence time on chemical and structural properties of hydrochar obtained by hydrothermal carbonization of water hyacinth. Energy 58, 376–383. https://doi.org/10.1016/j.energy.2013.06.023.

Glaser, B., Lehmann, J., Zech, W., 2002. Ameliorating physical and chemical properties of highly weathered soils in the tropics with charcoal – a review. Biol. Fertil. Soils 35, 219–230.

Gomez-Gomez, L., Rubio-Moraga, A., Ahrazem, O., 2011. Molecular cloning and characterisation of a pathogenesis-related protein CsPR10 from Crocus sativus. Plant Biol. 13, 297–303.

He, M., Xu, Z., Sun, Y., Chan, P.S., Lui, I., Tsang, D.C.W., 2021. Critical impacts of pyrolysis conditions and activation methods on application-oriented production of wood. Waste-derived Bioresource Technology 341 (2021), 125811.

Huang, D., Liu, L., Zeng, G., Xu, P., Huang, C., Deng, L., Wang, R., Wan, J., 2017. The efects of rice straw biochar on indigenous microbial community and enzymes activity in heavy metal-contaminated sediment. Chemosphere 174, 545–553. https://doi.org/10.1016/j.chemosphere.2017.01.130.

Ippolito, J.A., Stromberger, M.E., Lentz, R.D., Dungan, R.S., 2016. Hardwood biochar and manure co-application to a calcareous soil. Chemosphere 142, 84–91. https://doi.org/10.1016/j.chemo sphere.2015.05.039.

Karunanayake, A.G., Todd, O.A., Crowley, M.L., Ricchetti, L.B., Pittman, C.U., Anderson, R., Mlsna, T.E., 2017. Rapid removal of salicylic acid, 4-nitroaniline, benzoic acid and phthalic acid from wastewater using magnetized fast pyrolysis biochar from waste Douglas fir. Chem. Eng. J. 319, 75–88. https://doi.org/10.1016/j.cej.2017.02.116.

Kaza, S., Lisa, Y.C., Bhada-Tata, P., Van Woerden, F., 2018. What a waste 2.0: a global snapshot of solid waste management to 2050.

Lai, W.Y., Lai, C.M., Ke, G.R., Chung, R.-S., Chen, C.T., Cheng, C.-H., Pai, C.W., Chen, S.Y., Chen, C.C., 2013. The effects of woodchip biochar application on crop yield, carbon sequestration and greenhouse gas emissions from soils planted with rice or leaf beet. J. Taiwan Inst. Chem. Eng. 44, 1039–1044. https://doi.org/10.1016/j.jtice.2013.06.028 53.

Laird, D.A., Fleming, P., Davis, D.D., Horton, R., Wang, B.Q., Karlen, D.L., 2010. Impact of biochar amendments on the quality of a typical midwestern agricultural soil. Geoderma 158, 443–449, 2010.

Lawal, A.A., Hassan, M.A., Zakaria, M.R., Yusoff, M.Z.M., Norrahim, M.N.F., Mokhtar, M.N., Shiraj, Y., 2021. Effect of oil palm biomass cellulosic content on nanopore structure and adsorption capacity of biochar. Bioresour. Technol. 332, 125070.

Lehmann, J., Rillig, M.C., Thies, J., Masiello, C.A., Hockaday, W.C., Crowley, D., 2011. Biochar effects on soil biota–a review. Soil Biol. Biochem. 43, 1812–1836. https://doi.org/10.1016/j.soilbio.2011.04.022.

Lei, Q., Zhang, R., 2013. Effects of biochars derived from different feedstocks and pyrolysis temperatures on soil physical and hydraulic properties. J. Soils Sediments 13, 1561–1572.

Liao, N., Li, Q., Zhang, W., Zhou, G., Ma, L., Min, W., Ye, J., Hou, Z., 2016. Effects of biochar on soil microbial community composition and activity in drip-irrigated desert soil. Eur. J. Soil Biol. 72, 27–34. https://doi.org/10.1016/J.EJSOBI.2015.12.008.

Liu, Z., Chen, X., Jing, Y., Li, Q., Zhang, J., Huang, Q., 2014. Ffects of biochar amendment on rapeseed and sweet potato yields and water stable aggregate in upland red soil. Catena 123, 45–51. https://doi.org/10.1007/s11368-013-0738-7E.

MellerHarel, Y., Elad, Y., Rav-David, D., Borenstein, M., Shulchani, R., Lew, B., Graber, E.R., 2012. Biochar mediates systemic response of strawberry to foliar fungal pathogens. Plant Soil 357, 245–257. https://doi.org/10.1007/s11104-012-1129-3.

Mukherjee, A., Lal, R., Mukherjee, A., Lal, R., 2013. Biochar impacts on soil physical properties and greenhouse gas emissions. Agronomy 3, 313–339. https://doi.org/10.3390/agronomy30 20313.

Nandy, B., Sharma, G., Garg, S., Kumari, S., George, T., Sunanda, Y., Sinha, B., 2015. Recovery of consumer waste in India—a mass fow analysis for paper, plastic and glass and the contribution of households and the informal sector. Resour. Conserv. Recycl. 101, 167–181. https://doi.org/10.1016/J.RESCO NREC.2015.05.012.

Nong, H.T.T., Whangchai, K., Unpaprom, Y., Thararux, C., Ramaraj, R., 2020. Development of sustainable approaches for converting the agro-weeds Ludwigiahyssopifolia to biogas production. Biomass Convers. Biorefinery. https://doi.org/10.1007/s13399-020-01083-4.

NPMCR. Available online: http://agricoop.nic.in/sites/default/files/NPMCR_1.pdf. (Accessed 6 March 2019).

Ossai, I.C., Ahmed, A., Hassan, A., Hamid, F.S., 2020. Remediation of soil and water contaminated with petroleum hydrocarbon: a review. Environ. Technol. Innovat. 17, 100526. https://doi.org/10.1016/j.eti.2019.100526.

Rashid, M.I., Mujawar, L.H., Shahzad, T., Almeelbi, T., Ismail, I.M., Oves, M., 2016. Bacteria and fungi can contribute to nutrients bioavailability and aggregate formation in degraded soils. Microbiol. Res. https://doi.org/10.1016/j.micres.2015.11.007.

Ren, C., Zhang, W., Zhong, Z., Han, X., Yang, G., Feng, Y., Ren, G., 2018. Differential responses of soil microbial biomass, diversity, and compositions to altitudinal gradients depend on plant and soil characteristics. Sci. Total Environ. 610–611, 750–758. https://doi.org/10.1016/j.scitotenv.2017.08.110.

Rondon, M.A., Lehmann, J., Ramírez, J., Hurtado, M., 2007. Biological nitrogen fxation by common beans (Phaseolus vulgaris L.) increases with bio-char additions. Biol. Fertil. Soils 43, 699–708. https://doi.org/10.1007/s00374-006-0152-z.

Ros, M., Rodríguez, I., García, C., Hernández, T., 2010. Microbial communities involved in the bioremediation of an aged recalcitrant hydrocarbon polluted soil by using organic amendments. Bioresour. Technol. 101, 6916e6923.

Sahoo, S.S., Vijay, V.K., Chandra, R., Kumar, H., 2021. Production and characterization of biochar produced from slow pyrolysis of pigeon pea stalk and bamboo. Cleaner Engineering and Technology 3, 100101.

Sierra, M.J., Millan, R., Lopez, F.A., Alguacil, F.J., Canada, I., 2016. Sustainable remediation of mercury contaminated soils by thermal desorption. Environ. Sci. Pollut. Res. 23, 4898e4907. https://doi.org/10.1007/s11356-015-568-8.

Singh, S., 2020. Solid Waste Management in Urban India: Imperatives for Improvement. ORF Occasional Paper No. 283. Observer Research Foundation.

Song, Y., Lia, Y., Cai, Y., Fu, S., Luo, Y., Wang, H., Liang, C., Lin, Z., Hua, S., Li, Y., Chang, S.X., 2019. Biochar decreases soil N2O emissions in Moso bamboo plantations through decreasing labile N concentrations, N-cycling enzyme activities and nitrification/denitrification rates. Geoderma 348, 135–145.

Uras, Ü., Carrier, M., Hardie, A.G., Knoetze, J.H., 2012. Physico-chemical characterization of biochars from vacuum pyrolysis of South African agricultural wastes for application as soil amendments. J. Anal. Appl. Pyrolysis 98, 207–213. https://doi.org/10.1016/j.jaap.2012.08.007.

Vallad, G.E., Goodman, R.M., 2004. Systemic acquired resistance and induced systemic resistance in conventional agriculture. Crop Sci. 44, 1920–1934.

Van Der Heijden, M.G., Bardgett, R.D., Van Straalen, N.M., 2008. The unseen majority: soil microbes as drivers of plant diversity and productivity in terrestrial ecosystems. Ecol. Lett. 11 (3), 296–310.

Verheijen, F., Jeffery, S., Bastos, A.C., van der Velde, M., Diafas, I., 2009. Biochar Application to Soils. A Critical Scientific Review of Effects on Soil Properties, Processes and Functions. Office for the Official Publications of the European Communities, Luxemburg.

Villaverde, J., Rubio-Bellido, M., Lara, A., Merchan, F., Morillo, E., 2017. Combined use of microbial consortia isolated from different agricultural soils and cyclodextrin as a bioremediation technique for herbicide contaminated soils. Chemosphere 193, 118–125.

Wang, C., Chen, D., Shen, J., Yuan, Q., Fan, F., Wei, W., Li, Y., Wu, J., 2021. Biochar alters soil microbial communities and potential functions 3–4 years after amendment in a double rice cropping system. Agric. Ecosyst. Environ. 311, 107291.

Wang, D., Fonte, S.J., Parikh, S.J., Six, J., Scow, K.M., 2017. Biochar additions can enhance soil structure and the physical stabilization of C in aggregates. Geoderma 303, 110–117. https://doi.org/10.1016/j.geoderma.2017.05.027.

Wang, X., Songa, D., Liang, G., Zhang, Q., Ai, C., Zhou, W., 2015. Maize biochar addition rate influences soil enzyme activity and microbial community composition in a fluvo-aquic soil. Appl. Soil Ecol. 96, 265–272.

Yang, Q., Sun, Y., Sun, W., Qin, Z., Liu, H., Ma, Y., Wang, X., 2021. Cellulose derived biochar: preparation, characterization, and Benzo[a]pyrene adsorption capacity. Grain & Oil Science and Technology. https://doi.org/10.1016/j.gaost.2021.10.001.

Yu, Z., Chen, L., Pan, S., Li, Y., Kuzyakov, Y., Xu, J., Brookes, P.C., Luo, Y., 2018. Feedstock determines biochar-induced soil priming effects by stimulating the activity of specific microorganisms. Eur. J. Soil Sci. 69, 521–534.

Zhang, H., Wang, S., Zhang, J., Tian, C., Luo, S., 2021. Biochar application enhances microbial interactions in mega-aggregates of farmland black soil. Soil Tillage Res. 213, 105145.

Zhang, Q., Song, Y., Wu, Z., Yan, X., Gunina, A., Kuzyakov, Y., Xiong, Z., 2020. Efects of six-year biochar amendment on soil aggregation, crop growth, and nitrogen and phosphorus use efciencies in a rice-wheat rotation. J. Clean. Prod. https://doi.org/10.1016/j. jclepro.2019.118435.

Zhao, Y.H., Zhao, L., Mei, Y.Y., Li, F.Y., Cao, X.D., 2018. Release of nutrients and heavy metals from biochar-amended soil under environmentally relevant conditions. Environ. Sci. Pollut. Res. 25, 2517–2527. https://doi.org/10.1007/s11356-017-0668-9.

Zhou, Z., Liu, Y., Liu, S., Liu, H., Zeng, G., Tan, X., Yang, C., Ding, Y., Yan, Z., Cai, X., 2017. Sorption performance and mechanisms of arsenic(V) removal by magnetic gelatin-modified biochar. Chem. Eng. J. 314, 223–231. https://doi.org/10.1016/j.cej.2016.12.113.

Zheng, H., Wang, X., Luo, X., Wang, Xing, B., 2018. Biochar-induced negative carbon mineralization priming effects in a coastal wetland soil: roles of soil aggregation and microbial modulation. Sci. Total Environ. 610, pp. 951–960. https://doi.org/10.1016/j.scitotenv.2017.08.166.

Further reading

Bruun, E.W., Ambus, P., Egsgaard, H., Hauggaard-Nielsen, H., 2012. Effects of slow and fast pyrolysis biochar on soil C and N turnover dynamics. Soil Biol. Biochem. 46, 73–79.

Pappu, A., Saxena, M., Asolekar, S.R., 2007. Solid wastes generation in India and their recycling potential in building materials. Build. Environ. 42, 2311–2320. https://doi.org/10.1016/J.BUILD ENV.2006.04.015 21. Mtui GYS (2009) Rec.

Sheng, Y., Zhu, L., 2018. Biochar alters microbial community and carbon sequestration potential across different soil pH. Sci. Total Environ. 622, 1391–1399. https://doi.org/10.1016/j.scitotenv.2017.11.337.

Zhang, L.Y., Jing, Y.M., Xiang, Y.Z., Zhang, R.D., Lu, H.B., 2018. Responses of soil microbial community structure changes and activities to biochar addition: a meta- analysis. Sci. Total Environ. 643, 926–935. https://doi.org/10.1016/j. scitotenv.2018.06.231.

CHAPTER

11 Microbe-assisted heavy metal phytoremediation

Pranoti Belapurkar, Shivani Dubey and Seema Kelotra

Department of Biosciences, Acropolis Institute, Indore, Madhya Pradesh, India

1. Introduction

All living organisms including microorganisms dwell on natural resources like soil, water, and air. Indiscriminate release of heavy metals resulting from different modernization and anthropogenic activities is presenting a great threat not only to these natural resources but also to the food safety and security. The ever-increasing population of humankind adds to the anthropogenic pressure resulting in change in traditional agricultural practices and fast industrialization causing disruption of the natural ecosystems (Sarwar et al., 2017). Globally, around five million sites have been identified with heavy metal/metalloid pollution where their concentration in soil well exceeds the regulatory levels (Li et al., 2019). The major heavy metal pollutants are arsenic, chromium, cadmium, lead, copper, zinc, and mercury (Liu et al., 2013; Henao and Herrera, 2021). These heavy metal pollutants are extremely toxic to living forms even at extremely low concentration (Atobatele and Olutone, 2015). They are nonbiodegradable and hence become an integral part of the food chain and adversely affect human health by acting as carcinogen, mutagen, and teratogen (Ali et al., 2013; Ahemad and Kibret, 2013) as well as affect soil fertility and soil microbial diversity (Huang et al., 2009). Thus a lot of remediation strategies have been developed to relieve the soil from the heavy metal stress and preserve the environment to protect life which includes physical, chemical, and biological technologies. The physicochemical techniques are costly and are with many deficiencies as they produce secondary pollutants which further add to the adversity on soil fertility (Glick, 2010; Henao and Herrera, 2021). Biological techniques on the other hand are sustainable, environment friendly, and benign (Henao and Herrera, 2021). Bioremediation is a technique where microorganisms remove heavy metal pollutants via different strategies, while phytoremediation involves heavy metal pollutant removal using plants (Kumar Mishra, 2017; Doble and Kumar, 2005). Both of these bioremediation methods are economic, eco-friendly, and effective. Major bioremediation techniques include biosorption, biomineralization, and biooxidation (Jin et al., 2018). Phytoremediation involves plant—soil—atmosphere interactions such as phytoextraction, phytostabilization, phytovolatilization, and rhizofiltration. All of these techniques are dependent on soil, water, and climatic conditions making phytoremediation a slow and seasonal process (Laghlimi et al., 2015). Studies have shown that heavy metal-resistant microbe and plant interactions are very effective in heavy metal remediation. This synergism results in very fast, safe, and effective remediation making it an alternative method of choice for heavy metal remediation (Khan and Doty, 2011; Glick, 2014; Ahemad, 2015; Ma et al., 2016; Mishra et al., 2017; Henao and Herrera, 2021).

Integrative Strategies for Bioremediation of Environmental Contaminants, Volume Two. https://doi.org/10.1016/B978-0-443-14013-6.00016-0

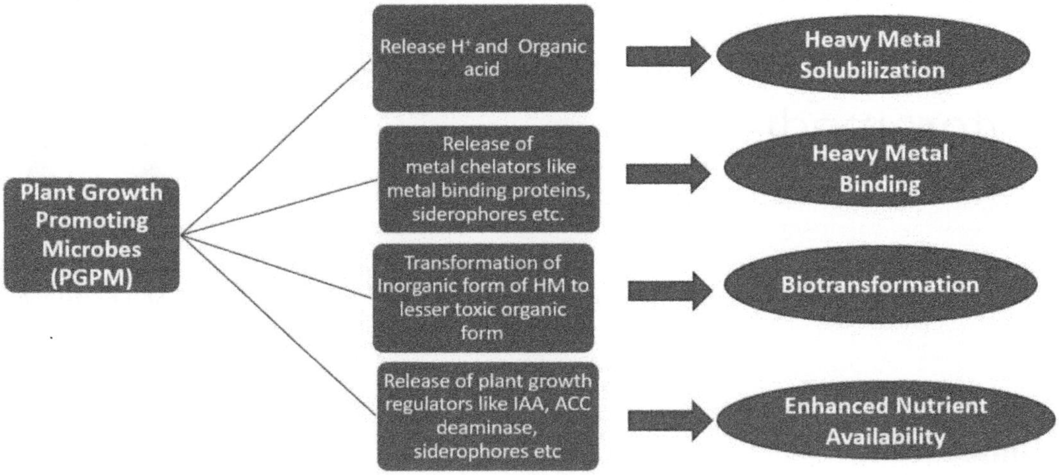

FIGURE 11.1

Mechanisms involved in remediation of heavy metal contaminated soil by plant growth-promoting microbes.

1.1 Microbe-assisted heavy metal remediation by plants

The plant microbe metal interactions affect biogeochemical cycling of heavy metals via processes such as metal immobilization and/or mobilization and their translocation or transformation for enhanced heavy metal bioremediation. Plant growth-promoting microorganisms (PGPMs) develop resistance against heavy metal-stressed environments by activities like heavy metal bioaccumulation, bioleaching, and bioexclusion. They use techniques such as acidification, chelation, and protonation that cause heavy metal mobilization; at the same time, they can perform techniques like precipitation, alkalinization, and complexation leading to immobilization of metals (Bruins et al., 2000; Ma et al., 2016) (Fig. 11.1 and Table 11.1).

Table 11.1 Examples showing microbe-assisted phytoremediation.

Mechanism	Plant	Bacteria	Metal	References
Bioaccumulation	*Chinese milkvetch (Astragalus sincus)*	*Mesorhizobium huakuii* subsp. rengei B3	Cadmium	Glick (2010)
Acidification	*Sedum alfredii*	*Sphingomonas*	Cadmium	Ma et al. (2016)
Chelation	*Thlaspi goesingense*	*Rhizosphere and endophytic bacteria*	Nickel	Glick (2010)
	Solanum nigrum	*Pseudomonas* sp.*LK9*	Cadmium, copper	Ma et al. (2016)

Table 11.1 Examples showing microbe-assisted phytoremediation.—cont'd

Mechanism	Plant	Bacteria	Metal	References
Precipitation	*Chickpea (Cicer arietinum)*	*P. aeruginosa*	Chromium	Ma et al. (2016)
Alkalinization	*Sedum alfredii*	*Glomus caledonium*	Cadmium	Ma et al. (2016)
Metal transformation	*Green chili*	*Cellulosimicrobium cellulans*	Chromium	Ma et al. (2016)
Metal transport distribution	*S. plumbizincicola*	*P. myrsinacearum*	Cadmium, zinc, lead	Ma et al. (2016)

The chapter henceforth focuses on different mechanisms of microbe-assisted heavy metal phytoremediation.

2. Metal detoxification

The important prerequisite for microbe-assisted phytoremediation is heavy metal resistance in plants and microorganisms. Plants exhibit adaptation to heavy metal-stressed environments via different physiological and molecular mechanisms. They show metal resistance through different mechanisms such as binding of heavy metal to plant cell wall, active transport of heavy metal ions into plant cell vacuole, intracellular complexation of heavy metals with the help of peptide ligands such as phytochelatins, metallothioneins, etc., or via sequestration of heavy metals through siderophore complexes that are formed in root cells or soil (Miransari, 2011; Ma et al., 2016). On the other hand, heavy metal resistance in plants results due to root exudates which contain different compounds such as low molecular weight organic acids (LMWOAs) that stimulate microbial growth, solubilize mineral nutrients, and also detoxify heavy metals (Liu et al., 2013; Ma et al., 2016). LMWOAs play a significant role in alleviating heavy metal toxicity as they have the ability to chelate heavy metals and metalloids present in the rhizosphere and prevent their entry inside the cell (Meier et al., 2012; Ma et al., 2016). The mechanisms of microbial metal resistance are summarized below:

2.1 Bioaccumulation

This is a process where heavy metals are accumulated intracellularly. It's a two-step process where the first step is metabolism-independent passive biosorption resulting from processes like adsorption, precipitation, chelation, complexation, and metal ion exchange. Second step involves metabolism-dependent active bioaccumulation where metal ion complexes are transported within microbial cells via processes like carrier mediated ion pumps or even endocytosis (Ma et al., 2016) (Fig. 11.2).

Bioaccumulation of heavy metals is being demonstrated by various microorganisms and results in reduction in plant metal toxicity and uptake. Studies even show that biosorption happens in surface molecules like S layer proteins where metal ions get trapped in living as well as dead cells. While bioaccumulation takes place through helper proteins and participates in providing essential nutrients like sulfur and phosphorus (Suarez and Reyes, 2002; Ma et al., 2016) and/or in enzyme driven metal reduction (Elangovan et al., 2005; Ma et al., 2016).

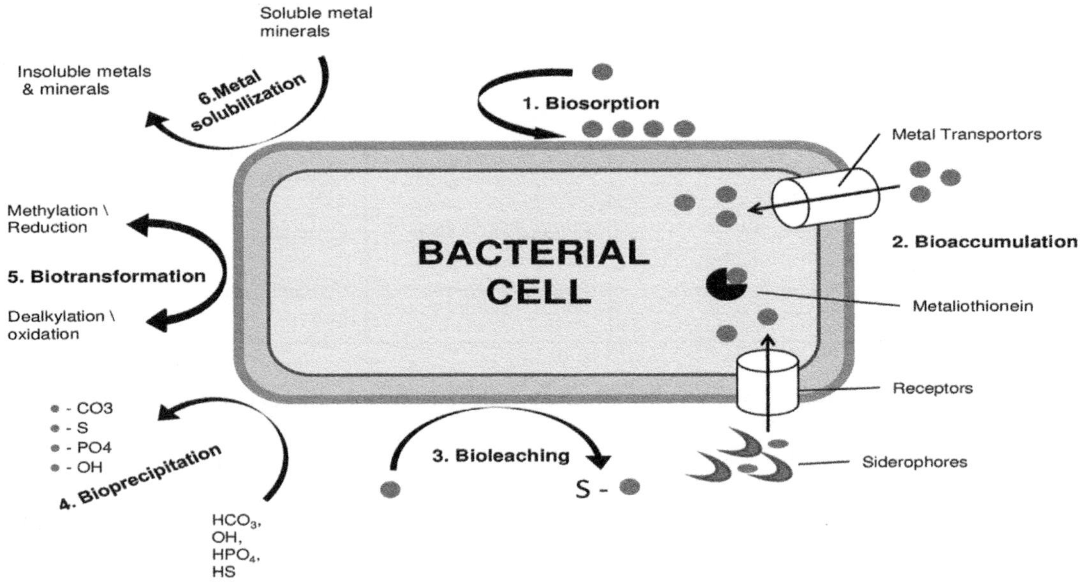

FIGURE 11.2

Various bacterial interactions with heavy metals in metalliferous soil.

2.2 Bioleaching

Different categories of bacteria belonging to mesophiles such as sulfur-oxidizing bacteria like *Acidithiobacillus thiooxidans* and iron-oxidizing bacteria like *Leptospirillum ferrooxidans* (Wong et al., 2004; Ma et al., 2016), thermophilic, thermophiles like archaebacteria, for example, *Sulfobacillus, Ambivalens* (Kletzin, 2006), heterotrophic bacteria like *Acetobacter, Pseudomonas*, etc., and fungi like *Penicillium, Aspergillus*, etc. (Mulligan and Galvez-Cloutier, 2003; Ma et al., 2016). Bioleaching of heavy metals from the environments occurs via metabolic processes like oxidation, reduction, complexation, adsorption, and dissolution to reduce heavy metal phytotoxicity (Pathak et al., 2009; Ma et al., 2016). (Fig. 11.2)

2.3 Bioexclusion

This process involves efflux of toxic heavy metals present inside the microbial cells via active mechanism. Heavy metals like Cd and As gain entry inside the microbial cells via ATPase-linked or non-ATPase nutrient transport systems. However, the efflux mechanism of heavy metals involves ATPase pumps or chemiosmotic ion/proton pumps, which makes them heavy metal resistant (Bruins et al., 2000; Ahemad, 2015; Ma et al., 2016).

3. Metal mobilization

Plants take up those heavy metals which are freely available in the soil to perform phytoextraction. However, heavy metal ions bind to soil particles and become insoluble and unavailable for uptake by plant. Therefore, metal mobility and solubilization in the soil play an important factor for efficient phytoextraction. Microbes play an important role in this regard. They mobilize heavy metals via processes like acidification, protonation, and chelation to modify the habitat of the rhizosphere helping in metal uptake by the plant (Glick, 2010; Sessitsch et al., 2013; Henao and Herrera, 2021). (Fig. 11.2)

3.1 Acidification

Solubility and mobility of metal ions in soil are greatly dependent on pH of the soil which decreases with increasing pH (Richards et al., 2000). The pH of the soil is the resultant of plant and microbial activity. The heavy metal cations adsorbed on the surface of soil particles are replaced by H ions excreted by plant roots causing lowering of pH in the rhizosphere. Therefore, root exudates help in soil heavy metal ion mobility and their bioavailability. Endophytic bacterium *Sphingomonas* SaMR 12 has shown to regulate the release of organic acids in root exudates which in turn enhance Cd bioavailability and uptake by plant (Ma et al., 2016).

3.2 Protonation

In order to replace heavy metal cations at sorption sites, soil microbes export protons to acidify their environments. For instance, *Rhodobacter sphaeroides* possess carboxylic moieties on its surface that help in extracellular biosorption of Ni^{2+} ions forming weak bonds (Giotta et al., 2011; Ma et al., 2016).

3.3 Chelation

Plants and microbes can scavenge heavy metal ions using organic compounds and prevent their resorption (Gadd, 2004). Metal binding peptides, organic acids, siderophores, biosurfactants, and metallophores are known organic chelators (Sessitsch et al., 2013; Ma et al., 2016).

3.3.1 Metal-binding peptides

Metal-binding peptides like metallothionein and phytochelatins play significant roles in metal uptake as well as sequestration of heavy metals and help in alleviating its phytotoxic effects. Phytochelatins are produced from tripeptide glutathione by PC synthase enzyme, whereas MTs are small cysteine-rich metal-binding peptides, which play a significant role in metal detoxification and scavenging of reactive oxygen species (ROS) in animals, plants, as well as microbes. *Tuber melanosporum* is a mycorrhizal fungus that produces polypeptides similar to MTs and PCs synthase and helps in enhancing tolerance against heavy metals like Cu, Zn, Cd, As, and Hg (Ma et al., 2016). (Fig. 11.2).

3.3.2 Organic acids

They are released by both host plant and microbes, participate in different processes taking place in the rhizosphere like nutrient uptake, heavy metal detoxification, and mobilization/immobilization. As discussed earlier, these organic acids, through complexation reactions, make metals in soil available for plant uptake (Ma et al., 2016). (Fig. 11.2).

Pseudomonas sp. Lk9 is an endophytic bacterium, which produces organic acids that participate in iron and phosphorus uptake by *Solanum nigrum* evident by its enhanced shoot biomass and accumulation of heavy metals in aerial parts of the plants (Ma et al., 2016).

3.3.3 Siderophores

Plants require Fe for its growth, which it uptakes from the soil; however, availability of Fe in soil is often in low concentration. Therefore to procure Fe, plant applies three strategies:

- Dicots and monocots solubilize Fe by acidification of the rhizosphere.
- They secrete phytosiderophores which facilitates the uptake of Fe^{3+}.
- They take up Fe^{3+} through microbial siderophores.

Phytosiderophores are secreted into the rhizosphere by potassium-mugenic acid symporter where they solubilize metal ions and cause their chelation facilitating their transport inside the plant cell (Fig. 11.2). Microbial siderophores show great affinity for metals in comparison to phytosiderophores. For example, an endophytic bacterium named *Rahnella* sp. JN 27 secretes siderophores that solubilize cadmium and facilitates its uptake by *Amaranthus hypochondriacus* and *Amaranthus mangostanus*, both of which are cadmium hyperaccumulators (Luo et al., 2014; Yuan et al., 2014; Ma et al., 2016).

3.3.4 Biosurfactants

These are amphiphilic substances which a microbial cell secretes extracellularly. It contains hydrophilic as well as hydrophobic moieties that help in reducing surface as well as interfacial tensions. These surfactants are able to form metal complexes at soil interface making them available to plant due to increased metal insolubility in heavy metal-stressed soil. *Bacillus subtilis*, *Torulopsis bombicola*, and *Pseudomonas aeruginosa* are these days being widely used for remediation of heavy metal-contaminated soil as they produce biosurfactants called surfactin, sophorolipids, and rhamnolipids, respectively (Pacheco et al., 2011; Ma et al., 2016).

3.3.5 Metallophores

Microorganisms produce low molecular weight organic ligands to regulate their intracellular ion concentration in order to keep their toxicity in check and to utilize them for their growth. Nitrogen-fixing bacterium *Azotobacter vinelandii* has shown to produce metallophores called protochelin and azotochelin that bind to metal cations and oxoanions (Fe and Mo) extracellularly and enhance their bioavailability to the bacterium. These two metals Fe and Mo are essential cofactors for working of nitrogenase enzyme (Deicke et al., 2013; Ma et al., 2016).

4. Metal immobilization

Microorganisms can also reduce heavy metal uptake also known as translocation into the plants by making them unavailable for the plant. For this, they use three processes namely precipitation, alkalinization, and complexation.

4.1 Precipitation

Some microorganisms can enzymatically catalyze precipitation of toxic metals like Cr and Se by reduction processes. *P. aeruginosa* OSG 4, a chromium-reducing bacterium, when inoculated into chick pea plants grown in Cr^{6+} contaminated soil, shows decreased Cr uptake in roots, shoots, and seeds, thus helping plants in normal robust growth. This happens because of the ability of bacteria to protect the plant against the inhibitory effect of Cr^{6+} concentration in the soil by reducing it to nontoxic immobile Cr^{3+} (Oves et al., 2013; Ma et al., 2016).

Microbes can also immobilize metal ions by enzymatic reactions or Fe oxidation or by interactions with inorganic acids (hydrogen sulfide, bicarbonate, and phosphate). In fact, sulfate-reducing bacteria are being used to treat S-containing, metal-rich industrial effluents. Similarly some bacteria promote plant growth by immobilizing and solubilizing metals and minerals in soil. For example, phosphate-solubilizing bacteria can reduce Pb availability in Pb-loaded soil by releasing P from insoluble phosphate compounds (Park et al., 2011; Ma et al., 2016).

4.2 Alkalinization

Both fungi (arbuscular mycorrhizal fungi, AMF) and bacteria (sulfur-reducing bacteria and cyanobacteria) show metal-absorbing ability by alkalinization. AMF secrete OH- ions causing alkalinization resulting in active uptake of NO3 by microbes, thus making metal unavailable in the rhizosphere by glomalin (insoluble metal-sorbing glycoprotein) secretion. Further, AMF acts as metal sinks which reduce metal mobility and availability to the plant to promote its growth in heavy metal-stressed soils (Ma et al., 2016).

4.3 Complexation

The microbes associated with plants excrete an exopolysaccharide (EPS) that creates a protective barrier against the adverse effects of metal biosorption (Zhou et al., 2013; Ma et al., 2016). EPS enables metal biosorption via processes such as complexation with negatively charged functional groups, metal ion exchange, and precipitation and protects microbes from metal stress. Similarly, AMF has shown to produce an insoluble metal-sorbing glycoprotein called glomalin responsible for heavy metal sequestration and immobility (Vodnik et al., 2008; Ma et al., 2016).

5. Metal transformation

Studies have shown that microbes perform redox processes to transform heavy metals from toxic state to nontoxic state. Metals like As, Cr, Hg, Fe, Mn, and Se are reduced via microbial redox reactions that cause changes in their speciation and soil mobility resulting in reduced plant toxicity. Commonly associated bacteria belong to the genus *Bacillus* and *Geobacillus* (Ma et al., 2016).

6. Metal transport and distribution

Metal ion translocation in different parts of the plant depends considerably on the plant species and metals. Since uptake is via roots, the root cell walls show high heavy metal concentration. Chelation of metal ions with ligands like amino acids, organic acids, thiols, etc., facilitates its movement from roots to the shoot. Organic acids show metal ion chelation with Cd and cause its translocation in *Brassica juncea.* Histidine shows better uptake of long distance Ni in hyperaccumulator *Alyssum lesbiacum.*

Microbes secrete a variety of organic acids which act as ligands and form heavy metal complexes eventually changing the form and distribution of heavy metals in plants facilitating their phytoremediation (Chiang et al., 2011; Ma et al., 2016).

7. Role of PGPM in plant growth promotion

Apart from the mechanisms discussed above, certain rhizospheric bacteria belonging to genera *Azospirillum, Azotobacter, Agrobacterium, Pseudomonas, Chromobacterium, Micrococcus, Erwinia, Flavobacterium, Caulobacter,* and *Serratia* (Bhattacharyya and Jha, 2012) promote plant growth either by producing phytohormones such as indole-3-acetic acid (IAA) or solubilize phosphate and make it available to plant or produce aminocyclopropane-1-carboxylate deaminase to protect the plant against pathogens and soil contaminants (Paul and Lade, 2014; Ahemad, 2015; Nehra and Choudhary, 2015). These bacteria produce various substances which may increase the phytoremediation process in plants by increasing their growth under heavy metal stress. These include:

7.1 Indole-3-acetic acid production

IAA is an important auxin which regulates several morphological and physiological functions in plants. It stimulates plant growth through extensive root systems, thereby increasing root surface area, resulting in better water and minerals absorption from soil. Rhizospheric bacteria produce this phytohormone and help in enhancing root elongation, and increases surface area, thus making nutrients available to the plants. Plants in turn enhance root exudate production which provides nutrition to rhizospheric bacteria (Spaepen and Vanderleyde, 2011). (Fig. 11.3).

7.2 Phosphate solubilization

Phosphorus is an essential macronutrient required by plants, but in soil mainly it is present in insoluble form making it unavailable to plants. Certain bacteria are capable of converting

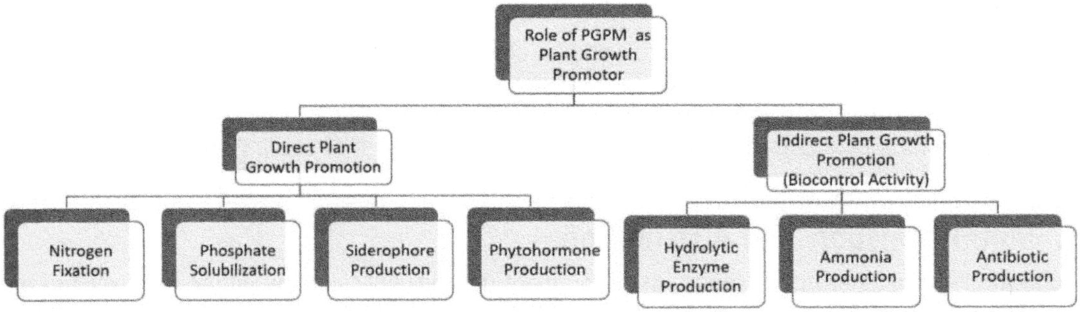

FIGURE 11.3

Role of PGPM as plant growth promoters.

phosphorus to a soluble form for plants and are referred to as phosphate-solubilizing bacteria which include genera *Beijerinckia*, *Azotobacter*, *Burkholderia*, *Microbacterium*, *Erwinia*, *Enterobacter*, *Flavobacterium*, *Rhizobium*, *Bacillus*, *Pseudomonas*, and *Serratia* (Bhattacharyya and Jha, 2012). Many species of these bacterial genera make phosphorus available to the plant by enhancing biological nitrogen fixation under metalliferous soil condition (Ojuederie and Babalola, 2017) (Fig. 11.3).

7.3 Aminocyclopropane-1-carboxylate deaminase production

Many rhizospheric bacteria including phosphate solubilizing bacteria promote plant growth-promoting enzyme such as 1-aminocyclopropane-1-carboxylate (ACC) deaminase, which facilitate plant growth in environmental stress conditions such as salinity, drought, heavy metal stress, etc. (Chandra and Singh, 2016). ACC deaminase catalyzes breakdown of ACC to alpha-ketobutyrate and ammonia and hence decreases the ethylene biosynthesis in plant tissues which otherwise at increased level would impair root growth and development (Glick, 2014) (Fig. 11.3).

8. Conclusion

Heavy metal contamination of the environment needs to be dealt with, to be economical, environment friendly, and sustainable. The plant-microbe synergism proposes to be a promising strategy for relieving the heavy metal stress from the soil.

To summarize, the synergistic interactions between heavy metal resistant microbes and plants are as follows:

- PGPMs associate with the rhizosphere and detoxify heavy metals and enhance their bioavailability.
- PGPMs and root exudates release metal chelators.
- PGPMs perform biotransformation and translocation of heavy metals.
- PGPMs release plant growth regulators.

References

Ahemad, M., 2015. Phosphate-solubilizing bacteria-assisted phytoremediation of metalliferous soils: a review. 3 Biotech 5 (2), 111–121.

Ahemad, M., Kibret, M., 2013. Recent trends in microbial biosorption of heavy metals: a review. Biochem. Mol. Biol. 1, 19–26.

Ali, H., Khan, E., Anwar, S.M., 2013. Phytoremediation of heavy metals concepts and applications. Chemosphere 91, 869–881.

Atobatele, O.E., Olutona, G.O., 2015. Distribution of three non-essential trace metals (cadmium, mercury and lead) in the organs of fish from aiba reservoir, iwo, Nigeria. Toxicol. Rep. 2, 896–903.

Bhattacharyya, P., Jha, D., 2012. Plant growth-promoting rhizobacteria (PGPR): emergence in agriculture. World J. Microbiol. Biotechnol. 28, 1327–1350.

Bruins, M.R., Kapil, S., Oehme, F.W., 2000. Microbial resistance to metals in the environment. Ecotoxicol. Environ. Saf. 45, 198–207.

Chandra, P., Singh, E., 2016. Applications and mechanisms of plant growth-stimulating rhizobacteria. In: Choudhary, D., Varma, A., Tuteja, N. (Eds.), Plant-Microbe Interaction: An Approach to Sustainable Agriculture. Springer, Singapore, pp. 37–62.
Chiang, P.N., Chiu, C.Y., Wang, M.K., Chen, B.T., 2011. Low-molecular weight organic acids exuded by Millet (Setaria italica (L.) Beauv.) roots and their effect on the remediation of cadmium-contaminated soil. Soil Sci. 176, 33–38.
Deicke, M., Bellenger, J.P., Wichard, T., 2013. Direct quantification of bacterial molybdenum and iron metallophores with ultra-high-performance liquid chromatography coupled to time-of-flight mass spectrometry. J. Chromatogr. A 1298, 50–60.
Doble, M., Kumar, A., 2005. Biotreatment of Industrial Effluents. Elsevier Science.
Elangovan, R., Avispa, S., Rohit, B., Ligy, P., Chandraraj, K., 2005. Reduction of Cr(VI) by a Bacillus sp. Biotechnol. Lett. 28, 247–252.
Gadd, G.M., 2004. Microbial influence on metal mobility and application for bioremediation. Geoderma 122, 109–119.
Giotta, L., Mastrogiacomo, D., Italiano, F., Milano, F., Agostiano, A., Nagy, K., et al., 2011. Reversible binding of metal ions onto bacterial layers revealed by protonation-induced ATR-FTIR difference spectroscopy. Langmuir 27, 3762–3773.
Glick, B.R., 2010. Using soil bacteria to facilitate phytoremediation. Biotechnol. Adv. 28, 367–374.
Glick, B.R., 2014. Bacteria with ACC deaminase can promote plant growth and help to feed the world. Microbiol. Res. 169, 30–39.
Henao, G.S., Herrera, G.T., 2021. Heavy metals in soils and the remediation potential of bacteria associated with the plant microbiome. Front. Environ. Sci. 9 (604216), 1–17.
Huang, S., Peng, B., Yang, Z., Chai, L., Zhou, L., 2009. Chromium accumulation, microorganism population and enzyme activities in soils around chromium-containing slag heap of steel alloy factory. Trans Nonferrous Met Soc China 19, 241–248.
Jin, Y., Luan, Y., Ning, Y., Wang, L., 2018. Effects and mechanisms of microbial remediation of heavy metals in soil: a critical review. Appl. Sci. 8 (8), 1336.
Khan, Z., Doty, S., 2011. Endophyte-assisted phytoremediation. Curr. Top. Plant Biol. 12, 97–105.
Kletzin, A., 2006. Metabolism of inorganic sulfur compounds in archea. In: Garret, R.A., Klenk, H.P. (Eds.), Archea: Evolution, Physiology and Molecular Biology. Blackwell Publishing), Oxford, pp. 262–274.
Kumar Mishra, G., 2017. Microbes in heavy metal remediation: a review on current trends and patents. Recent patents Biotechnol. 11 (3), 188–196.
Laghlimi, M., Baghdad, B., Hadi, H.E., Bouabdli, A., 2015. Phytoremediation mechanisms of heavy metal contaminated soils: a review. Oje 05 (08), 375. https://doi.org/10.4236/oje.2015.58031.
Li, C., Zhou, K., Qin, W., Tian, C., Qi, M., Yan, X., et al., 2019. A review on heavy metals contamination in soil: effects, sources, and remediation techniques. Soil Sediment Contam.: Int. J. 28 (4), 380–394.
Liu, M., Huang, B., Bi, X., Ren, Z., Shenga, G., Fu, J., 2013. Heavy metals and organic compounds contamination in soil from an e-waste region in South China. Environ. Sci. Process Impacts 15, 919–929.
Luo, Q., Sun, L., Hu, X., Zhou, R., 2014. The variation of root exudates from the hyperaccumulator Sedum alfredii under cadmium stress: metabonomics analysis. PLoS One 9, e115581.
Ma, Y., Oliveira, R.S., Freitas, H., Zhang, C., 2016. Biochemical and molecular mechanisms of plant-microbe-metal interactions: relevance for phytoremediation. Front. Plant Sci. 7, 918.
Meier, S., Alvear, M., Borie, F., Aguilera, P., Ginocchio, R., Cornejo, P., 2012. Influence of copper on root exudate patterns in some metallophytes and agricultural plants. Ecotoxicol. Environ. Saf. 75, 8–15.
Miransari, M., 2011. Soil microbes and plant fertilization. Appl. Microbiol. Biotechnol. 92, 875–885.
Mishra, J., Singh, R., Arora, N.K., 2017. Alleviation of heavy metal stress in plants and remediation of soil by rhizosphere microorganisms. Front. Microbiol. 8, 1706.

Mulligan, C.N., Galvez-Cloutier, R., 2003. Bioremediation of metal contamination. Environ. Monit. Assess. 84, 45–60.

Nehra, V., Choudhary, M., 2015. A review on plant growth promoting rhizobacteria acting as bioinoculants and their biological approach towards the production of sustainable agriculture. J. Appl. Nat. Sci. 7, 540–556.

Ojuederie, O., Babalola, O., 2017. Microbial and plant-assisted bioremediation of heavy metal polluted environments: a review. Published 1 december 2017 environmental science. Med. Int. J. Environ.

Oves, M., Khan, M.S., Zaidi, A., 2013. Chromium reducing and plant growth promoting novel strain *Pseudomonas aeruginosa* OSG41 enhance chickpea growth in chromium amended soils. Eur. J. Soil Biol. 56, 72–83.

Pacheco, G.J., Ciapina, E.M.P., Gomes, E.B., Pereira Jr., N., 2011. Biosurfactant production by Rhodococcus erythropolis and its application to oil removal. Braz. J. Microbiol. 41, 685–693.

Park, J.H., Bolan, N., Megharaj, M., Naidu, R., 2011. Isolation of phosphate solubilizing bacteria and their potential for lead immobilization in soil. J. Hazard Mater. 185, 829–836.

Pathak, A., Dastidar, M.G., Sreekrishnan, T.R., 2009. Bioleaching of heavymetals from sewage sludge: a review. J. Environ. Manag. 90, 2343–2353.

Paul, D., Lade, H., 2014. Plant-growth-promoting rhizobacteria to improve crop growth in saline soils: a review. Agron. Sustain. Dev. 34, 737–752.

Richards, B.K., Steenhuis, T.S., Peverly, J.H., McBride, M.B., 2000. Effect of sludge-processing mode, soil texture and soil pH on metal mobility in undisturbed soil columns under accelerated loading. Environ. Pollut. 109, 327–346.

Sarwar, N., Imran, M., Shaheen, M.R., Ishaque, W., Kamran, M.A., Matloob, A., et al., 2017. Phytoremediation strategies for soils contaminated with heavy metals: modifications and future perspectives. Chemosphere 171, 710–721.

Sessitsch, A., Kuffner, M., Kidd, P., Vangronsveld, J., Wenzel, W.W., Fallmann, K., et al., 2013. The role of plant-associated bacteria in the mobilization and phytoextraction of trace elements in contaminated soils. Soil Biol. Biochem. 60, 182–194.

Spaepen, S., Vanderleyden, J., 2011. Auxin and plant-microbe interactions. Cold Spring Harbor Perspect. Biol. 3, a001438.

Suarez, P., Reyes, R., 2002. Heavy metal incorporation in bacteria and its environmental significance. Interciencia 27, 160–172.

Vodnik, D., Grcman, H., Ma ˇcek, I., van Elteren, J.T., Kova ˇcevi ˇc, ˇ, M., 2008. The contribution of glomalin-related soil protein to Pb and Zn sequestration in polluted soil. Sci. Total Environ. 392, 130–136.

Wong, J.W.C., Xiang, L., Gu, X.Y., Zhou, L.X., 2004. Bioleaching of heavy metals from anaerobically digested sewage sludge using FeS2 as an energy source. Chemosphere 55, 101–107.

Yuan, M., He, H.D., Xiao, L., Zhong, T., Liu, H., Li, S.B., et al., 2014. Enhancement of Cd phytoextraction by two Amaranthus species with endophytic Rahnella sp. JN27. Chemosphere 103, 99–104.

Zhou, Q., Chen, Y., Yang, M., Li, W., Deng, L., 2013. Enhanced bioremediation of heavy metal from effluent by sulfate-reducing bacteria with copper-iron bimetallic particles support. Bioresour. Technol. 136, 413–417.

CHAPTER

12 Mechanisms involved in the removal of contaminants by biochar from an aqueous medium

Piyush Gupta

Department of Chemistry, Faculty of Engineering and Technology, SRM Institute of Science and Technology, Ghaziabad, Uttar Pradesh, India

1. Introduction

Every day, a big volume of wastewater is generated and discharged by industries like coal and steel, metal surface processing businesses such as iron picking and electroplating, and nonmetallic minerals industry, which has a significant impact on the environment. In order to meet with the sustainable development goals and conserving the ecosystems, precipitation, filtration with coagulation, sedimentation with coagulation, ion exchange, reverse osmosis, ozonation, advanced oxidation processes (AOPs), and membrane processes have all been used to treat industrial wastewater. Majority of these procedures demands large operating and capital expenses as well as good energy supplies. This has been observed as the principal impediment to their use in emerging and developing nations for the abatement/removal of potentially harmful chemicals from contaminated waterways (Ambaye et al., 2020). Biochar is a carbonaceous solid made by pyrolyzing biomass at temperatures below 700°C in the absence or low presence of oxygen. The resultant solid is carbon-rich and has promising adsorption properties, allowing it to remove both organic and inorganic pollutants from wastewater. Various waste materials including, straw, feces, and sludge have been investigated as biochar source materials. Park et al. (2016) shown that biochar can be produced by hydrothermal carbonization, pyrolysis, and gasification. The performance of biochar is mostly influenced by type of feedstock, temperature, residence time, mechanisms involved, nature of contaminant, and modification methods, among other factors. Biochar's adsorption properties have been utilized for waste management, environmental remediation, pollution abatement, water treatment, and produce sustainable biofuels. Many compounds like lignin, cellulose, hemicellulose, and polyphenolic groups are present in waste material like sawdust, which play an important role in attaching organic compounds to biochar adsorbent through diverse mechanisms such as complexation, ion exchange, and hydrogen bonding. These adsorbents were discovered to have a limited surface area and poor porosity. Activation of sludge-based adsorbents with chemicals may result in large surface area (>1800 $m^2\ g^{-1}$). Biochar is an environmentally beneficial absorbent; therefore, studying its physicochemical characteristics and a comprehensive understanding of its adsorption mechanisms are of current interest. Electrostatic interaction, pore filling, ion exchange, and

Integrative Strategies for Bioremediation of Environmental Contaminants, Volume Two. https://doi.org/10.1016/B978-0-443-14013-6.00015-9

precipitation are some of the adsorption mechanisms used by biochar to remove organic and inorganic contaminants from aqueous environments. This is dependent on biochar's physiochemical properties, adsorbent dose, wastewater complexity, ion concentrations, medium/effluent pH, etc. Its economic and environmental benefits are also examined, with a focus on large-scale implementation. Since biochar has numerous benefits, such as availability of a diverse range of raw materials, cheap cost, and recyclability, when used for environmental remediation, it may accomplish the target of converting trash into treasure. This chapter expounds on biochar's adsorption ability and mechanisms involved in the removal of contaminants from aqueous phase. Regeneration studies for biochar are also well described (Ambaye et al., 2021).

2. Adsorption mechanisms involved in the removal of contaminants by biochar from an aqueous medium

Adsorption occurs when the adsorbate becomes linked with the surface of the adsorbed until equilibrium is reached. The adsorption process includes the following steps: (a) physical adsorption, in which the adsorbate settles on the adsorbent's surface; (b) precipitation and complexation, in which the adsorbate deposits on the adsorbent's surface; and (c) pore filling, in which the adsorbate is condensed into the adsorbent's pore. This process happens in three stages, each of which qualifies three zones: the first is the clean zone, in which no adsorption occurs. The second step is known as the mass transfer zone, and it is where the adsorption takes place. The last stage is the tired zone, which is when equilibrium is reached. During the process, the saturated or exhausted zone expands, while the clean zone shrinks. If the adsorbate concentration is raised, the mass transfer zone becomes infected; otherwise, it is not infected. This pattern continues until the adsorbent reaches saturation, at which time the adsorbent is saturated (Ambaye et al., 2021). The elucidation of the underlying processes of the adsorption process is required for assessing the removal effectiveness of pollutants by biochars. The adsorption behavior of biochar for various contaminants (such as heavy metals, organic pollutants, and other pollutants) varies and is well connected with the characteristics of the contaminants. Furthermore, the adsorption process may be affected by biochar's features such as surface functional groups, specific surface area, porous structure, and mineral components.

2.1 Adsorption mechanisms involved in the removal of organic pollutants using biochar

The method of adsorption for eliminating organic pollutants is determined by the type of the contaminants and the chemical characteristics of the adsorbent surface (Rosales et al., 2017). Differences in the organic structure, surface electrical characteristics, and surface functional groups of biochar are significant explanations for the various adsorption methods. The study of biochar's adsorption process for organic pollutants is extremely beneficial to agricultural productivity and engineering applications. The mechanism of adsorption of organic pollutants by biochar may be largely categorized into diffusion, electrostatic interactions, pore filling, $\pi-\pi$ electron−donor−acceptor (EDA) interaction, H-bonding, partitioning, and hydrophobic interactions are the several mechanisms involved in organic pollutant adsorption as per the literature available.

2.1.1 Pore filling

The pore-filling mechanism is determined by the kind of biochar, nature, and the polarity of organic contaminant. This is a process in which organic pollutants accumulate on the surface of biochar, which has micropores (<2 nm) and mesopores (2–50 nm). Smaller molecules of adsorbate fill the micro- and mesopores of adsorbent easily due to strong affinity. In general, for this pore-filling process to be efficient, the biochar must encompass minimal quantity of volatile matter and occurring at low concentrations of organic pollutants. According to Kasozi et al. (2010), biochar made from oak, gamma grass, and loblolly for the sorption of catechol utilized micropore-filling mechanism was more dominant than other sorption mechanisms.

2.1.2 Partitioning

The adsorbate material diffuses into the pores of the noncarbonized component of the biochar during this phase. This part is easily able to interact with the organic adsorbate, resulting in its sorption. The adsorption of organic pollutant chemicals, on the other hand, is dependent on the properties of the noncarbonized biochar (crystalline/amorphous carbon) and the carbonized crystalline and graphene fractions of biochar. When biochar has high volatile matter content and large amounts of organic pollutants, this mechanism is highly visible and efficient (Keiluweit et al., 2010). According to Zhang et al. (2013), biochar made from pyrolyzing swine and dairy dung at 200°C and 350°C exhibited high sorbate partitioning of atrazine pollution utilizing organic carbon (OC) components of biochar. Sun et al. (2011) discovered similar results, demonstrating that organic components of biochar, generated from wood and grass, can improve fluridone and norflurazon adsorption via partitioning.

2.1.3 Electron–donor–acceptor interactions

The EDA interaction mechanism is commonly used in the adsorption of aromatic compounds on graphene-like structured biochar prepared at 1100°C to ensure complete graphitization. However, electron density of biochar to produce deficient or enriched π-electron depends on the pyrolysis temperature of the biochar, that is, if the temperature of the biochar is <500°C, the aromatic system of the biochar acts as an electron acceptor, and if the temperature is >500°C, the biochar acts as a donor (Sun et al., 2014; Zheng et al., 2013). Zheng et al. (2013) attempted to adsorb sulfamethoxazole utilizing reed biochar graphene surface enriched with π-electron. A high adsorption was noticed between biochar's graphene surface and aniline protonated rings of sulfamethoxazole. They also stated that π-EDA interaction on biochar surface between the electron-withdrawing substituent of chlorine and aromatic carbon promoted the adsorption of atrazine.

2.1.4 Hydrophobic interactions

This mechanism can be employed for the adsorption of hydrophobic and neutral organic pollutants. It requires less energy than partitioning. Furthermore, the hydrophobic interaction is the primary pathway for the adsorption of organic contaminants on graphene structured biochar (Ambaye et al., 2021). Li et al. (2018) found that hydrophobic interaction is the main mechanism involved in the adsorption of ionizable organic contaminants such as *o*-chlorobenzene acid, benzoic acid, and *p*-chlorobenzene acid. Chen et al. (2011) examined the sorption of perfluorooctanesulfonate on biochar made from maize straw. Because of the organic pollutant's strong hydrophobicity, elimination occurred via hydrophobic contact. The adsorption of perfluorooctanesulfonate molecules increased as the pyrolytic temperature for biochar preparation was increased. This occurred due to the high pyrolysis temperature causing a reduction in the number of polar groups on the biochar surface. Functional groups (hydroxyl, carboxyl, carbonyl, and amine) promote organic contaminant affinity and adsorption on the surface of biochar. This sort of adsorption mechanism is an EDA type that is based on the unequal distribution of electrons between the organic contaminant and the adsorbent (biochar) functional groups. At this level, it is important to note that the interaction between the adsorbent and organic contaminants with nitro- and chloro-substituents is greatly diminished. This increases the electrostatic interaction between the organic contaminant and the biochar as it is core-lated with high electron acceptor character of the substituent group in the contaminant (Ambaye et al., 2021).

2.1.5 Electrostatic interactions

This is the utmost mode of interaction, which paves the way for the adsorption of ionizable organic compounds to the positively charged surface of the biochar. Its efficiency to attract or repel contaminants is determined by the pH and ionic strength of the aqueous solution (Ahmad et al., 2014; Zheng et al., 2013). Mukherjee et al. (2011) studied the effect of pH on the electrostatic interaction between organic pollutants and biochar surface, which indicated a positive charge when the pH of the aqueous solution was low, but a negative charge when the pH of the aqueous solution was high. This suggests that the pH of the solution/effluent controls the net charge of the biochar's surface, that is, positive at low pH and negative at high pH. The ionic strength of the aqueous solution influences the electrostatic interaction between organic pollutants and biochar. Inyang et al. (2014) demonstrated that increasing the ionic strength of the sorbate solution from 0.01 to 0.1 M NaCl reduced methylene blue adsorption from 4.5 to 3 mg g^{-1} in their experiment on the removal of methylene blue by biochar generated from bagasse composite with carbon nanotubes (CNT). The reason for a decrease in adsorption capacity was attributed to an increase in the sorbent–sorbate repulsive electrostatic interactions.

2.1.6 Diffusion

Organic volatile matters diffuse from higher concentration regions to adsorbent pore structure.

2.1.7 Hydrogen bonding

Electron-rich polar functional groups of biochar form hydrogen bonding with inorganic/organic contaminants having electronegative atoms (Saha and Sengupta, 2021). The nitro, carboxylic acid, and ketonic groups on biochar surface function as electron acceptors, developing $\pi-\pi$ EDA interactions with aromatic compounds and so increased aromatic compounds adsorption. Biochar's various hydroxyl and amine groups can potentially be employed as π-electron-donor sites (Ahmed et al., 2018). Biochar's surface is often negatively charged, resulting in electrostatic attraction between biochar and positively charged organic molecules. The strength of electrostatic attraction is determined by the size of each atomic charge and distance between two atoms (Rosales et al., 2017). Biochar with little surface oxidation is often hydrophobic. This type of biochar reacts with hydrophobic organic molecules via hydrophobic interaction to remove organic pollutants (Inyang and Dickenson, 2015). The investigation of MO adsorption by ferric oxide-biochar nanocomposites discovered that complexation may aid in the removal of MO molecules in a single solute solution (Chaukura et al., 2017). When the surface of the biochar is negatively charged, the positively charged cationic organic molecule is easily electrostatically attracted (Dai et al., 2019).

The numerous processes postulated for the interaction of biochar with organic pollutants are outlined in Fig. 12.1 (Tan et al., 2015). Because of the coexistence of carbonized and noncarbonized fractions, the surface of biochar is heterogeneous, and the carbonized and noncarbonized phases of biochar usually represent separate adsorption mechanisms. The adsorption of organic compounds was determined by both partitioning into noncarbonized organic matter and adsorption onto carbonized matter. Qiu et al. (2009) investigated the mechanism of dye adsorption on a biochar based on straw. They claimed that dye adsorption processes included electrostatic attraction/repulsion, intermolecular hydrogen bonding, and $\pi-\pi$ interactions between graphene layers of biochar and dye molecules (Tan et al., 2015).

The adsorption of positively charged methylene blue dye (MBD) was examined using municipal sewage sludge and tea waste-based biochar. The effect of pH on adsorption revealed that adsorption was controlled by pH; for pH less than 2, the removal effectiveness of MBD was only 80%, but it was about 100% at pH 11. This clearly shows that electrostatic contact had a role in the adsorption process. The numerous H^+ ions filled the restricted bonding sites on biochar and resisted the positively charged MBD molecules at low pH, but when pH increased, the bonded sites were deprotonated and free for the MBD molecules to interact, dramatically boosting the adsorption capacity. The alternative mechanism

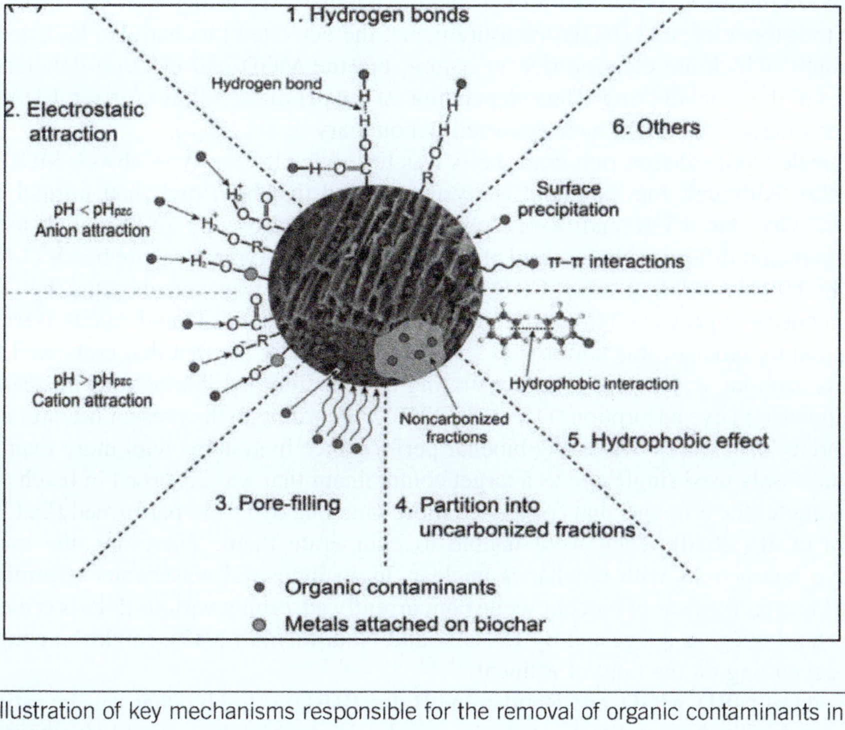

FIG. 12.1

Conceptual illustration of key mechanisms responsible for the removal of organic contaminants in aqueous solutions by biochars.

Reproduced from Tan, X., Liu, Y., Zeng, G., Wang, X., Hu, X., Gu, Y., Yang, Z., 2015. Application of biochar for the removal of pollutants from aqueous solutions. Chemosphere. 125, 70–85, Copyright © 2014, with permission from Elsevier B.V., License Number: 5196490788622.

could involve ion exchange; the concentration of released metal cations (Ca^{2+}, Na^{+}, K^{+}, and Mg^{2+}) by the biochar in the equilibrium solution was measured, and it was discovered that the concentration of these ions, particularly Na^{+} and K^{+} ions, was greatly increased after MBD adsorption, indicating that these ions were involved in the ion exchange process (Fan et al., 2016).

Using biochar generated from cashew nut shells, a study of surface and structural characteristics of biochar and their effects on Q_m (maximum adsorption capacity) of MBD was undertaken. It was discovered that Q_m changed linearly with the volume of pores with diameters ranging from 1.6 to 2 nm, and that adsorption capacity was mostly reliant on these pores. This is due to the fact that the diameter of the pore closely matches to the size of the MBD molecule; consequently, biochar adsorption may be extremely selective in wastewater treatment by tightly managing the volumes and diameter of surface pores. The study also discovered a strong relationship between Q_m and the presence of weakly acidic phenolic groups on the surface of biochar. These groups bind MBD molecules via dipole–dipole interactions or hydrogen bonding, both of which are aided by the phenolic groups indicated above (Spagnoli et al., 2017).

Malachite green dye (MGD) adsorption on activated biochar produced from *Opuntia ficus-indica* with a mesoporous structure revealed improved physical adsorption capacities, indicating that the pore-filling effect was the key mechanism. The adsorption process might have been aided by hydrogen bonding between the MGD functional groups and the biochar. According to X-ray diffraction (XRD) and Fourier transform infrared (FTIR) measurements, the activated biochar also included amine and hydroxyl groups, which are electron-rich in nature, but the MGD had electron-deficient functional groups as part of the aromatic ring. Thus, depending on the pH of the solution, π–π EDA interactions and cation–π interactions might have occurred (Choudhary et al., 2020).

Biochar made from calcium-rich crab shells was tested for its ability to absorb MGD and Congo red dye (CRD). Although zeta potential analysis revealed that both dyes had normal electrostatic interaction activity, but FTIR and X-ray photoelectron spectroscopy (XPS) analysis of biochar revealed a significant difference before and after adsorption, with typical calcite bands (1440, 873, and 710 cm^{-1}) in biochar missing after CRD adsorption, which was corroborated by XPS spectra (decreased intensity of peaks at 348 and 438 eV). This implied that CaO and calcite were involved in CRD adsorption by biochar; this notion was supported further by the fact that crab shell biochar was more effective without acid washing, demonstrating that calcium ion interactions in adsorption had a beneficial influence in dye adsorption (Dai et al., 2018). According to the present literature assessment, there is a scarcity of research evaluating biochar performance in systems with more than one dye. So far, studies have only used single dye as a target contaminant that was adsorbed in batch experiments. Studies utilizing textile effluents that comprised more than one dye were performed (Del Bubba et al., 2020; Santra et al., 2020). They were unable to enumerate them. Therefore, the mechanism of numerous dye interactions with biochar is unclear. In an industrial wastewater treatment scenario, determining the effectiveness of biochar while concurrently adsorbing various dyes is critical. Multiple treatment technologies, as suggested by De Gisi and Notarnicola (2017), might be coupled to treat wastewater depending on the kind of effluent.

The removal of MBD, MGD, and Rhodamine B dye (RBD) using biochar was investigated, and it was discovered that the intramolecular diffusion mechanism, which is involved in adsorption, is rate-limiting. The initial step is quick, and the subsequent steps get slower as adsorption intensity decreases and the number of vacant active sorption sites on the biochar surface decreases. FTIR examination revealed the presence of hydroxyl and carbonyl groups, indicating hydrogen bonding and $\pi-\pi$ stacking adsorption mechanisms (Yao et al., 2020; Srivatsav et al., 2020).

Pesticides such as carbofuran have been removed from aqueous solutions using steam-activated rice husk biochar with excellent adsorption (Mayakaduwa et al., 2017). Because carbofuran has a benzofuranol moiety and a methyl-carbamate moiety, it may form H-bonds between the >NH, $-C{=}O$, >O, phenols, alcohols, and carboxylic acid groups in the biochar (Mayakaduwa et al., 2017). Metribuzin (herbicide) and sulfathiazole (antibiotic) adsorption methods with biochar have also been shown using H-bonding between biochar and the adsorbate (Ahmed et al., 2017; Essandoh et al., 2017). Other important processes for the adsorption of organic pollutants onto biochar include electrostatic attractions and $\pi-\pi$ EDA interactions (Jang and Kan, 2019). As previously stated, the surface of biochar is negatively charged due to the presence of several functional groups. As a result of electrostatic attraction, positively charged organic pollutants can be attracted and retained on the biochar surface. Furthermore, $\pi-\pi$ interaction-mediated adsorption of hydrophobic molecules such as humic acids and phenanthrene was reported onto various biochars (Chu et al., 2017; Jin et al., 2017; Shankar et al., 2017). Complex formation and pore filling are also crucial in biochar-organic contaminant retention. According to Ahmed et al. (2017) and Jin et al. (2017), the retention of phenanthrene and sulfonamide antibiotics is mostly due to pore filling via the diffusion mechanism. Furthermore, using FTIR research, Mayakaduwa et al. (2017) discovered that biochar removes carbofuran by complexation (Palansooriya et al., 2020).

2.2 Adsorption mechanisms involved in the removal of inorganic pollutants using biochar

In the case of heavy metals, the conceivable adsorption processes often entailed the integrative effects of many types of interactions such as physisorption, chemisorption, electrostatic attraction, ion exchange, complexation, surface sorption, and precipitation/coprecipitation. The precise processes of various heavy metals differ, and the suitable characteristics of biochars contribute significantly to heavy metal adsorption. Fig. 12.2 depicts the potential pathways of biochar in respect to inorganic pollutant removal (Tan et al., 2015).

2.2.1 Physisorption/physical adsorption

The electrostatic and van der Waals interactions between the numerous free-moving electrons associated with aromatic functional groups on biochar surface cause physical adsorption, which finally leads to noncovalent interactions with $C{=}C$ aromatic bonds (Amusat et al., 2021).

2.2.2 Chemisorption/chemical adsorption

Many investigations have reported that chemical sorption occurs when functional groups on the surface of biochar, such as carboxylic, phenolic, and hydroxylic groups, exchange negatively charged ions for positively charged cationic pollutants. However, the conditions under which biochar is produced, as well as the types of biomass feedstocks used, influence the material's adsorption capability. For example, a moderate production temperature of 350–400°C would result in a high sorption capacity of biochar, but high temperatures would result in a poor sorption capacity. This impact might be caused by the degradation of surface functional groups, such as oxygen-containing functional groups, which are easily degraded at higher temperatures (Amusat et al., 2021).

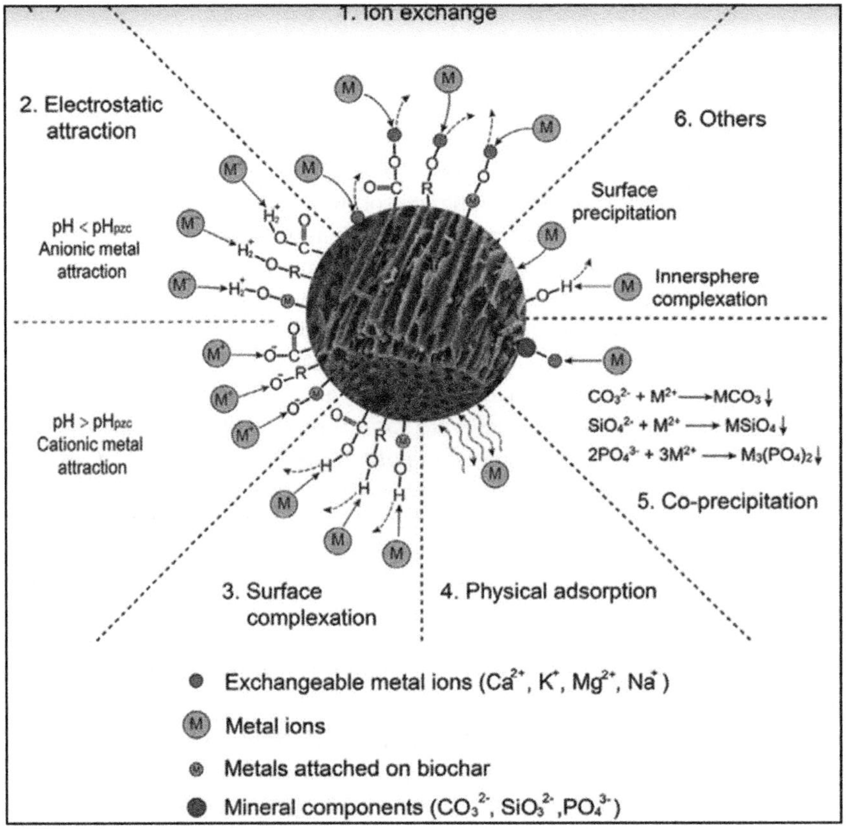

FIG. 12.2

Conceptual illustration of key mechanisms responsible for the removal of inorganic contaminants in aqueous solutions by biochars.

Reprinted from Tan, X., Liu, Y., Zeng, G., Wang, X., Hu, X., Gu, Y., Yang, Z., 2015. Application of biochar for the removal of pollutants from aqueous solutions. Chemosphere. 125, 70–85, Copyright © 2014, with permission from Elsevier B.V., License Number: 5196490788622.

2.2.3 Surface sorption

It is a physical process encompassing chemical bonds formation via metal ion diffusion in the pores of the sorbent. The volume of the pores and the surface area of the sorbent (biochar) are affected by the carbonization temperature. Kumar et al. (2017) investigated uranium adsorption onto pinewood biochar at temperatures ranging from 300 to 700°C. Their findings revealed that biochar generated at high temperatures may entirely remove uranium more effectively than biochar prepared at low temperatures. They ascribed this to the fact that increased carbonization increases the surface area and volume of the pores in the biochar. Wang et al. (2015a) conducted another study to test the adsorption of hazardous metals on biochar made from hickory wood and treated with $KMnO_4$ (Ambaye et al., 2021).

Their findings revealed that the biochar showed 153.1 mg g^{-1} adsorption of Pb, 28.1 mg g^{-1} adsorption of Cd, and 34.2 mg g^{-1} adsorption of Cu. This difference in adsorption might be attributed to the affinity of these metals, which have different valences toward biochar (Ambaye et al., 2021).

2.2.4 Electrostatic interaction

The electrostatic interaction between the charged biochar and metal ions in this process limits the mobilization of potentially hazardous metals (Mukherjee et al., 2011). According to Qiu et al. (2009), the removal of Pb from an aqueous solution using biochar made from rice and wheat results in a high removal of Pb due to the attraction of the positively charged Pb and negatively charged biochar. Furthermore, they claimed that raising the pyrolysis temperature beyond 400°C can improve biochar carbonization and increase the electrostatic interaction of the biochar to adsorb the pollutant. This immobilization process, however, is dependent on the point of zero charge of biochar and solution pH.

2.2.5 Ion exchange

Ion exchange is an effective technique for removing heavy metals from wastewater. It is a reversible chemical reaction in which heavy metals are removed by substituting resin ions for those in the effluent. Ion exchange replaces ions of comparable charge in solids with ions of positive and negative charges (cations and anions) from the solvent (aqueous solution). The binding surface sites of biochar can readily allow cationic pollutant adsorption to be separated via an ion-exchange process. The mechanism of cation adsorption (Cu^{2+} and Ag^{+}) and biochar adsorption in an aqueous solution was examined, and a drop in solution pH was observed throughout the process. This validated the exchange of ions between the biochar's two acidic functional groups and the ions of the other species in the solution throughout the reaction (Amusat et al., 2021).

The primary basis of this process is also the exchange of protons and ionized cations with dissolved salts on the surface of biochar. Its ability to remove heavy metals is determined by the contaminated size and surface functional group of the biochar. According to Ali and his group, the higher the cation exchange capacity of the biochar, the greater the metal adsorption. However, as pyrolysis temperatures rise over 350°C, the cation exchange capacity decreases. El-Shafey (2010) examined the removal of Hg^{2+} and Zn^{2+} from contaminated water using rice husk biochar produced at a pyrolysis temperature of 180°C. They demonstrated that Hg ions adsorption was superior than Zn ions. Trakal et al. (2016) studied the removal of Cd and Pb from biochar made from several feedstocks including wheat straw, graph stalk, grape husk, plum stone, and nutshell. The authors found that feedstocks containing iron oxides had greater Pb and Cd removal efficiencies. The addition of iron in the biochar feedstock was shown to improve the biochar's cation exchange capacity (Ambaye et al., 2021).

2.2.6 Precipitation/coprecipitation

It is one of the primary ways for removing inorganic pollutants from biochar. It entails the production of mineral precipitates in the solution or on the surface of the sorbing material, particularly for biochar, which is formed by the pyrolysis of cellulose and hemicelluloses at temperatures more than 300°C and has an alkaline characteristic. According to Puga and his coresearchers, biochar made from sugarcane and straw dust can improve Cd and Zn precipitation. However, they stated that the efficiency of biochar surface precipitation is dependent on the pyrolysis temperature, suggesting that additional study is required in this area (Ambaye et al., 2021). Precipitation/coprecipitation is a specialized mechanism, considered mainly for the adsorption of metal cations by biochar and its composites. During the adsorption process, it causes the formation of a solid on the surface of the adsorbent (biochar) or in the solvent. When the temperature is approximately 300°C, alkali-modified biochar produced by thermal treatment of hemicellulose and cellulose has a high yield. Such biochar produced under the same heat treatment settings improves the effectiveness of inorganic pollutant precipitation in various forms of insoluble precipitates. The presence of active sites on the surface of biochar or electrostatic interactions between cationic species and negatively charged functional groups on the surface of biochar has been linked to the coprecipitation efficacy of biochars derived from plants and animal feedstocks (Amusat et al., 2021).

2.2.7 Complexation

This metal complexation method comprises the arrangement of multiatom formation via the interaction of certain metal ligands to create complexes. Because of the functional groups that include oxygen in their structure, such as lactonic, phenolic, and carboxyl, biochar formed at low temperatures can bond with heavy metals. This oxygen content can promote the surface oxidation of biochar, resulting in increased metal complexation. When compared with biochar made from animals such as dairy manure and chicken litter, biochar prepared from plant biomass has a high efficiency in the binding of heavy metals such as Cd, Cu, Ni, and Pb to form metal complexes with carboxylic and phenolic functional groups (Zhang et al., 2017). They finally stated that plant-derived biochar has strong surface complexation and ion exchange capacities. More study is needed, however, to identify the generation of biochar metal complexes using advanced spectroscopic methods like as XPS and FTIR (Zhou et al., 2014; Ambaye et al., 2021).

Dong et al. (2011) proposed that sugar beet tailing biochar successfully eliminated Cr(VI) by electrostatic attraction of Cr(VI) through Cr(VI) reduction to Cr(III) and Cr(III) complexation. They divided the adsorption process into three stages: first, the negatively charged Cr(VI) species were migrated to the positively charged surfaces of biochar (at low pH) by electrostatic driving forces; second, Cr(VI) was reduced to Cr(III) by the participation of hydrogen ions and electron donors from biochar; and third, Cr(III) was reduced to Cr(III) by the participation of hydrogen ions and electron donors from biochar. Finally, a portion of the Cr(III) reduced from Cr(VI) was released into aqueous solution, while the remainder of the Cr(III) complexed with the function groups on biochar. According to Lu et al. (2012), functional groups played an important role in Pb adsorption on a sludge-derived biochar, including metal exchange with K^+ and Na^+ due to electrostatic outer-sphere complexation, surface complexation with free carboxyl and hydroxyl functional groups, and inner-sphere complexation with free hydroxyl groups. Similar processes were observed in the adsorption of Hg(II) onto soybean stalk-based biochar (Kong et al., 2011) and the adsorption of Cd(II) on maize straw biochar (Sun et al., 2014). Furthermore, the mineral components in biochar played an important part in the adsorption process. Xu et al. (2013) examined the removal of Cu, Pb, Zn, and Cd from aqueous solutions by rice husk biochar and dairy manure biochar at the same time. The results showed that the removal ability varied with different biochar feedstock sources, and mineral components derived from the feedstock, such as CO_3^{2-} and PO_4^{3-}, play a major role in the adsorption ability of biochar. Another investigation of Zn, Cu, and Cd removal by dairy manure-derived biochar discovered that the biochar formed from dairy manure was enriched with CO_3^{2-} and PO_4^{3-}. These mineral components acted as additional adsorption sites, leading to the biochar's high heavy metal adsorption capacity (Xu et al., 2013). However, according to the published data, the surface area and porous structure of biochar appear to have less of an influence on heavy metal adsorption than oxygen-containing functional groups. According to Ding et al. (2014), oxygen functional groups were most likely responsible for high Pb sorption onto low temperature biochars (250 and 400°C), whereas intraparticle diffusion was mostly responsible for poor Pb sorption onto high temperature biochars (500 and 600°C). Samsuri et al. (2014) employed oil palm biochar and rice husk biochar to adsorb various heavy metals; the findings revealed that the former had a better adsorption capacity for the heavy metals than the latter, implying that surface area was less essential than oxygen-containing functional groups (Tan et al., 2015).

Biochar's abundance of negatively and positively charged sites allows for the sorption of cations and anions from water. Carboxylic and hydroxyl groups on the biochar surface are often implicated in metal immobilization via three important mechanisms: complex formation, electrostatic attraction, and H^+/light metal ion exchange (He et al., 2017a,b; Karunanayake et al., 2017). Pb complexation with oxygen-containing functional groups of biochar and CNTs was detected in several investigations (He et al., 2017a,b; Ifthikar et al., 2017; Inyang et al., 2014). He et al. (2017a,b), on the other hand, proposed that this complexation is followed by H^+ ion exchange with metal cations. The pH of the solution appears to be an important factor in metal sorption. According to Karunanayake et al. (2017), the deprotonation of acidic functional groups on the surface of biochar at high solution pH attracts positively charged metal cations such as Cd^{2+} and Pb^{2+}. Metalloid anions, in addition to metal cations, can interact with the biochar surface. In aqueous environments, metalloids such as As(V) exist mostly as anions (i.e., $HAsO_4{}^{2-}$ and H_2AsO^{4-}) (Niazi et al., 2018a; Shakoor et al., 2019). The production of H-bonds between oxygen in the carbonyl/carboxyl groups of biochar and the hydroxyl groups of $HAsO_4{}^{2-}$ and H_2AsO^{4-} is principally responsible for As(V) adsorption (Niazi et al., 2018a,b). High-temperature biochar may contain organosulfur, sulfide, and calcium species. As(III)−S-like complexes and insoluble Ca−As precipitates are therefore formed, allowing As to be immobilized in water (Niazi et al., 2018a). Modified biochar, as opposed to original biochar, aids in the sorption of metals/metalloids using various mechanisms. Adsorption of As(V) on magnetic biochar was described by Zhang et al. (2016). Through ligand exchange, Fe oxide-loaded biochar was shown to form an inner-sphere surface complex with As(V) and a hydroxy-lated surface of Fe_3O_4 (Zhang et al., 2016). Furthermore, Hu et al. (2015) demonstrated that Fe-impregnation on hickory chips biochar increased As(V) sorption via the chemisorption process. However, as demonstrated by Karunanayake et al. (2017) and Zhou et al. (2017), the pH of the solution plays a crucial role in the removal of As from aqueous media. At low pH (i.e., pH 4), magnetic gelatin-modified biochar demonstrated increased As(V) removal (Zhou et al., 2017). This process was caused primarily by the protonation of numerous biochar functional groups and the pH-dependent charge on iron oxides (Zhou et al., 2017). The synthesis of new iron phosphate oxides, such as $Fe_5(PO_4)_4(OH)_3$, enabled the removal of $PO_4{}^{3-}$ by iron-modified corn straw biochar (Liu et al., 2015). Another potential process for $PO_4{}^{3-}$ adsorption in aqueous environments is nanopore swelling. The presence of more micropores in magnetic biochar than in original biochar resulted in increased $PO_4{}^{3-}$ adsorption owing to nanopore swelling (Chen et al., 2011). According to Chen et al. (2016), the high affinity and selectivity for F^-, the production of aluminum-fluoride complexes (AlF_x), and the valence forces between F^- and the biochar all contributed to F^- adsorption with aluminum hydroxide-coated biochar (Palansooriya et al., 2020).

Choudhary et al. (2020) noticed same mechanism for Cu^{2+} and Ni^{2+} adsorption onto biochar, but at the same timings, interaction of both Cu^{2+} and Ni^{2+} with Ca^{2+} and Mg^{2+} ions present in biochar via ion exchange was also proposed. Bandara et al. (2020) demonstrated the adsorption mechanism of Cu(II) and Cd(II) on biochar depends on forming carbonate and phosphate precipitation from electrostatic interaction, interaction between metal and CO_3^{2-}, PO_4^{3-}, and π−π surface interaction between negatively charged surface functional groups (-O⁻, -COO⁻) of biochar and positively charged heavy metals (Cd^{2+}, Cu^{2+}). Younis et al. (2020) concluded that the adsorption of Ba(II) and Sr(II) on biochar occurred due to Lewis acid−base interaction between oxygen functional groups (−OH, −COOH, −CHO, and Si−OH) (Lewis base) of biochar and metal (Lewis acid) as well as ion exchange between biochar's Na^+ ions and Ba(II) +Sr(II). Huang et al. (2020) revealed that the Cd^{2+} adsorption on biochar took place through ion exchange with K^+, Ca^{2+}, Na^+, and π−π interaction with the aromatic rings (C-H, C=C) of biochar. Ali et al. (2020) reported that As(III) in form of negatively charged ions $HAsO_4^{2-}$, $AsO4^{2-}$ in aqueous solution interacted with the positively charged K^+, Ca^{2+}, Na^+ metal ions and surface functional groups such as lactone, carboxyl, hydroxyl, or aromatic ring (O=C−O−, C=C, C−H, −OH, −COOH) of biochar through electrostatic, π−π and ion exchange mechanism. Zhang et al. (2019b) reported that Pb^{2+} ions were adsorbed on biochar surface through ion exchange with Ca^{2+}, Mg^{2+}, and π−π surface complexation with hydroxyl, aromatic groups of adsorbent. Another study showcased the cause of better adsorption of Pb and Cu and attributed to the formation of precipitation of $PbCO_3$ and CuO through coprecipitation mechanism with abandoned hydroxyl and carbonate groups of biochar (Zhang et al., 2019a). The mechanism of uranium (U(VI)) adsorption on biochar was explained by Mishra et al. (2017), and they observed that π−π interaction between surface functional groups such as hydroxyl, carboxyl groups of adsorbent, and uranium was the pathway of good adsorption of radioactive element. Ho et al. (2017) found that due to the presence of negatively charged surface functional groups (−OH, −COOH) on biochar surface, Pb^{2+} was attracted through electrostatic interaction. The adsorption of Cr(VI) on biochar was possible due to π−π interaction between metal (Cr) and carboxyl, hydroxyl groups, ion exchange with Ca^{2+}, Mg^{2+}, and formation of chromium phosphate/carbonate precipitation through interaction between Cr(VI) and CO_3^{2-}/PO_4^{3-} (Shakya and Agarwal, 2019). Jang et al. (2018) reported strontium (Sr^{2+}) ion removal due to interacting with negatively charged carbonyl, carboxylate groups of biochar through electrostatic interaction (Saha and Sengupta, 2021).

2.3 Adsorption mechanisms involved in the removal of microorganisms/pathogens using biochar

Most prior investigations found that biochar, particularly used in biofilters, effectively removed *Escherichia coli* from water. Unlike batch trials, column filters outperformed batch experiments in eliminating microbiological pollutants from both synthetic and natural waters. In a bioretention system, forestry wood waste biochar generated at 700°C was able to eliminate *E. coli* by 96.6%. Nonetheless, when these biochars were modified with H_2SO_4 and H_3PO_4, the removal % was substantially higher than with the original biochar (Lau et al., 2017).

A mechanism concept for microbial/biological retentions is illustrated in Fig. 12.3 (Gwenzi et al. 2017). Several researches have employed biochar-augmented sand filters to remove microbial pollutants from water. When utilizing biochar for water filters, it is necessary to examine the combined impacts of biochar and sand. Although there has been less study on the removal of pathogens/pathogenic indicators in aqueous systems by biochar as compared to organic and inorganic pollutants, information on the removal of microbial contaminants from soil/soil solution is widely available in the literature. Based on the literature, it may be supposed that the removal of microbial pollutants from aqueous solutions using biochar could be proportional to biochar-amended soils. As a result,

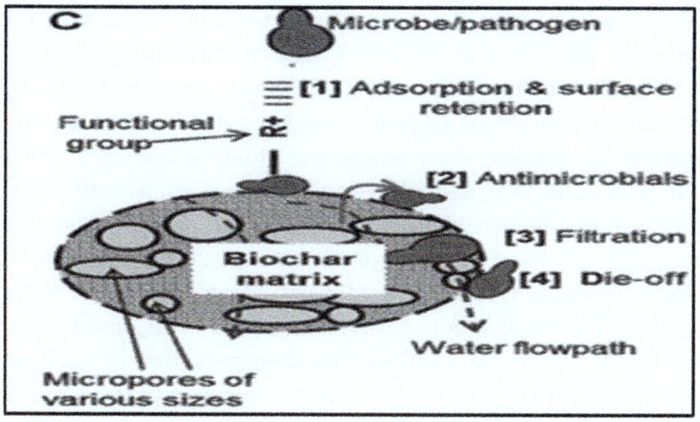

FIG. 12.3

Conceptual illustration of key mechanisms responsible for the removal of pathogens/microbial contaminants in aqueous solutions by biochars.

Adopted from Gwenzi et al. (2017), Copyright © 2017, with permission from Elsevier B.V., License Number: 5196510352111.

pathogenic clearance and die-off in biochar-augmented sand filters may be analogous to what occurs in biochar-amended soils. Several studies have documented the use of a biochar-augmented sand biofilter to remove *E. coli* and *Enterococci* under variable conditions. Due to improved overall performance of the sand filter reinforced with biochar, the combination of sand and biochar has demonstrated considerable *E. coli* removal (Mohanty and Boehm, 2015). It has been observed that biofilters may remove both *E. coli* and *Enterococci* from natural storm water (Afrooz and Boehm, 2017). These researchers observed that *E. coli* was more efficiently removed than *Enterococci*. It was therefore indicated that the microorganism type and the surface features of biochar were responsible for microbial retention capacity on biochar surfaces (Afrooz and Boehm, 2017). Reddy et al. (2014) discovered that waste wood biochar had a poor removal effectiveness of *E. coli* in a column filter system. This was mostly owing to the low input concentration and preexisting microbial levels (Reddy et al., 2014). Besides *E. coli*, water sources can also include extracellular DNA, RNA, and resistance genes (for example, antibiotic resistance genes) which may affect human health (Bergeron et al., 2017). Wang et al. (2014) studied that high-temperature-produced HCl-treated willow wood biochar removes more DNA than low-temperature-produced biochar. With an adsorption capacity of >600 mg g^{-1}, zerovalent iron (ZVI) modified bamboo biochar was able to extract Ag from an aqueous solution fast and efficiently (Zhou et al., 2014). More intriguingly, the silver nanoparticles-ZVI biochar composite fully prevented *E. coli* growth, whereas the original ZVI biochar (without silver nanoparticles) stimulated it. This implies that the silver nanoparticles-ZVI biochar composite showed high antimicrobial properties and might be beneficial as a value-added antibacterial/antifungal agent in water treatment operations (Zhou et al., 2014). Overall, the primary parameters examined in biochar-based water treatment for microbial pollutants removal are the type of microbial contaminant, biochar characteristics, quantity of biochar, microbial levels/concentrations in the water, and the methods employed. However, researches on harmful microorganism elimination approaches based on biochar are limited. More research is needed to determine the efficacy of biochar to reduce pathogens/pathogenic indicators including *Cryptosporidium, Legionella, Giardia lamblia, Coliphage, Enterococci,* parasites (e.g., helminth eggs), spores of sporulent microbiota, viral components, and total coliforms from water (Palansooriya et al., 2020).

The presence of lipid-containing substances on the membranes of some bacterium species, such as phospholipids and lipopolysaccharides, causes these organisms to be hydrophobic (Wu et al., 2014;

Dowhan et al., 2017). As a result, hydrophobic attraction at the surface is a frequent mechanism for *E. coli* deposition on biochar (Afrooz and Boehm, 2017; Lau et al., 2017). Lau et al. (2017) discovered that H_2SO_4, H_3PO_4, and KOH-modified biochar removed more *E. coli* than the amino-modified biochar. The results showed that introducing a substantial quantity of O-containing functional groups during amino-modification enhanced the hydrophilicity of biochar. As a result, the amino-modified biochar eliminated less hydrophobic bacteria than the H_2SO_4, H_3PO_4, and KOH-modified biochar, as well as the original biochar (Lau et al., 2017). Phosphate groups on bacterial cell walls generate negative electrostatic cell surfaces. Gram-negative bacteria (*E. coli*) have a net negative surface charge due to the presence of lipopolysaccharides, for example, phosphoryl and 2-keto-3-deoxyoctonate carboxylate groups in the outer membrane (Wu et al., 2014). Teichuronic acid residues, teichoic acids, and unsubstituted carboxylate groups on gram-positive bacteria promote a net negative charge on the cell surface. Chen and Walker (2012) discovered that *E. coli* had a slightly positive charge because of the varying ionic strengths and pH of the solution. Afrooz and Boehm (2017) discovered that *E. coli* has a smaller negative charge than *Enterococci*. Clearly, microorganisms species and solution chemistry are the determinants of electrostatic cell surface and charge density. In context of bacteria's negative or positive electrostatic cell surfaces, electrostatic deposition is another plausible method for adhering bacteria to charged sites on biochar surfaces (Afrooz and Boehm, 2017; Chen and Walker, 2012). According to Derjaguin-Landau-Verwey-Overbeek (DLVO) theory, bacterial adherence to biochar surfaces might be the result of combined attractive van der Waals and repulsive electrostatic interactions (Lau et al., 2017). Furthermore, gram-negative bacteria like *E. coli* may form bonds between the macromolecules on the cell surface and the biochar surface via "steric interaction" or "polymer bridging" under low ionic strength conditions (Afrooz and Boehm, 2017; Chen and Walker, 2012). Wang et al. (2014) described the adsorption of microbial compounds, such as DNA, onto willow wood biochar in addition to the microorganisms. DNA molecules were having negatively charged surfaces due to the presence of phosphate groups, van der Waals forces, hydrophobic, and electrostatic interaction, which were the common mechanisms for the retention of DNA on biochar surfaces due to different solution chemistry and biochar properties (Wang et al., 2014; Palansooriya et al., 2020).

A summary of mechanisms involved in organic, inorganic, and microbial contaminants is compiled in Table 12.1 (Saha and Sengupta, 2021).

3. Regeneration studies

A competitive adsorbent should be reusable and recyclable for industrial applications. Biochar regeneration is an inverted adsorption process which embraces two principles: (i) adsorbate desorption and (ii) adsorbate decomposition. It must follow the low-cost sorption-desorption cycles. After three hours of grinding the superfine magnetic charcoal adsorbent, the tetracycline itself deteriorated more than 97%, while the carbamazepine degradation rate reached 98.4% at a dose of 0.3 $g\ g^{-1}$ (Shan et al., 2016). A schematic illustration of applications of biomass-based biochar, desorption/regeneration of biochars, and management of spent biochar has been given in Fig. 12.4 (Tan et al., 2015).

There are no noticeable band changes in FTIR spectra of magnetic biochar made with eucalyptus leaf residue after five to six regeneration cycles. The regenerated adsorbents retained their good magnetic separation and adsorption properties. Regenerated magnetic biochars, on the other hand, had increased S_{BET} and pore volume (Wang et al., 2015b). Biochar regeneration and cost reduction may be achieved by

Table 12.1 Summary of mechanisms involved in adsorption of organic and inorganic contaminants (Saha and Sengupta, 2021).

S. No.	Target contaminant	Adsorbent	Adsorption mechanism
1.	Pb^{2+}, Cd^{2+}	Mn–Zn–Fe/BC	Electrostatic interaction, ion exchange, and π–π surface complex formation
2.	Pb(II)	H_3PO_4/BC	π–π surface complexation
3.	Cd^{2+}	NaOH/BC	Electrostatic interaction
4.	Cr(VI)	HNO_3/BC	π–π surface complexation
5.	U(VI)	HNO_3/BC	π–π surface complexation
6.	Pb(II)	H_2SO_4/BC	π–π surface complex and electrostatic interaction
7.	Sb(V)	Mn/BC	π–π surface complexation
8.	Hg(II)	Sulfur/BC	Lewis soft acid–base interaction
9.	Zn^{2+}	Sulfur/BC	Electrostatic interaction
10.	Cu(II)	Mn–Al–LDH/BC	Isomorphic substitution
11.	Cr(VI)	Fe_3O_4–SiO_2–NH_2/BC	Electrostatic interaction and Cr(III) forms chelation with amine group
12.	As(V)	Red mud/BC	Electrostatic interaction
13.	Cd^{2+}, Pb^{2+}	Soil/BC	Electrostatic interaction
14.	Pb^{2+}	$KMnO_4$/BC	Ion exchange and coprecipitation
15.	U(VI)	$KMnO_4$/BC	Lewis acid–base interaction
16.	As(III)	Fe–Mn–La/BC	Electrostatic interaction and π–π surface complexation
17.	Cd(II), As(V)	TiO_2/BC	Ion exchange
18.	Cu(II)	NH_2/BC	π–π surface complexation
19.	Ni(II)	$KMnO_4$–KOH/BC	Ion exchange mechanism
20.	Cr(VI)	N–Fe/BC	Electrostatic interaction
21.	Cr(VI)	S–Fe/BC	Lewis acid–base interaction and electrostatic interaction

Continued

Table 12.1 Summary of mechanisms involved in adsorption of organic and inorganic contaminants (Saha and Sengupta, 2021).—cont'd

S. No.	Target contaminant	Adsorbent	Adsorption mechanism
22.	As(V)	Al/BC	Electrostatic interaction
23.	Cr(VI)	ZnO/BC	Coprecipitation mechanism
24.	Pb^{2+}, CrO_4^{2-}	Mg−Al−LDH/BC	Ion exchange mechanism and electrostatic interaction
25.	Phosphate	Fe/BC	Electrostatic interaction, ion exchange, and diffusion
26.	Caffeine	BC	Hydrogen bonding and $\pi-\pi$ surface complexation
27.	Naproxen	BC	Pore filling and $\pi-\pi$ interaction
28.	Doxycycline hydrochloride	Cu/BC	Electrostatic interaction and $\pi-\pi$ surface complexation
29.	Tetracycline	BC	$\pi-\pi$ interaction, hydrophobic interaction, and hydrogen bonding
30.	Sulfamethoxazole Sulfapyridine	BC	$\pi-\pi$ interaction, hydrogen bonding, and hydrophobic interaction
31.	Diclofenac	Fe/BC	Hydrophobic interaction, $\pi-\pi$ interaction, and hydrogen bonding
32.	Furazolidone	Fe/BC	Hydrogen bonding
33.	Ciprofloxacin	BC and montmorillonite composite	$\pi-\pi$ interaction and electrostatic interaction
34.	Levofloxacin Amoxicillin	BC	$\pi-\pi$ interaction
35.	Atrazine	BC	Pore filling and $\pi-\pi$ interaction
36.	*E. coli*	Sulfuric acid-treated BC as a filter	Surface attachment
37.	a. Methyl orange b. Eriochrome black T c. Methylene blue d. Crystal violet	BC	Electrostatic, ion exchange, and $\pi-\pi$ interaction
	e. Congo red f. Malachite green	BC	Pore-filling mechanism, electrostatic attraction, hydrogen bonding, and $\pi-\pi$ interaction

Table 12.1 Summary of mechanisms involved in adsorption of organic and inorganic contaminants (Saha and Sengupta, 2021).—cont'd

S. No.	Target contaminant	Adsorbent	Adsorption mechanism
38.	Phosphate	La/BC	Electrostatic and π−π interaction
39.	Phosphate	Polyethyleneimine/BC	Electrostatic interaction
40.	Phosphate	Ca/BC	Coprecipitation
41.	Phosphate	Dolomite/BC	π−π interaction and electrostatic interaction
42.	Phosphate	Zn−Al−OH/BC	Hydrogen bonding, electrostatic interaction, and coprecipitation
43.	Bisphenol A	Fe/BC	Pore filling, hydrophobic interaction, π−π interaction, and hydrogen bonding
44.	Phenol, 2−4 dinitrophenol	BC	π−π interaction and hydrogen bonding
45.	NH_4^+	BC	Electrostatic interaction
46.	Nitrate	BC and montmorillonite composite	Electrostatic interaction

using different methods to handle different biochars. The desorption method for recovering biochar is critical in determining their viability for reuse. Different methods (complexation, precipitation, or ion exchange) can be used to recover metal from a solution. The inorganic chemicals employed in regeneration are inexpensive, the regeneration conditions are gentle, but the treatment duration is lengthy, and the regeneration effectiveness is often less than 80% (Salvador et al., 2015). Thermal regeneration is frequently employed in industrial and wastewater treatment plants, and it is generally regarded as the most cost-effective technology (Roman et al., 2013). The adsorption of extremely volatile substances is well suited to supercritical regeneration. Supercritical regeneration employing water and CO_2 as fluids, as compared to organic solvent regeneration, disposes of solvent pollution problems and risks toward the environment (Salvador et al., 2015). Microwave irradiation piques the interest of researchers due to its shorter regeneration time and easier control of heating temperature (Foo and Hameed, 2012). It is also effective in energy saving and lessening danger of secondary contamination.

Fig. 12.5 indicates the different methods utilized for the regeneration of exhausted biochar adsorbent (Fdez-Sanroman et al., 2020). Some regeneration methods reviewed by Dai et al. (2019) are discussed below.

3.1 Solvent regeneration

The equilibrium relationship between biochar, solvent, and adsorbate is used in solvent regeneration to break the adsorption equilibrium by altering the temperature and pH value of the solvent to desorb the adsorbate from activated carbon. Solvent regeneration is a suitable approach to consider for high

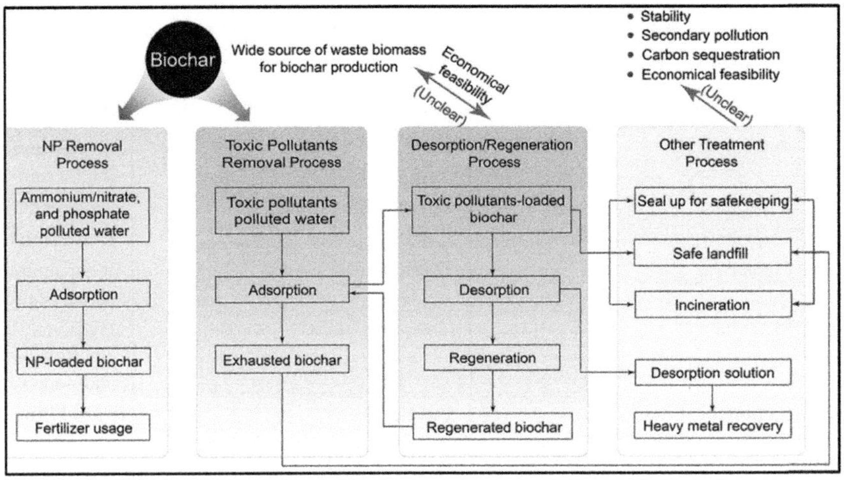

FIG. 12.4

Schematic representation of applications of biomass-based biochar, desorption/regeneration of biochars, and management of spent biochar.

Reproduced from Tan, X., Liu, Y., Zeng, G., Wang, X., Hu, X., Gu, Y., Yang, Z., 2015. Application of biochar for the removal of pollutants from aqueous solutions. Chemosphere. 125, 70–85, Copyright © 2014, with permission from Elsevier B.V., License Number: 5196490788622.

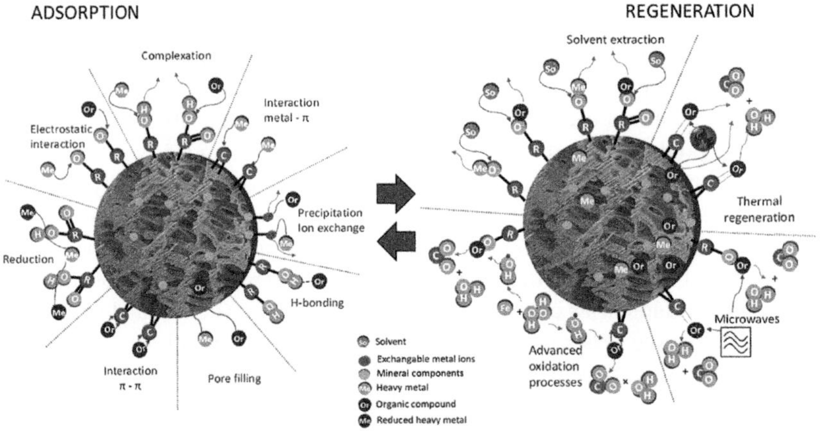

FIG. 12.5

Regeneration of exhausted biochar adsorbent (Fdez-Sanroman et al., 2020).

concentration and low boiling point organic matter adsorbents. It refers to the removal of adsorbate using inorganic acids like HCl, H_2SO_4, HNO_3 or alkalis such as NaOH, KOH, etc., which is also known as acid-base regeneration approach. For example, biochar adsorbed with a high concentration of phenol is washed with sodium hydroxide solution, and the desorbed phenol is recovered in the form of sodium phenolate (Hu, 2018). Alkalis like NaOH, $NaHCO_3$, and NaH_2PO_4 have been used to desorb As from Fe-impregnated biochar. Furthermore, 0.1 M NaOH could desorb 72% of the originally adsorbed As(V) on biochar, degraded from 72% (first cycle) to 60% (third cycle). Similarly, after five recycling cycles, Fe_3O_4-orange peel can be recycled using 0.1 M HNO_3 with a desorption capacity of 98.5% (Fdez-Sanroman et al., 2020). Several researches have been conducted to investigate the effectiveness of recycling biochar using various organic solvents. Organic solvents such as acetone, benzene, and methanol were used to desorb biochar. Adsorption principles include integration and similarity (Hu, 2018). Using carbon adsorbate, the best candidate solvent for desorption of red dye in water was 60% acetone with 52% desorption efficiency, and 40% isopropanol with 54% desorption efficiency for yellow dye in water (Lu et al., 2011).

3.2 Thermal regeneration

One of the most successful strategies for achieving desorption is thermal regeneration. At high temperatures, adsorbate is carbonized and degraded, finally becoming molecules smaller than the pore size of biochar and escaping. The regeneration efficiency improves as the pyrolysis temperature rises. Regenerated *Enteromorpha prolifera* biochar at 80°C, 150°C, and 200°C temperatures, retained 35%, 45%, and 48% of initial pyrene adsorption, respectively, although the regeneration rates of benzoapyrene were 31%, 41%, and 40%, respectively (Qiao et al., 2017). Thermal treatment was shown to be effective in removing dissolved OC by He et al. (2017a,b).

3.3 Supercritical fluid regeneration

By altering the operating pressure, the supercritical fluid is employed as an extractant to reach separation. CO_2 is most often utilized as supercritical fluid. The advantages of supercritical fluid regeneration include a short running cycle, low operating temperature, and reduced biochar loss. It may recover adsorbate without affecting its physicochemical characteristics. However, it has high equipment cost and pressure resistance, thus is in experimental stage (Hu, 2018).

3.4 Microwave irradiation regeneration

The microwave irradiation approach uses microwave intending polar molecules in biochar adsorbent to generate dipole toward polarization. The electromagnetic field is converted into thermal energy at the same time. The trapped organic material in the pore is heated and volatilized. Microwave irradiation heats biochar uniformly from its inner side to outside. Heating can decompose and desorb organic matter adsorbed in the pores of biochar. This heating technology demonstrated a more efficient heating procedure for the ease of heat control (Jia et al., 2016).

4. Conclusion

The comprehensive development model of "high energy consumption, high emissions, and high pollution" has resulted in increased environmental contamination with the continued expansion of industry. Inorganic and organic contamination has wreaked havoc on the environment of water and soil.

Biochar is gaining popularity as a low-cost, high-efficiency, regenerative adsorbent. The calcined material, temperature, pH, and other parameters impact the surface chemistry of biochar, and the removal process for various organic pollutants changes as well. Electrostatic attraction, π–π EDA interaction, pore filling, H-bonding, hydrophobic interactions, diffusion, and partitioning are the major adsorption mechanisms for eliminating organic contaminants. Hydrophobic interactions, surface attachment, van der Waals forces, and electrostatic attractions were observed as significant pathways for microbial/biological contaminant retention from wastewaters. The available literature data are enormous for adsorption mechanisms involved in inorganic pollutants, specifically on heavy metal remediation from aqueous solutions. The published data suggested efficient chemistry behind the adsorption of inorganics apparently as physisorption, chemisorption, surface sorption, electrostatic attractions, ion exchange, precipitation or coprecipitation, and complexation. Biochar application research in engineering is mostly focused on fixed bed and continuous countercurrent separation devices; regeneration is primarily focused on solvent regeneration, microwave radiation regeneration, thermal regeneration, and supercritical fluid regeneration. These two scientific areas are progressively gaining popularity.

References

Afrooz, A.N., Boehm, A.B., 2017. Effects of submerged zone, media aging, and antecedent dry period on the performance of biochar-amended biofilters in removing fecal indicators and nutrients from natural stormwater. Ecol. Eng. 102, 320–330.

Ahmad, M., Rajapaksha, A.U., Lim, J.E., Zhang, M., Bolan, N., Mohan, D., Vithanage, M., Lee, S.S., Ok, Y.S., 2014. Biochar as a sorbent for contaminant management in soil and water: A review. Chemosphere 99, 19–33.

Ahmed, M.B., Zhou, J.L., Ngo, H.H., Guo, W., Johir, M.A.H., Sornalingam, K., 2017. Single and competitive sorption properties and mechanism of functionalized biochar for removing sulfonamide antibiotics from water. Chem. Eng. J. 311, 348–358.

Ahmed, M.B., Zhou, J.L., Ngo, H.H., Johir, M.A.H., Sun, L.Y., Asadullah, M., Belhaj, D., 2018. Sorption of hydrophobic organic contaminants on functionalized biochar: Protagonist role of p-p electron-donor-acceptor interactions and hydrogen bonds. J. Hazard Mater. 360, 270–278.

Ali, S., Rizwan, M., Shakoor, M.B., Jilani, A., Anjum, R., 2020. High sorption efficiency for As(III) and As(V) from aqueous solutions using novel almond shell biochar. Chemosphere 243, 125330.

Ambaye, T.G., Vaccari, M., Castro, F.D., Prasad, S., Rtimi, S., 2020. Emerging technologies for the recovery of rare earth elements (REEs) from the end-of-life electronic wastes: A review on progress, challenges, and perspectives. Environ. Sci. Pollut. Res. 1–23.

Ambaye, T.G., Vaccari, M., van Hullebusch, E.D., Amrane, A., Rtimi, S., 2021. Mechanisms and adsorption capacities of biochar for the removal of organic and inorganic pollutants from industrial wastewater. Int. J. Environ. Sci. Technol. 18 (10), 3273–3294.

Amusat, S.O., Kebede, T.G., Dube, S., Nindi, M.M., 2021. Ball-milling synthesis of biochar and biochar–based nanocomposites and prospects for removal of emerging contaminants: A review. J. Water Proc. Eng. 41, 101993.

Bandara, T., Xu, J., Potter, I.D., Franks, A., Chathurika, J., Tang, C., 2020. Mechanisms for the removal of Cd(II) and Cu(II) from aqueous solution and mine water by biochars derived from agricultural wastes. Chemosphere 254, 126745.

Bergeron, S., Raj, B., Nathaniel, R., Corbin, A., LaFleur, G., 2017. Presence of antibiotic resistance genes in raw source water of a drinking water treatment plant in a rural community of USA. Int. Biodeterior. Biodegrad. 124, 3–9.

Chaukura, N., Murimba, E.C., Gwenzi, W., 2017. Synthesis, characterisation and methyl orange adsorption capacity of ferric oxide-biochar nano-composites derived from pulp and paper sludge. Appl. Water Sci. 7, 2175–2186.

Chen, B., Chen, Z., Lv, S., 2011. Novel magnetic biochar efficiently sorbs organic pollutants and phosphate. Bioresour. Technol. 102, 716–723.

Chen, G., Walker, S.L., 2012. Fecal indicator bacteria transport and deposition in saturated and unsaturated porous media. Environ. Sci. Technol. 46 (16), 8782–8790.

Chen, G.J., Peng, C.Y., Fang, J.Y., Dong, Y.Y., Zhu, X.H., Cai, H.M., 2016. Biosorption of fluoride from drinking water using spent mushroom compost biochar coated with aluminum hydroxide. Desalination Water Treat. 57 (26), 12385–12395.

Choudhary, M., Kumar, R., Neogi, S., 2020. Activated biochar derived from Opuntia ficus-indica for the efficient adsorption of malachite green dye, Cu^{+2} and Ni^{+2} from water. J. Hazard Mater. 392, 122441.

Chu, K.H., Shankar, V., Park, C.M., Sohn, J., Jang, A., Yoon, Y., 2017. Evaluation of fouling mechanisms for humic acid molecules in an activated biochar-ultrafiltration hybrid system. Chem. Eng. J. 326, 240–248.

Dai, L., Zhu, W., He, L., Tan, F., Zhu, N., Zhou, Q., He, M., Hu, G., 2018. Calcium-rich biochar from crab shell: An unexpected super adsorbent for dye removal. Bioresour. Technol. 267, 510–516.

Dai, Y., Zhang, N., Xing, C., Cui, Q., Sun, Q., 2019. The adsorption, regeneration and engineering applications of biochar for removal of organic pollutants: A review. Chemosphere 223, 12–27.

Del Bubba, M., Anichini, B., Bakari, Z., Bruzzoniti, M.C., Camisa, R., Caprini, C., Checchini, L., Fibbi, D., El Ghadraoui, A., Liguori, F., Orlandini, S., 2020. Physicochemical properties and sorption capacities of sawdust-based biochars and commercial activated carbons towards ethoxylated alkylphenols and their phenolic metabolites in effluent wastewater from a textile district. Sci. Total Environ. 708, 135217.

De Gisi, S., Notarnicola, M., 2017. In: Abraham, M. (Ed.), Industrial Wastewater Treatment. Elsevier, Oxford, UK, pp. 23–42, 978-0-12-804792-7.

Ding, W., Dong, X., Ime, I.M., Gao, B., Ma, L.Q., 2014. Pyrolytic temperatures impact lead sorption mechanisms by bagasse biochars. Chemosphere 105, 68–74.

Dong, X., Ma, L.Q., Li, Y., 2011. Characteristics and mechanisms of hexavalent chromium removal by biochar from sugar beet tailing. J. Hazard Mater. 190, 909–915.

Dowhan, W., Bogdanov, M., Mileykovskaya, E., Vitrac, H., 2017. Functional roles of individual membrane phospholipids in *Escherichia coli* and *Saccharomyces cerevisiae*. In: Geiger, O. (Ed.), Biogenesis of Fatty Acids, Lipids and Membranes. Springer International Publishing, Cham, pp. 1–22.

El-Shafey, E.I., 2010. Removal of Zn(II) and Hg(II) from aqueous solution on a carbonaceous sorbent chemically prepared from rice husk. J. Hazard Mater. 175, 319–327.

Essandoh, M., Wolgemuth, D., Pittman, C.U., Mohan, D., Mlsna, T., 2017. Adsorption of metribuzin from aqueous solution using magnetic and nonmagnetic sustainable low-cost biochar adsorbents. Environ. Sci. Pollut. Res. 24 (5), 4577–4590.

Fan, S., Tang, J., Wang, Y., Li, H., Zhang, H., Tang, J., Wang, Z., Li, X., 2016. Biochar prepared from co-pyrolysis of municipal sewage sludge and tea waste for the adsorption of methylene blue from aqueous solutions: Kinetics, isotherm, thermodynamic and mechanism. J. Mol. Liq. 220, 432–441.

Fdez-Sanromán, A., Pazos, M., Rosales, E., Sanromán, M.A., 2020. Unravelling the environmental application of biochar as low-cost biosorbent: A review. Appl. Sci. 10 (21), 7810.

Foo, K.Y., Hameed, B.H., 2012. Microwave-assisted regeneration of activated carbon. Bioresour. Technol. 119, 234–240.

Gwenzi, W., Chaukura, N., Noubactep, C., Mukome, F.N., 2017. Biochar-based water treatment systems as a potential low-cost and sustainable technology for clean water provision. J. Environ. Manag. 197, 732–749.

He, J., Song, Y., Chen, J.P., 2017a. Development of a novel biochar/PSF mixed matrix membrane and study of key parameters in treatment of copper and lead contaminated water. Chemosphere 186, 1033–1045.

He, X., Elkouz, M., Inyang, M., Dickenson, E., Werta, E.C., 2017b. Ozone regeneration of granular activated carbon for trihalomethane control. J. Hazard Mater. 326, 101–109.

Ho, S., Chen, Y., Yang, Z., Nagarajan, D., Chang, J., Ren, N., 2017. High-efficiency removal of lead from wastewater by biochar derived from anaerobic digestion sludge. Bioresour. Technol. 246, 142–149.

Hu, X., Ding, Z., Zimmerman, A.R., Wang, S., Gao, B., 2015. Batch and column sorption of arsenic onto iron-impregnated biochar synthesized through hydrolysis. Water Res. 68, 206–216.

Hu, Y., 2018. Research and development of regeneration technology of activated carbon. Coal. Chem. Ind. 41, 136–139.

Huang, F., Zhang, L., Wu, R., Zhang, S., Xiao, R., 2020. Adsorption behaviour and relative distribution of Cd^{2+} adsorption mechanisms by the magnetic and nonmagnetic biochars derived from chicken manure. Int. J. Environ. Res. Publ. Health 17 (5), 1602.

Ifthikar, J., Wang, J., Wang, Q., Wang, T., Wang, H., Khan, A., Jawad, A., Sun, T., Jiao, X., Chen, Z., 2017. Highly efficient lead distribution by magnetic sewage sludge biochar: Sorption mechanisms and bench applications. Bioresour. Technol. 238, 399–406.

Inyang, M., Dickenson, E., 2015. The potential role of biochar in the removal of organic and microbial contaminants from potable and reuse water: A review. Chemosphere 134, 232–240.

Inyang, M., Gao, B., Zimmerman, A., Zhang, M., Chen, H., 2014. Synthesis, characterization, and dye sorption ability of carbon nanotube-biochar nanocomposites. Chem. Eng. J. 236, 3946.

Jang, H.M., Kan, E., 2019. Engineered biochar from agricultural waste for removal of tetracycline in water. Bioresour. Technol. 284, 437–447.

Jang, J., Miran, W., Divine, S.D., Nawaz, M., Shahzad, A., Woo, S.H., Lee, D.S., 2018. Rice straw-based biochar beads for the removal of radioactive strontium from aqueous solution. Sci. Total Environ. 615, 698–707.

Jia, Y., Zong, Q., Zhang, M., Wang, Z.M., Zhang, L.H., 2016. Research progress in regeneration technology of activated carbon from flue gas desulfurization. Bull. China Ceram. Soc. 35, 815–818.

Jin, J., Sun, K., Wang, Z., Han, L., Du, P., Wang, X., Xing, B., 2017. Effects of chemical oxidation on phenanthrene sorption by grass-and manure-derived biochars. Sci. Total Environ. 598, 789–796.

Karunanayake, A.G., Todd, O.A., Crowley, M.L., Ricchetti, L.B., Pittman, C.U., Anderson, R., Mlsna, T.E., 2017. Rapid removal of salicylic acid, 4-nitroaniline, benzoic acid and phthalic acid from wastewater using magnetized fast pyrolysis biochar from waste Douglas fir. Chem. Eng. J. 319, 75–88.

Kasozi, G.N., Zimmerman, A.R., Nkedi-Kizza, P., Gao, B., 2010. Catechol and humic acid sorption onto a range of laboratory-produced black carbons (biochars). Environ. Sci. Technol. 44 (16), 6189–6195.

Keiluweit, M., Nico, P.S., Johnson, M.G., Kleber, M., 2010. Dynamic molecular structure of plant biomass-derived black carbon (biochar). Environ. Sci. Technol. 44, 1247–1253.

Kong, H., He, J., Gao, Y., Wu, H., Zhu, X., 2011. Cosorption of phenanthrene and mercury(II) from aqueous solution by soybean stalk-based biochar. J. Agric. Food Chem. 59, 12116–12123.

Kumar, A., Kumar, A., Sharma, G., Naushad, M., Stadler, F.J., Ghfar, A.A., Dhiman, P., Saini, R.V., 2017. Sustainable nano-hybrids of magnetic biochar supported g-C3N4/FeVO4 for solar powered degradation of noxious pollutants-synergism of adsorption, photocatalysis & photo-ozonation. J. Clean. Prod. 165, 431–451.

Lau, A.Y., Tsang, D.C., Graham, N.J., Ok, Y.S., Yang, X., Li, X.D., 2017. Surface-modified biochar in a bio-retention system for *Escherichia coli* removal from stormwater. Chemosphere 169, 89–98.

Li, H., Cao, Y., Zhang, D., Pan, B., 2018. pH-dependent KOW provides new insights into understanding the adsorption mechanism of ionizable organic chemicals on carbonaceous materials. Sci. Total Environ. 618, 269–275.

Liu, F., Zuo, J., Chi, T., Wang, P., Yang, B., 2015. Removing phosphorus from aqueous solutions by using iron-modified corn straw biochar. Front. Environ. Sci. Eng. 9 (6), 1066–1075.

Lu, H., Zhang, W., Yang, Y., Huang, X., Wang, S., Qiu, R., 2012. Relative distribution of Pb^{2+} sorption mechanisms by sludge-derived biochar. Water Res. 46, 854–862.

Lu, P.J., Lin, H.C., Yu, W.T., Chern, J.M., 2011. Chemical regeneration of activated carbon used for dye adsorption. J. Taiwan Inst. Chem. Eng. 42, 305–311.

Mayakaduwa, S., Herath, I., Ok, Y.S., Mohan, D., Vithanage, M., 2017. Insights into aqueous carbofuran removal by modified and non-modified rice husk biochars. Environ. Sci. Pollut. Res. 24 (29), 22755–22763.

Mishra, V., Sureshkumar, M.K., Gupta, N., Kaushik, C.P., 2017. Study on sorption characteristics of uranium onto biochar derived from eucalyptus wood. Water Air Soil Pollut. 228 (8), 1–14.

Mohanty, S.K., Boehm, A.B., 2015. Effect of weathering on mobilization of biochar particles and bacterial removal in a stormwater biofilter. Water Res. 85, 208–215.

Mukherjee, A., Zimmerman, A.R., Harris, W., 2011. Surface chemistry variations among a series of laboratory-produced biochars. Geoderma 163, 247–255.

Niazi, N.K., Bibi, I., Shahid, M., Ok, Y.S., Burton, E.D., Wang, H., Shaheen, S.M., Rinklebe, J., Lüttge, A., 2018a. Arsenic removal by perilla leaf biochar in aqueous solutions and groundwater: an integrated spectroscopic and microscopic examination. Environ. Pollut. 232, 31–41.

Niazi, N.K., Bibi, I., Shahid, M., Ok, Y.S., Shaheen, S.M., Rinklebe, J., Wang, H., Murtaza, B., Islam, E., Nawaz, M.F., Lüttge, A., 2018b. Arsenic removal by Japanese oak wood biochar in aqueous solutions and well water: investigating arsenic fate using integrated spectroscopic and microscopic techniques. Sci. Total Environ. 621, 1642–1651.

Palansooriya, K.N., Yang, Y., Tsang, Y.F., Sarkar, B., Hou, D., Cao, X., Meers, E., Rinklebe, J., Kim, K.H., Ok, Y.S., 2020. Occurrence of contaminants in drinking water sources and the potential of biochar for water quality improvement: A review. Crit. Rev. Environ. Sci. Technol. 50 (6), 549–611.

Park, J.H., Ok, Y.S., Kim, S.H., Cho, J.S., Heo, J.S., Delaune, R.D., Seo, D.C., 2016. Competitive adsorption of heavy metals onto sesame straw biochar in aqueous solutions. Chemosphere 142, 77–83.

Qiao, K., Tian, W., Bai, J., Dong, J., Zhao, J., Gong, X.X., Liu, S.H., 2017. Preparation of biochar from enteromorpha prolifera and its use for the removal of polycyclic aromatic hydrocarbons (PAHs) from aqueous solution. Ecotoxicol. Environ. Saf. 149, 80–87.

Qiu, Y., Zheng, Z., Zhou, Z., Sheng, G.D., 2009. Effectiveness and mechanisms of dye adsorption on a straw-based biochar. Bioresour. Technol. 100, 5348–5351.

Reddy, K.R., Xie, T., Dastgheibi, S., 2014. Evaluation of biochar as a potential filter media for the removal of mixed contaminants from urban stormwater runoff. J. Environ. Eng. 140 (12), 04014043.

Roman, S., Ledesma, B., Gonzalez, J.F., Al-Kassir, A., Engo, G., Alvarez-Murilloa, A., 2013. Two stage thermal regeneration of exhausted activated carbons. Steam gasification of effluents. J. Anal. Appl. Pyrolysis 103, 201–206.

Rosales, E., Meijide, J., Pazos, M., Sanroman, M.A., 2017. Challenges and recent advances in biochar as low-cost biosorbent: From batch assays to continuous-flow systems. Bioresour. Technol. 246, 176–192.

Saha, B., Sengupta, S., 2021. Role of biochar in water treatment. In: Advanced Materials and Technologies for Wastewater Treatment. CRC Press, pp. 217–246.

Salvador, F., Martin-Sanchez, N., Sanchez-Hernandez, R., Sanchez-Hernandez, R., Sanchez-Montero, J., Izquierdo, C., 2015. Regeneration of carbonaceous adsorbents. Part II: Chemical, microbiological and vacuum regeneration. Microporous Mesoporous Mater. 202, 277–296.

Samsuri, A., Sadegh-Zadeh, F., Seh-Bardan, B., 2014. Characterization of biochars produced from oil palm and rice husks and their adsorption capacities for heavy metals. Int. J. Environ. Sci. Technol. 11, 967–976.

Santra, B., Ramrakhiani, L., Kar, S., Ghosh, S., Majumdar, S., 2020. Ceramic membrane-based ultrafiltration combined with adsorption by waste derived biochar for textile effluent treatment and management of spent biochar. J. Environ. Health Sci. Eng. 18 (2), 973–992.

Shakoor, M.B., Niazi, N.K., Bibi, I., Shahid, M., Saqib, Z.A., Nawaz, M.F., Shaheen, S.M., Wang, H., Tsang, D.C., Bundschuh, J., Ok, Y.S., 2019. Exploring the arsenic removal potential of various biosorbents from water. Environ. Int. 123, 567–579.

Shakya, A., Agarwal, T., 2019. Removal of Cr(VI) from water using pineapple peel derived biochars: Adsorption potential and re-usability assessment. J. Mol. Liq. 293, 111497.

Shan, D., Deng, S., Zhao, T., Wang, B., Wang, Y.J., Huang, J., Yu, G., Winglee, J., Wiesner, M.R., 2016. Preparation of ultrafine magnetic biochar and activated carbon for pharmaceutical adsorption and subsequent degradation by ball milling. J. Hazard Mater. 305, 156–163.

Shankar, V., Heo, J., Al-Hamadani, Y.A., Park, C.M., Chu, K.H., Yoon, Y., 2017. Evaluation of biochar-ultrafiltration membrane processes for humic acid removal under various hydrodynamic, pH, ionic strength, and pressure conditions. J. Environ. Manag. 197, 610–618.

Spagnoli, A.A., Giannakoudakis, D.A., Bashkova, S., 2017. Adsorption of methylene blue on cashew nut shell based carbons activated with zinc chloride: The role of surface and structural parameters. J. Mol. Liq. 229, 465–471.

Srivatsav, P., Bhargav, B.S., Shanmugasundaram, V., Arun, J., Gopinath, K.P., Bhatnagar, A., 2020. Biochar as an eco-friendly and economical adsorbent for the removal of colorants (dyes) from aqueous environment: A review. Water 12 (12), 3561.

Sun, K., Ro, K., Guo, M., Novak, J., Mashayekhi, H., Xing, B., 2011. Sorption of bisphenol A, 17α-Ethinyl estradiol and phenanthrene on thermally and hydrothermally produced biochars. Bioresour. Technol. 102, 5757–5763.

Sun, J., Lian, F., Liu, Z., Zhu, L., Song, Z., 2014. Biochars derived from various crop straws: Characterization and Cd(II) removal potential. Ecotoxicol. Environ. Saf. 106, 226–231.

Tan, X., Liu, Y., Zeng, G., Wang, X., Hu, X., Gu, Y., Yang, Z., 2015. Application of biochar for the removal of pollutants from aqueous solutions. Chemosphere 125, 70–85.

Trakal, L., Veselská, V., Šafařík, I., Vítková, M., Číhalová, S., Komárek, M., 2016. Lead and cadmium sorption mechanisms on magnetically modified biochars. Bioresour. Technol. 203, 318–324.

Wang, C., Wang, T., Li, W., Yan, J., Li, Z., Ahmad, R., Herath, S.K., Zhu, N., 2014. Adsorption of deoxyribonucleic acid (DNA) by willow wood biochars produced at different pyrolysis temperatures. Biol. Fertil. Soils 50 (1), 87–94.

Wang, S., Gao, B., Zimmerman, A.R., Li, Y., Ma, L., Harris, W.G., Migliaccio, K.W., 2015a. Removal of arsenic by magnetic biochar prepared from pinewood and natural hematite. Bioresour. Technol. 175, 391–395.

Wang, S.Y., Tang, Y.K., Chen, C., Wu, J.T., Huang, Z.N., Mo, Y.Y., Zhang, K.X., Chen, J.B., 2015b. Regeneration of magnetic biochar derived from eucalyptus leaf residue for lead (II) removal. Bioresour. Technol. 186, 360–364.

Wu, E.L., Fleming, P.J., Yeom, M.S., Widmalm, G., Klauda, J.B., Fleming, K.G., Im, W., 2014. *E. coli* outer membrane and interactions with OmpLA. Biophys. J. 106 (11), 2493–2502.

Xu, X., Cao, X., Zhao, L., Wang, H., Yu, H., Gao, B., 2013. Removal of Cu, Zn, and Cd from aqueous solutions by the dairy manure-derived biochar. Environ. Sci. Pollut. Res. 20, 358–368.

Yao, X., Ji, L., Guo, J., Ge, S., Lu, W., Chen, Y., Cai, L., Wang, Y., Song, W., 2020. An abundant porous biochar material derived from wakame (*Undaria pinnatifida*) with high adsorption performance for three organic dyes. Bioresour. Technol. 318, 124082.

Younis, S.A., El-Salamony, R.A., Tsang, Y.F., Kim, K., 2020. Use of rice straw-based biochar for batch sorption of barium/strontium from saline water: Protection against scale formation in petroleum/desalination industries. J. Clean. Prod. 250, 119442.

Zhang, F., Wang, X., Xionghui, J., Ma, L., 2016. Efficient arsenate removal by magnetite-modified water hyacinth biochar. Environ. Pollut. 216, 575–583.

Zhang, H., Wang, Z., Li, R., Guo, J., Li, Y., Zhu, J., Xie, X., 2017. TiO2 supported on reed straw biochar as an adsorptive and photocatalytic composite for the efficient degradation of sulfamethoxazole in aqueous matrices. Chemosphere 185, 351–360.

Zhang, J., Hu, X., Zhang, K., Xue, Y., 2019b. Desorption of calcium-rich crayfish shell biochar for the removal of lead from aqueous solutions. J. Colloid Interface Sci. 554, 417–423.

Zhang, L., Li, W., Cao, H., Hu, D., Chen, X., Guan, Y., Tang, J., Gao, H., 2019a. Ultraefficient sorption of Cu^{2+} and Pb^{2+} ions by light biochar derived from Medulla etrapanacis. Bioresour. Technol. 291, 121818.

Zhang, W., Mao, S., Chen, H., Huang, L., Qiu, R., 2013. Pb(II) and Cr(VI) sorption by biochars pyrolyzed from the municipal wastewater sludge under different heating conditions. Bioresour. Technol. 147, 545–552.

Zheng, H., Wang, Z., Zhao, J., Herbert, S., Xing, B., 2013. Sorption of antibiotic sulfamethoxazole varies with biochars produced at different temperatures. Environ. Pollut. 181, 60–67.

Zhou, Y., Gao, B., Zimmerman, A.R., Cao, X., 2014. Biochar-supported zerovalent iron reclaims silver from aqueous solution to form antimicrobial nanocomposite. Chemosphere 117, 801–805.

Zhou, Z., Liu, Y.G., Liu, S.B., Liu, H.Y., Zeng, G.M., Tan, X.F., Yang, C.P., Ding, Y., Yan, Z.L., Cai, X.X., 2017. Sorption performance and mechanisms of arsenic(V) removal by magnetic gelatin-modified biochar. Chem. Eng. J. 314, 223–231.

CHAPTER

Microbial assisted restoration of contaminated agricultural soil for sustainable environment

13

Neha Rana[1] and Piyush Gupta[2]

[1]*Department of Pharmaceutical Sciences, KIET School of Pharmacy, KIET Group of Institutions, Ghaziabad, Uttar Pradesh, India;* [2]*Department of Chemistry, Faculty of Engineering and Technology, SRM Institute of Science and Technology, Ghaziabad, Uttar Pradesh, India*

1. Introduction

Soil is an important component of the common habitat, like plants, creatures, rocks, landforms, lochs, and waterways. It is a living environment for a diverse range of living beings. Furthermore, it serves as a source and storage location for gases (such as oxygen and carbon dioxide) within the climate. Soil is one of the corpus foundations of life on Earth, serving as a source of water and nutrients, a mechanism for waste filtration and breakdown, and an operating member in the cycling of carbon and other components through the environment.

Several highly toxic compounds have been released into the environment in recent years as a result of various industrial and agricultural activities. Pesticides, fuels, solvents, alkanes, polycyclic aromatic hydrocarbons (PAHs), nitrogen and phosphorus compounds, explosives, and dyes are examples of typical organic chemicals, whereas toxic heavy metals are examples of inorganic contaminants. Natural events such as earthquakes, landslides, hurricanes, and floods can also contribute to soil pollution. Organic pollutant contamination of soil, surface, and ground water is a major environmental concern in many countries. Furthermore, complex effluents containing multiple pollutants and toxic metals, and pesticides also harm plants and animals by contaminating the soil, as these chemicals penetrate deep into the soil and poison the groundwater system. It is necessary to investigate the viability of bioremediation technologies for contaminated soil remediation.

The effects of soil pollution are not limited to soil and its biota but spread to every aspect of the environment and affect every organism, from earthworms to humans. Toxins bioaccumulate in our bodies, causing a variety of diseases, including reproductive health, birth and developmental defects, neurologic effects, malnutrition, and mutations in body cells that lead to cancer (Johansen, 2003). Long-term exposure to such soil can alter the genetic make-up of the body, causing congenital illnesses and being carcinogenic. Soil pollution can have an impact on plant growth. Contamination of soil can disrupt the ecological balance and harm the microorganisms found in it. Large sections of land become infertile, unfit to support any life.

Soil is now recognized as a vital resource, and because of its slow formation, it can be considered nonrenewable. Furthermore, it has implications for environmental, economic, and cultural activities. As a result, soil restoration is required to avoid further negative consequences. Microbial remediation

Integrative Strategies for Bioremediation of Environmental Contaminants, Volume Two. https://doi.org/10.1016/B978-0-443-14013-6.00017-2

is the best choice among different bioremediation techniques, such as chemical and physical methods, due to its environmental friendliness and cost effectiveness. Microorganisms are widely distributed throughout the biosphere due to their ability to grow in a variety of environmental conditions. Bioremediation is the term used to describe this type of process. It is carried based on the ability of certain microorganisms to convert, modify, and utilize toxic pollutants in order to produce energy and biomass (Tang et al., 2007). Rather than simply collecting and storing pollutants, bioremediation is a microbiologically well-organized procedural activity used to break down or transform contaminates into less toxic elemental and compound forms. Bacteria, archaea, and fungi are examples of primary bioremediators. Microorganisms are restoring natural environments and preventing further pollution. The current chapter describes the various types of organic and inorganic soil pollutants, as well as the current use of microorganisms in bioremediation, which leads to the restoration of contaminated soil. It is currently a hot research area because microorganisms are eco-friendly and offer valuable genetic material for solving environmental problems.

2. Soil contamination

2.1 Inorganic toxic compounds

Inorganic residues in industrial waste pose significant disposal challenges. They contain metals with a high toxicity potential. Large amounts of arsenic fluorides and sulfur dioxide are also emitted by industrial activity (SO_2). Fluorides are found in the atmosphere due to the industries of superphosphate, phosphoric acid, aluminum, steel, and ceramics. Sulfur dioxide emitted by factories and thermal plants has the potential to make soils extremely acidic. These metals damage leaves and kill vegetation. Copper, mercury, cadmium, lead, nickel, and arsenic are some of the elements that can accumulate in soil if they enter by sewage, industrial waste, or mine washings. Some fungicides containing copper and mercury also pollute the soil. Automobile smoke contains lead, which is absorbed by soil particles and is toxic to plants. Many fertilizers containing an excessive amount of N and P pollute the soil. Radioactive materials such as uranium, thorium, strontium, and others are also causing soil pollution.

2.2 Organic wastes

Organic wastes of various types are a source of pollution. Domestic garbage, municipal sewage, and industrial wastes, when left in heaps or improperly disposed of, have a negative impact on the health of humans, plants, and animals. Borates, phosphates, and detergents are abundant in organic waste. If left untreated, they will have an impact on plant vegetative growth. Phenols and coal are the most common organic contaminants. Other liquid wastes, such as sewage, sewage sludge, and so on, are also significant sources of soil problems.

Organic pesticides are widely used nowadays to control a variety of pests. Pesticides can be harmful to microorganisms, which can affect plant growth. Pesticides that do not degrade quickly may cause such issues. Pesticide residues accumulated in higher concentrations are toxic. Pesticide persistence in soil and movement into water streams may also result in pesticide entry into foods, posing health risks. Currently, DDT and a number of organochloride pesticides have been declared harmful and banned in the United States and England (Apitz, 2008). Organochlorides have been replaced by organophosphate pesticides, which are more toxic but leave no residue, due to their drawbacks. Rodenticides and fungicides also contribute to soil pollution.

3. Bioremediation techniques

Bioremediation is a mechanism that employs living microorganisms to reduce or prevent natural contamination (Cristaldi et al., 2017). Instead of simply relocating the problem, bioremediation can provide a permanent in situ solution to contamination. This technique can be used to remove heavy metals, metalloids, and other inorganic pollutants from soil or water. When compared to other engineering techniques, it is a cost-effective, efficient, novel, eco-friendly, and solar-powered technology with high public acceptance. Bioremediation techniques based on strategies can be divided into two categories:

- Ex situ bioremediation.
- In situ bioremediation.

3.1 Ex situ bioremediation

Ex situ, as the name implies, refers to the removal of contamination mat to a remote treatment location. This method is not widely used because it entails the large task of excavating polluted soil and transporting it offsite. The basic idea behind ex situ remediation is to introduce proper soil oxygen, moisture, and nutrient conditions offsite. However, the ex situ bioremediation process risks spreading contamination or causing an accidental spill during transportation. This can be accomplished in two ways:

3.1.1 Slurry phase

This method entails combining contaminated soil with water and other additives in a large bioreactor and mixing it to keep the indigenous microorganisms in contact with the contaminants. Essential nutrients and oxygen are added, and the conditions in the bioreactor are optimized for the microorganisms to degrade contaminants.

3.1.2 Solid phase

Soils in the above-ground treatment area are treated with solid-phase treatment. Despite the fact that solid-phase systems are easier to process and maintain, they require more space and treatment time than slurry-phase processes. The following techniques can be used to achieve this treatment.

- Land farming

This technique essentially promotes biodegradation through the use of indigenous microorganisms and facilitates aerobic degradation of contaminants. It is completed through a simple methodology technique in which contaminated soil is excavated and spread over a prepared bed on a regular basis until pollutants are degraded. Nutrients and minerals are also added to promote the growth of indigenous species.

- Soil biopiles

This biodegradation technique is used to remediate excavated soil contaminated with petroleum. Soil biopiles, also known as biocells, are a type of soil biopile. This technology entails piling contaminated soil and stimulating microbial activity either aerobically or by adding nutrients, minerals, or moisture. Biopiles can range in height from three to ten feet. This technology also employs oxygen to stimulate bacterial growth. Aerated biopiles are created by forcing air to move through perforated piping installed throughout the pile.

- Composting

Composting is the process of combining contaminated soil with a biomass such as straw, hay, or corncobs to provide the microorganisms with the optimal levels of air and water. Composting involves placing contaminated soil in treatment vessels and mixing it there for aeration. Window composting is a type of composting process in which soil is placed in long piles called windows and mixed on a regular basis by tractors. For composting, a 75% contaminated soil to 25% compost ratio is used. Compost remediation is known as a faster remediation method because it can be completed in a matter of weeks.

3.2 In situ bioremediation

The in situ method is defined by the bioremediation process performed at the contamination site. In situ bioremediation is the preferred method because it requires less mechanical effort to eliminate spreading contaminants and prevent pollutant spread through transportation or pumping away to other treatment locations. In situ bioremediation refers to biological processes in which microorganisms convert organic contaminants to inorganic materials such as carbon dioxide, methane, water, and inorganic salts. This procedure can be carried out in either natural or engineered conditions. This can be attained through the use of the following in situ bioremediation techniques:

3.2.1 Intrinsic bioremediation

Intrinsic bioremediation is the process of converting environmental pollutants to nontoxic forms using the inherent abilities of the site's naturally occurring microbial population. This method is commonly used in underground locations, such as underground petroleum tanks. Intrinsic bioremediation harnesses the innate abilities of naturally occurring microbial communities to degrade environmental pollutants without modifying or speeding up the process. This technique involves feeding nutrients and oxygen to the indigenous microbial population in order to increase their metabolic activity.

3.2.2 Enhanced (engineered) In situ bioremediation

As the name implies, this technique involves introducing specific microorganisms to the contaminated site. Engineered in situ bioremediation accelerates the degradation process by improving the physicochemical conditions for microorganism growth.

3.2.3 Bioventing

Bioventing is an in situ remediation technique that utilizes microorganisms to degrade organic constituents that have become adsorbent on soils. This technique involves regulating airflow to increase oxygen to the unsaturated zone, which improves bioremediation by increasing the activity of indigenous microbes. Amendments are made to the bioventing process by adding nutrients and moisture to increase bioremediation and achieve microbial transformation of pollutants to a nontoxic state. This technique has received attention among other in situ bioremediation techniques, particularly in the restoration of sites polluted by light spilled petroleum products. Bioventing is primarily used to degrade adsorbed fuel residuals, but it can also be used to degrade volatile organic compounds (VOCs).

3.2.4 Bioslurping

The bioslurping technique combines bioventing and vacuum-enhanced pumping of soil and groundwater remediation by indirect oxygen provision and contaminant biodegradation stimulation. This technique employs a "slurp" that extends into the free product layer and draws liquids (free products and soil gas) from this layer in the same way that a straw does from any vessel. The bioslurping system consists of a well connected to an adjustable length called "slurp tube," which is connected to a vacuum pump, which is lowered into the light nonaqueous phase liquids (LNAPL) layer, and pumping begins to remove free product along with some groundwater. This method is used to clean up soil that has been contaminated with volatile and semi-VOCs.

3.2.5 Biosparging

Biosparging is a biological method for removing aromatic compounds such as benzene, toluene, ethylbenzene, xylene, and naphthalene from an area. This process requires the addition of specific aerobic bacteria to break down mineral oil and aromatic compounds into simpler and more useful forms. This technique is similar to bioventing in that air is introduced into the soil subsurface to stimulate microbial activity and thus improve pollutant removal from polluted areas. In biosparging, air is injected into the saturated zone, causing VOCs to rise to the unsaturated zone and promote biodegradation.

3.2.6 Bioaugmentation

Bioaugmentation is a method of enriching the existing microorganism population and making it more effective in reducing contamination. This technique involves the addition of organic culture to contaminated soil in order to create an environment similar to that of a bioreactor. Bioaugmentation is primarily used for bioremediation at oil-contaminated sites. In comparison to other methods of treating wastewater and soil contamination, bioaugmentation is a low-cost method.

3.2.7 Phytoremediation

Phytoremediation is the direct use of green plants and their associated microorganisms to stabilize or reduce contamination in soils, sludges, sediments, surface water, or ground water. To mitigate the toxic effects of pollutants, this technique applies plant interactions (physical, biochemical, biological, chemical, and microbiological) to contaminated sites. It is a complementary technology that can be used alongside or in place of mechanical conventional clean-up technologies, which frequently require large capital investments and consume a lot of energy. Depending on the pollutant type (elemental or organic), phytoremediation involves a number of mechanisms (accumulation or extraction, degradation, filtration, stabilization, and volatilization). The removal of elemental pollutants (toxic heavy metals and radionuclides) is primarily accomplished through extraction, transformation, and sequestration. This can be accomplished using the methods listed below:

a. Phytostabilization: It involves the use of plants to reduce heavy metal bioavailability in soil.
b. Phytoextraction: This method entails extracting and removing heavy metals from soil using plants.
c. Phytovolatilization: This technique involves the use of plants to absorb heavy metals from soil and release them as volatile compounds into the atmosphere.
d. Phytofiltration: It involves the use of hydroponically grown plants to absorb or adsorb heavy metal ions from groundwater and aqueous waste.

3.2.7.1 Phytostabilization

The phytostabilization process involves plants that have been established and primarily function to accumulate metals into root tissues or aid in their precipitation in the root zone. This method is based on the chemical stabilization of heavy metals through the use of various nonorganic and/or organic soil additives in conjunction with appropriately selected plant species. Phytostabilization reduces contaminant mobility and helps to reduce the risk of inorganic contaminants within the site. This technology does not generate contaminated secondary waste that must be treated further. This technique essentially limits heavy metal bioavailability while also restoring adequate soil quality.

3.2.7.2 Phytoextraction

Phytoextraction is a phytoremediation technique that involves the use of plants to absorb and remove metals and other contaminants from soil or water. This technology can be used to remove organic and inorganic pollutants from soil, water, and air. This should be accomplished at a lower cost than any other technology because it only requires the identification and planting of plants with the ability to hyperaccumulate.

3.2.7.3 Phytovolatilization

Phytovolatilization is the process by which contaminants are taken up by plants, converted into volatile compounds, and then released into the atmosphere. Because of its metabolic and transpiration pull, the plant absorbs organic pollutants and water as it grows and travels from root to other parts of the plant in the same or altered form.

3.2.7.4 Phytofiltration

The phytofiltration technique is commonly used to treat polluted water. This technique employs high metal-accumulating plants that act as biofilters, and it has the potential to be effective in sequestering metals from polluted waters. In this case, the roots absorb waste water and dissolved contaminants. Because many plant species naturally absorb heavy metals and other contaminants, it is a cost-effective method of remediation.

3.2.8 Phytodegradation

The term phytodegradation refers to the degradation of organic contaminants by plant enzymatic activities. Enzymes are released by the plant from its roots or through metabolic activities within plant tissues. Organic contaminants are taken up by roots and metabolized in plant tissues to less toxic substances during phytodegradation. Hydrophobic organic contaminants can be degraded more efficiently using the phytodegradation process.

3.2.9 Mycoremediation

Mycoremediation is a bioremediation technique that employs fungus as a bioremediator. This biotechnique employs specific fungi that produce enzymes capable of degrading a variety of pollutants and have been found to be a promising strategy for contaminant removal within a site. Mycoremediation is also a cost-effective and efficient technique.

3.2.10 Rhizofiltration

Rhizofiltration is the removal of toxic substances or pollutants from ground water through use of filtration by plant roots. The mechanism of rhizospheric accumulation by plants reinforces the rhizofiltration process. Plants that can absorb and resist high concentrations of toxic metals, such as hyperaccumulators, are ideal for rhizofiltration. The addition of PGPR to a contaminated site reduces metal toxicity in plants because the bioavailability of such metals decreases, increasing the capacity of plants to rid themselves of heavy metal contaminants and protect themselves from environmental stress.

3.2.11 Genetically engineered microorganisms

Because of their ability to degrade environmental pollutants through use of biochemical pathways related to the organism's activity and growth, microorganisms are used in bioremediation. Microorganisms can degrade hazardous substances found in polluted environments to harmless end products through the cometabolism process. The use of indigenous microorganisms to clean up polluted environments has not yielded many positive results. Heavy metals such as Hg, for example, cannot be removed from the environment by indigenous bacteria. However, recombinant DNA technology has a significant role to play in heavy metal contamination bioremediation because it improves the remediation process. With the discovery of restriction enzymes and DNA ligases in the 1970s, recombinant DNA technology made it possible to alter the genomes of living organisms. Microorganism's metabolic potentials have since been studied, and microbes have been genetically modified for specific purposes. The goal of genetic engineering for bioremediation is to modify plants, microorganisms, and enzymes so that they can be used to degrade harmful substances. Genetic engineering has enabled the development of bacteria capable of removing heavy metals such as As, Cd, Cu, Fe, Hg, and Ni (Azad et al., 2014). However, the rate of degradation is determined by the catalytic efficiency of enzymes found in cells or those induced to degrade a specific substrate.

Using recombinant DNA technology, genetically engineered microorganisms (GEMs) have foreign genes inserted into their genome from another organism of the same or different species. These engineered microbes have been used to generate competent strains for the bioremediation of contaminated environments due to their enhanced ability to degrade a wide range of contaminants. There have been reports of genetically engineered *Escherichia coli* strain M109 and *Pseudomonas putida* containing the merA gene being used to effectively remove Hg from contaminated soils and sediments (Deckwer et al., 2004). Azad et al. (2014) provided a comprehensive review of the use of genetically engineered bacteria and plants in the bioremediation of heavy metal and other organic pollutant-contaminated sites. The merA gene for Hg uptake, phenol catabolic genes (pheA, pheB, pheC, pheD, and pheR) (Marconi et al., 1997), and the ArsM gene for As removal from contaminated soils are examples of widely used genes (Liu et al., 2011). Similarly, *Cupriavidus metallidurans* strain MSR33 was genetically engineered with a pTP6 plasmid that provided genes (merB and merG) regulating Hg biodegradation, as well as the synthesis of organomercurial lyase protein (merB) and mercuric reductase protein (merA), giving the microorganism the ability to reduce Hg contamination from polluted sites (Rojas et al., 2011). *Pseudomonas* strains have also been made resistant to Hg by introducing novel genes into the strain using the pMR68 plasmid (Sone et al., 2013). The n-alkane-degrading microorganisms have specific genes such as alkB, alkB1, alkB2, alkM, aromatic hydrocarbons: xylE, and PAHs: ndoB and nidA, which are used as markers for microbial biodegradation identification.

The use of GEMs to aid in the remediation process is critical in the conflict against toxic substances in the environment. To use GEMs for bioremediation, the stability of the microbes must be maintained prior to field application, as the catabolic activity of the released GEM is primarily linked to the stability of the recombinant plasmid introduced into the organism. The use of genetic engineering increases the chances of obtaining effective pollutant-degrading microorganisms, which may have a higher potential for environmental cleanup than indigenous microbes. To successfully use GEMs for bioremediation in harsh environmental conditions, the recombinant bacterial population must be preserved in the soil, with appropriate environmental conditions prepared, and the recombinant bacteria must be able to withstand opposition from indigenous bacterial populations.

3.2.12 Microbial fuel cells

The bioelectrochemical remediation of contaminated soils using bioelectrochemical systems (BESs) is gaining significant attention in terms of soil ecosystem restoration. In this regard, a microbial fuel cell (MFC, a BES application) was used to treat total petroleum hydrocarbons (TPHs) in a soil microenvironment. A novel method known as BESs, which deals with the interface of a biofilm and an electrode environment, such as MFCs and microbial electrolysis cells (MECs), has demonstrated superior functions for the degradation of petrogenic pollutants. Pollutant degradation or oxidation in an anodic environment has aided in the generation of green energy. BES can also treat other pollutants such as sulfates, nitrates, and chemical oxygen demand (COD) at the same time (Srikanth et al., 2016). The electrochemically active biofilm on the anode of MFCs generates bioelectricity from the treatment of organics, whereas MECs treat specific pollutants under mild externally applied potentials for various types of pollutants. In BES systems, the biological and electrochemical effects combine to form hybrid mechanisms that facilitate complex bioelectrochemical treatment processes. Petroleum hydrocarbons in the soil act as electron donors, while carbon electrodes act as electron acceptors, resulting in the development of a biopotential in the soil environment. In conjunction with bioelectroremediation, the combination of a generated biopotential and associated microbial activity aids in the generation of biological electricity. Sulfate is a major pollutant found in waste/wastewater generated by the petroleum industry. High sulfate concentrations in wastewater are evidence of the ecosystem's negative effects. BESs have been proposed as an effective sulfate removal method.

4. Microorganisms in remediation of organic pollutants

Organic pollutants can be degraded biologically (microbially or enzymatically) in soil. However, the degradative efficiency of biological processes is determined by the biodegradability of the contaminants. Pollutants such as simple hydrocarbons C1−C15, alcohols, phenols, amines, acids, esters, and amides biodegrade quickly. Polychlorinated biphenyls (PCBs), PAHs, and pesticides, on the other hand, are extremely difficult to biodegrade. Under aerobic and anaerobic conditions, various bacteria and fungi can be used for biodegradation (Korda et al., 1997). Specific biochemical pathways have been developed by certain classes of bacteria and fungi that have adapted their enzymatic and metabolic complements to use these complex molecules as energy and carbon sources. Organic pollutants can be degraded biologically (microbially or enzymatically) in soil. However, the degradative efficiency of biological processes is determined by the biodegradability of the contaminants. Pollutants such as simple hydrocarbons C1−C15, alcohols, phenols, amines, acids, esters, and amides biodegrade quickly. PCBs, PAHs, and pesticides, on the other hand, are extremely difficult to biodegrade. Under aerobic and anaerobic conditions, various bacteria and fungi can be used for biodegradation (Korda et al., 1997). Specific biochemical pathways have been developed by certain classes of bacteria and fungi that have adapted their enzymatic and metabolic complements to use these complex molecules as energy and carbon sources.

One of the most serious environmental issues affecting aquatic and terrestrial environments is crude oil contamination. At the moment, approximately 80% of lands are affected by petroleum-derived products. Crude oil coats the soil's surface, causing carbon dioxide produced by soil organisms to be retained. It also reduces soil porosity by binding soil particles together. The amount of loss is determined by the amount and quality of oil spilled. Because microbes and plant roots can easily access these tiny pores, bioremediation is the best method for removing PAH from the soil. Some of the anaerobic bacteria species in the genus *Desulfococcus, Thauera, Dechloromonas,* and *Azoarcus* have the ability to degrade hydrocarbons (Rabus et al., 2016). The major bacterial genera that demonstrated crude oil-degrading ability are *Alcaligenes*, *Sphingomonas*, *Pseudomonas*, *Bacillus*, *Nocardia*, *Acinetobacter*, *Micrococcus*, *Achromobacter*, *Rhodococcus*, *Alcaligenes*, *Moraxella*, *Mycobacterium*, *Aeromonas*, *Xanthomonas*, *Arthrobacter*, *Flavobacterium*, *Micrococcus*, and *Azospirillum* (Das and Chandran, 2011). Some members of the following fungal genera have demonstrated crude oil degradation: *Candida, Stropharia, Rhodotorula, Pleurotus, Penicillium, Phanerochaete*, and *Fusarium*. Some microorganisms used in the remediation of various organic pollutants are depicted in Table 13.1.

4.1 Removal of organic nitrogen by microorganisms

Nitrogen enters the soil as a result of mineral and organic fertilization (manure and crop residues), as well as biological fixation of elemental nitrogen by symbiotic and free-living bacteria. Nitrogen is absorbed by plants in two forms: ammonium (NH^{4+}) and nitrate (NO_3^-). According to Lamb et al. (2014), nitrogen circulation in the environment involves six microbiological processes: atmospheric nitrogen fixation, mineralization, immobilization, nitrification, denitrification, and anaerobic ammonium oxidation (anammox), all of which are significantly influenced by soil physicochemical properties and climatic conditions. Microorganisms play an important role in the nitrogen transformation cycle because they contain genes that encode enzymes involved in nitrogen metabolism, which are used as indicators of the cycle's potential. An unbalanced nutrient balance can be harmful to plants because neither nitrogen deficiency nor excess nitrogen is beneficial to them. Nitrogen fertilizers used in large quantities in farmlands have an impact on the processes of nitrogen compound transformation in soil, particularly nitrification and denitrification.

The first stage in the decomposition of nitrogen-based organic matter is proteolysis. It is made up of the enzymatic hydrolysis of proteins into peptides, which are then converted into amino acids. Proteolytic activity is found in all soil microorganisms. Alkaline serine endopeptidases and subtilisin peptidases produced by bacteria are two of the most important proteases produced by microorganisms, for example, *Bacillus* and *Pseudomonas* genera, and by the *Aspergillus* sp., *Penicillium* sp., *Pythium* sp., *Trichophyton* sp., and *Cladosporium herbarum* fungi. Aside from that, these are cysteine endopeptidases made by bacteria of the *Clostridium, Micrococcus, Microsporum,* and *Streptococcus* genera, in addition to aspartyl endopeptidases produced by fungi and bacteria of the *Aspergillus, Candida, Mucor, Saccharomyces,* and *Shewanella* genera (Ayyasamy et al., 2009; Yegin et al., 2011). Acid proteinases are a class of proteolytic enzymes derived from fungi (*Saccharomyces* sp.,

Table 13.1 Microorganisms used in organic pollutants remediation.

Type of pollutant	Name of microorganism	References
Azo dyes	*Pycnoporus sanguineus*	Pointing and Vrijmoed (2000)
Polycyclic aromatic hydrocarbons	*Pycnoporus chrysosporium*	Bumpus (1998)
Hydrocarbon	*Aspergillus niger, Aspergillus fumigatus, Fusarium solani, and Penicillium funiculosum*	Al-Jawhari (2014)
Polychlorinated biphenyls	*Coriolopsis polyzona and T. versicolor*	Novotny et al. (1997)
Pentachlorophenol (PCP)	*P. chrysosporium and T. versicolor*	Alleman et al. (1995)
Monocyclic aromatic hydrocarbons	*Pseudomonas putida*	Safi et al. (2015); Sarang et al. (2013)
Phenols	*White rot fungi*	Nicell (2001)
Naphthalene	*Cyanobacteria, green algae, diatoms,* and *Bacillus licheniformis*	Sivakumar et al. (2012); Lin et al. (2010)
Organophosphates	*Agrobacterium, Pseudomonas* sp., *Flavobacterium* sp., *and Bacillus cereus*	Coppella et al. (1990); Mulbry (1998)
Crude oil	*A. niger, Candida glabrata, Candida krusei, and Saccharomyces cerevisiae*	Burghal et al. (2016)
Diesel oil	*Bacillus cereus A*	Maliji et al. (2013)
Industrial dyes	*Pycnoporus sanguineous, Phanerochaete chrysosporium, and Trametes trogii*	Yan et al. (2014)
Textile azo dyes	*Micrococcus luteus, Listeria denitrificans, and Nocardia atlantica*	Hassan et al. (2013)

Aspergillus sp., *Penicillium* sp., *Rhizopus* sp., and *Endothia* sp.), which are most active in substrates with a pH less than 5 (Sumantha et al., 2006). Furthermore, exopeptidases like glycine and alanine aminopeptidase produced by *Pseudomonas fluorescens* and *Lactobacillus lactis* are also considered important for organic nitrogen mineralization in soil (Allison and Vitousek, 2005; Dong et al., 2007; Grandy et al., 2007).

The next stage of nitrogen mineralization of organic matter is ammonification. Low molecular weight organic nitrogen compounds (amino acids, urea, purine bases, and nucleic acids) enter soil with animal and plant residues during this process, and the proteolysis products are further transformed when ammonia is released. Soil deamination is primarily a biochemical process caused by the presence of deaminases secreted by soil microorganisms, such as asparaginase, urease, and cytosine deaminase. These reactions are visible in *Aspergillus* sp., *Bacillus* sp., *P. fluorescens, Azotobacter agilis,* and *Streptomyces griseus.*

Ammonification process occurs in soil with a wide tolerance of environmental factors due to the large diversity of ammonifying microorganisms, which include aerobes and anaerobes, psychrophiles, mesophiles, and thermophiles, as well as organisms adapted to acidic and alkaline environments and a wide range of substrate moisture. Numerous bacteria are involved in the ammonification process (*Bacillus* sp., *Clostridium* sp., *Proteus* sp., *Pseudomonas* sp., *Serratia* sp., and *E. coli*), actinobacteria (*Streptomyces* sp., *Micromonospora* sp., and *Nocardia* sp.), and fungi (*Aspergillus* sp., *Mucor* sp., *Cladosporium* sp., *Botrytis* sp., and *Trichoderma* sp.) (Libudzisz et al., 2007; Fujii et al., 2020).

4.2 Removal of organic phosphorus by microorganisms

Phosphorus (P) is a macronutrient that is essential for plant growth and is involved in numerous metabolic reactions. P is emitted into the atmosphere by the combustion of fossil fuels and biofuels, and the largest reservoirs of P in soil and the marine environment are phosphate rock (PR) and sedimentary rock, respectively. Human activities such as the development and use of organophosphorus chemicals, the extraction of geological P reserves to produce P-fertilizers, and the disposal of animal excreta into the environment have all had a significant impact on soil P geochemical balances and ecosystem functions. Phosphate solubilizing microorganisms (PSMs), a large microbial community that mediates bioavailable soil P, play an important role in the soil P cycle by mineralizing organic P, solubilizing inorganic P minerals, and storing large amounts of P in biomass (Liang et al., 2020). PSMs can dissolve soil P into soluble and plant available orthophosphate forms by releasing phosphatase enzymes and organic acids, lowering soil pH, and increasing chelation activities with additional P adsorption sites (mostly PO_4^{3-}, HPO_4^{2-}, and $H_2PO_4^-$).

P exists in soil in a variety of forms, the most common of which are inorganic P (P_i) and organic P (P_o). Soil Po, which accounts for a significant portion of total soil P, is derived primarily from biological tissues where P is found as an integral component of organic compounds such as nucleotides, phospholipids, phosphoproteins, and coenzymes. The widespread use of P_o-containing products such as plasticizers, fire retardants, antifoam agents, and pesticides has resulted in their frequent occurrence in the environment as new P_o sources, increasing the quantity and variety of P_o forms in soils (Hoffman et al., 2017). PSMs isolated from bulk soils and rhizospheres have been shown to hydrolyze P_o mostly through phosphatase release (Bi et al., 2018). Phosphatases are enzymes that control P_o decomposition and mineralization. PSM-released nonspecific acid phosphatases (phosphohydrolases or NSAPs) dephosphorylate phosphodiester or phosphoanhydride bonds in P_o matters, and they play an important role in P_o mineralization, for example, *Deinococcus radiodurans*, *E. coli*, and *Microcystis aeruginosa.* Yeasts (including *Pichia acacia* and *Candida argentea*) were also shown to produce phytase while using phytic acid as their sole source of P (Liang et al., 2016).

5. Microorganisms in remediation of inorganic pollutants

Heavy metals, metalloids, and radionuclides comprise the majority of this group. These chemicals are linked to the processes (mining, fossil fuel combustion, municipal solid waste, industrial effluent, and phosphate fertilizer application) that allow them to enter the food chain.

5.1 Removal of heavy metals

Microorganisms are essential in the remediation of heavy metal-contaminated environments because they can tolerate metal toxicity in a variety of ways. The use of microorganisms to sequester, precipitate, or change the oxidation state of a variety of heavy metals has received a lot of attention (Kang et al., 2016). Heavy metal bioremediation will be successful if a consortium of bacterial strains is used rather than a single strain culture. Kang et al. (2016) investigated the synergistic effect of bacterial mixtures on the bioremediation of a Pb, Cd, and Cu mixture from contaminated soils using four strains: *Viridibacillus arenosi* B-21, *Sporosarcina soli* B-22, *Enterobacter cloacae* KJ-46, and *E. cloacae* KJ-47 were investigated. After 48 h, they discovered that the bacterial mixtures had greater resistance and efficiency for heavy metal remediation than single strain cultures, with remediation efficiencies of 98.3% for Pb, 85.4% for Cd, and 5.6% for Cu recorded. Microorganisms such as *Agrobacterium* spp., *Alcaligenes faecalis*, *Xanthomonas campestris*, *Bacillus* spp., *Zymomonas mobilis*, *Leuconostoc*, *Pseudomonas* spp., and *Acetobacter xylinum* have been identified as EPS-producing microorganism genera for heavy metal removal (Donot et al., 2012). Some commercial bacterial EPS with the required anionicity includes: alginate (*Pseudomonas aeruginosa* and *Azotobacter vinelandii*), gellan (*Sphingomonas paucimobilis*), hyaluronan (*Pseudomonas aeruginosa*, *Pasteurella multocida*, and Streptococci attenuated strains), xanthan (*X. campestris*), galactopol (*Pseudomonas oleovorans*), and fucopol (Enterobacter A47) (Gupta and Diwan, 2016). Table 13.2 summarizes some of the common microorganisms used in heavy metal removal (Achal et al., 2011, 2012; Balamurugan et al., 2014; Coelho et al., 2015; Dong et al., 2013; Farhan and Khadom, 2015; Livia and Benedito, 2015; Srivastava, 2015; Sukumar, 2010; Ta Stan et al., 2010; Vullo et al., 2008).

5.2 Removal of inorganic nitrogen

The following steps are followed to remove inorganic nitrogen from soil:

5.2.1 Immobilization

Microbial nitrogen immobilization is the uptake of nitrogen into microbial cells in the form of ammonium, nitrate, or amine in order to use this element in the synthesis of high molecular weight intracellular nitrogen compounds. Bottomley et al. (2012) state that both bacteria and fungi participate in the immobilization of inorganic nitrogen forms in soil but at different rates.

5.2.1.1 Nitrification

Another important process in the transformation of nitrogen compounds in soil is nitrification. It consists of the aerobic transformation of ammonium ions into nitrates. Nitritation is the first stage of the nitrification process. It is made up of nitrosobacteria (AOB-ammonia oxidizing bacteria) oxidizing ammonia (NH_3) to nitrite (NO_2^-). This transformation is initiated by these bacteria through the

Table 13.2 Microorganisms used in heavy metal remediation.

S.No.	Class of microorganism	Name of microorganism	Heavy metal
1.	Bacteria	*Bacillus cereus strain XMCr-6*	Cr(VI)
		Kocuria flava	Cu
		Sporosarcina ginsengisoli	As(III)
		Pseudomonas veronii	Cd, Zn, and Cu
		Pseudomonas putida	Cr(VI)
		Bacillus subtilis	Cr(VI)
2.	Fungi	*Aspergillus versicolor*	Ni and Cu
		Gloeophyllum sepiarium	Cr(VI)
		Rhizopus oryzae (MPRO)	Cr(VI)
3.	Yeast	*Saccharomyces cerevisiae*	Pd and Cd
4.	Algae	*Hydrodictylon, Oedogonium,* and *Rhizoclonium* spp.	As
		Spirogyra spp. and *Spirulina* spp.	Cr, Cu, Fe, Mn, and Zn

intermediate hydroxylamine (NH_2OH). The second stage of nitrification is nitrification, which is the oxidation of nitrites (NO_2) to nitrates (NO_3). A distinct group of prokaryotes is involved in this process; these nitrobacteria are referred to as nitrite-oxidizing bacteria (NOB). According to Bernhard (2010), the NOB group includes following genera: *Nitrospira, Nitrobacter, Nitrococcus,* and *Nitrospina.*

5.2.2 Denitrification

Denitrification, as compared to nitrification, is an anaerobic process in which denitrifying bacteria catabolize carbon substrates into energy (ATP), with reactive nitrogen forms serving as the ultimate electron acceptors in respiratory metabolism. Under anaerobic conditions, the same microorganisms reduce nitrates (or nitrites), taking the oxygen required for energy processes. Most denitrifying bacteria, according to Schulze and Mooney (2002), are heterotrophs of the following genera: *Alcaligenes, Bacillus, Paracoccus, Thiobacillus, Burkholderia, Thaurea, Rubrivax,* and *Ralstonia*. Denitrification is also feasible in the following species: *Achromobacter* sp., *Agribacterium* sp., *Alcaligenes* sp., *Flavibacterium* sp., *Serratia* sp., *Spirillum* sp., and *Micrococcus* sp. Some species of molds (*Aspergillus nidulans, Fusarium oxysporum, Fusarium decemcellulare, Trichoderma hamatum, Talaromyces flavus*, and *Chaciomhun* sp.) are also capable of denitrification (Libudzisz et al., 2007; Dilfuza, 2011; Fuka et al., 2007; Rana and Pandey, 2019). Cabello et al. (2004) found that hyperthermophilic (*Pyrobaculum aerophilum* and *Ferroglobus placidus*) and halophilic species (*Haloferax* sp. and *Haloarcula* sp.) of archaea were also capable of denitrification.

5.2.3 Codenitrification

Another process that occurs in the soil environment is codenitrification, also known as biotic nitrosation. The formation of N_2O and N_2 is the result of the process. Long et al. (2013); Spott et al. (2011) found that both fungi (*F. oxysporum, Fusarium solani,* and *Cylindrocarpon tonkinense*) and bacteria (e.g., *Streptomyces antibioticus, Corynebacterium nephridia, Pseudomonas denitrificans, Paracoccus denitastrificans, Rhodoclastobacter sphacterides,* and *Mesorhizobium* sp.) were involved in biotic nitrosation.

5.2.4 Anammox

Strous (1999) named anammox a new type of ammonia oxidation that occurs under anaerobic conditions (anaerobic ammonia oxidation). After denitrifiers, anammox bacteria are the second largest group of microorganisms that live in an anaerobic environment. Under anaerobic conditions, they oxidize ammonia (electron donor) with nitrite as an electron acceptor and CO_2 as a carbon source to produce gaseous N_2. Han and Gu (2013); Schmid et al. (2005) showed that seven genera of prokaryotes belonging to the Planctomycetes phylum (*Brocadia, Anammoxoglobus, Kuenenia, Scalindua, Jettenia, Brasillis,* and *Anammoximicrobium*) were involved in the anammox process.

5.2.5 Nitrogen fixing

Despite the fact that molecular nitrogen constitutes nearly 80% of the earth's atmosphere, plants and animals are unable to assimilate it due to its low reactivity. As a result, only a subset of prokaryotes known as diazotrophs, which includes heterotrophic bacteria, archaea, and cyanobacteria, can perform the energetically demanding process of binding molecular nitrogen (N_2) and incorporating it into organic compounds. Free-living bacteria are some nitrogen-fixing organisms (*Azotobacter* sp., *Azomonas* sp., *Azospirillum* sp., *Bacillus* sp., *Beijerinckia* sp., *Calothrix* sp., *Clostridium* sp., *Derxia* sp., *Desulfovibrio* sp., *Nostoc* sp., *Rhodobacter* sp., and *Rhodospirillum* sp.), whereas others make symbiotic systems with plants (rhizobia—*Rhizobium* sp., *Allorhizobium* sp., *Azorhizobium* sp., *Bradyrhizobium* sp., *Mesorhizobium* sp., and *Sinorhizobium* sp.) (Dilfuza, 2011).

5.3 Removal of inorganic phosphate

Soil P_i is typically present as relatively insoluble and stable forms of primary and secondary P minerals, as compared to P_o, which is more easily leached due to weak interactions with soil particles. P_i is a necessary but nonrenewable resource for all plants on earth. Because plant-available P_i is primarily derived from the soil environment, rising global food demand has significantly increased the use of PR in the production of P fertilizer (Cong et al., 2020). Furthermore, modern agricultural use of PR and P_i fertilizers can result in an excessive accumulation of soil P. A large number of heterotrophic and autotrophic microorganisms, including phosphate-solubilizing bacteria (PSB), phosphate-solubilizing fungi (PSF), phosphate-solubilizing actinomycetes (PSAs), and cyanobacteria, are commonly identified as PSM among heterogeneously distributed soil microflora (Turan et al., 2006; Zaidi et al., 2010). PSB, such as *Bacillus* sp., *Pseudomonas* sp., *Rhizobium* sp., and *Escherichia* sp., form the largest microbial communities with P solubilization abilities in soil. *Penicillium* sp., *Aspergillus* sp., *Mucor* sp., and *Rhizopus* sp. are the commonly isolated and demonstrated PSF microflora in soil. Some previous studies have found that several arbuscular mycorrhizal fungi (AMF), such as *Rhizophagus irregularis* (Mackay et al., 2017), *Glomus aggregatum*, and *Glomus mosseae* (Zai et al., 2017), can also solubilize P_i. Moreover, actinobacteria, especially *Streptomyces* and *Micromonospora*, are of increasing interest since these sporulating bacteria are capable to solubilize insoluble P minerals in P_i weathering process and the soil P cycle (Hamdaliet al., 2008; Chaiharn et al., 2018; Farhat et al., 2015). Yandigeri et al. (2011) evaluated the rock phosphate and tricalcium phosphate-solubilizing abilities of two diazotrophic cyanobacteria, *Westiellopsis prolifica*, and *Anabaena variabilis*.

5.4 Removal of radionuclides

The prevalent use of radioactive materials in research and development, biomedical, and industrial settings has resulted in a significant buildup of radioactive waste. The commonly encountered radionuclides include cobalt-60 (^{60}Co), plutonium-239 (^{239}Pu), radium-226 (^{226}Ra), radon- 222 (^{222}Rn), technetium-99 (^{99}Tc), thorium-232 (^{232}Th), and uranium-238 (^{238}U). Long-term exposure to radionuclides has been linked to an increased risk of leukemia, leukopenia, kidney damage, and genetic damage, all of which can result in fatal health problems that can be passed down to the next generation (Mohner et al., 2006). Currently, excavation and shipping to a distant waste disposal location is the most common method of eradicating radionuclide-contaminated soil, but it is costly. Given the high costs of physiochemical approaches, there has been unprecedented interest in microbes and plants containing radionuclides for decontamination of nuclear waste-affected sediments and waters (Kumar et al., 2007).

Microorganisms such as *Rhodanobacter* sp. and *Desulfuromus aferrireducens* were observed to be able to interact with these contaminants (Green et al., 2012). The interaction of site-specific microorganisms initiates the solubility of transformed radionuclides through electron addition or removal, increasing the mobility of the contaminant and allowing it to be easily flushed from the environment. Wildung and colleagues (2000) reported enzymatic reduction of U(VI) on the surface of the microorganism *Shewanella putrefaciens.* The direct microbial enzymatic reduction of Tc(VII) was first observed by Lloyd and Macaskie (1996) using *S. putrefaciens* and *Geobacter metallireducens*. *Desulfovibrio desulfuricans* and related organisms were observed to utilize formate as an efficient electron donor for Tc(VII) reduction. Although U and Tc are still the most important radionuclide pollutants in most medium- and low-level radioactive wastes, other actinides such as Th, Np, Pu, and Am can also be found at contaminated sites (Lloyd and Macaskie, 1996; Tamponnet and Declerck, 2008). Iron-reducing bacteria such as *Geobacter* sp. and *Rhodoferax ferrireducens* have the metabolic potential to reduce these radionuclides enzymatically (Kim et al., 2012). Several microorganisms, including *Microbacterium flavescens*, produced unidentified compounds such as organic acids, siderophores, and extracellular metabolites capable of dissolving and mobilizing radionuclides into the soil when grown in the presence of other radionuclides (e.g., U, Th, Am, and Pu). Premuzic and colleagues (1985) reported the production of extracellular chelating agents in *P. aeruginosa* that can bioaccumulate uranium. The microorganism *D. radiodurans* has been studied to detoxify Cr(VI), U(VI), and Tc(VII) from soil (Fredrickson et al., 2000). A recombinant *E. coli* strain with five times the sorption ability for U(VI) was developed (Beckwith et al., 2001).

6. Factors affecting microbial bioremediation

Bioremediation is the process of degrading, removing, altering, immobilizing, or detoxifying various chemicals and physical wastes from the environment using bacteria, fungi, and plants. Through their enzymatic pathways, microorganisms act as biocatalysts, facilitating the progression of biochemical reactions that degrade the desired pollutant. Many factors influence the efficiency of bioremediation, including the chemical nature and concentration of pollutants, the physicochemical characteristics of the environment, and their availability to microorganisms (El Fantroussi and Agathos, 2005). Because bacteria and pollutants do not come into contact with each other, the rate of degradation is slowed.

Furthermore, microbes and pollutants are not evenly distributed in the environment. Because of the numerous factors involved, controlling and optimizing bioremediation processes is a complex system. The presence of a microbial population capable of degrading pollutants, the availability of contaminants to the microbial population, and environmental factors (type of soil, temperature, pH, the presence of oxygen or other electron acceptors, and nutrients) play a significant role.

6.1 Biological factors

Organic compound degradation is influenced by biotic factors such as competition between microorganisms for limited carbon sources, antagonistic interactions between microorganisms, and predation of microorganisms by protozoa and bacteriophages. The rate of contaminant degradation is frequently determined by the contaminant concentration and the amount of "catalyst" present. The amount of "catalyst" in this context represents the number of organisms capable of metabolizing the contaminant as well as the amount of enzymes(s) produced by each cell. Furthermore, specific enzymes must be involved in contaminant metabolism, as well as their "affinity" for the contaminant and availability of the contaminant. Mutation, horizontal gene transfer, enzyme activity, interaction (competition, succession, and predation), its own growth until critical biomass is reached, population size, and composition are all included.

6.2 Environmental factors

The metabolic properties of the microorganisms and the physicochemical properties of the targeted contaminants determine the possibility of interaction during the process. The actual successful interaction between the two, on the other hand, is dependent on the environmental conditions of the interaction site. pH, temperature, moisture, soil structure, solubility in water, nutrients, site characteristics, redox potential, and oxygen content all have an impact on microorganism growth and activity, as do a lack of trained human resources in this field and the physicochemical bioavailability of pollutants (contaminant concentration, type, solubility, chemical structure, and toxicity). Biodegradation can occur in a wide range of pH; however, in most aquatic and terrestrial systems, a pH of 6.5–8.5 is generally optimal for biodegradation. Moisture affects the rate of contaminant metabolism by influencing the type and amount of soluble materials available, as well as the osmotic pressure and pH of terrestrial and aquatic systems.

6.2.1 Availability of nutrients

The addition of nutrients changes the essential nutrient balance for microbial growth and reproduction, as well as the rate and effectiveness of biodegradation. By optimizing the bacterial C: N: P ratio, nutrient balancing, particularly the supply of essential nutrients such as N and P, can improve biodegradation efficiency. Microorganisms require a variety of nutrients to survive and continue their microbial activities, including carbon, nitrogen, and phosphorous. The extent of hydrocarbon degradation is also limited in low concentrations.

6.2.2 Temperature

Temperature is the most important physical factor in determining the survival of microorganisms and the composition of hydrocarbons (Das and Chandran, 2011). Oil degradation by natural processes is very slow in cold environments such as the Arctic, putting microbes under more pressure to clean up the spilled petroleum. Biological enzymes involved in the degradation pathway have an optimum temperature and do not have the same metabolic turnover at all temperatures. Furthermore, the degradation process for a specific compound necessitates a specific temperature. Temperature can also speed up or slow down the bioremediation process because it has a large impact on microbial physiological properties.

6.2.3 Concentration of oxygen

Different organisms require oxygen, while others do not. Depending on their needs, the biodegradation rate can be ramped up. Because oxygen is a gaseous requirement for most living organisms, biological degradation occurs in both aerobic and anaerobic conditions. In most cases, the presence of oxygen can improve hydrocarbon metabolism.

6.2.4 Moisture content

Microorganisms require a sufficient amount of water to grow. The moisture content of the soil has a negative impact on biodegradation agents.

6.2.5 pH

The pH of a compound, which represents its acidity, basicity, and alkalinity, has an effect on microbial metabolic activity as well as the rate of removal. The pH of soil could indicate the potential for microbial growth (Asira et al., 2013).

6.2.6 Metal ions

Metals are important in small amounts for bacteria and fungi, but in large amounts, they inhibit cell metabolic activity. Metal compounds have an impact on the rate of degradation in both direct and indirect ways.

6.2.7 Toxic compounds

Some contaminants, when present in high concentrations, can cause toxic effects on microorganisms and slow down decontamination. The degree and mechanisms of toxicity vary depending on the toxicant, its concentration, and the microorganisms exposed.

7. Benefits and challenges

Microorganisms improve soil structure by improving aggregation and stability, as well as increasing nutrient bioavailability through nitrogen fixation and mobilization of key nutrients (phosphorus, potassium, and iron) to crop plants (Owen et al., 2015; Nguyen and Bruns, 2015). Under field conditions, the success rate of such inocula is determined by their antagonistic or synergistic interaction with indigenous microbes, as well as their inoculation with organic fertilizers. Coinoculation of bacteria and fungi with or without organic fertilizer is more beneficial than single inoculum for restoring soil fertility and organic matter content (Caesar-Tonthat et al., 2014). Such considerations are critical when considering bacteria and fungi inocula for soil restoration. This will aid agriculturists in developing a sustainable management strategy for restoring the fertility of degraded soil and assisting them in reducing the negative environmental impact of artificial fertilizers.

These microorganisms also promote shoot and root growth by producing growth hormones, control or suppression of disease, and improved soil structure. However, the use of microorganisms necessitates specific oxygen, temperature, pH, and moisture conditions that must be monitored on a regular basis. Furthermore, before employing microbes, their compatibility with soil microorganisms should be thoroughly investigated. This technique necessitates proper expertise to ensure that no negative effects on the environment occur. To meet these challenges, proper microorganism research should be conducted, as well as the use of newer methods such as genetic engineering and MFCs can be explored.

8. Conclusion

Agricultural soil contamination and its consequences cause biodiversity loss and pose serious threats to soil health and productivity. To reduce the uptake of contaminants by agricultural crops from contaminated soils, some practical and cost-effective measures should be developed. Bioremediation with microbes has proven to be the most cost-effective and environmentally friendly method of removing soil contaminants. GEMs and MFCs have the potential to effectively remediate contamination by removing organic and inorganic pollutants; however, their use should be subject to stringent biosafety procedures to ensure that there are no health or environmental risks. More efficient methods of using transgenic microbes should be developed, allowing for effective remediation of polluted environments without horizontal transfer of recombinant plasmids or pollens to indigenous organisms, which is currently a major challenge.

This chapter discusses various bioremediation techniques and emphasizes the use of microorganisms to remediate soil pollutants effectively. It also demonstrates the utility of bioremediation as a better substitute for the removal of organic and inorganic pollutants when compared to less efficient and more expensive physicochemical methods. However, adequate environmental conditions must be met and monitored in order for bioremediation to be effective. Furthermore, microorganisms not only remediate pollutants but also aid in soil restoration by maintaining the soil's nutrient balance. As a result, microorganisms can be used as an alternative to many synthetic fertilizers.

References

Achal, V., Kumari, D., Pan, X., 2011. Bioremediation of chromium contaminated soil by a brown-rot fungus, Gloeophyllum sepiarium. Res. J. Microbiol. 6, 166.

Achal, V., Pan, X., Fu, Q., Zhang, D., 2012. Biomineralization based remediation of As (III) contaminated soil by Sporosarcina ginsengisoli. J. Hazard. Mater. 201, 178–184.

Al-Jawhari, I.F.H., 2014. Ability of some soil fungi in biodegradation of petroleum hydrocarbon. J. Appl. Env. Microbiol. 2, 46–52.

Alleman, B.C., Loan, B.E., Gilbertson, R.L., 1995. Degradation of pentachlorophenol by fixed films of whiterot-fungi in rotating tube bioreactors. Water Res. 29, 61–67.

Allison, S.D., Vitousek, P.M., 2005. Responses of extracellular enzymes to simple and complex nutrient inputs. Soil Biol. Biochem. 37, 937–944.

Apitz, S.E., 2008. Is risk based, sustainable sediment management consistent with European policy. J. Soils Sediments 8, 461–466.

Asira, E., 2013. Factors that determine bioremediation of organic compounds in the soil. Aca. J. Interdiscipl. Stud. 2, 125–128.

Ayyasamy, P.M., Chun, S., Lee, S., 2009. Desorption and dissolution of heavy metals from contaminated soil using *Shewanella sp*. (HN-41) amended with various carbon sources and synthetic soil organic matters. J. Hazard Mater. 161, 1095–1102.

Azad, M.A.K., Amin, L., Sidik, N.M., 2014. Genetically engineered organisms for bioremediation of pollutants in contaminated sites. Chin. Sci. Bull. 59, 703–714.

Balamurugan, D., Udayasooriyan, C., Kamaladevi, B., 2014. Chromium (VI) reduction by *Pseudomonas putida* and *Bacillus subtilis* isolated from contaminated soils. Int. J. Environ. Sci. 5, 522.

Beckwith, C.S., McGee, D.J., Mobley, H.L., Riley, L.K., 2001. Cloning, expression, and catalytic activity of *Helicobacter hepaticus* urease. Infect. Immun. 69, 5914–5920.

Bernhard, A., 2010. The nitrogen cycle: processes, players, and human impact. Nat. Educ. Knowl. 3, 25.

Bi, Q.F., Zheng, B.X., Lin, X.Y., Li, K.J., Liu, X.P., Hao, X.L., Zhang, H., Zhang, J.B., Jaisi, D.P., Zhu, Y.G., 2018. The microbial cycling of phosphorus on long-term fertilized soil: insights from phosphate oxygen isotope ratios. Chem. Geol. 483, 56–64.

Bottomley, P., Taylor, A.E., Myrold, D.D., 2012. A consideration of the relative contributions of different microbial subpopulations to the soil cycle. Front. Microbiol. 3, 373.

Bumpus, J.A., 1998. Biodegradation of polycyclic aromatic hydrocarbons by *Phanerochaete chrysosporium*. Appl. Environ. Microbiol. 55, 154–158.

Burghal, A.A., Najwa, M.J.A., Al-Tamimi, W.H., 2016. Mycodegradation of crude oil by fungal species isolated from petroleum contaminated soil. Int. J. Innov. Res. Sci. Eng. Technol. 5, 1517–1524.

Cabello, P., Roldan, M.D., Moreno-Vivian, C., 2004. Nitrate reduction and the nitrogen cycle in archaea. Microbiology 150, 3527–3546.

Caesar-Tonthat, T.C., Stevens, W.B., Sainju, U.M., Caesar, A.J., West, M., Gaskin, J.F., 2014. Soil-aggregating bacterial community as affected by irrigation, tillage, and cropping system in the northern great plains. Soil Sci. 179, 11–20.

Chaiharn, M., Pathom-aree, W., Sujada, N., Lumyong, S., 2018. Characterization of phosphate solubilizing *Streptomyces* as a biofertilizer. Chiang Mai J. Sci. 45, 701–716.

Coelho, L.M., Rezende, H.C., Coelho, L.M., de Sousa, P.A., Melo, D.F., Coelho, N.M., 2015. Bioremediation of polluted waters using microorganisms. In: Shiomi, N. (Ed.), Advances in Bioremediation of Wastewater and Polluted Soil. InTech, Shanghai, China.

Cong, W.F., Suriyagoda, L.D., Lambers, H., 2020. Tightening the phosphorus cycle through phosphorus-efficient crop genotypes. Trends Plant Sci. 25 (10), 967–975.

Coppella, S.J., Cruz, N.D., Payne, G.F., Pogell, B.M., Speedie, M.K., Karns, J.S., Sybert, E.M., Connor, M.A., 1990. Genetic engineering approach to toxic waste management case study for organophosphate waste treatment. Biotechnol. Prog. 6, 76–81.

Cristaldi, A., Conti, G.O., Jho, E.H., Zuccarello, P., Grasso, A., Copat, C., Ferrante, M., 2017. Phytoremediation of contaminated soils by heavy metals and PAHs. A Brief Review. Env.Tech and Innv. 8, 309–326.

Das, N., Chandran, P., 2011. Microbial degradation of petroleum hydrocarbon contaminants: an overview. Biotechnol. Res. Int. 941810.

Deckwer, W.D., Becker, F., Ledakowicz, S., Wagner-Döbler, I., 2004. Microbial removal of ionic mercury in a three-phase fluidized bed reactor. Environ. Sci. Technol. 38, 1858–1865.

Dilfuza, E., 2011. Role of microorganisms in nitrogen cycling in soils. In: Minarsari, M. (Ed.), Soil Nutrients. Nova Science Publishers Inc. Hauppauge., NJ, USA, pp. 159–176.

Dong, S., Brooks, D., Jones, M.D., Grayston, S.J., 2007. A Method for linking in situ activities of hydrolytic enzymes to associated organisms in forest soils. Soil Biol. Biochem. 39, 2414–2419.

Dong, G., Wang, Y., Gong, L., Wang, M., Wang, H., He, N., Zheng, Y., Li, Q., 2013. Formation of soluble Cr (III) end-products and nanoparticles during Cr (VI) reduction by *Bacillus cereus* strain XMCr-6. Biochem. Eng. J. 70, 166–172.

Donot, F., Fontana, A., Baccou, J., Schorr-Galindo, S., 2012. Microbial exopolysaccharides: main examples of synthesis, excretion, genetics and extraction. Carbohydr. Polym. 87, 951–962.

El Fantroussi, S., Agathos, S.N., 2005. Is bioaugmentation a feasible strategy for pollutant removal and site remediation? Curr. Opin. Microbiol. 8, 268–275.

Farhan, S.N., Khadom, A.A., 2015. Biosorption of heavy metals from aqueous solutions by *Saccharomyces cerevisiae*. Int. J. Ind. Chem. 6, 119–130.

Farhat, M.B., Boukhris, I., Chouayekh, H., 2015. Mineral phosphate solubilization by *Streptomyces sp.* CTM396 involves the excretion of gluconic acid and is stimulated by humic acids. FEMS Microbiol. Lett. 362.

Fredrickson, J.K., Kostandarithes, H.M., Li, S.W., Plymale, A.E., Daly, M.J., 2000. Reduction of Fe(III), Cr(VI), U(VI), and Tc(VII) by *Deinococcus radiodurans* R1. Appl. Environ. Microbiol. 66, 2006–2011.

Fujii, K., Yamada, T., Hayakawa, C., Nakanishi, A., Funakawa, S., 2020. Decoupling of protein depolymerization and ammonification in nitrogen mineralization of acidic forest soils. Appl. Soil Ecol. 153, 103572.

Fuka, M.M., Braker, S.H.G., Philippot, L., 2007. Molecular tools to assess the diversity and density of denitrifying bacteria in their habitats. In: Biology of the Nitrogen Cycle. Elsevier, Amsterdam, The Netherlands, pp. 313–330.

Grandy, A.S., Neff, J.C., Weintraub, M.N., 2007. Carbon structure and enzyme activities in alpine and forest ecosystems. Soil Biol. Biochem. 39, 2701–2711.

Green, S.J., Prakash, O., Jasrotia, P., Overholt, W.A., Cardenas, E., Hubbard, D., 2012. Denitrifying bacteria from the genus *Rhodanobacter* dominate bacterial communities in the highly contaminated subsurface of a nuclear legacy waste site. Appl. Environ. Microbiol. 78, 1039–1047.

Gupta, P., Diwan, B., 2016. Bacterial exopolysaccharide mediated heavy metal removal: a review on biosynthesis, mechanism and remediation strategies. Biotechnol. Rep. 13, 58–71.

Hamdali, H., Bouizgarne, B., Hafidi, M., Lebrihi, A., Virolle, M.J., Ouhdouch, Y., 2008. Screening for rock phosphate solubilizing *Actinomycetes* from Moroccan phosphate mines. Appl. Soil Ecol. 38, 12–19.

Han, P., Gu, J.-D., 2013. More refined diversity of anammox bacteria recovered and distribution in different ecosystems. Appl. Microbiol. Biotechnol. 97, 3653–3663.

Hassan, M.M., Alam, M.Z., Anwar, M.N., 2013. Biodegradation of textile azo dyes by bacteria isolated from dyeing industry effluent. Int. Res. J. Biol. Sci. 2 (8), 27–31.

Hoffman, K., Butt, C.M., Webster, T.F., Preston, E.V., Hammel, S.C., Makey, C., Lorenzo, A.M., Cooper, E.M., Carignan, C., Meeker, J.D., 2017. Temporal trends in exposure to organophosphate flame retardants in the United States. Environ. Sci. Technol. Lett. 4, 112–118.

Johansen, B.E., 2003. The Dirty Dozen: Toxic Chemicals and the Earth's Future. Praeger Publishers, Westport, CT.

Kang, C.-H., Kwon, Y.-J., So, J.-S., 2016. Bioremediation of heavy metals by using bacterial mixtures. Ecol. Eng. 89, 64–69.

Kim, S.J., Koh, D.C., Park, S.J., Cha, I.T., Park, J.W., Na, J.H., 2012. Molecular analysis of spatial variation of iron-reducing bacteria in riverine alluvial aquifers of the Mankyeong. River. J. Microbiol. 50, 207–217.

Korda, A., Santas, P., Tenente, A., Santas, R., 1997. Petroleum hydrocarbon bioremediation: sampling and analytical techniques, in situ treatments and commercial microorganisms currently used. Appl. Microbiol. Biotechnol. 48, 677–686.

Kumar, R., Singh, S., Singh, O.V., 2007. Bioremediation of radionuclides: emerging technologies. OMICS 11, 295–304.

Lamb, J.A., Fernandez, F.G., Kaiser, D.E., 2014. Understanding Nitrogen in Soils. University of Minnesota, Minneapolis, MN, USA, pp. 1–5.

Liang, J.L., Liu, J., Jia, P., Yang, T.T., Zeng, Q.W., Zhang, S.C., Liao, B., Shu, W.S., Li, J.T., 2020. Novel phosphate-solubilizing bacteria enhance soil phosphorus cycling following ecological restoration of land degraded by mining. ISME J. 14, 1600–1613.

Liang, X., Csetenyi, L., Gadd, G.M., 2016. Lead bioprecipitation by yeasts utilizing organic phosphorus substrates. Geomicrobiol. J. 33, 294–307.

Libudzisz, Z., Kowal, K., Zakowska, Z., 2007. Technical Microbiology. Microorganisms and Their Environment. Scientific Publisher PWN, Warsaw, Poland.

Lin, C., Gan, L., Chen, Z.L., 2010. Biodegradation of naphthalene by strain *Bacillus fusiformis* (BFN). J. Hazard Mater. 182, 771–777.

Liu, S., Zhang, F., Chen, J., Sun, G., 2011. Arsenic removal from contaminated soil via biovolatilization by genetically engineered bacteria under laboratory conditions. J. Environ. Sci. 23, 1544–1550.

Lívia de, C.F., Benedito, C., 2015. Potential application of modified *Saccharomyces cerevisiae* for removing lead and cadmium. J. Bioremed. Biodegrad. 6, 2.

Lloyd, J.R., Macaskie, L.E., 1996. A novel Phosphor Imager-based technique for monitoring the microbial reduction of technetium. Appl. Environ. Microbiol. 62, 578–582.

Long, A., Heitman, J., Tobias, C., Philips, R., Song, B., 2013. Co-occurring anammox, denitrification, and codenitrification in agricultural soils. Appl. Environ. Microbiol. 79, 168–176.

Mackay, J.E., Cavagnaro, T.R., Müller Stöver, D.S., Macdonald, L.M., Grønlund, M., Jakobsen, I.A., 2017. Key role for arbuscular mycorrhiza in plant acquisition of P from sewage sludge recycled to soil. Soil Biol. Biochem. 115, 11–20.

Maliji, D., Olama, Z., Holail, H., 2013. Environmental studies on the microbial degradation of oil hydrocarbons and its application in Lebanese oil polluted coastal and marine ecosystem. Int. J. Curr. Microbiol. App. Sci. 2, 1–18.

Marconi, A.M., Kieboom, J., de Bont, J.A., 1997. Improving the catabolic functions in the toluene-resistant strain pseudomonas putidas. Biotechnol. Lett. 19, 603–606.

Mohner, M., Lindtner, M., Otten, H., Gille, H.G., 2006. Leukemia and exposure to ionizing radiation among German uranium miners. Am. J. Ind. Med. 49, 238–248.

Mulbry, W.W., 1998. Selective deletions involving the organophosphorus hydrolase gene adpB from *Nocardia* strain B-1. Microbiol. Res. 153, 213–217.

Nguyen, N., Bruns, T., 2015. The microbiome of *Pinus muricata* ectomycorrhizae: community assemblages, fungal species effects, and burkholderia as important bacteria in multipartnered symbioses. Microb. Ecol. 1–8.

Nicell, J., 2001. Environmental applications of enzymes. Interdiscipl. Environ. Rev. 3, 14–41.

Novotny, C., Vyas, B.R.M., Erbanova, P., Kubatova, A., Sasek, V., 1997. Removal of PCBs by various white rot fungi in liquid cultures. Folia Microbiol. 42 (2), 136–140.

Owen, D., Williams, A.P., Griffith, G.W., Withers, P.J.A., 2015. Use of commercial bio-inoculants to increase agricultural production through improved phosphrous acquisition. Appl. Soil Ecol. 86, 41–54.

Pointing, S.B., Vrijmoed, L.L.P., 2000. Decolorization of azo and tri-phenylmethane dyes by *Pycnoporus sanguineus* producing laccase as the sole phenoloxidase. World J. Microbiol. Biotechnol. 16, 317–318.

Premuzic, E.T., Francis, A.J., Lin, M., Schubert, J., 1985. Induced formation of chelating agents by *Pseudomonas aeruginosa* grown in presence of thorium and uranium. Arch. Environ. Contam. Toxicol. 14, 759–768.

Rabus, R., Boll, M., Heider, J., Meckenstock, R., Buckel, W., 2016. Anaerobic microbial degradation of hydrocarbons: from enzymatic reactions to the environment. J. Mol. Microbiol. Biotechnol. 26, 5–28.

Rana, A., Pandey, R.K., Ramakrishnan, B., 2019. Enzymology of the nitrogen cycle and bioremediation of toxic nitrogenous compounds. In: Smart Bioremediation Technologies. Elsevier, Amsterdam, The Netherlands, pp. 45–61.

Rojas, L.A., Yáñez, C., González, M., Lobos, S., Smalla, K., Seeger, M., 2011. Characterization of the metabolically modified heavy metal-resistant *Cupriavidus metallidurans* strain MSR33 generated for mercury bioremediation. PLoS One 6, e17555.

Safi, yanu I., Isah, A.A., Abubakar, U.S., Singh, R.M., 2015. Review on comparative study on bioremediation for oil spills using microbes. Res. J. Pharmaceut. Biol. Chem. Sci. 6, 783–790.

Sarang, B., Richa, K., Ram, C., 2013. Comparative study of bioremediation of hydrocarbon fuel. Int. J. Biotechnol. Bioeng Res. 4, 677–686.

Schmid, M.C., Maas, B., Dapena, A., Van de Pas-Schoonen, K., Van de Vossenberg, J., Kartal, B., Van Niftrik, L., Schmidt, I., Cirpus, I., Kuenen, J.G., 2005. Biomarkers for in situ detection of anaerobic ammonium-oxidizing (anammox) bacteria. Appl. Environ. Microbiol. 71, 1677–1684.

Schulze, E.-D., Mooney, H.A., 2002. Biodiversity and Ecosystem Function. Springer, Berlin/Heidelberg, Germany.

Sivakumar, G., Xu, J., Thompson, R.W., Yang, Y., Randol-Smith, P., 2012. Integrated green algal technology for bioremediation and biofuel. Bioresour. Technol. 107, 1–9.

Sone, Y., Mochizuki, Y., Koizawa, K., Nakamura, R., Pan-Hou, H., Itoh, T., Kiyono, M., 2013. Mercurial-resistance determinants in *pseudomonas* strain k-62 plasmid pmr68. Amb. Express 3, 41.

Spott, O., Russow, R., Stange, C.F., 2011. Formation of hybrid N_2O and hybrid N_2 due to codenitrification: first review of a barely considered process of microbially mediated N-nitrosation. Soil Biol. Biochem. 43, 1995–2011.

Srikanth, S., Kumar, M., Singh, D., Singh, M.P., Das, B.P., 2016. Bioresour. Technol. 221, 70–77.

Srivastava, S., Anil Dwivedi, K., 2015. Biological wastes the tool for biosorption of arsenic. J. Bioremed. Biodegrad. 7, 2.

Strous, M., Fuerst, J.A., Kramer, E.H.M., Logemann, S., Muyzer, G., Van de Pas-Schoonen, K.T., Webb, R., Kuenen, J.G., Jetten, M.S.M., 1999. Missing lithotroph identified as new Planctomycete. Nature 400, 446–449.

Sukumar, M., 2010. Reduction of hexavalent chromium by *Rhizopus oryzae*. Afr. J. Environ. Sci. Technol. 4, 412–418.

Sumantha, A., Larroche, C., Pandey, A., 2006. Microbiology and industrial biotechnology of food-grade proteases: a perspective. Food Technol. Biotechnol. 44, 211.

Ta stan, B.E., Ertu grul, S., Donmez, G., 2010. Effective bioremoval of reactive dye and heavy metals by *Aspergillus versicolor*. Bioresour. Technol. 101, 870–876.

Tamponnet, C., Declerck, S., 2008. Radionuclide (RN) pollution is a worldwide problem that arises from human activities. J. Environ. Radioact. 99, 773–774.

Tang, C.Y., Criddle, C.S., Leckie, J.O., 2007. Effect of flux (trans membrane pressure) and membranes properties on fouling and rejection of reverse osmosis and nano filtration membranes treating perfluorooctane sulfonate containing waste water. Environ. Sci. Technol. 41, 2008–2014.

Turan, M., Ataőglu, N., ¸Sah_N, F., 2006. Evaluation of the capacity of phosphate solubilizing bacteria and fungi on different forms of phosphorus in liquid culture. J. Sustain. Agric. 28, 99–108.

Vullo, D.L., Ceretti, H.M., Daniel, M.A., Ramírez, S.A., Zalts, A., 2008. Cadmium, zinc and copper biosorption mediated by *Pseudomonas veronii* 2e. Bioresour. Technol. 99, 5574–5581.

Wildung, R.E., Gorby, Y.A., Krupka, K.M., Hess, N.J., Li, S.W., Plymale, A.E., 2000. Effect of electron donor and solution chemistry on products of dissimilatory reduction of technetium by *Shewanella putrefaciens*. Appl. Environ. Microbiol. 66, 2451–2460.

Yan, J., Niu, J., Chen, D., Chen, Y., Irbis, C., 2014. Screening of *Trametes* strains for efficient decolorization of malachite green at high temperatures and ionic concentrations. Int. Biodeterior. Biodegrad. 87, 109–115.

Yandigeri, M.S., Yadav, A.K., Srinivasan, R., Kashyap, S., Pabbi, S., 2011. Studies on mineral phosphate solubilization by *cyanobacteria Westiellopsis* and *Anabaena*. Microbiology 80, 558–565.

Yegin, S., Fernandez-Lahore, M., Salgado, A.J.G., Guvenc, U., Goksungur, Y., Tari, C., 2011. Aspartic proteinases from *Mucor spp*. in cheese manufacturing. Appl. Microbiol. Biotechnol. 89, 949–960.

Zai, X.M., Zhang, H.S., Hao, Z.P., 2017. Effects of arbuscular mycorrhizal fungi and phosphate-solubilizing fungus on the rooting, growth and rhizosphere niche of beach plum (prunus maritima) cuttings in a phosphorus-deficient. Soil. J. Am. Pomol. Soc. 71, 226–235.

Zaidi, A., Ahemad, M., Oves, M., Ahmad, E., Khan, M.S., 2010. Role of phosphate-solubilizing bacteria in legume improvement. In: Microbes for Legume Improvement. Springer, Cham. Switzerland, pp. 273–292.

CHAPTER

14 Recent advances in dyes removal technologies from wastewater using biochar

Piyush Gupta[1] and Neha Rana[2]

[1]*Department of Chemistry, Faculty of Engineering and Technology, SRM Institute of Science and Technology, Ghaziabad, Uttar Pradesh, India;* [2]*Department of Pharmaceutical Sciences, KIET School of Pharmacy, KIET Group of Institutions, Ghaziabad, Uttar Pradesh, India*

1. Introduction

Competition for water resources, water shortages, poor water quality, and unpredictability of hydrological events are mostly caused by climate change, population growth, rising living standards, and uneven water distribution. Water is essential for sustainable development, but it is not equally available, and clean water is out of reach in many parts of the world (Mekonnen and Hoekstra, 2016). The impact of environmental stress on water bodies may be seen not only in terms of quantity but also in terms of quality. Worldwide, the globalization of industrialization has resulted in a high level of contamination of water sources. Dyeing industries, tanneries, paper industries, metal-plating industries, mining activities, agricultural waste, fertilizer industries, and pesticides are the major polluters of water reservoirs. During the first half of the 21st century, industrial water consumption is predicted to rise by 283%, resulting in increased industrial wastewater discharges. Colors appeal to people's aesthetic senses; thus, they will continue to be used in different products. Colors are usually employed in a number of industries like food industries, textile industries, tanneries, plastic industries, pharmaceutical industries, etc. Many of these enterprises' products contain a range of colors from different classes of dyes, resulting in a complicated wastewater as shown in Fig. 14.1 (Dutta, 2021). Industrial dyes, especially those employed in the textile sector, contain complex molecular structures that are synthetic in origin and resistant to degradation. The majority of chemicals are used in the dyeing process, which involves adding a color to the dye pools and immersing the cloth in the dye baths until the dye is fixed. There is a need to add salt to the bath in some situations, and tonnes of synthetic dye are created each year, with 10%–15% of it not making it into the finished product, damaging the environment further. Because the dyes are not completely attached to the fiber (wool, nylon, cotton, viscose, polyester, and acrylic) of different textiles during the dyeing process, the effluent also contains a substantial quantity of refractory unfixed dyes (such as basic dyes, acid dyes, chrome dyes, sulfur dyes, optical/fluorescent brightener, and azoic dyes).

The textile sector uses a lot of water in its manufacturing process, and it's also one of the biggest generators of wastewater with carcinogenic and mutagenic substances (Rubeena et al., 2018).

Integrative Strategies for Bioremediation of Environmental Contaminants, Volume Two. https://doi.org/10.1016/B978-0-443-14013-6.00008-1

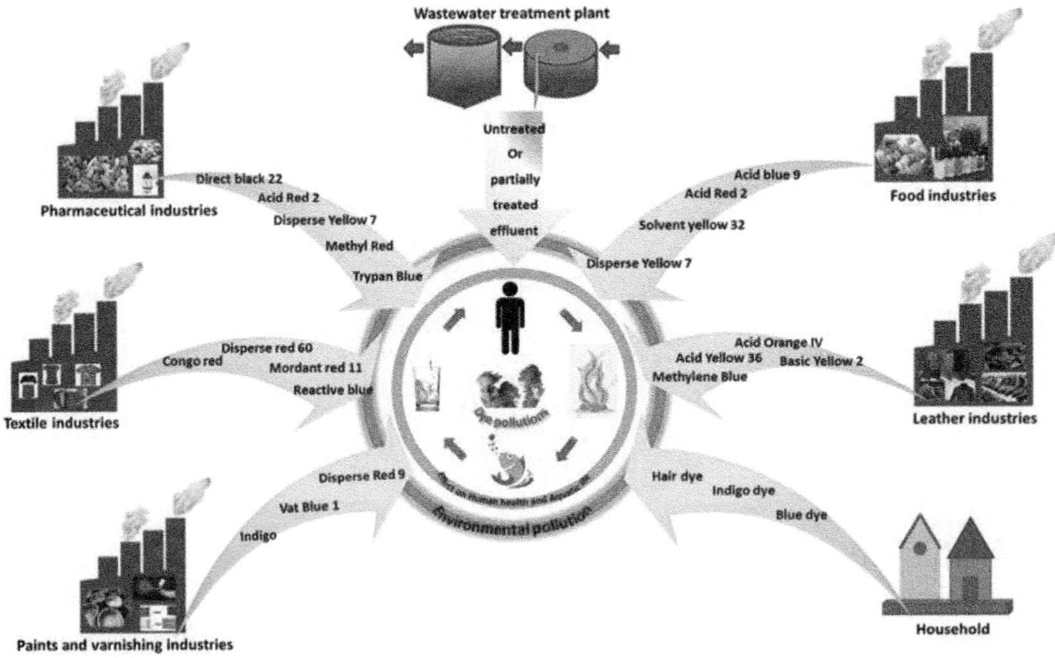

FIGURE 14.1

Dyes: Sources and pathways in the environment (Dutta, 2021).

Textile sector wastewater is typically released straight into lakes and rivers without treatment, and these water sources are widely used by people for household purposes (Barredo-Damas et al., 2010). Due to wastewater complexity, such as high solubility, nondegradability, and often changing speciation in water, dyes cannot be removed using traditional treatment unit operations. Conventional and advanced approaches for removing or degrading these persistent and emerging contamination due to dyes are limited due to high capital requirements, a lack of adaptable technical tools, societal hurdles, and a focus on centralized systems. As a result, novel solutions and extensive decentralized systems in this sector that are appropriate for rural regions and capital-constrained nations are urgently needed to bridge the gap. Different methods can be based upon physical approaches (filtration, sedimentation, coagulation, floatation, distillation, reverse osmosis, solvent extraction, and adsorption), chemical approaches (neutralization, oxidation, reduction, catalysis, ion exchange, and electrolysis), and biological approaches (aerated lagoons, trickling filters, activated sludge, stabilization, and anaerobic digestion) (Mostafa, 2015). Physical techniques, such as adsorption, filtration, and membrane filtration, are highly widespread in dye wastewater treatment due to their excellent color removal effectiveness. Other treatment methods, including nanofiltration, reverse osmosis, and multiple effect evaporators, are effective but costly, whereas typical treatment methods are nondestructive, less expensive, time-consuming, and inefficient (Piaskowski et al., 2018). Adsorption by porous materials is one of the most promising and cost-effective methods for removing dissolved contaminants and may be used instead of energy-intensive methods (Anjum et al., 2017). Biochar (BC), which is inexpensive, plentiful, and may have equivalent adsorption capability to activated carbon, can be employed as a cost-effective alternative (Inyang and Dickenson, 2015). The efficiency, local manufacture, availability, and prices, on the other hand, must all be considered. Low-temperature pyrolysis of carbon-rich biomass (from agricultural and forestry wastes) in low-oxygen circumstances produces biochar (Palansooriya et al., 2020). Recent research trends are shown in Fig. 14.2A, B, C using different keywords like dye wastewater treatment, dyeing technologies, biochar, and other materials (Dutta, 2021; Mamane, 2020).

2. Dyes removal technologies from wastewater using biochar

A dye is defined as a material capable of providing color to a substrate through physical/chemical binding. Synthetic dyes and natural dyes are two types of dyes. Synthetic dyes are more extensively used than natural dyes since they are easier to produce in many hues and are characterized by their fastness. Synthetic dyes are classified into two groups: (a) by their application mode (direct, reactive, basic, dispersion, and vat dying) and (b) by their chemical structure (anthraquinone, azo, phthalocyanine, sulfur, and triarylmethane dyes). Aromatic molecular structures derived from hydrocarbons such as benzene, toluene, naphthalene, anthracene, and xylene are present in these dye molecules. Dyes may also be classified as cationic, anionic, and nonionic dyes comprising different chromophoric and auxochromic groups. The azo dyes are distinguished by their strong polarization and recalcitrance. Amphoteric characteristics can be used to classify azo dyes that contain additional acidic groups such as carboxyl, hydroxyl, or sulfonyl substituents. In comparison to hydrocarbons, the presence of an amino group promotes greater water solubility and mobility, as well as a higher boiling point and a lower Henry's law constant. However, vital applications of dyes in everyday life result in their discharge into the environment together with other harmful organic/inorganic compounds. More than 100,000 different dyes are used in industrial dyeing, according to statistics, and the majority of industrial effluent is released into the water systems. In water, even tiny amounts of color are visible and dangerous to the ecosystems (Guan et al., 2017; Hao et al., 2000). Other components, such as metals and auxiliaries used in dye manufacturing and dyeing processes, are also discharged in addition to dyes. These dissolved dye molecules prevent sunlight from passing through, increase biochemical oxygen demand (BOD), and lower the quantity of dissolved oxygen (DO) of the affected water system. As a result, they degrade the environment and cause illnesses in humans, animals, and plants. The type of the dyes, contaminants, and wastewater composition all influence whether or not a technique is capable of treating the specific wastewater. A number of wastewater treatment methods is given in Fig. 14.3 (Shindhal et al., 2021).

Anionic dyes are water-soluble and difficult to remove using traditional procedures. For the total elimination of acidic and reactive dyes, biological treatments are insufficient. Nonionic dyes, also known as dispersion dyes, do not ionize in aqueous solutions and are very resistant to degradation due to their fused aromatic ring structure. A few cationic dyes, such as methyl blue, can, however, be easily removed using advanced oxidation techniques and adsorption. Fig. 14.4 comprises a flow chart showing different types of dye molecules, ecotoxicological effects of dyes, various adsorbents used for adsorption of dyes, and adsorption mechanisms with influencing factors (Dutta, 2021).

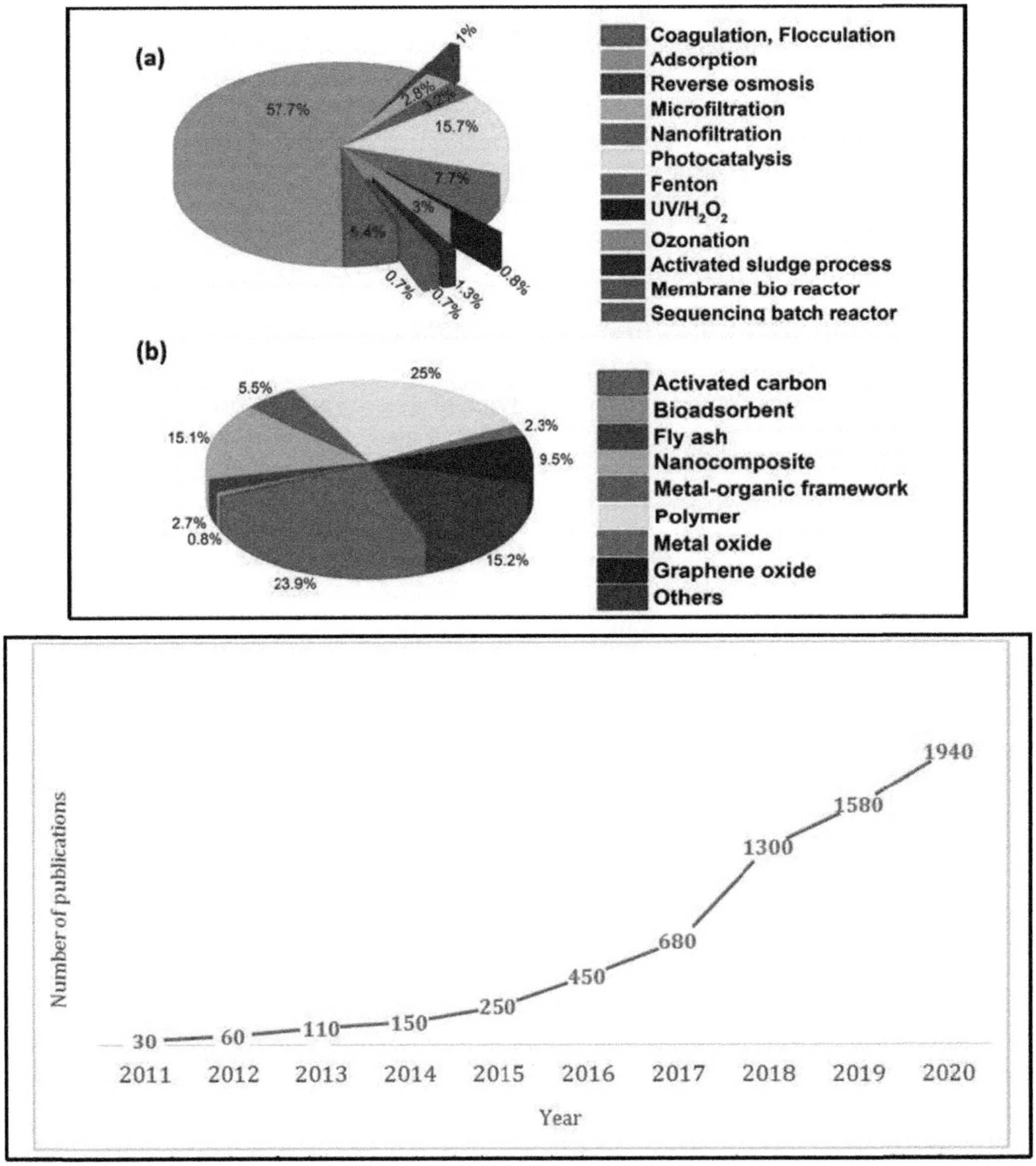

FIGURE 14.2

Pie charts illustrating the literature available (%) on dye removal based on (A) various physicochemical and biological techniques and (B) adsorption techniques using different adsorbents (Dutta, 2021). (C). Frequency of publications on dye removal using keywords: dye sorption, dyeing wastewater, and biochar (Mamane, 2020).

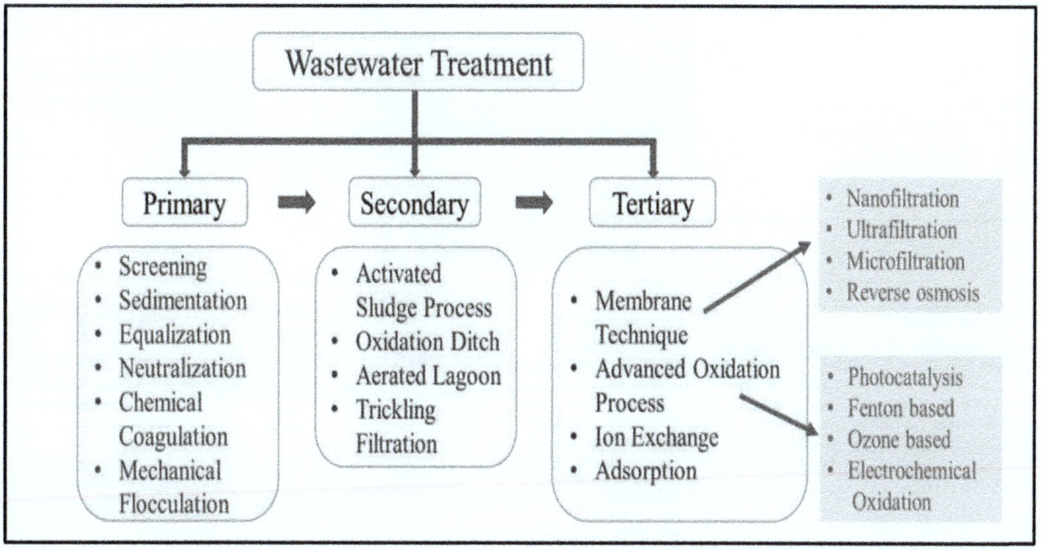

FIGURE 14.3

Classification of wastewater treatment (Shindhal et al., 2021).

A pictorial representation of treatment of dye wastewater using different adsorbents is represented in Fig. 14.5 (Dutta, 2021). As a result, present effluent treatment approaches are insufficient for the dyeing businesses, and one must implement innovative and more effective treatment strategies, such as biochar adsorption, biochar-based layered double hydroxides (LDHs), and biochar-based photocatalysts (BCPs) to prevent continued insemination of surface water with such mutagenic and carcinogenic chemicals.

2.1 Biochar as adsorbents

BC is made from biomass through pyrolysis, gasification, and hydrothermal carbonization. Wood, forestry products, sludge, manure, agricultural residues, and organic wastes are all biomass feedstocks which are used in the preparation of BC. BC has a well-developed mesoporous structure and numerous surface functional groups as compared to most typical activated carbon and is efficiently used as an adsorbent (Fig. 14.6) for the treatment of dye wastewater (Dutta, 2021).

The production of biochar from sewage sludge (SS) is in line with the objective of long-term resource recovery and encourages the use of wastewater in a circular economy. Thermochemical conversion of SS to BC addresses two main challenges at once: it reduces disposal costs and serves as a resource for removing harmful contaminants from water and wastewater. BC is an economically feasible material for wastewater treatment because of its reusability and wide-accessibility, regardless of the season. The use of SS-derived BC as an adsorbent for pollutants in wastewaters, the potential use of BC as a catalyst and support material in advanced oxidation processes, and the use of BCs as an electrode material are all drawn the researcher's attention (Gopinath et al., 2021). Methylene blue

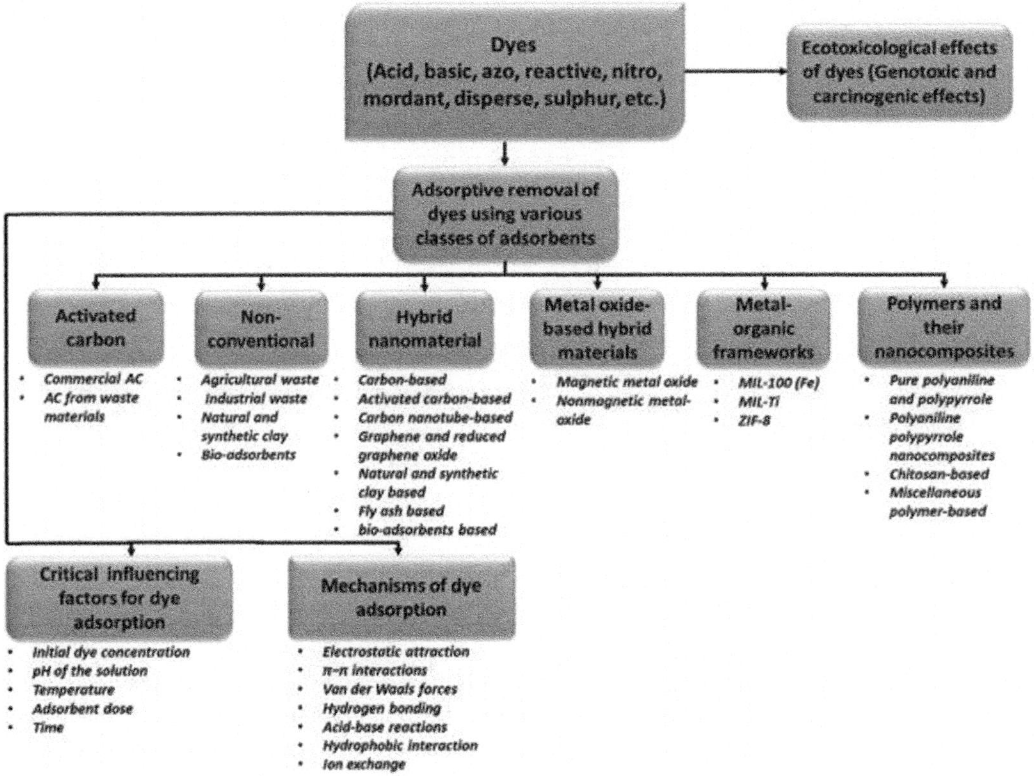

FIGURE 14.4

Flow chart showing different types of dye molecules, dyes removal using various adsorbents, and adsorption mechanisms with influencing factors (Dutta, 2021).

(MB) adsorption on sewage-derived BC was demonstrated by Fan et al. (2016). Hydrogen bonding, ion exchange, and electrostatic attraction are all part of the MB adsorption process on BC. BC was made by mixing dry, powdered SS and tea waste in a 1:1 ratio and pyrolyzing it at 300°C for 2 h, with the resulting biochar being employed for MB adsorptive removal (Fan et al., 2016).

Chen et al. (2019) developed BC by copyrolyzing SS and rice husk (RH) (1:1) at 500°C for 2 h and employed it to remove dye molecules such as Reactive Blue 19 (RB), Acid Orange II (AO), Direct Red 4BS (DR), and MB. Physisorption of the dye molecules on active adsorption sites of BC was confirmed by XRD examination. The impact of pH added to the evidence that electrostatic interactions are important in dye removal by this material, which was corroborated by the zeta potential measurements. External diffusion and intraparticle diffusion both affected the rate-limiting steps, according to kinetic studies. Adsorption isotherm modeling studies demonstrated that both monolayer and multilayer coverage favored dye removal by BC. The removal mechanism varied in accordance with the type of dye molecule and their individual unique interaction with BC. The following order for BC's adsorption capability for the various dyes is: DR > AO > RB > MB. Thermodynamic experiments found that RB

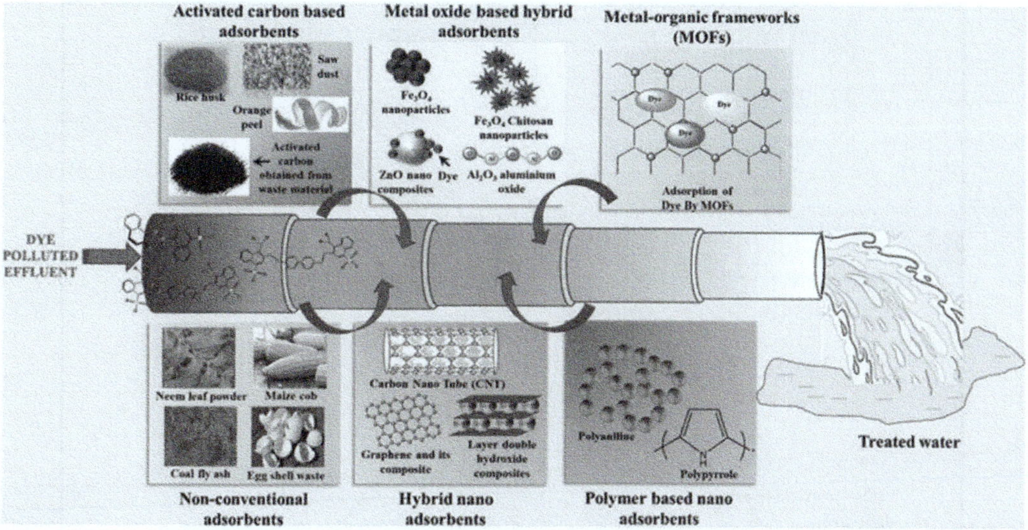

FIGURE 14.5

Pictorial representation of removal of dyes-contaminated effluent using different adsorbents (Dutta, 2021).

and MB adsorption onto biochar was endothermic, but DR and AO adsorption was exothermic (Chen et al., 2019). Leng et al. (2015) employed a liquefaction process with solvents such as ethanol, methanol, and acetone at a temperature of 260–380°C to produce BC, which was then used to adsorb malachite green (MG) and malachite blue (MBE). Although the surface area and pore size of this biochar (12–17 $m^2\ g^{-1}$ and 18.1–20.1 nm, respectively) were found to be smaller than those normally found in BCs, the material was found to be rich in oxygen functional groups. Zhang et al. (2016) synthesized $SrFe_{12}O_{19}$-magnetic BC and eliminated MG. The adsorption capacity was observed to be 388.65 $m^2\ g^{-1}$ for an initial concentration of MG of 500 mg L^{-1} and a BC dose of 2.0 g L^{-1}.

BC was recently prepared using pretreated SS for the adsorption of cationic red X-GRL (Xiao et al., 2018). This approach raised the pore volume and surface area of BC by 33.3% and 43.5%, respectively, resulting in a 49.2% improvement in dye removal efficiency, with an adsorption capacity of 39.1–47.6 $m^2\ g^{-1}$ (Xiao et al., 2018). Parshetti et al. (2013) examined the adsorption of textile colors in wastewater using BC made from food waste. They discovered that an alkaline pH increased dye adsorption. The strong interaction between the negatively charged sites on the BC's surface and the positively charged dyes was attributed to adsorption. At pH 3, however, the efficacy of adsorbing organic dye was reduced due to the presence of additional H^+ ions that competed with the dye's positive charges. Similar findings on the influence of pH on BC's adsorption capacity were published by Tsai and Chen (2013) and Xu et al. (2011). As a result, by varying the charged sites, the pH of the solution changes the adsorption capacity of organic compounds and inorganic contaminants from industrial effluent on BC. According to one study reviewed in Ambaye et al. (2021), BCs were prepared after pyrolyzing raw materials like peanut, cole, and rapeseed straw at 350°C which showed an adsorption capability of 123.5–195.4 mg g^{-1} for methyl violet (MV). At room temperature, rapeseed

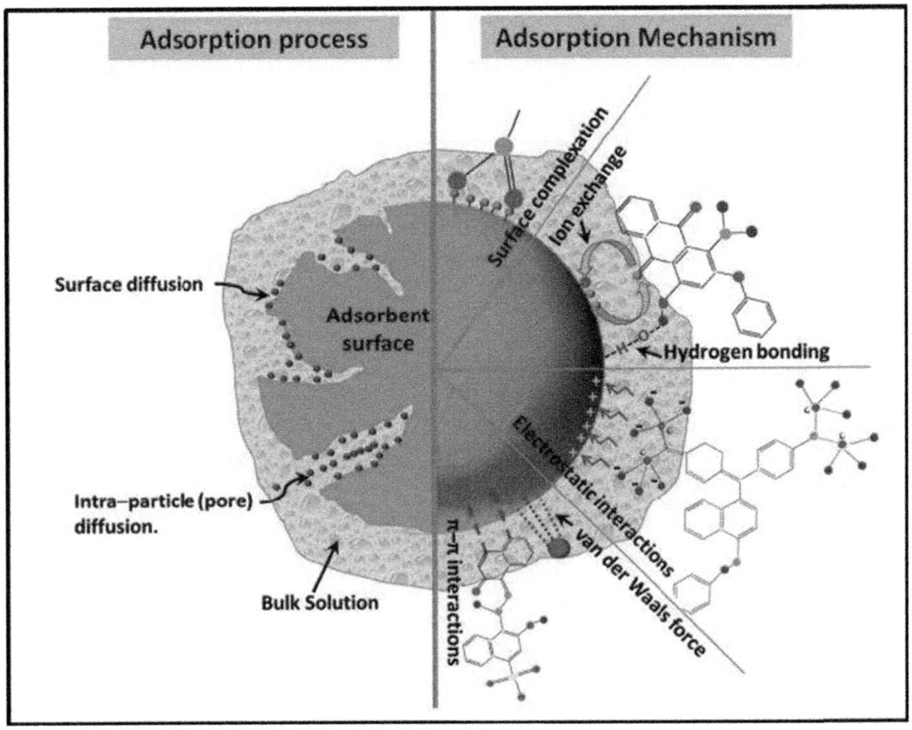

FIGURE 14.6

Diagrammatic illustration of adsorbent surface, adsorption process, and adsorption mechanisms (Dutta, 2021).

straw-BC showed highest adsorption capability. Zeta potential and Fourier transform infrared (FTIR) analyses revealed the presence of electrostatic attraction between MV and charcoal. MV was mostly adsorbed on the hydrophilic and COO^- sites, according to the findings (Ambaye et al., 2021). Adekola et al. (2019) developed, characterized an activated BC made from plantain peels, and observed RB adsorption data set. Mohan et al. (2014) reviewed the literature and compiled different investigations on adsorption of organic, inorganic pollutants, including heavy metals, dyes, pesticides, phenols, PAHs, solvents, etc. Ball-milling synthesis of BC and BC-based nanocomposites and possibilities for the removal of emerging pollutants was discussed by Amusat et al. (2021). Several studies demonstrated the synthesis, characterization of ferric oxide-BC (Fe_2O_3-BC) nanocomposites produced from pulp and paper sludge, and their adsorption capabilities for methyl orange (MO) (Chaukura et al., 2017). Lonappan et al. (2016) studied the sorption of MB on three types of BC made from pine wood (BC-PW), pig manure (BC-PM), and cardboard (BC-PD) under varied pyrolytic conditions. All three forms of BCs were found to have good adsorption capacity, with BC-PM having the best adsorption impact due to its high ash content (Lonappan et al., 2016). Wathukarage et al. (2017) observed the adsorption of CV using BC, designated as GBC300, GBC500, and GBC700, generated from pyrolyzing woody tree, *Gliricidia sepium* at 300, 500, and 700°C, respectively. Ji et al. (2016) conducted an experiment using MB and sunset yellow (SY) as adsorbed substances and straw BC prepared at 500 and 700°C. Liu et al. (2017) conducted a series of adsorption tests to determine the absorption efficiency of BC produced from poplar catkins. They examined the removal impact of BC and discovered that poplar catkin BC was a good adsorbent. BC made from pecan nutshell was tested as a low-cost pathway to eliminate Reactive Red 141 (RR 141) from aqueous solutions by Zazycki et al. (2018). Saygl and Guzel (2016) synthesized BC made with $ZnCl_2$ and active tomato processing waste as the raw material and observed the removal efficiencies for MB and MY (Metanil Yellow). According to the findings, this BC was found to be excellent for the adsorption of both cationic and anionic types of dyes. Furthermore, according to Zhu (2017), the current study cycle is mostly short-term, and the BC aging problem caused by long-term trials might affect the dye adsorption process. Dai et al. (2019) gave an overview on BC's adsorption, regeneration, and technical applications for the elimination of organic contaminants. Phuong et al. (2019) investigated adsorption of MB and Bromocresol Green (BG) using BC made from rice waste residues, that is, straw and husk of Japanese Koshihikari and Vietnamese IR50404 rice varieties. All BCs adsorbed cationic MB dye more efficiently than anionic BG dye. The capacity of adsorption of two dyes was greater in Vietnamese IR50404 BCs than in Japanese Koshihikari BCs, by roughly 1.5 times for MB and 1.7 times for BG. When the pH was changed from 2 to 10, the alkaline condition increased cationic MB dye adsorption by approximately 27.2%–33.5% for rice straw BCs and 86.2%–92.2% for RH BCs; however, the anionic BG dye adsorption decreased slightly by 27.2%–32.9% for rice straw BCs and 47%–70.6% for RH BCs. The pH data revealed the presence of negative charges in the BCs, as well as their electrostatic interaction with dyes. The dye intraparticle diffusion was corroborated by the adsorption kinetic analysis, which involved a complicated process including both surface adsorption and intraparticle transport within the pores of BCs. Physisorption by porous diffusion, hydrogen bonding, $\pi-\pi$ or $\pi^+-\pi$ interaction, common to both cationic and anionic dyes, and extra electrostatic interaction for cationic dyes with BCs in aqueous solution were examined in this work as plausible explanations for dye adsorption. Adsorption thermodynamic study revealed that the process was endothermic with positive ΔH^o values and spontaneous with negative ΔG^o values ($\Delta G^o = 35.27-41.94$ kJ mol; $\Delta H^o = 3.92-23.69$ kJ mol^{-1}). Within a pH range of 2–10, cationic MB dye adsorption favored $pH \geq 7$, while anionic BG dye adsorption favored $pH < 3$. The Langmuir model fitted the adsorption isotherm data nicely. The cationic MB dye adsorption capacity on BCs was much higher than the anionic BG dye adsorption capacity. When compared to the Japanese Koshihikari variety, BCs from Vietnamese IR50404 had a better adsorption efficiency, which was attributable to the higher surface area in their BCs. The adsorption rates of the adsorption systems were modified by intraparticle diffusion. The anionic dye adsorption process was primarily governed by physical interactions, whereas the cationic dye adsorption process was governed by both physical and chemical interactions. Hydrogen bonding between the H-bonds of −OH (or −COOH) of BCs and the nitrogen atom of the MB molecules (or the oxygen atom in the BG molecules), $\pi-\pi$ interaction between the aromatic rings of BCs and the aromatic rings of MB and BG dyes, or π^+-π interaction between the dye molecules' cation N^+ and aromatic rings of BCs are all possible physical adsorption mechanisms. The cationic adsorption process may be governed by electrostatic attraction between the oxygenated surface functional groups (i.e., −OH and −COOH) of BCs and N^+ (chemisorption) of MB (Phuong et al., 2019). The adsorption process allows contaminants to transition from solution to adsorbent phase without producing harmful intermediates (Ngwabebhoh et al., 2016). Due to increased efficiency, reduced cost, convenience and simplicity of operation, and lesser susceptibility to harmful contaminants, the adsorption technique has recently overcome its competitors (Fan et al., 2017; Madadi and Bester, 2021). MB, a toxic chemical dye, has a high adsorption capability in bamboo and kenaf (*Hibiscus cannabinus*) fiber char (Mahmoud et al., 2012; Yang et al., 2014). Fan et al. (2017) demonstrated that BC made from sludge may be utilized as a reusable absorbent to remove MB from wastewater. De Caprariis et al. (2018) employed activated biochars to remove dyes (MB) from water using deep eutectic solvents (DESs). When compared to normal BCs, activated BCs had a higher adsorption capability for removing colors, according to their research. The ability of straw BC and activated carbon to remove reactive brilliant blue (RBB) and RB dyes from water was compared in another study performed by Qiu et al. (2009). Despite having equal surface areas, the porosity, point of zero surface charge, and surface acidity of straw BC and activated carbon were reportedly varied. At pH 3.0 and 6.5, both dyes were readily absorbed by straw BC and activated carbon (Qiu et al., 2009). Xu et al. (2011) studied the ability of BC for adsorption of a triphenylmethane dye (MV), from aqueous solutions. The adsorption capacity of BC in the removal of MV from wastewater was shown to be dependent on feedstocks, with canola straw (CS), peanut straw (PS), soybean straw (SBS), and RHs and observed to be in descending order, that is, CS > PS > SS > RH. Nonetheless, BC may adsorb high concentrations of MV present in the medium.

A single and two-stage stirred adsorber were used to test the adsorption effectiveness of magnetic chicken bone BC (MCB) on the removal of RB dye and tetracycline (TC) in another investigation done by Oladipo and Ifebajo (2018). According to the findings, 75% clearance of RB dye and TC in a single-stage system takes 63.0 g MCB in 12 h. Although, in a two-stage stirred adsorber, 33.2 and 22.2 g of MCB can remove 96% TC and RB dye, respectively, from 100 mg L^{-1} effluent solutions in 180 min (Oladipo and Ifebajo, 2018).

2.1.1 BC/layered double hydroxide composites

The majority of current research, however, is concentrated on single BC adsorption dyes, and there are few studies on other types of BC composites (Li et al., 2018). Using the coprecipitation approach, LDH-BC composites in various Mg:Al molar ratios (2:1, 3:1, and 4:1) were synthesized. The composites were used to adsorb MB from aqueous solutions. The characterization techniques, for instance, Brunauer–Emmett–Teller (BET) measurements, thermogravimetric analysis (TG/DTG), FTIR spectroscopy, dispersive energy spectroscopy (DESS), X-ray diffraction (XRD), and scanning electron microscopy (SEM) were used to analyze the composites and pure bovine bone BCs. The adsorption tests for MB were performed in a finite bath. Results showed that pH 12 is better for dye adsorption, with a clearance rate of over 95% for all composites. The pseudo-second order (PSO) model best fitted the adsorption kinetics, which reached equilibrium in around 20 min. The adsorption equilibrium isotherms fitted well with Redlich–Peterson model. At 40°C, a maximum adsorption capacity of 406.47 mg g^{-1} was noticed. The −ve ΔG^o values suggested that the adsorption process was spontaneous. The + ve value of ΔH^o (30.72 kJ mol^{-1}) showed the physical character of the adsorption, whereas the positive value of ΔS^o (0.1863 kJ mol^{-1}) indicated that the structure of the adsorbent changed and increased randomly throughout the dye fixing (Meili et al., 2019).

A green BC/iron oxide composite was developed in this work combining banana peel extract and $FeSO_4$. In comparison to the unmodified banana peel-BC, the modified BC's ability to adsorb MB was significantly increased (Langmuir Q_{max} of 862 mg g^{-1} for MB at pH = 6.1, C_0 = 500 mg L^{-1}, and T = 313 K), and it performed well throughout a wide pH range (pH 2.05–9.21). The adsorption process fitted well with Langmuir isotherm model and the pseudo-second-order kinetic model. Various experimental approaches were used to examine the material characteristics and corresponding adsorption processes. Greater electronic attraction to MB molecules is related to increased MB adsorption by the BC/iron oxide composite, as revealed by X-ray photoelectron spectroscopy (XPS) measurement. After five regeneration cycles, the high adsorption capacity was maintained. This research revealed that BC may be modified following a green synthesis method to generate BC/iron oxide composite with high MB removal capability (Zhang et al., 2020).

BC/LDH composites have received a lot of interest recently as low-cost, long-lasting materials for water treatment. This research assessed the most recent developments in BC/LDH composite applications in water treatment, with a focus on adsorption and catalytic degradation of different contaminants (Zubair et al., 2021). The adsorption of several noxious pollutants onto BC/LDH composites, such as heavy metals, anions, dyes, and medicines, was discussed elaborately, including the adsorption process mechanisms and regeneration ability. Fig. 14.7 illustrates the dye wastewater treatment employing biochar as an adsorbent (Srivatsav, 2020). The synergistic impact of LDH with BC improved the stability, specific surface area, structural heterogeneity, surface functional groups, and adsorptive properties of the resultant BC/LDH composites significantly. The primary problems and obstacles connected with the synthesis and uses of BC/LDH composites in water remediation were highlighted in the study. A research plan for future study was also proposed to ensure the effective usage of BC/LDH composites in water purification (Zubair et al., 2021).

2.2 Biochar-based photocatalysts

BC as a carbonaceous material can be used as a catalyst or catalyst support in a variety of applications. Carbonaceous materials and mechanisms of BCPs in wastewater treatment are shown in Fig. 14.8 (Cui et al., 2020). BCPs have garnered a lot of interest in the environmental area in recent years (Sutar et al., 2021): TiO_2-coconut shell, TiO_2-reed straw, TiO_2-walnut shells, TiO_2-wood pellets, TiO_2-olive

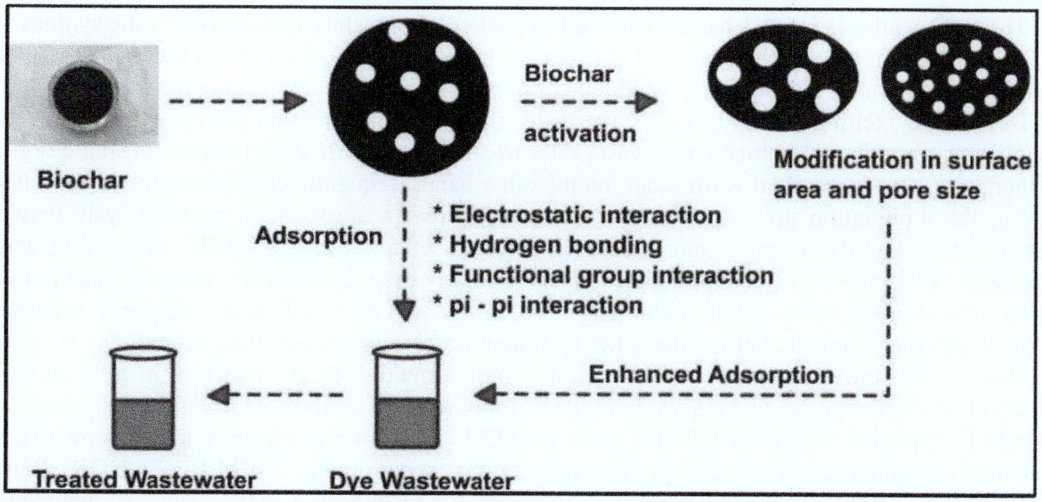

FIGURE 14.7

Dye wastewater treatment employing biochar as an adsorbent (Srivatsav, 2020).

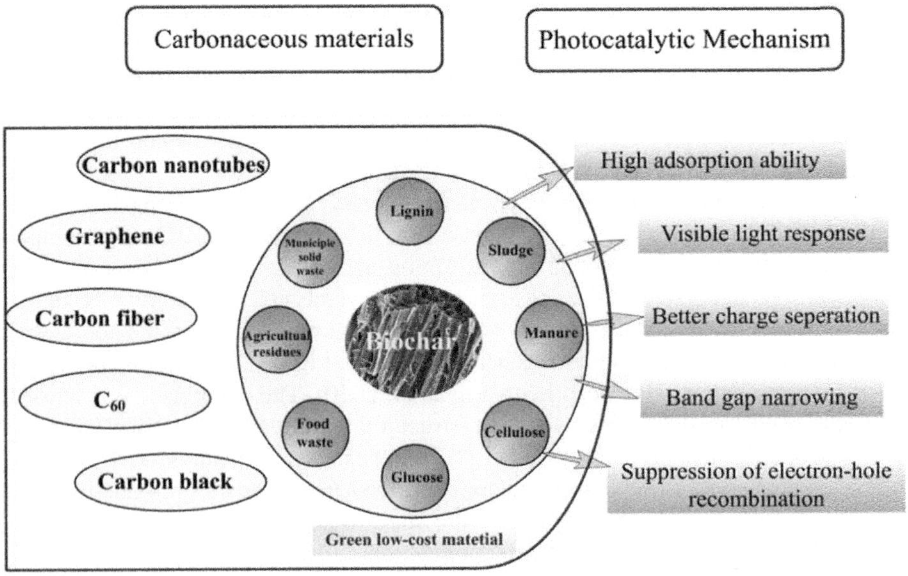

FIGURE 14.8

Carbonaceous materials and mechanisms of BC-based photocatalysts (BCPs) in wastewater treatment (Cui et al., 2020).

pits, TiO_2-plum stones, TiO_2-corn cob, TiO_2-wood shaving, TiO_2-bamboo, TiO_2-paper sludge and wheat husks, Bi_2O_3-corn cobs, g-C_3N_4/$FeVO_4$-pinus roxburghii, Bi_2O_3-corn cobs, etc. (Cui et al., 2020).

The BCP synthesis process has a significant impact on photocatalytic activity. For the synthesis of BCPs, several effective techniques and procedures have been described, including sol-gel, hydrothermal/solvothermal, ultrasonography, calcination, and in situ methods, among others (Fig. 14.9).

Despite the exciting growth of BCPs, there are still certain information gaps and difficulties. The majority of research on photocatalytic wastewater treatment using BCPs is focused on single organic or inorganic substances. Real wastewater, on the other hand, frequently contains many contaminants. During the elimination process, there may be competitive adsorption or photocatalysis between distinct materials. As a result, more research into the effectiveness of BCPs in treating actual wastewater is required. The cost of photocatalysis is greatly reduced when it is recycled. The reuse of BCPs from slurry systems is still a challenge, which restricts their utilization. Magnetic separation using an external magnetic field in the water treatment process is one possible technique to solve this problem. More study is needed to determine the most effective BCPs recovery on a broad scale. Different characterization techniques have been employed to analyze mechanical, physical, and chemical properties of BC and BCPs, such as SEM, transmission electron microscopy (TEM), scanning electron microscopy-energy dispersive X-ray spectrometry (SEM-EDS), FTIR, Raman spectroscopy, XRD, XPS, solid-state nuclear magnetic resonance (NMR), etc., as included in Fig. 14.10 (Cui et al., 2020).

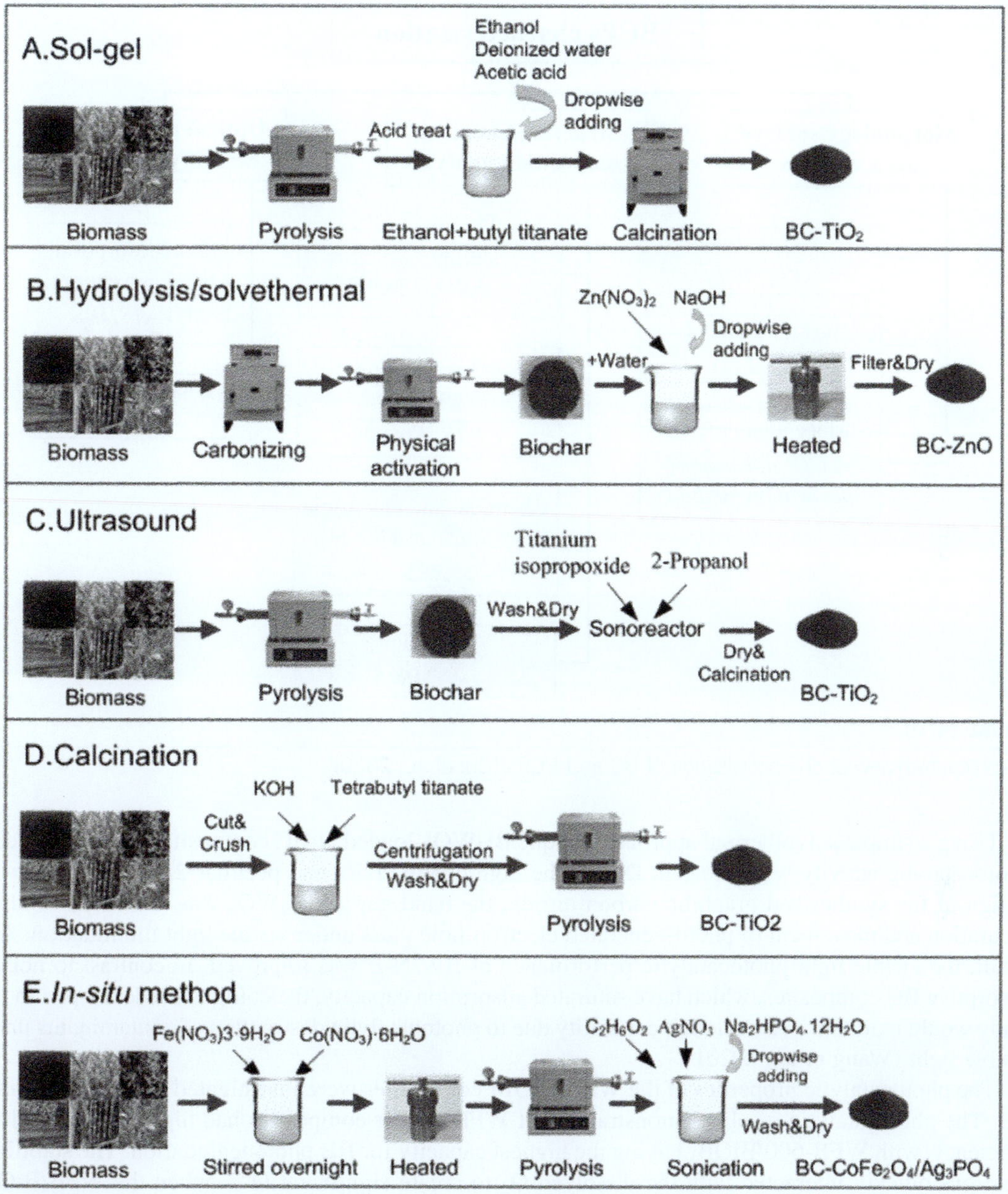

FIGURE 14.9

Schematic representation of preparation methods of BCPs (Cui et al., 2020).

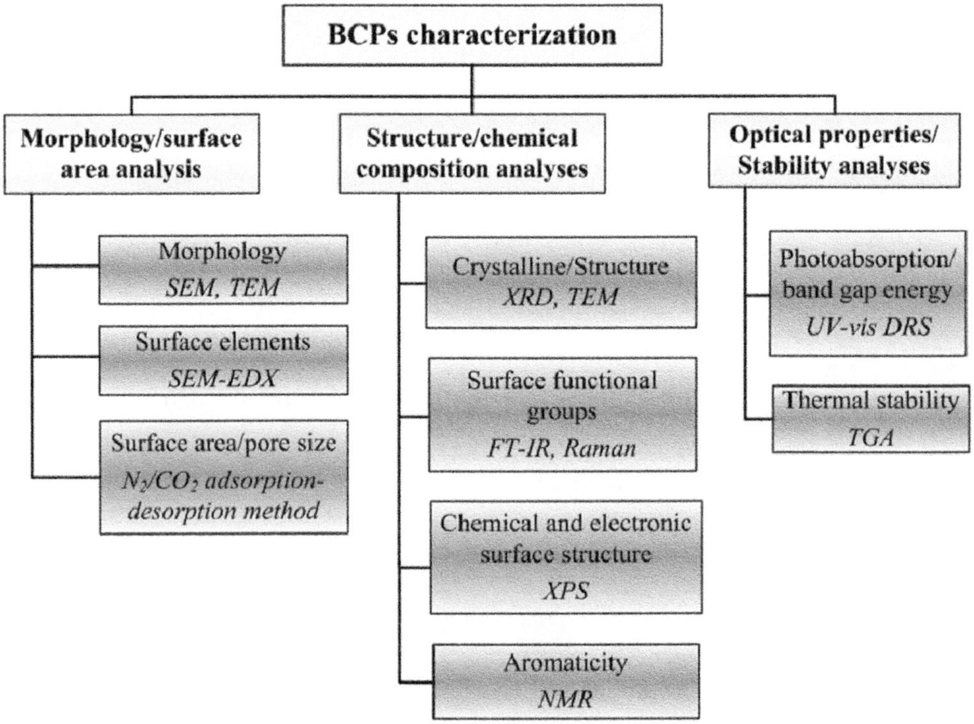

FIGURE 14.10

Graphical overview of characterization of BC and BCPs (Cui et al., 2020).

Using a simple solvothermal approach, unique Bi_2WO_6-loaded N-BC composites with outstanding photocatalytic activity were created. Due to the high conductivity and peculiar 2D sheet framework design of the synthesized graphitic carbon nitride, the band gap of Bi_2WO_6 was decreased to allow separation and movement of photo-generated electron-hole pairs under visible light illumination. As a result, the visible light photocatalytic performance of BW/N-B was improved. In contrast to normal adsorptive BC composites, which have saturated adsorption capacity, the composites developed in this study would maintain a remediation capability due to photocatalytic destruction of contaminants under visible light (Wang et al., 2020).

The photocatalytic properties of the WFB/BiOBr composites were investigated in order to degrade RB. The photocatalytic results demonstrated that WFB/BiOBr composites had higher photocatalytic efficiency, with WFB-600/BiOBr having the highest capacity for RB photodegradation. The stabilized electrons quickly interacted with the dissolved O_2 to create significant $O^{2-\bullet}$ when the WFB/BiOBr composite was utilized for RB degradation. During the whole photocatalytic degradation phase, $O^{2-\bullet}$ and h^+ were more important for the breakdown of RB dye (Geng et al., 2020).

Leichtweis et al. (2020) developed a novel, low-cost, and efficient BC-ZnO composite, and tested its photocatalytic efficacy by degrading RR97. On the BC's surface, SEM examination revealed a homogeneous dispersion of ZnO particles. According to the elemental mapping investigation, the

N20Z composite showed acceptable particle dispersion, suggesting that both ZnO and BC contribute to the production of photocatalyst composites. The photocatalytic efficiency demonstrated that all composites (N5Z, N10Z, and N20Z) had lower adsorption capacity and higher photocatalytic activity than pure BC, indicating that the BC-ZnO formation delivers the material essential properties. In terms of photocatalytic efficiency, the N20Z composite beat the competition, totally eliminating RR97 in just 67 min. The recombination process of the ZnO electron/hole pair of composites was blocked by BC, which resulted in a higher synthesis of reactive molecules. The dye molecules adsorbed on BC's outer surface were continually photodegraded, resulting in smaller molecules that were either absorbed into BC's pores or remained in the liquid phase; in the meantime, new dye molecules were adsorbed on BC. Most notably, the N20Z composite can perform efficiently in neutral pH environments, has great endurance, and can be easily withdrawn from the solution after nine recycles. There is a modest drop-in catalytic activity during the recycle test, as the scientists predicted. This might be due to the catalyst's active ingredients being leached and destroyed (Leichtweis et al., 2020).

Lu et al. (2019) used a direct hydrolysis technique and pyrolysis of discarded walnut shells to make a variety of BC-TiO_2 photocatalysts for the elimination of MO. After recycling five times, it was discovered that a catalyst with a weight ratio of 0.2/1 (BC/Ti) had the greatest decolorization efficiency of 92.45% and mineralization efficiency of 76.56% (Lu et al., 2019). The addition of BC-based carbon quantum dots (BCQDs) to a semiconductor might alter the band structure and increase light absorption. Using the solvothermal technique, Yao et al. (2018) created a BCQDs modified flower-like BiOI composite. The primary active species for MB photodegradation were identified to be $\bullet O^{2-}$ and h^+. The BCQDs worked as electron acceptors, allowing electron-hole pairs to be separated even more easily. Wang et al. (2019) used hydrothermal treatment of bamboo powder to make BCQDs and BCQDs/Ag/Bi4Ti_3O_{12}. They claimed that the use of BCQDs made photoelectron production and transit easier. In addition, the BCQDs served as an electron reservoir, enhancing photocatalytic activity.

Mian et al. (2019) successfully synthesized nanoparticles (NPs)-supported-BC composite using one-step NH_3-ambiance pyrolysis for MB breakdown and mineralization. The resultant composite N-doped a TiO_2/Fe_3O_4/BC structure, decreasing the band gap energy and enhancing visible light sensitivity. Adsorption, ultraviolet (UV) degradation, and Fenton reaction procedures were all used in this study. At 800°C, SEM examination indicated a homogeneous dispersion of NPs over the BC surface, whereas pyrolysis at 600°C resulted in the formation of massive NP crystals. The presence of mesoporous nature in BC and its composite was verified by N_2 adsorption-desorption isotherm data. The high rate of H_2O_2 hydrolysis, which resulted in accessible molecular O_2 in the solution, might be linked to the considerable MB breakdown found in this study at a high alkaline solution. Later, the released O_2 reacts with electrons in the conduction band to create $^{\bullet}OH$ and $O^{2-\bullet}$. By absorbing photon energy equal to or larger than the bandgap, visible light irradiation over C1 might theoretically produce an e^-/h^+ pair on TiO_2. Charge transfer from the TiO_2 electron-hole pair to the Fe^{3+}/Fe^{2+} redox pair and/or reactive impurities can subsequently occur through the BC's dense aromatic carbon layer, increasing the lifetime of the e^-/h^+ pair. In addition, the local trap phase of C1 coupled with Ti^{3+} can improve the charge separation efficiency of BC-TiO_2. By gaining an electron, the catalyst's Fe^{3+} can be reduced to Fe^{2+}, which subsequently reacts with H_2O_2 in the medium to create $^{\bullet}OH$, as described by the classical Haber–Weiss process. $^{\bullet}OH$ may also interact with Fe^{3+} to regenerate Fe^{2+}, which is necessary for the Fenton reaction. ROS (reactive oxygen species) like superoxide radical ($O^{2-\bullet}$), hydroxyl radical ($^{\bullet}OH$), and photo-generated hole (h^+) are involved in the breakdown of MB in the photocatalytic redox process, with superoxide ($O^{2-\bullet}$) playing the most important role, followed by $^{\bullet}OH$ and h^+. After five times recycling, the ability to extract MB decreases, and the reused catalyst takes longer to disintegrate MB (Mian et al., 2019).

Zhang and Lu (2018) successfully synthesized a BC-based TiO_2/BC composite produced from coconut shell at a low cost and without causing environmental harm. The findings of the high-resolution TEM revealed that thick BC aided TiO_2 film development, and anatase TiO_2 was made utilizing a sol–gel technique. Micro- and macropores with diameters of 15–20 and 2–4 m were discovered on the surface of the BC using a high-resolution SEM. The greatest removal % was achieved using strong acid and alkali conditions, which is principally controlled by the oxidative destruction of KN-R molecules by •OH and $O^{2-\bullet}$. The KN-R molecules and intermediates first approached the exterior of TiO_2 film, where the dye molecules' adsorption process on BC and its porosity enhanced the interaction of TiO_2 and KN-R molecules. Additionally, photon activity occurred for the photo-generation of h^+ and e^- when TiO_2 absorbed sufficient photon energy equal to or greater than its band gap, but the BC prevented h^+ and e^- from reuniting, resulting in accelerated electron-hole pair transport. Finally, photo-generated holes h^+ on TiO_2's surface interacted with H_2O and OH^- to produce •OH, whereas photo-generated electrons e^- approached O_2 to produce $O^{2-\bullet}$. As a result, both •OH and $O^{2-\bullet}$ oxidized KN-R molecules and their byproducts (Zhang and Lu, 2018).

2.3 Biochar as electrode material and wetland matrix combined with microbial fuel cells

In this investigation, the corncob biochar modified with sulfuric acid (SA-MCB) was employed in a microbial fuel cell (MFC) for degrading real dye wastewater (RDW). The whole MFC, which generates electricity, was developed by using waste materials to show that low-cost options are possible to mimic the excellent outcomes. Biochar made from waste corncobs following acid alteration has been tested as a supplement in an MFC for the treatment of RDW. The effect of two dosages of SA-MCB (0.5 and 1 g) and a blank (no dose) on MFC performance was observed. With a 0.5 g biochar dosage, the current density (0.28 A m^{-2}), maximum power density (49.92 W m^{-2}), chemical oxygen demand (COD) removal efficiency (88.39%), total dissolved solid (TDS) reduction (84.4%), and decolorization efficiency (81.6%) were achieved. The structural alterations on the SA-MCB following usage in the MFC were analyzed using FTIR spectroscopy and SEM. SEM study demonstrated that SA-MCB deposition resulted in improved biofilm development. When compared with the system comprising no SA-MCB dose, power density was found to be enhanced by more than 24 times with 0.5 g of SA-MCB dose. The deformation of azo bond during treatment was verified by UV spectrophotometer analysis. Overall, the findings imply that using SA-MCB biochar as an anode material and as a supplement in MFC systems might result in an effective anode-biofilm in wastewater treatment MFCs (Sonu et al., 2020).

The goal of the study conducted by Sonu et al. (2021) was integrating constructed wetland with MFC (CW-MFC) utilizing corn cob biochar as matrix and demonstrating the impacts on textile wastewater (TW) treatment in terms of color, COD, and TDS removal as well as bioelectricity generation. In batch mode with a retention duration of 10 days for 10 cycles, two vertical down flow CW-MFC systems were tested for TW treatment and system reusability (with and without corn cob biochar). In terms of TW treatment, bioelectricity generation, and matrix reusability, CW-MFC with corn cob biochar (CW-MFC A) performed better than the CW-MFC without corn cob biochar

(CW-MFC B). The power density (102.08 mW m^{-2}), COD removal efficiency (83%), decolorization efficiency (90%), and TDS reduction (84%) obtained with CW-MFC A were observed as greater than those obtained with CW-MFC B (78.24 mW m^{-2}, 66%, 65%, 67%, respectively). Furthermore, germination tests employing the treated effluent from CW-MFC A yielded 100% germination of *Vigna radiata* seedlings, demonstrating the use of corn cob biochar as a wetland matrix for wastewater treatments (Sonu et al., 2021).

Low and middle-income nations will find it simpler to implement cost-effective, efficient, and sustainable water treatment technologies that use current materials and technology, therefore enhancing public health. Biochar's capacity to mediate and facilitate microbial contamination degradation, as well as its carbon sequestration potential, has sparked interest in recent years. Biochar is a potential option for use in cost-effective and long-term biological water treatment, particularly in low and middle-income nations (India, Indonesia, China, etc.), where ample biomass in the form of crop residues and organic wastes is readily available. Upon burning in open areas, this biomass tends to spread pollution in the environment. Jayakumar et al. (2021) reviewed the scope, possible advantages (both economic and environmental), and limitations of adopting biologically enhanced biochar (BEB) for long-term biological water treatment. The importance of correlating biochar characteristics to biofilm growth, which can ultimately influence process efficiency, was emphasized. "Sequential biochar systems" are described as specially developed end-of-life approaches for BEB filters to reduce their environmental effect, with instances of its integration into biological water treatment that can meet zero waste standards for BEBs. Biochar incorporation in microbial degradation of pollutants makes biochar an attractive choice for biological water treatment. The carbon-rich products like biochar can be employed in tertiary step of water treatment which use adsorption and microbial activity both for remediating pollutants. Microbes (naturally present in water or supplied externally to target specific pollutants) can immobilize on the porous carbon surface, creating biofilms. In a bioregeneration process, biofilms metabolize and destroy the adsorbed pollutants. This bioregeneration process unclogs blocked pores in filter materials (such as charcoal), restoring their adsorptive ability and extending their lifetime dramatically. Biofilms can breakdown and eliminate a variety of organic, inorganic, and biological pollutants from water. Adsorption, biofilm development, biodegradation, desorption, and diffusion of pollutants and nutrients across the biochar biofilm—water interface are all steps in the removal of contaminants in a biological filter. The adsorptive capabilities of biochar in water treatment have been extensively investigated; however, studies that explore the combined adsorption and biological degradation of pollutants using biochar are a new field of research. Despite a small number of research, the applications of BEBs have shown its effectiveness in removing organic, inorganic, and biological contaminants from wastewater. Microbial biofilms and their biosorption properties for bioremediation are a vast field of study. Biochar units, for example, absorb arsenic from wastewater, then separate periphytic biofilm reactors (including both heterotrophic and phototrophic bacteria), and adsorb the residual arsenic in the water via biosorption by biofilms. They may also be used for treating complex dye wastewater comprising recalcitrant dyes. BEB filters, when properly built, provide a long-term solution for water treatment, particularly in low and middle-income nations, with extra economic and sociological advantages to local populations. BEB may definitely help in achieving many of the United Nations Sustainable Development Goals (SDGs), particularly SDG-6 (Clean Water and Sanitation) and SDG-7 (Affordable and Clean energy) by 2030 (Jayakumar et al., 2021).

3. Key challenges

3.1 Recovery and disposal of spent biochar

Ahmad et al. (2012) used n-hexane as a desorbing solvent to investigate the recycling of trichloroethylene adsorbed biochar. Only 20% of trichloroethylene was desorbed by the solvent from biochar, indicating that trichloroethylene has a significant affinity for biochar surface functional groups. It was attributed to diffusion mechanisms which dominated trichloroethylene desorption from biochar, and the presence of noncarbonized fraction in biochar may have guarded the biochar pores, resulting in reduced desorption. These researchers used biochar made from Chinese fan palm seed, Arjun seed, Casuarina (abundantly available as agro-waste in India), and Ashoka seed biochar to remove a variety of dyes, including MB (Ahmad et al., 2020), Congo red, MG, auramine dye, and agrochemicals such as pesticides and insecticides. They've also found that biochar can be recovered and reused six to eight times, with highly efficient adsorption capabilities of 90%−98%. These loaded biochars might also be employed as a biofertilizer to boost soil fertility in accordance with crop needs (Gautam et al., 2021).

3.2 Toxicity associated with biochar

Biochar has been widely used in water across the world to reduce pollution and has shown promising results in meeting the allowable standards set forth by the World Health Organization. However, excessive use of biochar in wastewater treatment may cause toxicity in animals, plants, microbes, and agricultural crops exposed to contaminant-laden biochar. The biochar may be absorbed by the microorganism's live cells and tissues, as well as agricultural crops and plants, throughout the treatment. As a result of the adsorbed poisonous elements on biochars, it may demonstrate harmful effects on the biotic environment. Pyrolysis of biomass at high temperatures increases surface area, which aids in the efficient adsorption of dyes, heavy metals, pesticides, pharmaceutical compounds, and other hazardous compounds from water and soil. However, the presence of organic matter in the environmental matrix may limit agrochemicals adsorption (Gautam et al., 2021).

3.3 Cost analysis and assessment

Because of high adsorption efficiency, inexpensiveness, low energy consumption, high yield, and ease of application in wastewater, biochar is increasingly being used for environmental rehabilitation. Due to minimal energy use and little activation of synthesized biochar, it is estimated that producing biochar is less expensive than producing commercial activated carbon. The cost of biochar has been estimated to be 246 $ per tonne, whereas activated carbon costs roughly 1500 $ per tonne. The basic commercial cost of biochar, according to one study, was between US 0.8- and 18 $ per kg. This biochar pricing demonstrates that low-cost biochar may be used in wastewater treatment. Biochar is also obtained as a byproduct from biooil manufacturing units and may be utilized to remediate a range of contaminants in the environment. Cost analysis is the most prominent part of every application, including pollution cleanup. The high cost and energy needs of physicochemical methods used to remove contaminants make them unsuitable. Microbiological research is similarly broad, time-consuming, and inefficient. As a result, continual attempts have been made to use adsorption-based technologies like biochar, which have been shown to be cost-effective, to solve such problems.

Goswami et al. (2020) reported the synthesis of Arjuna seeds using pyrolyzer with a batch capacity of 4.75 L. In batch production, 0.6 kg of biomass yields 40% yield and 440 mL of biooil. The pyrolysis unit consumed roughly 4.2 kW of energy or 0.75 $ per kg. The availability of raw material was free of any charge and, thus, not included in the final cost estimation. The pyrolysis unit is a one-time expenditure of approx. 720 $. Transportation, biomass cleaning, and labor charges, on the other hand, might cost up to 0.15 $ per kg of biochar produced. As a result, the overall cost was estimated to be roughly 0.91 $ per kg for biochar generation, along with the production of 440 mL biooil which can be used for other purposes (Gautam et al., 2021).

3.4 Future perspectives

Biochars produce significant interactions between the surface of biochar and various environmental pollutants present in aqueous environments due to their huge number of functional groups and increased surface area. However, contaminant-loaded biochar may provide a danger of improper disposal, posing a new challenge to the scientific community in terms of the likelihood of changed environmental mobility of these pollutants, potentially altering the framework for contaminant risk assessment in the environment. As a result, it is critical to obtain detailed information on the chemical and physical interactions between biochars and contaminants found in wastewater during the laboratory-scale study. Because of these factors, there is always a secondary risk connected with any new technology, and hence the use of loaded biochar for biooil conversion might be an intriguing topic to investigate. The production of biooil from used biochar materials has been attempted using catalyst-aided rapid pyrolysis or hydrothermal conversion. For comparable investigations and the application of processed biochar as a biofertilizer, a thorough procedure has been developed. Nanobiochar has grabbed the interest of the scientific community due to its effective physicochemical features. Various literature outlining its potential and uses have been published. Novel synthetic approaches for nanobiochar production have been investigated extensively. Furthermore, many aspects of nanobiochar preparation, synthesis, and application have been reported (Ramanayaka et al., 2020). Nanobiochar is prepared via mechanically grinding employing a milling process in parent biochar made from biomass pyrolysis. In addition, direct synthesis of graphic nanosheets employing flash heating has been described. The process conditions have a significant impact on the properties of nanobiochar (32 times more BET specific surface area at higher temperatures than parent biochar). Nanobiochar has been used in sensors, supercapacitors, electrode materials, health care, and wastewater treatment, among other applications. However, the assessment of harmful nanobiochar's impacts on human and environmental health requires more attention. Nanobiochar still has some limitations, such as low yield, high agglomeration potential, and difficulties recovering from environmental matrices, all of which need to be addressed. Despite the increased usage of nanobiochar, available research data are insufficient to infer nature and behavior of nanobiochar. As a result, future study should focus on identifying techniques for improving synthesis protocols, increasing nanobiochar production, reducing agglomeration, and recognizing toxicological impact. Nanobiochar has been utilized as a contaminant removal material, as a biomolecule carrier, as a catalyst, as an electrode material, as a sensing material, as a carbon black replacement, and as a battery industry material (Gautam et al., 2021).

4. Impact

Fashion designers, producers, and consumers should all be concerned about the dyeing industry's long-term viability. The generation of process residuals as waste and wastewater necessitates the development of innovative and cost-effective technologies and processes for the collection, reuse, recycling, and upcycling of textile waste (clothing and other textiles), for the reduction of water, energy, and chemical consumption, and for the treatment of TW in both treatment plants and small dye house industries. The environmental regulation for dyeing wastewater is often highly strict, necessitating the requirement for effective treatment procedures that must follow zero liquid discharge (ZLD) policy in common or noncommon treatment facilities; however, in reality, due to treatment costs, this is not usually done. Furthermore, even when facilities are designed for the ZLD, they still produce hundreds of tonnes of hazardous solid waste each day in the form of sludge (residual dyes and waste salts). In addition to effective water treatment using innovative nanotechnological methods and novel hybrid membranes, efforts should be undertaken to recover dyes and other organics from wastewater before they are discharged into the land and water bodies (Mamane, 2020). The dye removal technologies using BC were discussed in this chapter, and recommendations for reusing the water back to the dye houses and recovering the biochar after usage were made.

5. Conclusion

Dyes add color and vibrancy to our lives, but the bulk of them have unstable chemical properties and a complex structure that might have a considerable negative impact on human health. As a result, scientists should focus more on investigation and study in order to reduce the harm caused by dyes to people. Dyes are released by the textile, paper, pulp, tannery, cosmetic, and leather industries. These colors are frequently dangerous when they are released into the environment. Significant progress has been made in the treatment and management of dye wastewater, which can be treated using biological, physical, or chemical methods. Many researchers are already conducting dye removal studies, and BC's debut has opened up new research opportunities. Advanced eco-friendly methods appear to be the most appropriate because they offer the potential for resource recovery (in the form of energy or chemicals). Various BCs were found to be effective at adsorbing colors from textile/industrial wastewater effluents. Basic environments (pH greater than 7) have been shown to be suitable for cationic dyes, whereas acidic environments (pH less than 7) have been shown to be ideal for anionic and acid dyes, with an optimum pH of 3–4. Different postproduction modification modes of BCs can enhance the adsorption process' effectiveness and hence the overall treatment process. BC's long-term usage and reuse should be monitored. An attempt was made to summarize existing data and provided scientific facts in this chapter that may be used to spark fresh research and scientific breakthroughs in this field.

References

Adekola, F.A., Ayodele, S.B., Inyinbor, A.A., 2019. Activated biochar prepared from plaintain peels: characterization and rhodamine B adsorption data set. Chem. Data Collect. 19, 100170.

Ahmad, A., Khan, N., Giri, B.S., Chowdhary, P., Chaturvedi, P., 2020. Removal of methylene blue dye using rice husk, cow dung and sludge biochar: characterization, application, and kinetic studies. Bioresour. Technol. 306, 123202.

Ahmad, M., Lee, S.S., Dou, X., Mohan, D., Sung, J.K., Yang, J.E., Ok, Y.S., 2012. Effects of pyrolysis temperature on soybean stover- and peanut shell-derived biochar properties and TCE adsorption in water. Bioresour. Technol. 118, 536–544.

Ambaye, T.G., Vaccari, M., van Hullebusch, E.D., Amrane, A., Rtimi, S., 2021. Mechanisms and adsorption capacities of biochar for the removal of organic and inorganic pollutants from industrial wastewater. Int. J. Environ. Sci. Technol. 1–22.

Amusat, S.O., Kebede, T.G., Dube, S., Nindi, M.M., 2021. Ball-milling synthesis of biochar and biochar-based nanocomposites and prospects for removal of emerging contaminants: a review. J. Water Proc. Eng. 41, 101993.

Anjum, N.A., Gill, S.S., Tuteja, N. (Eds.), 2017. Enhancing Cleanup of Environmental Pollutants. Springer.

Barredo-Damas, S., Alcaina-Miranda, M.I., Bes-Piá, A., Iborra-Clar, M.I., Iborra-Clar, A., Mendoza-Roca, J.A., 2010. Ceramic membrane behavior in textile wastewater ultrafiltration. Desalination 250 (2), 623–628.

Chaukura, N., Murimba, E.C., Gwenzi, W., 2017. Synthesis, characterisation and methyl orange adsorption capacity of ferric oxide-biochar nano-composites derived from pulp and paper sludge. Appl. Water Sci. 7, 2175–2186.

Chen, S., Qin, C., Wang, T., Chen, F., Li, X., Hou, H., Zhou, M., 2019. Study on the adsorption of dyestuffs with different properties by sludge-rice husk biochar: adsorption capacity, isotherm, kinetic, thermodynamics and mechanism. J. Mol. Liq. 285, 62–74.

Cui, J., Zhang, F., Li, H., Cui, J., Ren, Y., Yu, X., 2020. Recent progress in biochar-based photocatalysts for wastewater treatment: synthesis, mechanisms, and applications. Appl. Sci. 10 (3), 1019.

Dai, Y., Zhang, N., Xing, C., Cui, Q., Sun, Q., 2019. The adsorption, regeneration and engineering applications of biochar for removal organic pollutants: a review. Chemosphere 223, 12–27.

De Caprariis, B., De Filippis, P., Petrucci, E., Scarsella, M., 2018. Activated biochars used as adsorbents for dyes removal. Chem. Eng. Trans. 65, 103–108.

Dutta, S., Gupta, B., Srivastava, S.K., Gupta, A.K., 2021. Recent advances on the removal of dyes from wastewater using various adsorbents: a critical review. Mater. Adv. 2, 4497–4531.

Fan, S., Tang, J., Wang, Y., Li, H., Zhang, H., Tang, J., Wang, Z., Li, X., 2016. Biochar prepared from co-pyrolysis of municipal sewage sludge and tea waste for the adsorption of methylene blue from aqueous solutions: kinetics, isotherm, thermodynamic and mechanism. J. Mol. Liq. 220, 432–441.

Fan, S., Wang, Y., Wang, Z., Tang, J., Tang, J., Li, X., 2017. Removal of methylene blue from aqueous solution by sewage sludge-derived biochar: adsorption kinetics, equilibrium, thermodynamics and mechanism. J. Environ. Chem. Eng. 5, 601–611.

Gautam, R.K., Goswami, M., Mishra, R.K., Chaturvedi, P., Awashthi, M.K., Singh, R.S., Giri, B.S., Pandey, A., 2021. Biochar for remediation of agrochemicals and synthetic organic dyes from environmental samples: a review. Chemosphere 272, 129917.

Geng, A., Xu, L., Gan, L., Mei, C., Wang, L., Fang, X., Li, M., Pan, M., Han, S., Cui, J., 2020. Using wood flour waste to produce biochar as the support to enhance the visible-light photocatalytic performance of BiOBr for organic and inorganic contaminants removal. Chemosphere 250, 126291.

Gopinath, A., Divyapriya, G., Srivastava, V., Laiju, A.R., Nidheesh, P.V., Kumar, M.S., 2021. Conversion of sewage sludge into biochar: a potential resource in water and wastewater treatment. Environ. Res. 194, 110656.

Goswami, M., Chaturvedi, P., Sonwani, R.K., Gupta, A.D., Singhania, R.R., Giri, B.S., Rai, B.N., Singh, H., Yadav, S., Singh, R.S., 2020. Application of Arjuna (Terminalia arjuna) seed biochar in hybrid treatment system for the bioremediation of Congo red dye. Bioresour. Technol. 307, 123203.

Guan, C., Liu, L., Liu, Y.J., Li, R.D., 2017. Effectiveness of acid-dye wastewater treated by $CaCO_3$ modified diatomite. Sci. Technol. Eng. 17, 328–331.

Hao, J.O., Kim, H., Chiang, P.C., 2000. Decolourization of wastewater. Environ. Sci. Technol. 30, 449–505.

Inyang, M., Dickenson, E., 2015. The potential role of biochar in the removal of organic and microbial contaminants from potable and reuse water: a review. Chemosphere 134, 232–240.

Jayakumar, A., Wurzer, C., Soldatou, S., Edwards, C., Lawton, L.A., Mašek, O., 2021. New directions and challenges in engineering biologically-enhanced biochar for biological water treatment. Sci. Total Environ. 148977.

Ji, X., Chen, F., Yang, C., University, H., 2016. Sorption properties and mechanisms of organic dyes by straw biochar. Acta Sci. Circumstantiae 36, 1648–1654.

Leichtweis, J., Silvestri, S., Carissimi, E., 2020. New composite of pecan nutshells biochar-ZnO for sequential removal of acid red 97 by adsorption and photocatalysis. Biomass Bioenergy 140, 105648.

Leng, L., Yuan, X., Huang, H., Shao, J., Wang, H., Chen, X., Zeng, G., 2015. Bio-char derived from sewage sludge by liquefaction: characterization and application for dye adsorption. Appl. Surf. Sci. 346, 223–231.

Li, X.P., Wang, C.B., Zhang, J.G., Liu, B., Liu, J.P., Chen, G.Y., 2018. Application of biochar in removal of organics and heavy metals from water. Acta Pet. Sin. 34, 1047–1056.

Liu, X., Sun, J., Duan, S., Wang, Y., Hayat, T., Alsaedi, A., Wang, C., Li, J., 2017. A valuable biochar from poplar catkins with high adsorption capacity for both organic pollutants and inorganic heavy metal ions. Sci. Rep. 7, 10033.

Lonappan, L., Rouissi, T., Das, R.K., Brar, S.K., Ramirez, A.A., 2016. Adsorption of methylene blue on biochar microparticles derived from different waste materials. Waste Manag. 49, 537–544.

Lu, L., Shan, R., Shi, Y., Wang, S., Yuan, H., 2019. A novel TiO_2/biochar composite catalysts for photocatalytic degradation of methyl orange. Chemosphere 222, 391–398.

Madadi, R., Bester, K., 2021. Fungi and biochar applications in bioremediation of organic micropollutants from aquatic media. Mar. Pollut. Bull. 166, 112247.

Mahmoud, D.K., Salleh, M.A.M., Karim, W.A.W.A., Idris, A., Abidin, Z.Z., 2012. Batch adsorption of basic dye using acid treated kenaf fibre char: equilibrium, kinetic and thermodynamic studies. Chem. Eng. J. 181–182, 449–457.

Mamane, H., Altshuler, S., Sterenzon, E., Vadivel, V.K., 2020. Decolorization of dyes from textile wastewater using biochar: a review. Acta Innov 37, 36–46.

Meili, L., Lins, P.V., Zanta, C.L.P.S., Soletti, J.I., Ribeiro, L.M.O., Dornelas, C.B., Silva, T.L., Vieira, M.G.A., 2019. MgAl-LDH/Biochar composites for methylene blue removal by adsorption. Appl. Clay Sci. 168, 11–20.

Mekonnen, M.M., Hoekstra, A.Y., 2016. Four billion people facing severe water scarcity. Sci. Adv. 2 (2), e1500323.

Mian, M.M., Liu, G., Yousaf, B., Fu, B., Ahmed, R., Abbas, Q., Munir, M.A.M., Ruijia, L., 2019. One-step synthesis of N-doped metal/biochar composite using NH_3-ambiance pyrolysis for efficient degradation and mineralization of methylene blue. J. Environ. Sci. (China) 78, 29–41.

Mohan, D., Sarswat, A., Ok, Y.S., Pittman Jr., C.U., 2014. Organic and inorganic contaminants removal from water with biochar, a renewable, low cost and sustainable adsorbent-a critical review. Bioresour. Technol. 160, 191–202.

Mostafa, M., 2015. Waste water treatment in textile Industries-the concept and current removal technologies. J. Biodivers. Environ. Sci. (JBES) 7 (1), 501–525.

Ngwabebhoh, F.A., Gazi, M., Oladipo, A.A., 2016. Adsorptive removal of multi-azo dye from aqueous phase using a semi-IPN superabsorbent chitosan-starch hydrogel. Chem. Eng. Res. Des. 112, 274–288.

Oladipo, A.A., Ifebajo, A.O., 2018. Highly efficient magnetic chicken bone biochar for removal of tetracycline and fluorescent dye from wastewater: two-stage adsorber analysis. J. Environ. Manag. 209, 9–16.

Palansooriya, K.N., Yang, Y., Tsang, Y.F., Sarkar, B., Hou, D., Cao, X., Meers, E., Rinklebe, J., Kim, K.H., Ok, Y.S., 2020. Occurrence of contaminants in drinking water sources and the potential of biochar for water quality improvement: a review. Crit. Rev. Environ. Sci. Technol. 50 (6), 549–611.

Parshetti, G.K., Hoekman, S.K., Balasubramanian, R., 2013. Chemical, structural and combustion characteristics of carbonaceous products obtained by hydrothermal carbonization of palm empty fruit bunches. Bioresour. Technol. 135, 683–689.

Phuong, D.T.M., Loca, N.X., Miyanishi, T., 2019. Efficiency of dye adsorption by biochars produced from residues of two rice varieties, Japanese Koshihikari and Vietnamese IR50404. Desalination Water Treat. 165, 333–351.

Piaskowski, K., Świderska-Dąbrowska, R., Zarzycki, P.K., 2018. Dye removal from water and wastewater using various physical, chemical, and biological processes. J. AOAC Int. 101 (5), 1371–1384.

Qiu, Y., Zheng, Z., Zhou, Z., Sheng, G.D., 2009. Effectiveness and mechanisms of dye adsorption on a straw-based biochar. Bioresour. Technol. 100, 5348–5351.

Ramanayaka, S., Vithanage, M., Alessi, D.S., Liu, W.J., Jayasundera, A.C.A., Yong Sik, O., 2020. Nanobiochar: production, properties, and multifunctional applications. Environ. Sci.: Nano 7, 3279–3302.

Rubeena, K.K., Reddy, P.H.P., Laiju, A.R., Nidheesh, P.V., 2018. Iron impregnated biochars as heterogeneous Fenton catalyst for the degradation of acid red 1 dye. J. Environ. Manag. 226, 320–328.

Saygili, H., Guzel, F., 2016. High surface area mesoporous activated carbon from tomato processing solid waste by zinc chloride activation: process optimization, characterization and dyes adsorption. J. Clean. Prod. 113, 995–1004.

Shindhal, T., Rakholiya, P., Varjani, S., Pandey, A., Ngo, H.H., Guo, W., Ng, H.Y., Taherzadeh, M.J., 2021. A critical review on advances in the practices and perspectives for the treatment of dye industry wastewater. Bioengineered 12 (1), 70–87.

Sonu, K., Sogani, M., Syed, Z., 2021. Integrated constructed wetland-microbial fuel cell using biochar as wetland matrix: influence on power generation and textile wastewater treatment. ChemistrySelect 6 (32), 8323–8328.

Sonu, K., Sogani, M., Syed, Z., Dongre, A., Sharma, G., 2020. Effect of corncob derived biochar on microbial electroremediation of dye wastewater and bioenergy generation. ChemistrySelect 5 (31), 9793–9798.

Srivatsav, P., Bhargav, B.S., Shanmugasundaram, V., Arun, J., Gopinath, K.P., Bhatnagar, A., 2020. Biochar as an eco-friendly and economical adsorbent for the removal of colorants (dyes) from aqueous environment: a review. Water 12 (12), 3561.

Sutar, S., Otari, S., Jadhav, J., 2021. Biochar based photocatalyst for degradation of organic aqueous waste: a review. Chemosphere 132200.

Tsai, W.T., Chen, H.R., 2013. Adsorption kinetics of herbicide paraquat in aqueous solution onto a low-cost adsorbent, swine-manure-derived biochar. Int. J. Environ. Sci. Technol. 10 (6), 1349–1356.

Wang, T., Liu, S., Mao, W., Bai, Y., Chiang, K., Shah, K., Paz-Ferreiro, J., 2020. Novel Bi_2WO_6 loaded N-biochar composites with enhanced photocatalytic degradation of rhodamine B and Cr(VI). J. Hazard Mater. 389, 121827.

Wang, T., Liu, X., Men, Q., Ma, C., Liu, Y., Ma, W., Liu, Z., Wei, M., Li, C., Yan, Y., 2019. Surface plasmon-resonance effect of Ag nanoparticles for improving the photocatalytic performance of biochar quantumdot/Bi4Ti 3O12 nanosheets. Chin. J. Catal. 40, 886–894.

Wathukarage, A., Herath, I., Iqbal, M.C.M., Vithanage, M., 2017. Mechanistic understanding of crystal violet dye sorption by woody biochar: implications for wastewater treatment. Environ. Geochem. Health 39, 1–15.

Xiao, B., Dai, Q., Yu, X., Yu, P., Zhai, S., Liu, R., Guo, X., Liu, J., Chen, H., 2018. Effects of sludge thermal-alkaline pretreatment on cationic red X-GRL adsorption onto pyrolysis biochar of sewage sludge. J. Hazard Mater. 343, 347–355.

Xu, R.K., Xiao, S.C., Yuan, J.H., Zhao, A.Z., 2011. Adsorption of methyl violet from aqueous solutions by the biochars derived from crop residues. Bioresour. Technol. 102 (22), 10293–10298.

Yang, Y., Lin, X., Wei, B., Zhao, Y., Wang, J., 2014. Evaluation of adsorption potential of bamboo biochar for metal-complex dye: equilibrium, kinetics and artificial neural network modeling. Int. J. Environ. Sci. Technol. 11, 1093–1100.

Yao, X., Ma, C., Huang, H., Zhu, Z., Dong, H., Li, C., Zhang, W., Yan, Y., Liu, Y., 2018. Solvothermal-assisted synthesis of biomass carbon quantum dots/bismuth oxyiodide microflower for enhanced photocatalytic activity. Nano 13 (03), 1850031.

Zazycki, M.A., Godinho, M., Perondi, D., Foletto, E.L., Collazzo, G.C., Dotto, G.L., 2018. New biochar from pecan nutshells as an alternative adsorbent for removing Reactive Red 141 from aqueous solutions. J. Clean. Prod. 171, 57–65.

Zhang, J., Liu, M., Yang, T., Yang, K., Wang, H., 2016. A novel magnetic biochar from sewage sludge: synthesis and its application for the removal of malachite green from wastewater. Water Sci. Technol. 74, 1971–1979.

Zhang, P., O'Connor, D., Wang, Y., Jiang, L., Xia, T., Wang, L., Tsang, D.C., Ok, Y.S., Hou, D., 2020. A green biochar/iron oxide composite for methylene blue removal. J. Hazard Mater. 384, 121286.

Zhang, S., Lu, X., 2018. Treatment of wastewater containing Reactive Brilliant Blue KN-R using TiO_2/BC composite as heterogeneous photocatalyst and adsorbent. Chemosphere 206, 777–783.

Zhu, W.Y., 2017. Study on Mechanism of the Efficient Adsorptive Removal of OrganicPollutants onto Biochars. Dissertation. North China University of Water Resources and Electric Power, MA.

Zubair, M., Ihsanullah, I., Aziz, H.A., Ahmad, M.A., Al-Harthi, M.A., 2021. Sustainable wastewater treatment by biochar/layered double hydroxide composites: progress, challenges, and outlook. Bioresour. Technol. 319, 124128.

CHAPTER

Restoration of contaminated agricultural soils by microbes

15

Ashwini A. Waoo[1] and Jyoti Pandey[2]

[1]*Department of Biotechnology, Faculty of Life Science and Technology, AKS University, Satna, Madhya Pradesh, India;*
[2]*Department of Biotechnology, APS University, Rewa, Madhya Pradesh, India*

1. Introduction

In agriculture, soil health is essential for producing nutritious food products. Today, soil health is under threat. With plowing, tilling, and excessive fertilizer, we have destroyed them over the past century. Agricultural soil is seriously contaminated. Worldwide, soils are being contaminated as a result of anthropogenic activities.

Plowing and over-tilling enhanced erosion of cropland 10 to 100 times faster than nature. The intensification of agriculture and expansion of industries has resulted in the increased release of a wide range of xenobiotics. Contaminants can harm human beings, livestock, wildlife, crops, and indigenous plants, resulting in an ecological imbalance in nature. Around the world, scientists are trying to solve these problems by utilizing physical, chemical, and thermal processes that resemble excavation and transportation of contaminated soil. Unfortunately, these methods are expensive, and they cannot always guarantee that the pollutants are completely removed or destroyed from the soil. Some of the common pollutants found in soil are polycyclic aromatic hydrocarbons (PAHs), heavy metals, or pesticides (Mirsal, 2018).

A pesticide is a chemical compound used to kill pests. Pesticides are chemicals that weaken and kill pests. Herbicides, insecticides, rodenticides, fungicides, and bactericides can be classified according to the kinds of pests they target. Extracts of plants were used as fungicides and herbicides during the previous centuries. Synthetic chemistry led to rising in pesticide use during the 1930s. During this time, inorganic compounds such as sulfur and arsenic compounds were used for crop protection (Matthews, 2018). In the environment, soil plays a crucial role. Nature is just as important as plants, animals, rocks, landforms, and rivers. It is a natural habitat for a variety of animals.

Some pesticides remain in the soil, and their accumulation affects the microbes that live there. Pesticides can be introduced to humans via ingestion of contaminated water and food, through the inhalation of contaminated air, or directly from industrial, agricultural, and domestic use (Kim et al., 2017). The pesticides can take entry into the human body via the mouth, the eyes, and the respiratory system. Chemical pesticides are toxic depending on their electronic characteristics and the molecular structure, dosage, and time of exposure relating to soil profiles (Hamadache et al., 2016; Heard et al., 2017). When this type of pesticide reaction occurs, pesticides in the soil decrease and effective

Integrative Strategies for Bioremediation of Environmental Contaminants, Volume Two. https://doi.org/10.1016/B978-0-443-14013-6.00010-X

remediation technologies must be adopted to achieve this. An eco-friendly and cost-effective replacement for toxic and expensive methods like chemical or physical methods is bioremediation—an eco-friendly and cost-effective option. In biodegradation, microbial activity can be explored to obtain complete removal. Both bacteria and fungi can produce pesticides that can reduce complex compounds, CO_2, water, oxides from plants that can use as sources of carbon or mineral elements as well as energy. As catalysts, enzymes play a vigorous role in these reactions (Senko, et al., 2017).

The provisioning services include food, fiber, raw materials, wood, and infrastructure support. Regulation services include flood mitigation, nutrient filtering, carbon storing, greenhouse gases reduction, waste detoxification, and biological and cultural identification (Dominati et al., 2014).

Soil is a link between different ecosystems, such as the biosphere, atmosphere, and hydrosphere. Soils are therefore crucial in preserving environmental quality at the local, regional, and global levels. Its buffering capacity, for example, contributes to water quality because of its capability to act as a pollutant's sink and can play a significant role in controlling the negative impacts of the environment as a polluter. Thermomechanical, biological, and thermal methods have all been used to achieve soil and sediment rehabilitation, which is still done today. As a result, there is growing interest in using intrinsic approaches for soil and sediment remediation to minimize the negative impacts of soil contaminants (Khan et al., 2018).

Bioremediation is the best method for protecting agricultural soils from contamination; there is no single method for treating various soil contamination and restoring polluted environments. To restore soil and water to a clean, nontoxic state, bacteria, fungi, and plants are used in bioremediation. As a result of the need to find alternatives to the traditional methods of soil restoration, bioremediation techniques were developed utilizing microbes (bacteria and fungi). A bioremediation procedure uses microbial activity to eliminate contaminants in soil by transferring microorganisms that degrade contaminants or stimulating native microorganisms to degrade and transform contaminants, thus providing substrates for the microorganisms. The microbial degradation of contaminants in soil is one of the primary mechanisms for the elimination of contaminants from soil, which is noninvasive and cost-effective in bioremediation (Picariello, 2023). Throughout this chapter, the focus will be on ways soil degradation leads to a clear economic cost in terms of lost services, including the impact of land-use change and agriculture management on the soil. Like this, Amundson et al. examined soils and climate security, human impacts on soils, and sustainable agricultural practices for continuing intensification of agriculture as necessary for food security.

A change in farming practices can restore the productivity of such soils; severely degraded soils lose their productivity (Oldeman et al., 1990; Utuk and Daniel, 2015). Through soil contamination, the decrease in organic matter and nutrient-rich soil layers has resulted in nutrient depletion, soil structure, water-holding capacity, and soil fertility (Montgomery, 2007). Growing crops on such soils is mostly dependent on the availability of nutrients and the organization of the soil. Bioremediation is the decomposition and degradation of environmental waste materials by microorganisms. Nowadays, bioremediation is the best method to treat contaminated agricultural soil. Polymers produced by microbes have low water-holding capacities. Bioremediation of contaminated agricultural soil requires the removal of pollutants, soil restoration, soil testing, and prevention of future contamination.

It is important to restore contaminated soils. The degradation of soil due to chemicals affects 12% of all the world's degraded soils (Madison, et al., 1995). Contamination of soil is not only an aesthetic and health issue, it also has a profound impact on the economy due to reduced productivity, quality, and evaluations of contaminated sites. Further, the costs of remediating contaminated soils are high (Bini, 2011).

2. Sources of soil contamination

Contaminated soil can reduce soil health if harmful artificial drain into water reaches streams, lakes, or seas. Furthermore, soil increases contamination by the addition of unstable mixtures into the environment. When natural materials disintegrate in soil, sulfur dioxide and other sulfur compounds are released, causing corrosive downpours. Types of sources of soil pollution are mentioned in Fig. 15.1.

Soil can become polluted for various reasons, including:

1. Leaking landfills
2. Industrial waste discharge
3. Contaminated water seeping into the ground
4. Underground storage tank rupture
5. Excess use of pesticides, herbicides, or fertilizers
6. Solid waste seepage
7. Improper installations, detonation, and dismantlement of munitions

Heavy metals and other toxic components are the main reason why soil toxins are released in sewage. Plant and animal health are adversely affected by soil pollution, which occurs when chemicals, salts, or disease-causing agents accumulate in soils.

2.1 Man-made pollutants

Various types of anthropogenic soil pollution occur, some intentionally (industrial) and some accidentally. Contamination levels in soil can be increased by human-caused soil pollution in association with natural processes.

- The accidental spilling and leakage of chemicals during storage and transport;
- Factories and manufacturing processes involving furnaces or processes that may disperse contaminants into the environment;

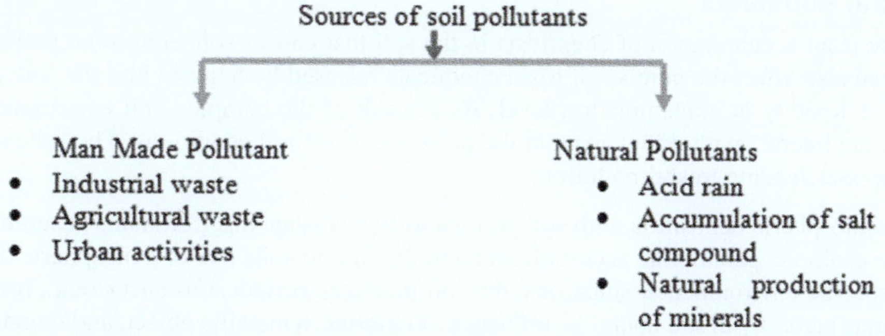

FIGURE 15.1

Sources of soil pollutants.

- The crushing and grinding of raw materials and heavy metals, which emit toxic substances;
- Dissemination of herbicides, pesticides, insecticides, fertilizers, and/or pesticides;
- Emissions from transportation activities;
- Unintentional or deliberate dumping of chemical waste, including illegal dumping;
- Garbage in landfills can leak into groundwater and generate polluted vapors;
- Lead-based paint chippings falling from walls.

2.1.1 Industrial waste

Chemical contaminants are spread in the soil with improper disposal of industrial waste products. As a result, plants and animals suffer, as well as the water supply and public health. Toxic fumes from regulated landfills contain toxic chemicals that can cause acid rain and damage the soil profile. As a result of industrial activities like dumping oil and fuel, heavy metals, toxic chemicals, and industrial waste, the soil is acidified and contaminated.

2.1.2 Agricultural pollution

Soil pollution can be occurred by agricultural processes. Fertilizers are used to enhance crop yield, but they also cause soil pollution. Pesticides also cause harm to plants and animals by polluting the soil. These chemicals seep deep into the soil and pollute the groundwater, and runoff pollutes local water systems and causes eutrophication. Phosphorus is one of the main contributors to eutrophication, as its high amount enhances cyanobacteria and algae growth, thereby reducing dissolved oxygen in the water.

2.1.3 Urban activities

Soil pollution can be due to direct or indirect ways, by human activities. Proper drainage and increased runoff can contaminate nearby land areas and streams. Several chemicals and pollutants are deposited into the soil when the trash is thrown away in an unorganized fashion. Those materials may seep into groundwater or runoff in local water systems, and excess waste storage can lead to the accumulation of bacteria in the soil, resulting in methane gas production from decomposition by bacteria, resulting in global warming and bad air quality. In addition, it produces foul odors and negatively impacts the quality of life.

2.2 Natural pollutants

Unless there is an accumulation of chemicals in the soil that causes soil to become polluted, natural processes can also affect the number of toxic chemicals released by humans into the soil, reducing or increasing its toxicity or contamination level. As a result of the complex soil environment, various chemicals may interact with pollutants and the presence of natural conditions. The following are the natural processes leading to soil pollution:

- Due to atmospheric deposition with soil permeability misbalancing, pollutants accumulate in the soil. For example, perchlorate accumulates naturally in arid soils due to atmospheric deposition.
- Under specific environmental situations, the soil produces certain substances (e.g., formation of perchlorate naturally in soil under the influence of chlorine, a metallic object, and thunderstorms).
- Sewer lines leaking into the subsurface (addition of chlorine, for example, can generate trihalomethanes, like chloroform).

The soil can also be polluted by natural events such as earthquakes, landslides, hurricanes, and floods. These events alter the composition of the soil, resulting in contamination. Natural sulfide-bearing rocks weather to form arsenopyrite (gossans) mineralized zones. In most cases, these minerals are spatially variable, with a higher concentration in deeper layers. As is slightly bioaccessible if it comes from natural sources. The radioactive gas radon (Rn) is also present in soil and rocks. Rocks rich in feldspar and illite are particularly radioactive in acidic igneous rocks.

Contamination of soil occurs when these xenobiotic substances adhere to the dirt, either directly from spills onto the dirt or soil that has been contaminated by the substance.

2.2.1 Acid rain

Sulfur dioxide, oxides of nitrogen, and ozone are the main contributors to acid rain. Acid rain occurs when pollutants in the air mix up with rain and come back to the ground. Acidity in rainwater is caused by sulfuric and nitric acid solutions. Acid rain lowers the pH of the soil, thereby increasing its acidity, which reduces the availability of important nutrients in the soil. Acid precipitation is most likely to occur in soils with low cation exchange capacity and low base saturation.

2.2.2 Accumulation of salt compound

Agriculture under irrigation faces an enormous problem of soil salinity. The soils in hot, dry regions of the world are frequently saline and have low agricultural potential. To make matters worse, most crops in these areas are maintained by irrigation, and inadequate irrigation management results in 20% of irrigated land being affected by secondary salinization (Glick et al., 2017). In arid and semiarid regions, land and water resources are often salinized as a result of irrigation, a major activity. Soil contains soil ions (electrically charged atoms or compounds). They are formed when soil minerals weather. Occasionally, they may migrate upward from shallow groundwater or be used as fertilizers or irrigation water.

By exploiting their unique attributes, microorganisms could play an important role. The properties of these plants include tolerance to salinity, genetic diversity, the ability to synthesize compatible solutes, hormone production, biocontrol efficiency, and their interaction with crops (Shrivastava et al., 2015).

2.2.3 Natural production of minerals

Over the past decades, increasing industrialization, the use of chemicals, mining and smelting residues, and traffic in many urban areas have resulted in environmental hazards for terrestrial and aquatic ecosystems, as well as food quality and socioeconomic problems. Plants rely on soils for trace elements that, in small amounts, are considered essential for human, plant, and animal growth. Soil/groundwater contamination can result from natural processes (industrial and agricultural activities, weathering/alteration of rocks and raw materials) and human activities (mining and smelting), often result in elevated contents of harmful elements (Cr, Cu, Hg, Pb, Zn, Sb, Co, Ni, Cd, and As).

3. Impacts of soil pollution

Pollution of soil affects not only soil and its biota but also by environment and every life-form, from earthworms to humans. Among these effects are:

- **Human health:** Human beings are reliant on the soil for our food, so soil pollution affects us as well. Various diseases can be caused by the bioaccumulation of toxins in the body. Today, reproductive health, birth defects, neurologic problems, malnutrition, and cancer-causing mutations in the body are all on the rise. The direct effect of soil on humans is attributable to the inhalation of polluted soil vapors and contamination.
- **Growth of plants:** Soil contamination directly affect the ecological balance. The plants cannot adapt to abrupt changes in the soil's chemistry, which in turn affects the microorganisms. This results in soil disintegration. Massive tracts of land can become infertile and unfit for living things. The plants that do grow on these soils will retain the toxins and adapt to the natural environment. As fruitfulness declines, the land becomes unsuitable for horticulture as well as vegetation. Contaminated soil makes large plots of land hazardous to health.
- **Diminished soil fertility:** The presence of poisonous synthetics in the dirt can lower soil fertility, resulting in a decline in soil productivity. Defiled soils are then used to produce leafy foods, which need quality supplements and contain toxic substances to cause chronic health conditions in individuals consuming them.
- **Odor contamination:** An area that is littered with garbage and debris destroys the serenity. Landfills emit harmful and foul gases that pollute the atmosphere and adversely affect the health of certain individuals. They are plagued by a terrible smell.
- **Changes in soil structure:** Numerous soil-living creatures (such as creepy crawlies, crawlers, and microorganisms) can cause soil structure to be altered. Consequently, their hunters may also have to move to different locations in search of food.
- **Impact on ecosystem and biodiversity:** The lack of biodiversity in an environment can be caused by soil contamination. Birds, creepy crawlies, well-developed creatures, and reptiles that live in dirt can be impacted by contamination. Dirt is a noteworthy environment.
- **Tainting of water sources:** Surface runoff causes water contamination when it rains because it transports degraded soil to water sources. Toxins can also penetrate deep into the ground and contaminate groundwater. Both animals and humans are unable to utilize the contaminated water. Amphibians will likewise be impacted since they will find their living spaces uninhabitable in these water bodies.

4. Treatments of contaminated agricultural soils

Agricultural soil can be treated chemically, enzymatically, mechanically, and by microorganisms, which is referred to as bioremediation, or remediation by living organisms. Remediation is getting rid of a problem; whenever it is associated with the care of an ecological problem, like soil or groundwater contamination, it is called bioremediation. Bioremediation is the use of living microorganisms to reduce natural contamination or to prevent it. Today, different types of treatment are available, with microbial treatment being the most used.

- **Mechanical treatment:** By using physical machinery.
- **Chemical treatment:** By using chemicals.
- **Enzymatic treatment:** By using enzymes.
- **Biological treatment:** By using microorganisms.

Microbiological bioremediation can be applied in a variety of ways for soil contaminant-specific site remediation by using specialized strains of bacteria with a particular zest for consuming contaminants (Volkering et al., 1997) (Fig. 15.2). Various steps in the bioremediation process are shown below in Fig. 15.3.

5. Types of Bioremediation

Depending on the type of microorganism used, pesticides can be biodegraded either aerobically or anaerobically. Depending on where the bioremediation treatment is performed, bioremediation techniques can be classified into three groups, such as ex situ and in situ. Bioremediation techniques based on strategies can be divided into two groups—ex situ bioremediation and in situ bioremediation (Fig. 15.4).

5.1 Ex situ bioremediation

Toxic waste can be collected from contaminated sites, and bioremediation can be carried out using the requisite microorganisms (often a consortium of organisms). Some places have successfully used this process in place of in situ bioremediation.

Ex situ means removing contamination to a remote treatment site. Since this classification involves the big job of excavating contaminated soil and transporting it to an off-site location, it is not very popular. In ex situ remediation, appropriate soil oxygen, nutrients, and moisture conditions are introduced off-site. During the transport process, the ex situ bioremediation process can spread contamination or result in an accidental spill. Below are two techniques that can be applied.

5.1.1 Slurry phase

A large bioreactor is used to combine polluted soil with water and nutrients supplement so that indigenous microbes can be in vicinity with the pollutants. The nutrients and oxygen are supplemented

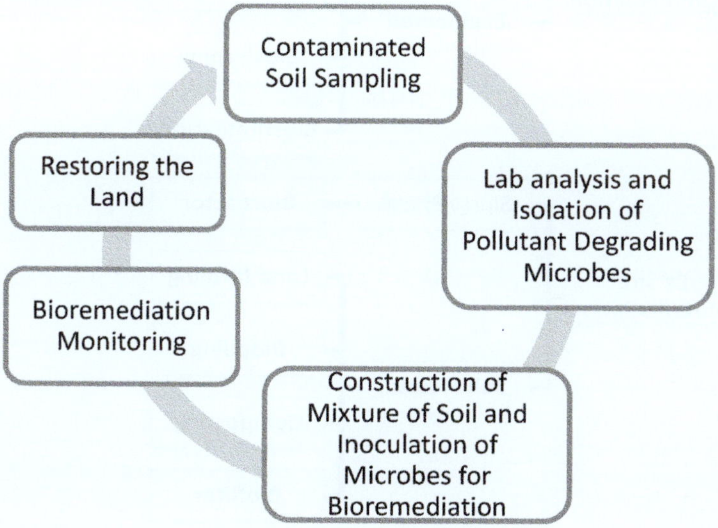

FIGURE 15.2

Restoration of contaminated agricultural soil by using microorganisms.

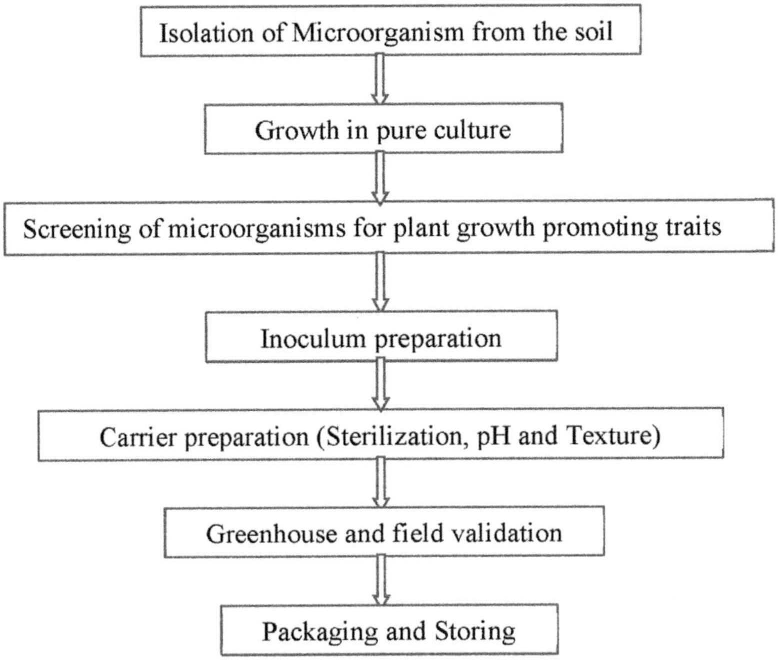

FIGURE 15.3

Steps of bioremediation of agricultural soil.

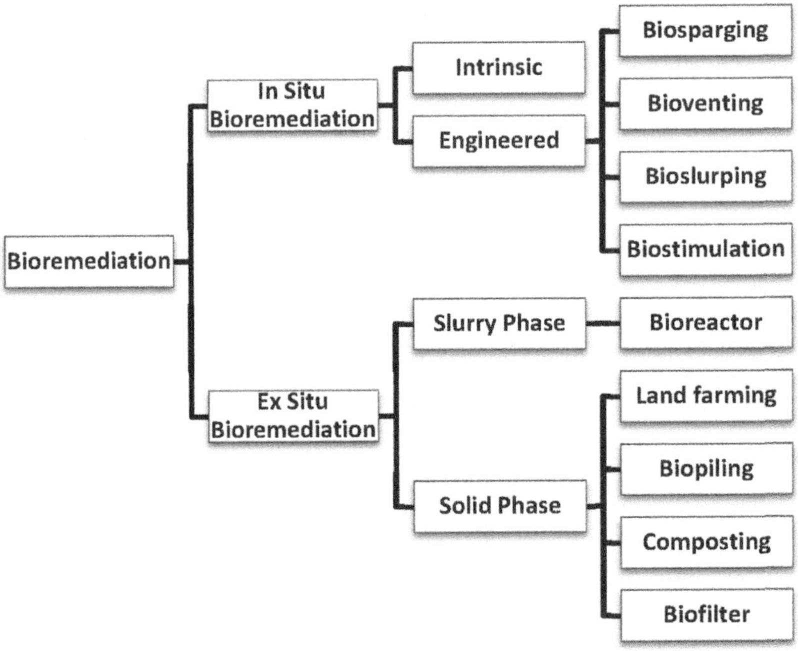

FIGURE 15.4

Types of bioremediation.

to the bioreactor, and conditions are set to allow the microorganisms to remove the contaminants. The water is eliminated from the solids after the treatment and disposed of, or further processed if still contaminated, if necessary. Slurry-phase treatment is relatively fast when compared with other biological treatment methods.

- **Bioreactors:** An important type of ex situ technique is slurry bioreactors (SBs). Under controlled environmental conditions, SBs provide the best option for bioremediating soils polluted by recalcitrant pollutants. SB involves excavating and conditioning soil before it is loaded into bioreactors. The main characteristic of SB is that soil inside the reactor is kept in the aqueous condition through some type of mixing so that biological treatment is done under saturated conditions with nearly homogeneous suspension.

SB operates in batch and semicontinuous modes. First, contaminated soil is crushed and screened. Soils that contain coarser fractions (pebbles and sands, 0.85–4 mm) are then discarded, while fine fractions (clay and organic matter, <0.85 mm) are retained and loaded into bioreactors. Pollutants tend to accumulate in fine soil particles. The contaminated fractions of soil are mixed with water or wastewater to create a slurry whose concentration ranges from 15% to 60%. Microorganisms must be properly aerated with continuous mixing to keep solid particles suspended.

The degradation process is very fast and effective due to close contact between the pollutants and microorganisms, as well as optimal conditions (nutrients, temperature, aeration, etc.). High costs, however, limit the use of slurry-phase bioreactors.

The polluted soil is either treated in solid form or slurry form. Solid-phase reactors work based on the mechanical decomposition of soil by attrition and intensive mixing of its components in a closed container. As a result, contaminants, microorganisms, nutrients, water, and air are constantly in contact with each other. The SB is defined as a system that mixes soil, water, and biomass (typically indigenous microorganisms) as a three-phase mixture to increase the bioremediation rate for soil-bound and water-soluble pollutants. Biodegradation in bioreactors generally occurs at a greater rate and extent than in situ or solid-phase systems as the contained environment is easier to manage and is, therefore, more controllable and predictable. The contaminated soil must either be pretreated (e.g., excavation) or removed by soil washing or physical extraction (e.g., vacuum extraction) before being placed in a bioreactor (Vidali, 2001).

5.1.2 Solid phase

Above-ground soils are treated with solid-phase treatment. There are collection systems in this area to prevent pollutants from escaping the treatment. Degradation is enhanced by controlling variables like moisture, nutrients, heat, and oxygen. While solid-phase systems are easy to maintain and operate, they require a lot of space and more time than slurry-phase systems. Treatment can be done using the following techniques.

- **Land farming:** Basic concepts of this technique include promoting degradation by indigenous microorganisms and facilitating aerobic degradation. The process involves excavating contaminated soil and spreading it over a prepared bed until the pollutants are degraded. Additionally, nutrients and minerals are added to enhance the growth of the native microbial species.

Bioremediation of soil contaminated with hydrocarbons can be achieved ex situ by land farming, an aboveground bioremediation method. An illustration of a land farming system (a solid phase soil reactor) is shown here. Typically, this process involves spreading excavated contaminated soil on a thin layer of sand that contains microorganisms and nutrients. Regular plowing is necessary for good soil mixing and aeration. By mixing compost into the soil and/or raising the temperature, biodegradation can be enhanced.

Enhanced microbial activity leads to the degradation of adsorbed petroleum product constituents through microbial respiration. Soils less than three feet deep that are contaminated are shallow (under three feet of the ground). Microbial action can be stimulated without excavating. It is necessary to excavate petroleum-contaminated soils deeper than five feet and reapply them to the ground surface.

Bioremediation of contaminated soil with chloromethane, benzene, toluene, and xylene can be achieved with land farming. BTX aromatics are found in these compounds.

In this process, excavated contaminated soil is spread in a thin layer on the ground surface and aerobic microbial activity is stimulated by aeration, mineral additions, and moisture. Because contaminated soil is treated in thin layers of up to 0.4 m in thickness, it requires a large treatment area. During regular intervals, plowing, harrowing, and milling are done to enhance oxygen supply as well as mixing. If there is sufficient land available, the treatment process is cost-effective (Fig. 15.5).

- **Soil biopiles:** The biodegradation technique is used to remediate soil that contains petroleum. Soil biopiles are also called biocells. As part of this technology, contaminated soil is piled up and microbial activity is stimulated either in the presence of oxygen or by adding minerals, nutrients, or moisture. Biopiles can be between three and ten feet high. Oxygen is also used in this technology to promote microbial proliferation. Biopiles are aerated by injecting air through perforated piping positioned throughout the pile.

Composting is a hybrid of land farming and composting. Excavated soils are then mixed with soil additives, spread on a treatment region, and then bioremediated using forced aeration. Contaminants are converted to carbon dioxide and water. There are four components to a bipile system: a treatment bed, aeration equipment, irrigation/nutrient equipment, and a leachate collecting system. Soil piles can be 20 feet tall and covered with plastic to regulate runoff, volatilization, and evaporation with solar

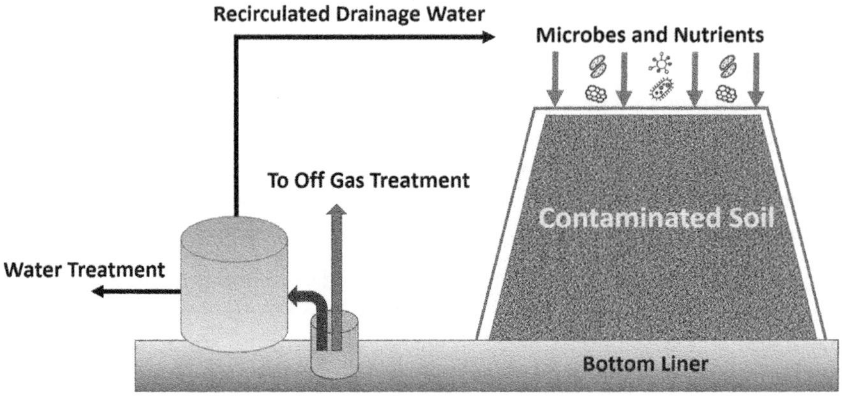

FIGURE 15.5

Land farming.

heating. Before volatile organic compounds enter the air stream, they are treated if necessary (Shukla et al., 2010). In biopiles, native aerobic and anaerobic microorganisms can flourish.

- **Composting:** The composting process involves mixing contaminated soil with biomass such as hay, straw, or corncobs, which provides just the right amount of air and water for the microorganisms. In composting, contaminated soil is placed in treatment vessels for aeration and then mixed. Composting in a window is a kind of composting in which the soil is arranged in long piles, called windows, and mixed by tractors. For composting, use 75% contaminated soil and 25% compost. Depending on the type of soil, characteristics, and contamination level, the ratio will vary. The compost remediation process is known as fast remediation because it can be completed in weeks. Bioremediation uses it to degrade toxic organic compounds and possibly lessen the toxicity of metallic contaminants in organic residues, wastes, and byproducts. Organic wastes are degraded by microorganisms in compost like those that occur naturally in soil. Composted materials often have higher temperatures than soils, so contaminants are solubilized and metabolic activity is higher.
- **Biofilter:** Organic contaminants can cometabolize with high levels of the substrate in composts. Metals, plastics, glass, and stones, which are nondegradable or unwanted, can be ground, mixed, and sieved out to allow compostable material to be biologically treated. A composting mechanism is affected by several factors including the organic contaminant, the conditions and procedures of composting, the microbial cell numbers, and time (Barker and Bryson, 2002).

5.2 In situ bioremediation

It is an in situ method where bioremediation is carried out at the site of contamination. This is the preferred method of bioremediation because it requires less effort to eliminate pollutants and prevent the spreading through transportation and pumping. The in situ bioremediation process utilizes microorganisms that metabolize organic contaminants into inorganic materials, such as methane, carbon dioxide, water, and salts. Natural or engineered conditions can be used to accomplish this.

5.2.1 Intrinsic in situ bioremediation

A process of intrinsic bioremediation involves converting environmental pollutants to nontoxic forms by using the natural abilities of the naturally occurring microbes on the site. This process is typically used underground, such as in underground petroleum storage tanks. Intrinsic bioremediation harnesses the natural abilities of natural microbial communities to degrade pollutants without modifying them or accelerating them in any way. This technique aims to increase the metabolic activity of indigenous microbes by providing them with nutrients and oxygen.

5.2.2 Enhanced (engineered) in situ bioremediation

In this technique, specific microbes are introduced into the polluted site. Enhanced physicochemical conditions facilitate in situ bioremediation, which accelerates degradation.

- **Bioventing:** The bioventing technique degrades organic matter adsorbing on soils using microorganisms. The technique involves regulating airflow to increase oxygen to unsaturated zones to boost bioremediation by increasing the number of indigenous microbes. Bioventing is a process in which nutrients and moisture are added to increase the microbial transformation of pollutants into nontoxic forms. In situ bioremediation techniques such as this one have gained popularity especially in sites polluted with petroleum spills. Bioventing is used mostly for the degradation of fuel residuals, but it can be used for the deprivation of volatile organic compounds in the active soil.

An in situ remediation method utilizes native microorganisms to degrade organic material adsorbing to unsaturated soils. The extraction method utilizes vacuum-enhanced soil vapors. Pressure differences in the subsurface cause the inflow of atmospheric air and, consequently, oxygen, allowing contaminants to be aerobically degraded. Petroleum products, like gasoline, kerosene, jet fuels, and diesel fuel, can be remedied using this technology. The extracted soil vapor must be treated if the contaminants to be eliminated are volatile through adsorption on activated carbon by biodegradation in a biofilter (Mohapatra, 2008).

- **Bioslurping:** Bioslurping using vacuum-enhanced pumping of soil and water, indirect oxygenation, and stimulation of biodegradation of contaminants are achieved. Liquids can be drawn from a layer of free product using a "slurp" that extends into the layer in a similar way to how a straw pulls liquid from a vessel. The process of slurping is accomplished by attaching an adjustable tube to a well and introducing the tube into the liquid layer of the nonaqueous phase and pumping to remove the free product and some groundwater. This method can be used to restore soils that contain volatile or semivolatile organic compounds.
- **Biosparging:** Biosparging consists of using a biological approach to remove aromatic compounds from an area such as benzene, xylene, toluene, and naphthalene. Mineral oil and aromatic compounds are broken down into simpler and more useful forms by using specific aerobic bacteria. Using this technique, the air is injected into the soil to stimulate microbial activity to promote pollutant elimination from polluted areas. During biosparging, the air is inserted into the saturated zone, causing volatile organic compounds to move to the unsaturated zone, thus promoting biodegradation. Biosparging is affected by two major factors:
  (1) Soil permeability (affecting pollutant bioavailability)
  (2) Pollutant biodegradability
  Utilizing indigenous microorganisms degrades organic contaminants in saturated soil. Through the drilling of boreholes, the air is injected into the saturated zone (below the water table) to activate the soil's indigenous microorganisms by increasing oxygen dissolution. Oxygen increases aerobic biodegradation of contaminants in soil or groundwater.
  By using biosparging, petroleum constituents can be reduced from the soil beneath the water table, within the capillary fringe, or in groundwater. The most common application of biosparging is in sites with midweight petroleum products, like diesel fuel; lighter petroleum products volatilize quickly and are removed very quickly by sparging. The effectiveness of the technology is dependent on soil permeability (Vidali, 2001).
- **Biostimulation:** Nitrogen and phosphorus are typically added because they increase biodegradation and microbial diversity. Betancur-Corredor et al. (2015) demonstrated the degradation of DDD, DDT, and DDE by enhancing the microbial population and addition of a surfactant. In studies conducted in soil using biostimulation processes, Bacmaga et al. (2019) studied the biodegradation of tebuconazole. This process requires control of the concentration of nutrients provided throughout the process to enhance microbial activity and diversity. For this reason, tebuconazole must be reduced in soil concentration since it negatively influences enzymatic activity and microbial growth and proliferation. Pesticides harm the microbial

population. Bird dropping-fertilized soils degraded tebuconazole more intensely than compost-fertilized soils. It can be done in the following ways:

(1) Nitrogen and phosphorus are added. Addition of cosubstrates, for example, methane, to degrade trichloroethylene.

(2) Hydrophobic compounds are dispersed in water with the addition of surfactants.

The addition of nutrients and cosubstrates initiates the growth of microorganisms, while the surfactants expose the hydrophobic molecules. Bioremediation of contaminated soil or land results in biostimulation in all these situations (Fig. 15.6).

6. Bioaugmentation—a bioremediation technique

Cleaning up contaminated sites with bioremediation is a highly effective method. Microorganisms and electron acceptors determine whether this is an aerobic or anaerobic process. Natural bioremediation can occur or it can be engineered (engineered bioremediation).

Natural attenuation refers to bioremediation that occurs without human intervention except for monitoring. In natural attenuation, indigenous microorganisms are responsible for the effects. As part of biostimulation, indigenous microbial populations are employed for the remediation of contaminated soils. An increasing supply of nutrients or the addition of some stimulating agents to soil can enhance microbial activity, thus catalyzing natural attenuation processes.

Biological augmentation is the process of enriching an existing population of microorganisms to reduce the contamination level. An organic culture is added to the contaminated soil to create an environment like that of a bioreactor. The two most common options are: adding an adapted pure bacterial strain or adapted consortium to a contaminated site. Oil-contaminated sites typically require bioaugmentation during bioremediation. The bioaugmentation method is a low-cost alternative to other techniques of treating soil and water contamination. Microorganisms either indigenous or exogenous are added to the contaminated area. Adding microbial cultures to a land treatment unit is limited by two factors:

a. Nonindigenous cultures rarely can compete effectively with native populations to achieve useful population levels.

b. As long as land treatment units are effectively managed, native microbes in the soil can degrade biodegradable waste (Vidali, 2001).

Microbial Cells	Biosoption
	Biotransformation
	Bioleaching
	Biomineralization
	Bioaccumulation
	Biodegradation of Chelating Agents
	Microbially Enhanced Chemisorption of Metals

FIGURE 15.6

Metal microbial interaction.

Bioaugmentation involves adding specific microorganisms to polluted soil. In some cases, the native soil microorganisms are not able to break down the pollutants effectively due to the complexity of their molecules. Polychlorobiphenyls (PCBs), polycyclic aromatic hydrocarbons (PAHs), trinitrotoluene (TNT), and certain pesticides are examples of these pollutants.

As part of the bioaugmentation process, a microbial consortium or one strain is inoculated into the soil to increase microbial diversity. As a result, microorganisms with specific metabolic activities enhance the biodegradation process. Pesticide concentration in the soil is an important parameter that shapes the process as high pesticide concentrations inhibit the functions of soil bacteria. Doolotkeldieva et al. (2018) studied the biological degradation of pesticide in soils in detention sites. Doolotkeldieva et al. in 2018 identified several bacterial strains in the studied soils during their preliminary study. Aldrin is a chlorinated hydrocarbon pesticide that is diffused. Bacterial strains with genes (cytochrome P450), such as *Bacillus polymyxa* and *Pseudomonas fluorescens*, have been demonstrated to be able to degrade aldrin in a relatively short period. Selection of specific bacteria, optimizing soil conditions like temperature, etc.

Using the pH and nutrients in the soil, the next experimental test was developed. Soil inoculated with the consortia was contaminated with pesticides and inoculated in mesocosms (Doolotkeldieva, et al., 2018). Since pesticides leach into the soil's subsurface and are adsorbed on the soil, the concentration of pesticide in contaminated soil differs at different depths. Odukkathil et al. (2016) analyzed the bioaugmentation method in a 4500 cm^3 glass column as an experimental test. During infiltration, pesticides drift downward; it appears that pesticide concentrations were high in the bottom soil, whereas pesticide concentrations were low in the central soil as a result of higher microbial activity.

7. Mycoremediation by white-rot fungi

Bioremediation using fungi is known as mycoremediation. These techniques use fungi-releasing enzymes that can degrade several pollutants and have been proven to be promising in removing contaminants from a site. Bioremediation applications are predominantly based on bacteria and few attempts on using fungi. Fungi contribute to the cycle of elements by decomposing and transforming organic and inorganic materials. These characteristics could be used in bioremediation to degrade organic compounds and lessen metal contamination risks. Fungi can sometimes outperform bacteria, not only in metabolic versatility but also in environmental resilience. They can oxidize a wide range of chemicals and survive under harsh environmental conditions like low moisture and high levels of pollutants.

It was discovered in 1985 that the white-rot fungus (*Phanerochaete chrysosporium)* could metabolize multiple key pollutants. A major characteristic of these fungi is their property to metabolize complex compounds, including lignin. Other species of white-rot fungi later showed similar characteristics. Furthermore, white-rot fungi have the advantage of degrading lignin extracellularly through their hyphal extension. In this way, they are having access to soil pollutants that other organisms can't and maximize the surface area for enzyme interaction. Extreme conditions, like extreme pH, high temperature, and moisture, can be tolerated by these inexpensive fungi. Most microbial organisms used for bioremediation need pretreatment of their environment to survive; however, white-rot fungi can be directly applied to many areas because they degrade upon nutrient deprivation.

The first fungus to be linked to organic pollution degradation is *P. chrysosporium*. Studies indicated that it can bioremediate pesticides, carbon tetrachloride, dioxins, and many other contaminants. The fungus *P. chrysosporium* has emerged as a model for bioremediation among fungi. *Trametes versicolor* and *Pleurotus ostreatus* are other species of white-rot fungi.

8. Bacterial bioremediation

Bioremediation is a process in which microorganisms degrade and lessen the poisonousness of pollutants in the environment. Bioremediation is thus dependent on the microbial and environmental interaction with pollutants. These microbes can either be naturally present at the site of remediation or artificially injected. The biodegradation process is a consequence of microbial metabolism, and some microbes break down pollutants directly to capture carbon and energy. Here we discuss bacteria and fungi, which are commonly used in bioremediation, and archaebacteria, a new group of organisms that have unique applications in bioremediation.

The diversity of bacteria makes them excellent elements in bioremediation. There are some universal toxins in bacteria, so any given substrate is likely to be digested if supplemented with the optimum conditions (anaerobic, aerobic environment). The following are specific bacteria species responsible for bioremediation.

- ***Pseudomonas putida:*** The gram-negative soil bacterium *P. putida* is involved in the remediation of toluene. A product of petroleum refinement, naphthalene, can also be degraded in contaminated soils.
- ***Dechloromonas aromatica:*** It is a rod-shaped microbe capable of oxidizing aromatic compounds including chlorobenzoate, benzoate, and toluene by reducing oxygen, nitrate, or chlorate. Benzene can only be oxidized by this organism anaerobically. Since ground and surface water are prone to benzene contamination, *D. aromatica* is useful for in situ technique of bioremediation.

8.1 Nitrifiers and denitrifiers

Wastewater can be cleaned using industrial bioremediation. Mineral nitrogen compounds (such as ammonia, nitrite, and nitrate) are generally eliminated by microbial activity. Nitrogen is removed from the atmosphere by nitrification and denitrification. During nitrification, *Nitrosomonas europaea* oxidizes ammonium to nitrite. Microbes such as bacteria then oxidize nitrite to nitrate. *Paracoccus denitrificans* use the nitrate produced during ammonium oxidation to accept electrons in anaerobic conditions. It produces N_2. Ammonium and nitrate can be used to treat eutrophication in natural waters.

Nowadays, artificial inorganic nitrogen sources replace the nitrogen mineralized by soil microbes. In addition, nitrogen fixation by soil microbes can be inhibited by a high amount of inorganic nitrogen compounds. Artificial nitrogen fixing, however, requires a lot of energy. To reduce energy waste and sustainably use natural resources, organic farming has become a trend in agriculture. An important issue is the nitrogen sources of agricultural fields. Ammonia is only produced by bacteria when nitrogen in the air is combined with hydrogen. The nitrogen cycle and organic farming rely on nitrogen-fixing bacteria. They are sometimes symbiotic with plants that have infections of their root systems. Infecting the roots of plants in the bean family (Fabaceae or Leguminosae) is a genus of nitrogen-fixing bacteria called Rhizobium.

8.2 Bioremediation of hydrocarbon pollutants

Hydroxycarbons are deep underground but were brought to the surface for transformation and utilization as fossil fuels. In the current era, most pollution is produced by PAHs, which appear when carbon materials are under anaerobic conditions, and nitrate is an electron acceptor for microbes like *P. denitrificans* during ammonium oxidation. Dinitrogen gas is produced as a result. Ammonium and nitrate, both pollutants that contribute to eutrophication in waters, are remediated in this process.

Marine rod-shaped bacteria *Alcanivorax borkumensis* consume hydrocarbons from fuel and produce carbon dioxide as a result. As a result of its ability to grow rapidly in oil-stained environments, it has been used in the cleanup of more than 830,00 gallons of oil from the oil spill in the Gulf of Mexico. Other microorganisms having the capacity of biodegradation of hydrocarbons are shown in Table 15.1.

8.3 Phosphate-solubilizing microorganisms

Numerous bacterial, fungal, and actinomycete genera can release biologically available phosphorous as orthophosphate by specific mechanisms involving the production of organic acids, phosphatase, and siderophore production. Plant growth could be enhanced either directly by phosphate-solubilizing microorganism (PSM) releasing organic acids into the environment or indirectly by PSM producing phytohormones, toxins-resistant compounds, antifungal compounds, and other valuable bioactive molecules.

Table 15.1 Microorganisms for bioremediation of hydrocarbons.

Microorganisms	Compound
Penicillium chrysogenum	Aromatic hydrocarbons, benzene, toluene, xylene, and phenol
Pseudomonas putida	Benzene and xylene
Aspergillus niger, Fusarium solani, and Penicillium funiculosum	Hydrocarbon
Microorganisms and heavy metals	
Saccharomyces cerevisiae	Lead, nickel, mercury
Pseudomonas fluorescens and *Pseudomonas aeruginosa*	Fe^{2+}, Zn^{2+}, Pb^{2+}, Mn^{2+}, and Cu^{2+}
Microorganisms and pesticides	
Bacillus and Staphylococcus	Endosulfan
Pseudomonas sp., Acinetobacter sp., Photobacterium sp., and *Enterobacter sp.*	Methyl parathion and chlorpyrifos

9. Bioremediation of pesticides in soil using biopesticides

Plants, animals, microorganisms, and minerals are the sources of biopesticides. Depending on the active ingredient used, they can be categorized into three main groups: biochemical pesticides, plant-incorporated protectants (PIPs), and microbiological pesticides (Fig. 15.7).

9.1 Biochemical pesticides

These natural compounds control pests in a nontoxic manner. They can be extracted naturally or synthesized with a similar structure as well as activity that existed in nature (Mishra et al., 2012). Semiochemicals are the compounds produced by plants and animals. Among these compounds are pheromones, kairomones, allomones, and attractants. Sarwar et al. (2015) suggest that they are essential for feeding, mating, and laying eggs (ovipositing) of pests. Therefore, controlling their concentration can influence their life cycle.

9.2 Plant-incorporated protectant

Pests that feed on plants can produce PIPs. Plants can be genetically altered to produce pesticidal proteins by inserting a gene that acts on a specific protein into the plant's genetic material. Thus, the plant can produce pesticidal compounds for certain pests (Parker et al., 2017).

9.3 Microbial pesticides degradation

Living organisms such as bacteria, algae, fungi, and viruses are used in microbial pesticides to control pests. In addition to forming toxic metabolites, they also hinder the growth of other microorganisms (Montesinos, 2003).

During bioremediation, microorganisms use pesticides as part of their metabolic reactions with other nutrients, so they are removed from the environment. The potential of these processes is determined by the features of pesticides, including their distribution, bioavailability, and persistence in soil. The low solubility of pesticides and adhesion to soil particles negatively affect the accessibility of pesticides to microbes (Ortiz-Hernandez et al., 2014). Additionally, soil characteristics and environmental factors, such as water content, pH, temperature, and microbial diversity, affect bioremediation effectiveness.

Biodegradation processes result in the breakdown of pesticides or their complete mineralization by microbes, which use these compounds as fuel for their metabolic processes. The biochemical reactions

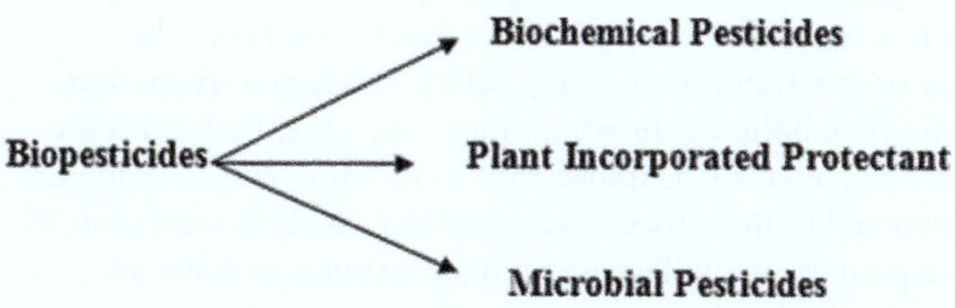

FIGURE 15.7 Types of active ingredients.

are catalyzed and influenced by enzymes, such as oxygenases, hydrolases, and peroxidases. Pesticides degrade in three stages, as follows:

- **Phase 1:** The pesticide is converted to a water-soluble and less toxic form through oxidation or reduction.
- **Phase 2:** Products in Phase 1 are transformed mostly into sugars and amino acids, which have more water solubility and less toxicity.
- **Phase 3:** Phase-2 metabolites are converted into secondary conjugates that are less toxic. Bacteria and fungi may produce intra- or extracellular enzymes as part of the degradation process.

Bioremediation activities must consider the degradation time as a relevant parameter (Khajezadeh, 2020), which is dependent on the concentration of pollutants at the beginning and end of the process. Due to several factors, such as temperature, water content, availability, microbial activity, and pesticide leaching in the soil, this method has limitations (Soulas and Lagacherie, 2021).

- **Bacterial degradation:** Many bacterial strains have been identified over the years that can degrade pesticides. Bacteria possess specific characteristics that make them particularly suitable for degradative processes. A bacterial strain's adaptation, proliferation, and role are affected by several factors, including pH, temperature, water content, and pollutants. Metabolites tend to be harder to remove than their original compounds; they can also cause environmental problems. Insecticide chlorpyrifos, an organophosphate, is hydrolyzed into 3,5,6-trichloro-2-pyridinol (TCP) by microbes. TCP is more soluble in water than chlorpyrifos, causing soil and water contamination (Abraham and Silambarasan, 2016).

When using a bacterial consortium, the degradation is typically easier than using a pure culture (Jariyal et al., 2018). Bacteria live together in nature and interact with each other for survival. Each bacterium in the metabolic pathway of pesticide degradation can produce metabolites that can be used as substrates.

- **Fungal degradation:** Agriculture soils are inhabited by many fungi that can biodegrade pesticides. Microorganisms of this class include yeast and filamentous fungi. Produces many enzymes that are involved in degradation; the fungus can promote degradation (Oliveira et al., 2015). Furthermore, microorganisms modify soil properties, such as ion exchange capacity and permeability. Fungi degrade pollutants more effectively than bacteria because of some characteristics such as growth morphology and resistance even when exposed to high concentrations of contaminants. For enhancing degradation, it is often recommended to use fungi and bacteria since fungi can oxidize pesticides into forms that are easier and more accessible to bacteria (Purnomo et al., 2020).
- **Enzymatic degradation:** Microorganisms and plants produce enzymes that are accountable for biodegradation by enzymes. Pesticide degradation is catalyzed by enzymes that are biological macromolecules capable of catalyzing biochemical reactions. These molecules lower the activation energy thus lowering the rate of reaction (Scott et al., 2008). Oxidation, reduction, hydrolysis, and conjugation are the main metabolic act in which they take part. Pesticides are degraded by oxidation, which occurs when an electron is transferred from reductants to oxidants. Laccase and oxygenase enzymes are involved in this process. Oxygenases catalyze oxidation by incorporating oxygen into aromatic compounds, while laccases reduce oxygen to water by

breaking the ring, resulting in free radicals. A reaction generates heat or energy, which microorganisms use for metabolic processes. Hydrolysis involves cleaving substrate bonds by the addition of −H or−OH groups from water molecules. Thus, pesticide molecules are broken down into smaller compounds than the originals. Lipases, cellulases, and esterases are typical enzymes involved in hydrolysis pathways. The reduction process is initiated by reductive enzymes (nitroreductase).

9.4 Factors affecting rates of biodegradation

Various factors affect the rate of biodegradation; by controlling these factors, enhanced biodegradation can be achieved.

- The rate of biodegradation is governed by pH, temperature, carbon sources, soil texture, moisture, aerobic or anaerobic conditions, and the amount of the pollutant. On the other hand, it is impossible to generalize the best conditions for bioremediation.
- In aerobic environments, slower degradation will be caused by substituents, while in anaerobic environments, they will cause faster degradation. As a result of steric hindrance, chlorine slows molecule degradation. In aerobic conditions, molecules with a higher chlorination rate degrade slower. High concentrations of pollutants degrade more rapidly. The enzymes cease to degrade it if the concentration goes down a threshold level.
- Biodegradation of soil with small particles, such as clay, can take several years due to diminishing its bioavailability. Chlorine creates a steric hindrance that prevents enzymes from accessing compounds with higher chlorination. As a result, compounds with higher chlorination degrade more slowly.
- The kinetics of a compound involves both its rate of transformation and the curves describing that transformation and is influenced by each of the factors above. Generally, first-order kinetics (exponential decay) describe biodegradation when the substrate concentration is low, whereas zero-order kinetics (linear degradation) describe biodegradation when the substrate concentration is high. Some chemicals cannot be transformed by microbes if the concentration falls below a certain critical threshold level.

10. Future considerations

The soil microbiota includes bacteria and fungi. By affecting soil nutrient cycles as well as improving the structure of the soil, microbes play a crucial role in the restoration of degraded soils. Despite encouraging evidence to suggest that microbial inoculums can improve nutrient bioavailability and soil aggregation, many aspects need to be explored further.

Several studies have demonstrated that bacterial or fungal inoculums alone or in combination improve soil fertility and aggregation. Coinoculation is performed better in most cases than their sole/lone counterpart, but there are still certain aspects that need attention, such as testing these approaches in the field. Experimental studies are typically conducted in laboratories or pots under controlled conditions, so the performance or adaptability of these microorganisms under field conditions may differ dramatically. The competition for resources increases under natural field conditions, as both biotic and abiotic factors are uncontrolled. The use of organic fertilizers in conjunction with microbial inoculum could be advantageous for restoring fertility and productivity to degraded land. The scientific community has not paid much attention to the selection of microbial inoculum to optimize their synergistic effect. Researchers need to explore this further.

The study of microbial inocula in crop production should focus on the whole cycle of nutrients rather than the singular fate of each nutrient. In future studies, the focus must be on understanding how microbial inoculum influences nutrient cycling. Using bacterial and fungal inoculums could successfully restore fertility to degraded soils. To maximize the fertility and structure of these soils, they would select bacteria and arbuscular mycorrhiza fungi that can form long-lasting associations with each other.

11. Application of microorganisms in soil restoration

For the restoration of contaminated soil, microorganisms can be utilized in various ways for bioremediation. Here, some applications of microorganisms are mentioned:

- It is a short, natural process, and it is acceptable for the treatment of contaminated materials like soil. The bacteria that degrade the pollutant and increase in number when the contaminant is present.
- When a contaminant is degraded, the biodegradative population declines. In most cases, the residue from the treatment is harmless, including water, carbon dioxide, and biomass.
- It often requires very little effort and can be done on-site, without causing major disruptions to normal activities. Moreover, it eliminates the need to transport waste off-site and the potential environmental and health risks associated with transportation.
- It is a cost-effective method since it loses less than other technologies (methods) used to clean up hazardous waste and also an important method for treating oil contamination.
- Furthermore, it helps in the destruction of pollutants, many of the toxic compounds can be changed into harmless chemicals, and as a result, future liability for treating and disposing of contaminated materials is eliminated.
- There are no hazardous chemicals used in it. Microbial growth is stimulated by supplementing the nutrients, especially fertilizers. These are typically applied to lawns and gardens. As harmful chemicals are converted into harmless gases and water during bioremediation, they are destroyed completely.
- Naturally, they are simple, labor-saving, and cheap.
- This product is eco-friendly and sustainable.
- Contaminants do not simply get transferred to another environment.
- Nonintrusive, potentially allowing the site to be used for years to come.

12. Conclusions

The removal efficiency of bioremediation must be maximized by maintaining nutrient levels, pH levels, water content, oxygen levels, and temperature. Food security will be impacted significantly by land degradation in the upcoming year. Microbial (fungal and bacterial) inoculums and their combination with organic fertilizers proved to be the better approach for the restoration of degraded soil and limit the enormous use of chemical fertilizers. These microbes can enhance soil aggregation and nutrient bioavailability through fixation, chelation, siderophores, and glomalin production. Thus, they could be applied to restore fertility to degraded soils. To ensure success in field applications, it is critical to understand the mechanisms through which microbial inoculums provide or mobilize nutrients in degraded soil. Bacteria and fungi in soil can produce these benefits even though the exact mechanisms are not well understood. It has become clearer that some or all these microbes can perform their associated functions. Utilizing these mechanisms, improved soil fertility and crop production capacity can be achieved with cost-effective and innovative management practices in degraded soils. It may not always be advantageous to use single inoculums or coinoculation between bacteria and bacteria, bacteria and fungi, or fungi and fungi.

Generally, the study found that pollution poses a threat to our health and damages the environment and soil, which affects crop production. As a result of bioremediation, pollution can be reduced and removed, and the air, water, and soil provided to future generations will be clean. Bioremediation is emerging science and technology that offers an alternative method of detoxifying contaminants. By using biological processes, contaminants or impairments are degraded, broken down, transformed, and essentially removed from soils or waters. As bacteria, fungi, and plants carry on with their lives, they alter contaminants by performing their normal functions. By using chemical pollutants as energy sources, these organisms can make them harmless or less toxic.

References

Abraham, J., Silambarasan, S., 2016. Biodegradation of chlorpyrifos and its hydrolysis product 3,5,6-trichloro-2-pyridinol using a novel bacterium Ochrobactrum sp. JAS2: a proposal of its metabolic pathway. Pestic. Biochem. Physiol. 126, 13–21, 2016.

Betancur-Corredor, B., Pino, N.J., Cardona, S., Peñuela, G.A., 2015. Evaluation of biostimulation and Tween 80 addition for the bioremediation of long-term DDT-contaminated soil. J. Environ. Sci. 28, 101–109. https://doi.org/10.1016/j.jes.2014.06.044. ISSN 1001-0742.

Bacmaga, M., Wyszkowska, J., Kucharski, J., 2019. Biostimulation as a process aiding tebuconazole degradation in soil. J. Soils Sediments 19, 3728–3741.

Barker, A.V., Bryson, G.M., 2002. Bioremediation of heavy metals and organic toxicants by composting. Sci. World J. 2, 407–420.

Bini, C., 2011. Soil Restoration: Remediation and Valorization of Contaminated Soils.

Dominati, E., Mackay, A., Green, S., Patterson, M., 2014. A soil change-based methodology for the quantification and valuation of ecosystem services from agroecosystems: a case study of pastoral agriculture in New Zealand. Ecol. Econ. 100, 119–129.

Doolotkeldieva, T., Konurbaeva, M., Bobusheva, S., 2018. Microbial communities in pesticide-contaminated soils in Kyrgyzstan and bioremediation possibilities. Environ. Sci. Pollut. Res. 25, 31848–31862.

Glick, B.R., Cheng, Z., Czarny, J., Duan, J., 2017. Promotion of plant growth by ACC deaminase-producing soil bacteria. Eur. J. Plant Pathol. 119, 329–339.

Hamadache, M., Benkortbi, O., Hanini, S., Amrane, A., Khaouane, L., Si Moussa, C., 2016. A quantitative structure activity relationship for acute oral toxicity of pesticides on rats: validation, domain of application and prediction. J. Hazard. Mater. 303, 28–40.

Heard, M.S., Baas, J., Dorne, J.L., Lahive, E., Robinson, A.G., Rortais, A., Spurgeon, D.J., Svendsen, C., Hesketh, H., 2017. Comparative toxicity of pesticides and environmental contaminants in bees: are honey bees a useful proxy for wild bee species? Sci. Total Environ. 578, 357–365.

Jariyal, M., Jindal, V., Mandal, K., Gupta, V.K., Singh, B., 2018. Bioremediation of organophosphorus pesticide phorate in soil by microbial consortia. Ecotoxicol. Environ. Saf. 159, 310–316.

Khan, N.T., Jamee, N., Khan, M.J., 2018. A brief overview of contaminated soil remediation methods. Biotechnol Ind J 14 (4), 171.

Khajezadeh, M., Abbaszadeh-Goudarzi, K., Pourghadamyari, H., Kafilzadeh, F., 2020. A newly isolated Streptomyces rimosus strain capable of degrading deltamethrin as a pesticide in agricultural soil. J. Basic Microbiol. 60, 435–443.

Kim, K.H., Kabir, E., Jahan, S.A., 2017. Exposure to pesticides and the associated human health effects. Sci. Total Environ. 575, 525–535.

Madison, W.I., Adriano, D.C., Chlopecka, A., Kaplan, D.I., Clijsters, H., Vangronsveld, J., 1995. Soil Contamination and Remediation: Philosophy, Science and Technology. INRA Les olloques, n° 85, Paris, pp. 465–504.

Matthews, G.A., 2018. History of Pesticides. CABI, Boston, MA, USA, ISBN 9781786394873.

Mirsal, I.A., 2018. Soil Pollution Origin, Monitoring & Remediation, second ed. Springer-Verlag: Berlin/Heidelberg, Germany, ISBN 9783540707752.

Mishra, D., Tiwari, S.K., Agarwal, S., Sharma, V.P., Chaturvedi, R.K., 2012. Prenatal carbofuran exposure inhibits hippocampal neurogenesis and causes learning and memory deficits in offspring. Toxicol. Sci. 127, 84–100.

Montesinos, E., 2003. Development, registration and commercialization of microbial pesticides for plant protection. Int. Microbiol. 6 (4), 245–252. https://doi.org/10.1007/s10123-003-0144-x. PMID: 12955583.

Montgomery, D.R., 2007. Soil erosion and agricultural sustainability. Proc. Natl. Acad. Sci. U. S. A. 104 (33), 13268–13272.

Odukkathil, G., Vasudevan, N., 2016. Residues of endosulfan in surface and subsurface agricultural soil and its bioremediation. J. Environ. Manage. 165, 72–80. https://doi.org/10.1016/j.jenvman.2015.09.020.

Oldeman, L., Hakkeling, R., Sombroek, W.G., 1990. World Map of the Status of Human-Induced Soil Degradation: An Explanatory Note. International Soil Reference, and Information Centre.

Oliveira, B.R., Penetra, A., Cardoso, V.V., Benoliel, M.J., Barreto Crespo, M.T., Samson, R.A., Pereira, V.J., 2015. Biodegradation of pesticides using fungi species found in the aquatic environment. Environ. Sci. Pollut. Res. 22, 11781–11791.

Ortiz-Hernandez, M.L., Rodriguez, A., Sanchez-Salinas, E., 2014. Castrejon-godinez, M.L. (Bioremediation of soils contaminated with pesticides: experiences in Mexico. In: Alvarez, A., Polti, M.A. (Eds.), Bioremediation in Latin America: Current Research and Perspectives. Springer, Cham, Switzerland, ISBN 9783319057385.

Parker, K.M., Sander, M., 2017. Environmental fate of insecticidal plant-incorporated protectants from genetically modified crops: knowledge gaps and research opportunities. Environ. Sci. Technol. 51, 12049–12057.

Picariello, E., Baldantoni, D., De Nicola, F., 2023. How soil microbial communities from industrial and natural ecosystems respond to contamination by polycyclic aromatic hydrocarbons. Processes 11 (1), 130. https://doi.org/10.3390/pr11010130.

Purnomo, A.S., Sariwati, A., Kamei, I., 2020. Synergistic interaction of a consortium of the brown-rot fungus Fomitopsis pinicola and the bacterium Ralstonia pickettii for DDT biodegradation. Heliyon 6.

Sarwar, M., 2015. Information on activities regarding biochemical pesticides: an ecological friendly plant protection against insects. Int. J. Eng. Adv. Res. Technol 1, 27–31.

Scott, C., Pandey, G., Hartley, C.J., Jackson, C.J., Cheesman, M.J., Taylor, M.C., Pandey, R., Khurana, J.L., Teese, M., Coppin, C.W., 2008. The enzymatic basis for pesticide bioremediation. Indian J. Microbiol. 48, 65–79.

Senko, O., Maslova, O., Efremenko, E., 2017. Optimization of the use of His_6-OPH-based enzymatic biocatalysts for the destruction of chlorpyrifos in soil. Int. J. Environ. Res. Public Health 14, 1438.

Shrivastava, P., Kumar, R., 2015. Soil salinity: a serious environmental issue and plant growth-promoting bacteria as one of the tools for its alleviation Saudi. J. Biol. Sci. 22 (2), 123–131. https://doi.org/10.1016/j.sjbs.2014.12.001.

Shukla, K.P., Singh, N.K., Sharma, S., 2010. Bioremediation: developments, current practices, and perspectives. Genet. Eng. Biotechnol. J. 1–19.

Soulas, G., Lagacherie, B., 2021. Modelling of microbial degradation of pesticides in soils. Biol. Fertil. Soils 33, 551–557.

Utuk, E.E.D., 2015. Land degradation: a threat to food security: a global assessment. J. Environ. Earth Sci. 5 (8), 13–21.

Vidali, M., 2001. Bioremediation. An overview. Pure Appl. Chem. 73 (7), 1163–1172.

Volkering, F., Breure, A.M., Rulkens, W.H., 1997. Microbiological aspects of surfactant use for biological soil remediation. Biodegradation 8 (6), 401–417.

Further reading

Amundson, R., Berhe, A.A., Hopmans, J.W., Olson, C., Sztein, A.E., Sparks, D.L., 2015. Soil and Human Security in the 21st Century, vol 348, p. 1261071.

Bouwer, E.J., Zehnder, A.J.B., 1993. Bioremediation of organic compounds-putting microbial metabolism to work. Trends Biotechnol. 11, 260–267.

Das, M. and Adholeya, A., Role of Microorganisms in Remediation of Contaminated Soil.

Kopittke, P.M., Neal, W.M., Peng, W., Brigid, A.M., Lombi, E., 2019. Soil and the intensification of agriculture for global food security. Environ. Int. 132, 105078.

Site https://www.environmentalpollutioncenters.org/soil/causes.

Verma, A., 2021. Bioremediation Techniques for Soil Pollution: An Introduction. In: IntechOpen. https://doi.org/10.5772/intechopen.99028. Available from: https://www.intechopen.com/online-first/78227.

Wang, X., Li, L., Yan, X., Meng, X., Chen, Y., 2020. Processes of chromium (VI) migration and transformation in chromate production site: a case study from the middle of China. Chemosphere 257, 127282.

Welsh Government, 2009. River Basin Districts Surface Water and Groundwater Classification (Water Framework Directive) (England and Wales) Direction. Available online: https://gov.wales/sites/default/files/publications/201907/091222direct53.pdf.

CHAPTER 16

Microbes for the remediation of heavy metal–contaminated soil

Ashwini A. Waoo and Charu Vyas

Department of Biotechnology, Faculty of Life Science and Technology, AKS University, Satna, Madhya Pradesh, India

1. Introduction

Research in bioremediation is a vital area of study that offers cleaner alternatives to conventional methods. Microorganisms convert toxic metals into less toxic forms by using various mechanisms, such as biosorption, bioaccumulation, chelating agents, bioleaching, biomineralization, and enzyme-catalyzed transformation. In addition, plants have a variety of strategies for absorbing and accumulating metals and forming metal-bound compounds. By integrating microorganisms and plants in bioremediation, heavy metals in polluted soils may be effectively removed. This chapter summarizes the microbial- and plant-mediated strategies for soil remediation of heavy metal contamination.

Chemicals and heavy metals inevitably contaminate the soil environment, resulting in problems with environmental contamination. Agriculture, urbanization, or industrialization are some of the major sources of heavy metal contamination in the environment. In landfills, the generation of leachate is inevitable since they are the ultimate waste disposal method. The leachate from landfills contains a high concentration of heavy metals and is highly heterogeneous. Leachate that moves into the soil, groundwater, or surface water may cause pollution problems and cause toxicity to humans and other living organisms. Biological organisms can solubilize metals and increase their bioavailability. For this siderophores are produced in their biomass, and subsequently, adsorb to outer membrane proteins and precipitate. The amount of municipal solid waste generated around the world every year is estimated at 1.3 billion tonnes, and the amount is expected to grow to approximately 2.2 billion tonnes by 2025. Landfills are an important facility for the waste management industry because they are the most common way to dispose waste. These landfills pose a serious threat to the local environment because they are typically not sanitary. The improper disposal of waste will pollute the environment and lead to waste-borne diseases. Leachate and landfill gases are also released during landfilling activities. Leachate is the result of rainwater passing through permeable materials and leaching. In the absence of proper collection, leachate will flow or migrate to other bodies of water. Over time, leachate is produced as the degradable fractions of the waste decompose, dispersing into the underlying soil if a site is not contained. It contains more than 200 different elements or heavy metals (arsenic, cadmium, cobalt, copper, mercury, manganese, nickel, lead, tin, and thallium), which have the potential to damage the environment and human health.

Integrative Strategies for Bioremediation of Environmental Contaminants, Volume Two. https://doi.org/10.1016/B978-0-443-14013-6.00014-7

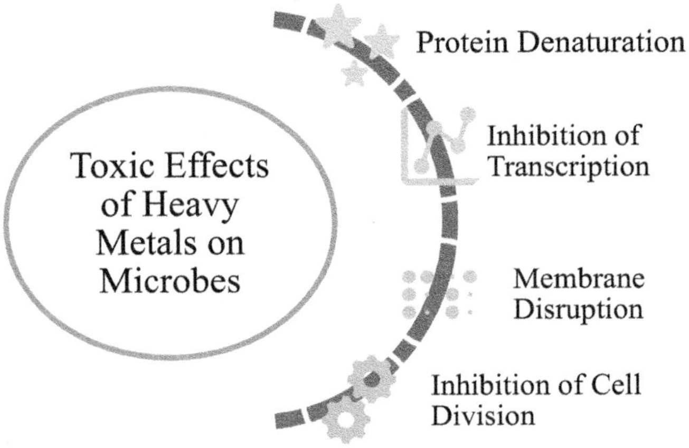

FIGURE 16.1

Heavy metal toxicity mechanism to microbes (Ahemad and Malik, 2012; Khan et al., 2008).

There is increasing concern about heavy metal toxicity due to its ecological, evolutionary, nutritional, and environmental hazards as shown in Fig. 16.1. Heavy metals disrupt human metabolism in two ways: they accumulate and thereby disrupt vital organ function, such as the heart, kidneys, brain, bone, and liver, and vital nutritional minerals are displaced from their original locations by them, hindering their biological function. The International Agency for Research on Cancer and US EPA classified these heavy metals as human carcinogens. For contaminated soil to be cleaned, remediation is necessary due to the toxicity effect.

The use of bioremediation techniques is an effective and efficient way to remove pollutants. Through chemical, physical, and biological means, microorganisms will enzymatically attack contaminants and transform them into less toxic products. As microbial growth and activity are dependent on environmental conditions, its application often involves manipulating environmental parameters to permit growth and degradation at a faster rate. Several factors affect the degradation of pollutants, such as the presence of a microbial population that can degrade the pollutants and the environmental factors (Fauziah et al., 2017).

Heavy metals and metalloids may accumulate in soil from emissions from rapidly expanding industrial areas, the disposal of metal wastes, lead-contaminated gasoline, and paints, mine tailings, land application of fertilizers, sewage sludge, wastewater irrigation, pesticides, and coal combustion residues (Khan et al., 2008; Zhang et al., 2010). Heavy metals represent a poorly defined group of inorganic chemical hazards, the most common of which are chromium (Cr), lead (Pb), arsenic (As), cadmium (Cd), zinc (Zn), copper (Cu), nickel (Ni), and mercury (Hg) (GWRTAC, 1997). A variety of anthropogenic activities release heavy metals into the environment. Unlike organic contaminants, which are degraded by microbes to carbon (IV) oxide (Kirpichtchikova et al., 2006), most metals are not degraded by microbes or chemicals (Adriano, 2003; Maslin and Maier, 2000). Changes in their chemical forms and bioavailability are possible, however. Organic contaminants can be severely inhibited by toxic metals in soil.

By ingesting or contacting soil contaminated with heavy metals, or through contact with soil, plants, or animals, soil-metal contamination can harm people and the environment, and drinking contaminated groundwater (McLaughlin et al., 2000a,b; McLaughlin et al., 2000a,b; Ling et al., 2007). Heavy metal—contaminated soil ecosystems need to be characterized and remedied to be adequately protected and restored. Environmental protection and public health laws, both nationwide and internationally, are primarily based on the chemical characteristics of natural phenomena, those found in the food chain (Kabata-Pendias and Pendias, 2001). Heavy metal speciation and bioavailability in the soil can be determined by soil characterization, but striving to restore heavy metal—contaminated soils would require understanding their source, their chemistry, and their possible health effects. Risk assessments are effective scientific tools that enable decision-makers to manage sites that have been contaminated cost-effectively while also protecting public and ecosystem health.

For the remediation of heavy metal—contaminated sites, immobilization, soil washing, and phytoremediation strategies are often cited as some of the best demonstrated available technologies (BDATs) (GWRTAC, 1997). Even though these technologies are cost-effective and environmentally friendly, only developed nations have reported field applications. These technologies have not yet become commercially available in the majority of developing nations, likely because they are not well aware of their inherent benefits and operating standards. Governments and the public have become more aware of the health implications of contaminated soils, which has led to increased interest in the adaptations in technologies to remediate contaminated sites (Bolan et al., 2008). To restore contaminated lands in growing nations with high population density and limited funding, low-cost and sustainable remedial measures are necessary to lessen the risks, make the resource available for agriculture, increase food security, and reduce problems related to land tenure. In this chapter, scattered literature is examined to provide perception into the heavy metal speciation, bioavailability, and thereby choice of a suitable remedial alternative for main heavy metals (Cr, Pb, As, Cd, Zn, Ni, Cu, etc.). As alternatives for soil cleanup, there is also a discussion of immobilization, and phytoremediation, including their principles, advantages, and disadvantages.

There is a great deal of natural and inorganic waste created annually in the materials cycle, and the way it is disposed of is going to be a global hassle (Zhao and Kaluarahchi, 2002). The harmful effects of heavy metals on ecosystems and organisms relying on them make it urgent to remove excessive pollutants caused by those poisonous factors as a result of those toxic elements (Maslin and Maier, 2000). Multiple strategies are employed for the treatment of media contaminated with heavy metals, and these fall into two broad categories: abiotic (chemical and physical) methods and biotic (biological) methods (Bolan et al., 2008; DeVolder et al., 2003). Most conventional methods for removing heavy metals from the environment are very expensive to implement at a large scale, in addition to being risky to track and control. To counter the negative effects of conventional physicochemical methods of removing heavy metals from the environment, eco-friendly methods using different organisms are being used (DeVolder et al., 2003; Sposito and Page , 1984). Several studies have been conducted for the bioremediation of heavy metals. Yet, there is little research on using synergistic heavy metal—tolerant microorganisms for heavy metal bioremediation.

2. Sources of heavy metal pollution in the environment

Industrialization is one of the leading causes of heavy metal pollution (Dwivedi et al., 2012). Heavy metals are metalloids or metallic elements that have relatively large atomic weights and are toxic at low concentrations (Butu and Iguisi, 2013). The presence of heavy metals is of great concern because they are nondegradable and therefore persistent. The term "soil heavy metal pollution" refers to the occurrence of heavy metals in the soil (Abd-Alla et al., 2012) that results in concentrations that exceed background levels (Jin et al., 2018). Heavy metal pollution is caused by two types of sources namely natural sources and anthropogenic sources. Natural sources may be pesticides insecticides, paints, pigments, etc., whereas anthropogenic sources include weathering of rocks, forest fires, volcanic activities, etc., as shown in Fig. 16.2.

2.1 Natural sources

Even though heavy metals may occur naturally (Selvi et al., 2019; Kumar and Bharadvaja, 2020) assert that different anthropogenic activities have resulted in a high number of heavy metals in soil. Heavy metal levels in the environment have increased since the industrial revolution, as stated by Rehan and Alsohim (Rehan and Alsohim, 2019). Surface water, groundwater, sediments, and soils contaminated with heavy metals are global issues (Mishra, 2017; Xie et al., 2016). Since heavy metals have a nonbiodegradable nature, they are more likely to accumulate in the environment through food chains (Dhankhar and Hooda, 2011). Metal pollution can indirectly affect biogeochemical cycles and ocean productivity in general (Bong et al., 2010). A soil's metal content adversely affects the metabolism, respiration, and activities of soil microbes, and the transformation of organic carbon into biocarbon. Additionally, heavy metals in the environment can cause physiological disorders and malnutrition in plants, and they can be transferred to higher trophic levels where they may cause irreversible harm to health (Jin et al., 2018). In living organisms, heavy metals can accumulate and cause a variety of metabolic and physiological disorders (Ahemad and Malik, 2012). Even at very low concentrations, heavy metals are toxic; as a result of their recalcitrance and tenacity, heavy metals pose a severe threat to the superiority of the environment and the life of organisms (Njoku et al., 2020a,b; Gupta et al., 2016) by carcinogenic and mutagenic effect. The functions and activities of vital cellular components

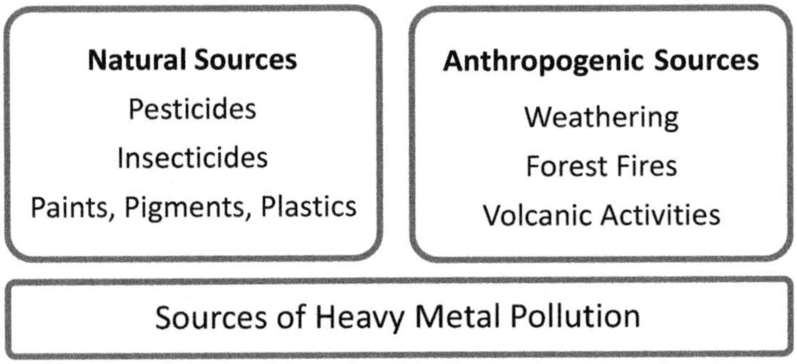

FIGURE 16.2

Heavy metal sources in the agricultural soil (Wuana and Okieimen, 2011).

can be disrupted when heavy metals are present (Rehan and Alsohim, 2019). In addition to their bioaccumulation and nonbiodegradability, heavy metal pollution is a significant environmental issue (Igiri et al., 2018) and poses health risks to both humans and ecosystems (Rehan and Alsohim, 2019). Heavy metals are generally recalcitrant (Njoku et al., 2020), which makes them pose a potential risk to environmental superiority and the life of organisms (Gupta et al., 2016).

During pedogenetic weathering of rocks, heavy metals occur in the soil environment at concentrations that are considered trace (< 1000 mg kg^{-1}) and rarely toxic (Kabata-Pendias and Pendias, 2001; Pierzynski et al., 2000). By disturbing and accelerating nature's slow-moving geochemical cycle of metals, soils of the rural and urban site may accumulate heavy metals above levels high enough to pose health risks to individuals, plants, animals, ecosystems, or other media (Amore et al., 2005).

Heavy metals are a direct contaminant in soil environments because:

(i) They are generated in manmade cycles at a higher rate than natural cycles.
(ii) They are moved from mines to random locations where direct exposure to them is greater.
(iii) Metal concentrations in waste products are higher than those in receiving environments.
(iv) The type of metal present in the receiving environment system (species) may make it more bioavailable.

2.2 Anthropogenic sources

Anthropogenic emissions of several heavy metals into the atmosphere are projected to be higher than natural fluxes (Sposito and Page, 1984). Anthropogenic heavy metals tend to be more mobile, more bioavailable, than pedogenic or lithogenic metals. In contaminated sites, metal-bearing solids may come from a wide variety of sources, including metal mine tailings, improperly protected landfills, leaded gasoline, lead-containing paints, fertilizer, compost, pesticides, coal combustion residues, petrochemicals, and atmospheric deposition (Khan et al., 2008; Zhang et al., 2010) are discussed here.

- **Fertilizers:** Farmers were the first to influence the soil significantly with fertilizers. Plants need both macronutrients and essential micronutrients to grow and complete their life cycle. It is possible to supplement crops with heavy metals that are missing from some soils. Cereal crops cultivated on Cu-deficient soils are occasionally supplemented with Cu, as are root crops and cereal crops with Mn. Fertilizers are often added in large amounts to soils in farming systems to deliver N, P, and K for plants. Heavy metals, such as cadmium and lead, may be found in the compounds used to supply such elements as impurities, causing an increase in their content after persistent fertilizer application. Metals, such as cadmium and lead, have no physiological activity. When certain phosphatic fertilizers are applied to the soil, they add Cd and other potentially poisonous factors such as F, Hg, and Pb to the soil.
- **Pesticides:** Several pesticides were used in agriculture in the past contained metals. About 10% of the insecticides and fungicides approved in the United Kingdom contain compounds containing Cu, Hg, Pb, Mn, or Zn. Pesticides such as Bordeaux mixture (copper sulfate) and copper oxychloride contain copper. Fruit orchards have used lead arsenate to control parasitic insects for many years. In New Zealand and Australia, arsenic-containing compounds are extensively used for controlling livestock ticks, and copper, chromium, and arsenic compounds (CCA) have been used for timber preservation. Contaminated sites can cause problems, especially if they are

redeveloped for agricultural or nonagricultural uses. Contrary to fertilizers, the use of such substances is more localized, restricted to sites or crops.

- **Biosolids and manures:** Numerous biosolids such as farm animal manures, composts, and municipal sewage sludge to land result in the buildup of heavy metals such as Cr, Cu, As, Cd, Pb, Ni, Se, Mo, Zn, Hg, and Sb in the soil (Yeung et al., 1997). Farm animals such as poultry, cattle, and pigs produce manure that is commonly used as solids or slurries on pastures and vegetation. Even though most manure is seen as valuable fertilizer, in the pig and poultry industries, metal contamination may also result from Cu and Zn introduced to feed as growth promoters, as well as arsenic found in poultry health products. As, Cu, and Zn are present in excessive concentrations in the manures of animals fed these diets, which, if repeatedly sprayed on confined areas of land, may accumulate in the soil over time. Wastewater treatment processes produce organic solids referred to as biosolids (sewage sludge) that can be recycled. Many countries that allow the reuse of biosolids generated by urban populations have a common practice of applying biosolids to the ground. Due to its wide recognition and regulatory definition, sewage sludge appears in many references. As a substitute for sewage sludge, biosolids are becoming more common because of their better representation of sewage sludge's beneficial characteristics.

Heavy metal contamination of soils by biosolids has caused great concern regarding their use in agricultural practices. Metal concentrations in biosolids are significantly affected by the type of process used in treating them, as well as their nature and intensity of industrial activity. The most common metals observed in biosolids are Pb, Ni, Cr, Cd, Cu, and Zn. It is possible for metals added to the soil during the application of biosolids to leach downward through the soil profile, potentially contaminating groundwater. Recently, studies on soils treated with biosolids in New Zealand showed increased Cd, Ni, and Zn concentrations in drainage leachates.

- **Wastewater:** Wastewater is currently used to irrigate several hectares of arable land worldwide. Some studies suggest that most vegetables consumed in urban areas in Asia and Africa are grown using completely wastewater irrigation. In general, farmers care less about the environment and are interested in maximizing their profits and yields. Though water effluents have exceptionally low metal concentrations, soil irrigation with them can eventually accumulate heavy metals.
- **Metal mining and industrial wastes:** Directly discharged tailings (heavier and larger debris particles settled in the lowest part of the flotation cell) from mining are directly released into natural slopes, including onsite wetlands, and result in elevated concentrations. Intensive mining and smelting of Pb and Zn ore have contaminated soil, posing hazards to human and ecological health. The reclamation of these sites tends to be prolonged and luxurious, and soil productivity may not be restored. People are at risk from soilborne heavy metals because of their bioavailability. Ingesting contaminated soil is a direct result of ingesting plant material grown in such soils (food chain) or ingestion of plants grown in such soils (oral bioavailability).

Textile, tanning, petrochemical, pesticide, and pharmaceutical industries create other types of substances that are highly variable in composition; such substances may come from accidental oil spills or originate from petroleum-based products. Furthermore, many contain heavy metals and toxic organic compounds and are rarely, if ever, transported off-shore. Many of these products are either very low in plant vitamins or do not have any soil-conditioning properties.

- **Airborne sources:** There are several sources of airborne metals, including stack or duct emissions, waste piles, and fugitive emissions. Particulates in gas streams generally release airborne metals. During high-temperature processing, some metals, such as As, Pb, and Cd may also volatilize. If the reducing atmosphere is kept, these metals will convert into oxides and condense into fine particles. Natural air currents may disperse stack emissions over a wide area until they are removed from the gas stream by precipitation. Because fugitive emissions are produced near the ground, they are dispersed over a much smaller area. In general, fugitive emissions have lower contaminant concentrations than stack emissions. Depending on the site-specific conditions, both types of sources may emit different types of metals.

Most fossil fuels include heavy metals, so this is a form of contamination that has persisted steadily since the industrial revolution. Solid particles from fires and other emissions from chimneys eventually end up on land or in the sea. There have been reports of very high concentrations of Cd, Pb, and Zn in plants and soil adjacent to smelting works, for instance. Pb also contaminates soils in urban areas and those adjacent to major roads through the emissions of petrol with tetraethyl lead; this contributes significantly to the level of lead in soils. There are also sources of Zn and Cd on the roadside, including tyres and lubricant oils (Wuana and Okieimen, 2011).

3. Potential risks of heavy metals

Anthropogenic pressure linked to population growth, coupled with soil pollution by heavy metals, poses a threat to the environment and food security. Several anthropogenic activities are responsible for environmental pollution and human exposure to heavy metals. These include mining, industrial production, and agricultural use of metal-containing compounds. Heavy metals and metalloids are present in the soil at five million sites with levels above the regulatory limit.

As a result of heavy metal contamination, the ecosystem and humans are at risk of many different health issues. Additionally, heavy metal contamination threatens food chain security, food quality, and the ability to use the land for agricultural production, which in turn exacerbates land tenure problems. There are several elements with metallic properties that are classified as heavy metals, including transition metals, metalloids, lanthanides, and actinides. For some enzymes, they are essential co-factors, while for others, they are not. There are two groups of trace elements: cobalt, copper, iron, manganese, molybdenum, selenium, and zinc. These elements are tightly regulated by interactions with binding proteins as they pose potential risks to cell function. The second group is composed of arsenic, cadmium, lead, mercury, tungsten, and vanadium, which are nonessential metals that possess toxic properties and enter tissues through their physicochemical properties, including ionic charge.

Particular attention should be paid to the latter since they have no essential functions in living beings, are highly hazardous even at low levels of exposure, and are considered a primary threat to life. The EPA identifies arsenic, cadmium, lead, and mercury as some of the most toxic metals found in the environment. There are over 20 heavy metals listed by the United States ATSDR (Agency for Toxic Substances and Disease Registry) that are toxic, but four are particularly harmful to humans. As a result, arsenic is ranked first on ATSDR's "top 20 list" of poisonous substances. Lead occupies the second spot, whereas cadmium is in seventh place.

Among the heavy metals that are most frequently found at contaminated sites, they include Pb, Cr, As, Cd, Cu, As, Zn, and Hg. Biomagnification in the food and Hg. Biomagnification in the food chain can cause those metals to decrease crop production. Understanding heavy metal speciation, bioavailability, and remedial options requires an understanding of the chemistry, environment, and health effects of these metals. A heavy metal's fate and transport in soil are heavily influenced by its chemical form and speciation. The soil is a reservoir for heavy metals that are first subject to fast reaction (minutes, hours) and then slow reaction. They are then redistributed in different chemical forms with varying bioavailability and toxicity. In soil, heavy metals are thought to be distributed due to a variety of processes including mineral precipitation and dissolution, ion exchange, adsorption, and desorption, aqueous complexation, biological immobilization, and plant uptake.

3.1 Lead

Lead (Pb) is a toxic nonessential heavy metal that has a long-term impact on living organisms regarding their growth, physiology, and biochemistry due to its high persistence in water and soil and accumulation in the top eight inches of soil. Despite its natural occurrence, anthropogenic activities like mining, fossil fuel combustion, and manufacturing increase its concentration in soils (Tchounwou et al., 2012). Pb(II) is a common and reactive form of lead, forming mono- and polynuclear oxides and hydroxides. Pb compounds insoluble in water are lead carbonates (form lead phosphates when pH exceeds 6) and lead hydroxides.

There are two routes of exposure, inhalation, and ingestion, and both cause similar effects. When Pb accumulates in the organs, for example, the brain, it may result in poisoning (plumbism) or even death. As a result of lead exposure, the kidneys, gastrointestinal tract, and central nervous system can also be affected. Under-six-year-olds are most at risk for mental deterioration, impaired development, less IQ, shortened attention, hyperactivity, and mental retardation from lead exposure. When exposed to lead, adults usually experience lowered reaction time, loss of memory, insomnia, nausea, and weakened joints. The brain, nervous system, RBC, and kidneys can suffer serious harm from lead exposure. Nowadays, lead poisoning that is severe enough to cause apparent illness is very rare.

Ingestion of contaminated soil or dust is the most serious source of soil lead exposure. Generally, plants do not absorb lead. However, in soils containing high levels of lead, some can be taken up. Several studies suggested that lead does not easily accumulate in the fruits of vegetables and crops (e.g., squash, tomatoes, corn, beans, strawberries, and apples). Leafy vegetables (e.g., lettuce) and root crops (e.g., carrots) contain higher concentrations. Since plants take up very little soil lead, lead levels in soil that are safe for plants will be higher than levels in soil that are considered unsafe.

3.2 Chromium

The most common form of Cr found at contaminated sites is chrome (VI). Depending on redox conditions and pH, chrome can also exist in the +III oxidation state. Chromite (VI) dominates in shallow aquifers that are aerobic. Under anaerobic conditions encountered in deeper groundwater, chromium (VI) can be reduced to chromium (III) by soil organic matter, S^{2-} and Fe^{2+} ions. Chromate and dichromate adsorb on soil surfaces, particularly iron and aluminum oxides. Cr (III) is the dominant form of chromium at low pH (< 4). Chrome (VI) is the more toxic and mobile form of chromium. The mobility of chrome (III) is decreased by the adsorption of clays and oxide minerals less than pH 5 and by the formation of Cr $(OH)_3$ above pH 5. In its precipitated or soluble form, chromium can be transported to surface waters by surface runoff. Soluble and unadsorbed chromium complexes can leach into groundwater. As soil pH increases, Cr (VI) leaches more readily. In natural waters, however, most of the Cr is particle-associated and accumulated into the sediment. Chromium is associated with allergic dermatitis problems in humans.

3.3 Arsenic

Natural and abundant, arsenic (As) is a metalloid element that has metallic and nonmetallic properties and can be found in large amounts in the Earth's crust. Both inorganic and organic forms of As are found in soil, with the latter being highly toxic. Surface soils with good drainage (oxygenated) typically accumulate as oxyanions of As^{5+} (arsenate), whereas waterlogged soils contain As^{3+} (arsenite). Compared to inorganic As^{5+} species, inorganic As^{3+} species are generally more mobile and toxic, influencing mobility, toxicity, and environmental behavior. In contrast to As^{3+}, As^{5+} forms complexes on mineral surfaces, especially on hydroxides and oxides of iron (III) and on calcite. Arsenic from those surfaces is released into groundwater under iron-reducing conditions and remains mobile after reduction to As^{3+}. In addition, arsenite inactivates more than 200 enzymes, while arsenate replaces phosphate, which plays a role in many biochemical pathways. As a result of exposure to arsenic, skin damage, cancer risks, and circulatory problems may occur.

3.4 Zinc

Zinc concentrations in drinking water can reach unhealthy levels due to industrial sources of toxic waste. Zinc is a trace element that's vital to human health. Zinc shortages can lead to birth defects. Global zinc production is still increasing, which means more Zn accumulated in the environment. The wastewater from industrial plants contains large amounts of Zn, which pollutes water. The result is that rivers continually deposit unpolluted sludge along their banks. Fish living in Zn-contaminated waters may accumulate Zn in their bodies. Zn can also increase the acidity of water. Zinc can biomagnify up the food chain when it enters the fish. Water-soluble zinc present in soils can pollute groundwater. Usually, plants have a high Zn uptake that their system cannot handle, as Zn accumulates in the soil. Moreover, Zn can interrupt the activity of soil microbes and earthworms and retard the breakdown of organic matter.

3.5 Cadmium

Despite its high toxicity and solubility in water, cadmium (Cd) is a nonessential element and a significant pollutant. Inorganic and organic ligands tend to form stable dissolved complexes with them, which inhibit their solubility and precipitation. The secondary anthropogenic sources of cadmium in soils are waste materials from mining, industry, and agricultural applications. Cadmium inhibits enzyme activity and DNA-mediated transformation in microorganisms. Cd occurs as the divalent Cd (II) ion in its compounds. As soils and surface waters become acidic as a result of acid rain, the geochemical mobility of Cd increases, and its concentration in lake water increases. Lead and zinc refining produces cadmium as a byproduct. Fertilizers and pesticides are inputs to agriculture that increase the concentration of Cd in soils, and whether plants can absorb Cd depends on its bioavailability.

After being absorbed by an organism, cadmium retains its properties for many years. Cadmium exposure to humans through food chains has been studied since the 1970s. Due to concerns about this latter method (crops), research has been conducted on the potential impact of applying sewage sludge to soils used for human consumption crops or using cadmium-enriched phosphate fertilizer.

The results of this research led to the setting of maximum permissible concentration levels for numerous food crops. Several enzymes in the body are affected by cadmium. Proteinuria is thought to occur as a result of Cd badly affecting enzymes in the kidney tubules that reabsorb proteins. Cadmium poisoning was most spectacularly reported when people in the Jintsu River Valley, Japan, were exposed to cadmium through their diet. The victims were affected by the itai-itai disease, which means ouch in Japanese. The symptoms resulted from painful osteomalacia (bone disease) coupled with kidney failure. Cadmium poisoning was detected in irrigated rice contaminated by a mine producing Pb, Cd, and Zn upstream. As a result of chronic accumulation in the kidneys, humans are at risk of kidney dysfunction. Tobacco smoking and food consumption are the main routes by which Cd enters the body.

3.6 Copper

The element copper is essential, but high doses can damage the liver, kidneys, stomach, and intestinal tracts. It is normal to find copper in drinking water from copper pipes and in chemicals formulated to control algal growth. Cu interacts complexly with the environment, but research shows that the majority of Cu released into the environment is stable, or quickly becomes stable, and does not pose a threat to the environment. Cu is not magnified or accumulated in the body as it is in some man-made materials. The complexation of Cu in soil implies there will only be a small fraction of copper in solutions of ionic copper, Cu (II). The solubility of copper increases drastically at pH 5.5, which is close to the ideal pH for farmland soils of 6.0–6.5. Cu and Zn are both essential elements for plants, microbes, animals, and humans. The relationship between soil and water pollution and metal absorption by plants is influenced by chemical, physical soil factors, and the physiological characteristics of the crops themselves. There are both direct and indirect risks from soil contamination with trace metals: direct, due to the negative effects of the metals on crops, and indirect, since the metals can enter the food chain and cause health problems. Adding just a few percent to crop yields could cause significant long-term losses in production and income. Food importers specify maximum levels of metals in food, which could limit the ability of farmers to export contaminated crops.

3.7 Nickel

There are very low levels of nickel in the environment, but it can be hazardous when it is used in excessive amounts. Animals that live near refineries are more likely to develop various types of cancer because of this. Steel and other metal products are the most common applications for Ni. Plating industries, fossil fuel combustion, nickel mining, and electroplating are the major sources of nickel in soil. These gases are released into the atmosphere by power plants and incinerators due to precipitation reactions. Removing nickel from the air usually takes a long time. When Ni compounds are released into the environment, they will tend to adsorb on sediments in soil and become immobile as a result. Acidic soils make Ni more mobile, however, and often penetrates the adjacent groundwater. The presence of Ni can also affect microorganisms' growth, but the organisms usually develop resistance to it after some time. Plants and animals do not accumulate nickel, so it is not known whether nickel will biomagnify through the food chain. Nickel is an important food for animals.

4. Bioremediation of heavy metal

There is an untreated flow of pollutants into the soil from the environment. Developing a low-adverse-impact solution and implementing a cost-efficient one should be the goal of research. It is ideal to rely on the metabolism of fungi and bacteria to break down contamination. Therefore, agricultural sites requiring decontamination must be assessed. A suitable bioremediation technique should be selected. Assessing soil quality and microbial diversity is part of monitoring system efficacy and potential. Microbes are nature's original recyclers. They can also form mixed populations or consortiums. Genetically modified microorganisms are very promising solutions for in situ agriculture soils' xenobiotics bioremediation. The genes are specifically designed for each promoter sequence to degrade specific contaminants, allowing us to create synthetic combinations not found in nature.

4.1 Mechanisms of heavy metal uptake by microorganisms

Microorganisms are crucial to the biogeochemical cycling of metals and the remediation of heavy metal–contaminated environments. Microorganisms can live in their presence despite the toxic poisonous impact they exert, because of different mechanisms that reduce their toxicity. The following five mechanisms are important in bacterial resistance to heavy metals:

1. Extracellular barriers: cell walls, plasma membranes, or capsules can prevent metal ions from entering the cell.
2. Active metal ions transport (efflux): includes three kinds of proteins: ATP-type ATPases, CDF (cation diffusion facilitator), and ROND (resistance, nodulation, and cell division).
3. Extracellular sequestration: occurs when metal ions accumulate in the periplasm or the outer membrane, or as insoluble compounds in cellular components.
4. Intracellular sequestration: occurs when metals are accumulated in nonbioavailable form within the cytoplasm to prevent their exposure to essential components of the cell. There are two examples of this form of metal resistance: production of metallothionein by Synechococcus sp. and Pseudomonas sp. contains cysteine-rich proteins.
5. Reoxidation of metal ions: Cu and As are oxidized to produce important detoxification mechanisms as well, for example, Cu (I) to Cu (II) by CueO and As (V) to As(III) to then eliminate A(III) from cells.

It is possible to categorize microorganisms into biochemical and molecular mechanisms (Ma et al., 2016; Khan et al., 2008). Biochemical mechanisms include those by which microorganisms interact with extracellular soluble metal ions and are responsible for enhancing microbe tolerance through metal detoxification, mobilization, immobilization, transformation, transport, and distribution (Ma et al., 2016). Molecular mechanisms represent genes of heavy metal resistance that may be located either on the chromosome or in extrachromosomal genetic elements.

4.2 Role of microorganisms (bacteria, algae, and fungi)

Heavy metal remediation in soils is vital to the preservation of the environment and the protection of dwelling organisms. Physical and chemical approaches include engineering repair, which can be quick but expensive and ineffective since they cause secondary pollution and adverse effects on soil properties. In contrast, biological methods are considered an effective approach for heavy metal remediation. These strategies include bioremediation, a field of work in which microorganisms eliminate soil pollutants, and phytoremediation, an approach that employs vegetation to remove heavy metals. Microbial bioremediation reduces the cost of the cleanup process associated with heavy metal contamination because it is green, efficient, economic, and environmentally friendly (González and Ghneim-Herrera, 2021).

Microbial remediation of heavy metals is characterized by the following mechanisms: biosorption surface precipitation, formation of organic ligand complexes, and biomineralization and biooxidation.

Many microorganisms have been recognized for their ability to bioremediate soil pollutants and to increase the production of agricultural vegetation with low input. Heavy metals are eliminated from the environment by microorganisms. By converting metals from one oxidation state to another, microbial cells can reduce their toxicity. Metals and soil particles containing heavy metals can be dissolved by microbial metabolic activities. Heavy metals can be degraded, detoxified, or transformed by microbes through precipitation, biosorption, and enzymatic transformation into more stable, inert forms. There have been reports of several microorganisms that are effective green bioremediators of heavy metals, either through bioaccumulation, biotransformation, or biosorption.

Different microorganisms have different capacities to degrade heavy metals using the degradative capability relying specifically on degradative plasmids and spores. The elimination of heavy metals by microorganisms includes the accumulation of the metal ions in the cell, the efflux of metal ions outside the cell, and their reduction to a nontoxic form. A variety of factors, such as temperature, pH, organic acids, and humic acids, are likely to affect the transformation, transportation, and valence state of heavy metals, thereby affecting microbial bioremediation of heavy metal—contaminated environments.

Even though numerous studies have been conducted on the bioremediation of heavy metals, little has been reported on using a synergistic combination of heavy metal—tolerant fungi and bacteria for the bioremediation of heavy metals. Therefore, heavy metal—tolerant fungal and bacterial consortiums are a promising tool for removing heavy metal pollution from polluted environments (Njoku et al., 2020a,b). Environmental, chemical, and biological factors affect microbial bioremediation as shown in Fig. 16.3.

4.3 Microbial bioremediation by bioaugmented composting

Recent research shows that bioaugmented manure can be used as bioinoculants, biocontrol agents, and bioremediation. In addition to agriculture, bioaugmented fertilizers are becoming an important asset for the bioremediation of soil contaminated with heavy metals and organic pollutants. The presence of humic substances and iron oxide in composts, as well as organics, limits the bioavailability of metals when applied to heavy metal—contaminated sites. As well, stabilized organic matter produces complexes with metals, restricting their mobility and their ability to be assimilated by plants. Additionally, composting microbes such as bacteria and fungi can accumulate metal ions in their cells by

Environmental Constraints	Chemical Constraints	Biological Constraints
• Availability of nutrients • Temperature • Concentration of oxygen • Moisture content • pH • Site characterization and selection • Metal ions • Toxic compounds • Redox potential • Mass transfer characteristics • Site hydrogeologic characteristics	• Bioavailability of contaminants • Biodegradability of contaminants • Contaminant properties • Nutrients • Oxygen, air, hydrogen peroxide • Alternative electron acceptors • Metal ions • Toxic compounds • Biogeochemical parameters	• Mutation • Horizontal Gene Transfer • Enzyme Activity • Interaction (Competition, Succession, and Predation)

FIGURE 16.3

Factors affecting bioremediation.

intracellular uptake and can chelate metal ions by the carboxyl, hydroxyl, or other active functional groups on cell (including the dead cell) walls.

Cocomposting soil as part of bioremediation is a cost-effective technique for cleaning natural organic contaminations such as xenobiotics, petroleum products, polycyclic aromatic hydrocarbons (PAHs), etc. Biological processes degrade contaminants in soils, surface waters, and air. Degradation of contaminants may involve a variety of mechanisms, including mineralization by microbial activity, transformation into nontoxic products, volatilization, and formation of humus and inert products, including carbon dioxide, water, and salts. Composting parameters such as aeration, C: N ratio, and duration are dependent on the type of contaminants and waste materials used. In bioremediation, the productivity of manure is determined by using procedure parameters such as pH, temperature, C: N ratio, and hardheadedness of the material that will be treated as compost. Composter bulking agents such as peat moss, pine shavings, and bran flakes facilitate the growth of aerobic microorganisms. Oil-contaminated soil can be cleaned economically through the application of compost or cocompost. It is often not possible for one organism to fully degrade PAHs, but rather, they are degraded by a consortium of organisms, each of which performs a different degradative step.

After 35 days of moderately thermophilic composting, Finstein et al. (1987) concluded that the degradation of sludge of oil refinery was comparable to that achieved after 1280 days on a simulated land farm. Beaudin et al. (1999) investigated the decomposition of grease and mineral oils in hydrocarbon-contaminated soil after 287 days of composting with alfalfa and maple leaves. Yeung et al. (1997) reported that forced aeration and heating of soils were effective in biodegrading fuel oil hydrocarbons in soil. After 96 h of exposure, composts enriched with propane or methane can remove 99% of pollutants like trichloroethylene (Sukesan and Watwood, 1998).

These complex substances must be degraded by diverse populations or consortiums of microorganisms. The effect of composting as a bioremediation technique for reducing the hydrocarbon content of oil refinery sludge was examined by Marín et al. (2006). The experiment showed 60% degradation in three months when composting was done with a bulking agent. According to Sayara et al. (2011), 89% of the total PAHs were degraded by the end of the composting period (30 days) as opposed to only 29.5% by the soil microbes with no cosubstratum.

The current xenobiotics used in harvesting and crop production are mostly natural aggravates that are exposed to organic decay as opposed to physical degradation. Sometimes, organic matter in soil that is simply decomposable provides an insufficient substrate for microbes to break down pesticides. Xenobiotics in the soil can be decomposed more quickly as a result of the enormous biological activity in composting (Büyüksönmez et al., 2000). Strom (1998) observed that the pesticide levels in compost were low, in part because the feedstock was not contaminated at the initial stages and in part. After all, pesticides were degraded during composting. In food wastes composting, diazinon [OO-diethyl-O-(2-isopropyl-6-methyl-5-pyrimidinyl) phosphorothionate] and parathion (OO-diethyl-O-4- nitrophenyl phosophorothionate) were quickly destroyed completely; Farrow et al. (1969) showed that dieldrin (polychlorinated dimethanonaphthalene) could be largely reduced, but DDT (dichlorodiphenyltri-chloroethane) did not biodegrade. Büyüksönmez et al. (2000) stated that most pesticides detected in feedstocks and composts were organochlorine insecticides. It has also been reported that compost addition can improve the degradation of two herbicides (benthiocarb [S-4-chlorobenzyl diethyl thi-ocarbamate] and MCPA [4-chloro-2-methylphenoxyacetic acid]). Cocomposting increased atrazine mineralization (Rao et al., 1996). Composts decreased the concentration of pesticides in soils to a level that was nonhazardous to crops. It should be noted, however, that the intermediates produced during the composting of pesticides are not often nontoxic; therefore, caution should be maintained when using composts after bioremediation.

4.4 Microbial consortium approach for bioremediation

Natural recyclers are microorganisms. A mixed population or consortium of microorganisms is metabolically advanced for metal biosorption and increasingly suited for field application. Research on microbial consortia is of great importance for understanding the biodegradation of complex pollutants, as it normally involves cooperation between more than one species. A microbial community with expansive enzymatic capabilities—a microbial community with a variety of pollutant substrates—can enable a higher level of degradation. Furthermore, some pollutants can only decompose through cometabolism. Using cocultivation, it is possible to enhance the degrading ability of the bacteria, indicating that they normally exist together and are dependent upon one another for the use of ecological substances, attributing to the synergistic effects of their interactions. Several members of a microbial network may play critical roles and may need each other to survive when the source of energy is limited and confined to complex carbons. Any microbial consortium's degradative capability limit is not necessarily the aftereffect of merely adding up and simply including the individual strains' capacities. Through the oxidation and hydrolysis pathways of cometabolism, each bacterium can produce metabolites that can be used by the enzyme of the next bacterium. Toxic metabolites of the species preceding it are evacuated by one species (Vyas and Waoo, 2019).

4.5 Phytoremediation and plant microbiome

However, phytoremediation depends entirely on the climate, water, and soil conditions, proving to be a sluggish and seasonal process. The process of phytoremediation consists of numerous procedures based upon the plant–soil–environment interactions. For heavy metal–contaminated soil, four processes are known: phytoextraction, phytostabilization, phytovolatilization, and hemofiltration. For the remediation of heavy metals from soils, microbial strategies represent a good option. Despite the presence of traces of bacteria that can remediate heavy metal—contaminated soils, three key components still need to be addressed:

a) Microbial resources need to be developed for bioremediation and bacteria-assisted phytoremediation.
b) Most research is conducted with strains resistant to few metals, and polluted soils exhibit heavy metal contamination.

Lead, arsenic, and cadmium, which are nonessential heavy metals or metalloids, pose health and environmental risks. Microbiological bioremediation has emerged as a potential method of reducing heavy metal concentrations in the environment due to bacteria's ability to sequester and transform heavy metals. It has been reported that bacterial strains can reclaim contaminated soils, but to date, none have been reported to be the most efficient bioremediation or bacterial-assisted phytoremediation techniques that can maximize plant growth and productivity. The amount of metallic pollutants in soil has increased dramatically in recent years due to geological and anthropogenic causes.

Analysis of the most evaluated genera *Agrobacterium, Klebsiella, Bacillus, Enterobacter, Pseudomonas, Microbacterium, Rhodococcus,* and *Mesorhizobium* showed significant differences in tolerance levels and revealed the involvement of different biochemical and molecular mechanisms involved in plants' growth promotion and resistance gene expression. Molecular and biochemical mechanisms support the ability of *Klebsiella* and *Enterobacter* to tolerate cadmium and lead. These strains are also able to mitigate the plant growth inhibition caused by phytotoxic metal concentrations. In conclusion, *Klebsiella* and *Enterobacter* are the best candidates for biodegradation or bacteria-assisted phytoremediation of soils containing arsenic, cadmium, and lead (Ma et al., 2016).

Microorganisms associated with plants such as plant growth-promoting rhizobacteria (PGPR) play a vital role in the remediation of polluted soil and the enhancement of plant growth. Biochemical nitrogen fixation, phosphate solubilization, phytohormone production, siderophore production, production of 1-aminocyclopropane-1-carboxylic acid (ACC) deaminase, among others. Heavy metal—contaminated soils can be greatly improved using PGPR, which represents an important metal-resistant bacteria and phytoremediation-enhancing agent (Ma et al., 2016).

4.6 Metagenomic profiling

Multiplicity in the microbial network arises from the interaction of anthropogenic and geogenic forces. The study of diverse microbes in an environment is incomprehensible by traditional cultural methods. A variety of advanced methods of metagenomic profiling have led to unprecedented advances and improvements in exploring, investigating, and profiling uncultivable organisms of the environment. Moreover, metagenomics provides direct access to the pool of genomes without the bias of culturing, allowing a deeper understanding of the range of degradation pathways of natural microorganisms that remain largely ineffective (if not unknown). The goal could be to develop more efficient and modified strains and consortiums for bioremediation applications, among other things. In recent years, metagenomics has been applied to better understand microbial ecology and diversity. Their use in ecological monitoring and bioremediation is also increasingly common. The detection of microbial communities can provide insight into the rehabilitation process by detecting natural or environmental changes in a short period.

The metagenomic process provides insights into environmental variations, enabling the detection of the diversity of microorganisms in rehabilitating and restoring habitats. An assessment of restoration achievement in an ecosystem might be improved by comparing the composition of microbial networks with rehabilitating communities at preserved reference sites. A subsequent assessment of the biological system's recovery may be based on this. It has been generally used to comprehend microbial networks in ecological frameworks using metagenomics, which is the study of a microbial community's entire genome. There are two types of metagenomic approaches-targeted metagenomics and shotgun metagenomics. As part of targeted metagenomics, the diversity of a gene is probed to identify its full complement of sequences in the environment, used both to determine its phylogenetic diversity and relative abundance in a sample. To examine the diversity of small subunit rRNA sequences (16S/18S rRNA) in a sample, this methodology is consistently used. Alternatively, it can be used to explore the impact of environmental contaminants on the structure of microbial communities.

Genomic sequencing is used in shotgun metagenomics to probe the total genome of an environmental community. A sequence library is prepared by extracting environmental DNA and then fragmenting it. The genomic content of those libraries is then determined by sequencing the libraries. The depth of sequencing is often the limiting factor in shotgun metagenomics. Shotgun metagenomics sometimes only focused on the majority group of microbes in a community and does not completely cover the genome. The phylogenetic marker gene sequencing method can also be used in metagenomics to assess taxonomic and functional profiling as shown in Fig. 16.4.

The utility of high-throughput 16S rRNA sequencing and metagenomics increases as sequencing costs continue to decline. In this way, it is possible to study in detail the effects of various bioremediation interventions on the native microbial community. These techniques can be optimized to enrich targeted groups of microbes, which are important for bioremediation. Furthermore, these methods allow the unbiased assessment of the microbial community, capturing culture-resistant microbes. Utilizing metagenomics for remediation could allow rapid, minimally invasive removal of contaminants.

Metagenomic approaches can also provide a mechanistic understanding of bioremediation processes, which will minimize efforts to optimize bioremediation efficiency. In bioremediation strategies, 16S rRNA sequencing and metagenomics have great potential and provide deep insights into microbial response to remediation techniques. Bioremediation will become more comprehensive with a combination of these approaches and pure culture analysis of environmentally relevant microbes. Applications of metagenomics include bioremediation and environmental monitoring (Vyas and Waoo, 2020).

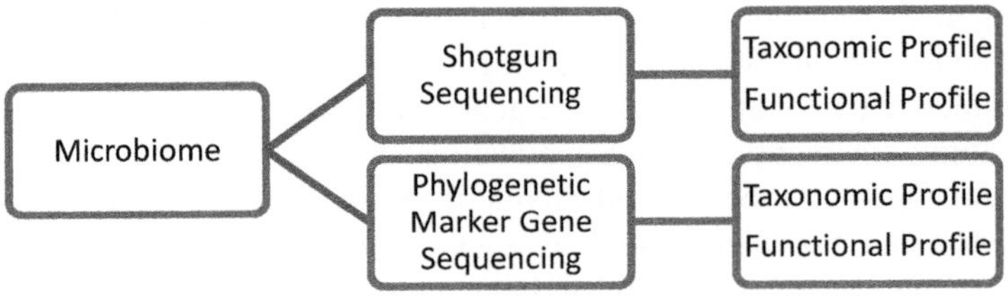

FIGURE 16.4

Steps in metagenomics.

5. Conclusion

Bioaugmented compost created by cocomposting is a sustainable and economical bioremediation method for agricultural waste soils, heavy metals, and organic pollutants, thereby increasing soil fertility and soil health, reducing environmental risks, and benefiting the environment. For example, bioremediation alone was able to transform defiled soil into gardening and agricultural soil at the trial site, and research is underway to make this procedure significantly more productive and effective for reusing and recycling soil with little effort. The bioremediation of heavy metal pollution using microbial consortiums is an eco-friendly approach, utilizing microbial strains and their properties to aid in heavy metal removal. From the perspective of efficient process scaling-up in the context of larger-scale applications, their genetic and biochemical history is used to explore different perspectives on bioremediation and new ideas.

The field of metagenomics is rapidly evolving and has profound effects on the microbial world. Soil metagenomics initiatives, by allowing direct investigation of bacteria, viruses, and fungi irrespective of their cultivability and taxonomy, might be an amazing resource to the scientific group and provide a much better understanding of microbial diversity and functions in the soil. Research on novel effective microorganisms and plants (vegetation) suitable for bioremediation to shorter durations and new technologies for mass production needs to be intensified. Though pollution is an inherent problem associated with the environment, these efforts at bioremediation can prove to be a small but beneficial step and lead to a bigger improvement in the socioeconomic situation of the area. This would be beneficial to both the farmers because their land would be more productive, thereby doubling their income, and to the residential population since a healthy environment would be created.

Microbes can reduce the growth of plant pathogens, increase nitrogen fixation, and cause a decrease in the bioavailability of inorganic and organic pollutants in soils. Research in the future should focus on increasing the efficiency of bioremediation technology to reduce environmental stress on terrestrial and aquatic ecosystems. Future studies on microbial remediation should address these challenges as a trial-and-error research niche.

References

Abd-Alla, M.H., Morsy, F.M., El-Enany, A.W.E., Ohyama, T., 2012. Isolation and characterization of a heavy metal-resistant isolate of *Rhizobium leguminosarum* bv. viciae potentially applicable for biosorption of Cd^{+} and Co^{2+}. Int. J. Biodeterior. Biodegrad. 67, 48–55.

Adriano, D.C., 2003. Trace Elements in Terrestrial Environments: Biogeochemistry, Bioavailability and Risks of Metals, 2nd edition. Springer, New York, NY, USA.

Amore, J.J.D., Al-Abed, S.R., Scheckel, K.G., Ryan, J.A., 2005. Methods for speciation of metals in soils: a review. J. Environ. Qual. 34 (5), 1707–1745.

Ahemad, M., Malik, A., 2012. Bioaccumulation of heavy metals by zinc resistant bacteria isolated from agricultural soils irrigated with waste water. J. Bacteriol. 2 (1), 12–21.

Beaudin, N., Caron, R.F., Legros, R., Ramsay, J., Ramsay, B., 1999. Identification of the key factors affecting composting of a weathered hydrocarbon-contaminated soil. Biodegradation 10, 127–133.

Bolan, N.S., Ko, B.G., Anderson, C.W.N., Vogeler, I., 2008. Solute interactions in soils in relation to bioavailability and remediation of the environment. In: Proceedings of the 5th International Symposium of Interactions of Soil Minerals with Organic Components and Microorganisms (Pucón, Chile).

Bong, C.W., Malfatti, F., Azam, F., Obayashi, Y., Suzuki, S., 2010. The effect of zinc exposure on the bacteria abundance and proteolytic activity in seawater. In: Hamamura, N., Suzuki, S., Mendo, S., Barroso, C.M., Iwata, H., Tanabe, S. (Eds.), Interdisciplinary Studies on Environmental Chemistry - Biological Responses to Contaminants. Terrapub, pp. 57–63.

Butu, A.W., Iguisi, E.O., 2013. Concentration of heavy metals in sediment of river Kubanni, Zaria, Nigeria. J. Environ. Earth Sci. 2 (1), 10–17.

Büyüksönmez, F., Rynk, R., Hess, T.F., Bechinski, E., 2000. Occurrence, degradation, and fate of pesticides during composting. Part II: occurrence and fate of pesticides in compost and composting systems. Compost. Sci. Util. 8, 61–81.

DeVolder, P.S., Brown, S.L., Hesterberg, D., Pandy, K., 2003. Metal bioavailability and speciation in a wetland tailings repository amended with biosolids compost, wood ash, and sulfate. J. Environ. Qual. 32 (30), 851–864.

Dhankhar, R., Hooda, A., 2011. Fungal biosorption –an alternative to meet the challenges of heavy metal pollution in aqueous solutions. Environ. Technol. 32 (5), 467–491.

Dwivedi, S., Mishra, A., Saini, D., 2012. Removal of heavy metals in liquid media through fungi isolated from waste water. Int. J. Sci. Res. 1 (3), 181–185.

Farrow, R.P., Elkins, F.R., Rose, W.W., Lamb, F.C., Ralls, J.W., Mercer, J.W., 1969. Canning operations that reduce insecticide levels in prepared foods and in solid food waste. Residue Rev. 29, 73–87.

Fauziah, S.H., Jayanthi, B., Emenike, C.U., Agamuthu, 2017. Remediation of heavy metal contaminated soil using potential. Microb. Iso. Closed Dispos. Site 7 (4).

Finstein, M., Miller, F., Hogan, J.A., Strom, P.F., 1987. Analysis of EPA guidance on composting sludge: Part I - biological heat generation and temperature. Biocycle 28, 20–26.

Gupta, A., Joia, J., Sood, A., Sood, R., Sidhu, C., Kaur, G., 2016. Microbes as potential tool for remediation of heavy metals: a review. J. Microb. Biochem. Technol. 8 (4), 364–372.

González, H.S., Ghneim-Herrera, T., 2021. Heavy metals in soils and the remediation potential of bacteria associated with the plant microbiome. Front. Environ. Sci. 9.

GWRTAC, 1997. Remediation of Metals-Contaminated Soils and Groundwater. Tech. Rep. TE-97-01. GWRTAC, Pittsburgh, Pa, USA. GWRTAC-E Series).

Igiri, B.E., Okoduwa, S.I.R., Idoko, G.O., Akabuogu, E.P., Adeyi, A.O., Ejiogu, I.K., 2018. Toxicity and bioremediation of heavy metals contaminated ecosystem from tannery wastewater: a review. J. Toxicol. 2018, 2568038.

Jin, Y., Luan, Y., Ning, Y., Wang, L., 2018. Effects and mechanisms of microbial remediation of heavy metals in soil: a critical review. Appl. Sci. 8, 1336–1353.

Kabata-Pendias, A., Pendias, H., 2001. Trace Metals in Soils and Plants, 2nd edition. CRC Press, Boca Raton, Fla, USA.

Khan, S., Cao, Q., Zheng, Y.M., Huang, Y.Z., Zhu, Y.G., 2008. Health risks of heavy metals in contaminated soils and food crops irrigated with wastewater in Beijing, China. Environ. Pollut. 152 (3), 686–692.

Kirpichtchikova, T.A., Manceau, A., Spadini, L., Panfili, F., Marcus, M.A., Jacquet, T., 2006. Speciation and solubility of heavy metals in contaminated soil using X-ray microfluorescence, EXAFS spectroscopy, chemical extraction, and thermodynamic modeling. Geochem. Cosmochim. Acta 70 (9), 2163–2190.

Kumar, L., Bharadvaja, N., 2020. Microbial remediation of heavy metals. In: Shah, M. (Ed.), Microbial Bioremediation and Biodegradation. Springer, Singapore, pp. 49–72.

Ling, W., Shen, Q., Gao, Y., Gu, X., Yang, Z., 2007. Use of bentonite to control the release of copper from contaminated soils. Aust. J. Soil Res. 45 (8), 618–623.

Ma, Y., Oliveira, R.S., Freitas, H., Zhang, C., 2016. Biochemical and molecular mechanisms of plant-microbe-metal interactions: relevance for phytoremediation. Front. Plant Sci. 7, 918.

Marín, J., Moreno, J., Hernández, T., García, C., 2006. Bioremediation by composting of heavy oil refinery sludge in semiarid conditions. Biodegradation 17, 251–261.

Maslin, P., Maier, R.M., 2000. Rhamnolipid-enhanced mineralization of phenanthrene in organic-metal co-contaminated soils. Ann. Finance 4 (4), 295–308.

McLaughlin, M.J., Zarcinas, B.A., Stevens, D.P., Cook, N., 2000a. Soil testing for heavy metals. Commun. Soil Sci. Plant Anal. 31 (11–14), 1661–1700.

McLaughlin, M.J., Hamon, R.E., McLaren, R.G., Speir, T.W., Rogers, S.L., 2000b. Review: a bioavailability-based rationale for controlling metal and metalloid contamination of agricultural land in Australia and New Zealand. Aust. J. Soil Res. 38 (6), 1037–1086.

Mishra, G.K., 2017. Microbes in heavy metal remediation: a review on current trends and patents. Recent Pat. Biotechnol. 11 (3), 188–196.

Njoku, K.L., Asunmo, M.O., Ude, E.O., Adesuyi, A.A., Oyelami, A.O., 2020a. The molecular study of microbial and functional diversity of resistant microbes in heavy metal contaminated soil. Environ. Technol. Innovat. 17.

Njoku, K.L., Akinyede, O.R., Obidi, O.F., 2020b. Microbial remediation of heavy metals contaminated media by *Bacillus megaterium* and *Rhizopus stolonifer.* Sci. African 1–8.

Pierzynski, G.M., Sims, J.T., Vance, G.F., 2000. Soils and Environmental Quality, 2nd edition. CRC Press, London, UK.

Rao, N., Grethlein, H., Reddy, C.A., 1996. Effect of temperature on composting of atrazine amended lignocellulosic substrates. Compost. Sci. Util. 4.

Rehan, M., Alsohim, A.S., 2019. Bioremediation of heavy metals. In: Saldarriaga-Noreña, H., Murillo-Tovar, M.A., Farooq, R., Dongre, R., Riaz, S. (Eds.), Environmental Chemistry and Recent Pollution Control Approaches. IntechOpen.

Sayara, T., Borràs, E., Caminal, G., Sarra, M., Sanchez, A., 2011. Bioremediation of PAH's contaminated soil through composting: influence of bioaugmentation and biostimulation contaminant biodegradation. Int. Biodeterior. Biodegrad. 65, 859–865.

Selvi, A., Rajasekar, A., Theerthagiri, J., Anathaselvam, A., Sathishkumar, K., Madhavan, J., Rahman, P.K.S.M., 2019. Integrated remediation processes toward heavy metal removal/recovery from various environments —a review. Front. Environ. Sci. 7.

Sposito, G., Page, A.L., 1984. Cycling of metal ions in the soil environment. In: Sigel, H. (Ed.), Metal Ions in Biological Systems, vol 18. Marcel Dekker, Inc., New York, NY, USA, pp. 287–332 of Circulation of Metals in the Environment.

Strom, P.F., 1998. Evaluating pesticide residues in yard trimmings compost. Biocycle 39, 80.

Sukesan, S., Watwood, M.E., 1998. Effects of hydrocarbon enrichment on trichloroethylene biodegradation and microbial populations in finished compost. J. Appl. Microbiol. 85, 635–642.

Vyas, C., Waoo, A., Sept. 2019. A "bioremediation: an innovative approach for reclamation of agricultural land to sustain farmers' income". In: Proceedings of BSADFI2022" at National Conference on "Biotechnology and Sustainable Agriculture for Doubling of Farmers' Income by 2022. AKS University, Satna., pp. 5–7 (Empyreal Publication House).

Vyas, C., Waoo, A.A., 2020. Prognostication of bioremediation requisite around industrially contaminated environment— A review. Curr. Biotechnol. 9 (1), 3–14.

Tchounwou, P.B., Yedjou, C.G., Patlolla, A.K., Sutton, D.J., 2012. Heavy metal toxicity and the environment. Exp. Suppl. 101, 133–164.

Wuana, R.A., Okieimen, F.E., 2011. Heavy Metals in Contaminated Soils: A Review of Sources, Chemistry, Risks and Best Available Strategies for Remediation, 2011. International Scholarly Research Network. *Article ID 402647.*

Xie, Y., Fan, J., Zhu, W., Amombo, E., Lou, Y., Chen, L., Fu, J., 2016. Effect of heavy metals pollution on soil microbial diversity and bermudagrass genetic variation. Front. Plant Sci. 31.

Yeung, P.Y., Johnson, R.L., Xu, J.G., 1997. Biodegradation of petroleum hydrocarbons in soil as affected by healing and forced aeration. J. Environ. Qual. 26, 1511–1516.

Zhang, M.K., Liu, Z.Y., Wang, H., 2010. Use of single extraction methods to predict bioavailability of heavy metals in polluted soils to rice. Commun. Soil Sci. Plant Anal. 41 (7), 820–831.

Zhao, Q., Kaluarachchi, J.J., 2002. Risk assessment at hazardous waste-contaminated sites with variability of population characteristics. Environ. Int. 28 (1–2), 41–53.

Further reading

DPR-EGASPIN, 2002. Environmental Guidelines and Standards for the Petroleum Industry in Nigeria (EGASPIN). Department of Petroleum Resources, Lagos, Nigeria.

Duah-Yentumi, S., Kuwatsuka, S., 1980. Effect of organic matter and chemical fertilizers on the degradation of benthiocarb and mcpa herbicides in the soil. Soil Sci. Plant Nutr. 26, 541–549.

NJDEP, 1996. *Soil cleanup criteria*, New Jersey department of environmental protection, proposed cleanup standards for contaminated sites. NJAC 7, 26D.

Riley, R.G., Zachara, J.M., Wobber, F.J., 1992. Chemical Contaminants on DOE Lands and Selection of Contaminated Mixtures for Subsurface Science Research. US-DOE, Energy Resource Subsurface Science Program, Washington, DC, USA.

Tessier, A., Campbell, P.G.C., Blsson, M., 1979. Sequential extraction procedure for the speciation of particulate trace metals. Anal. Chem. 51 (7), 844–851.

Ure, A.M., Quevauviller, P.H., Muntau, H., Griepink, B., 1993. Speciation of heavy metals in soils and sediments. An account of the improvement and harmonization of extraction techniques undertaken under the auspices of the BCR of Commission of the European Communities. Int. J. Environ. Anal. Chem. 51 (1), 35–151.

CHAPTER

Potential of microbes for the bioremediation of heavy metal—contaminated soil

17

Anshika Jain, Kalpana Katiyar, Virendra Kumar, Arya Sahu and Vaishali Mishra
Department of Biotechnology, Dr. Ambedkar Institute of Technology for Handicapped, Kanpur, Uttar Pradesh, India

1. Introduction

The indiscriminate application of heavy metals in various industrial processes has brought changes in the natural composition of the hydrosphere, lithosphere, as well as atmosphere. The significant changes in heavy metals are changes in oxidation state and concentration of heavy metals. Sometimes these alteration causes transformation to a toxic state and accumulation of one or more heavy metal above the permissible level in the hydrosphere, lithosphere, and biosphere. Heavy metal becomes carcinogenic, toxic, and mutagenic to living organisms even at low concentrations. The prolonged exposure to heavy metals leads to various disease development in living beings. The World Health Organization (WHO) also emphasized and claimed that several million humans across the globe are suffering from metal-induced severe diseases. Different pollution and natural imbalances increase rapidly due to population growth, modernization, and industrialization. The significant soil, water, land, and air contamination are mainly due to increased anthropogenic activities resulting from industrialization and modernization. This environmental contamination needs attention.

The industrial revolution recolonized human society in many ways. However, when we observe this revolution in the context of the environment, it was realized that lack of sustainability in the industrial revolution resulted in the accumulation of inorganic, organic, and substituted chemicals in the habitats of the living organism. The accidental technological disaster around the globe deteriorated the environment from time to time, as shown in chronological order in Table 17.1. These tragedies caused considerable environmental, economic, and health damages and lengthy legal disputes and stimulated new global regulations.

The soil is considered a nonrenewable natural resource. The formation of soil (pedology) is a slow and gradual process that takes thousands of years. The soil is a cornerstone for living society and the ecosystem. The soil is a habitat for 90% of living organisms. The soil is a major medium for food production. The intentionally and unintentionally anthropogenic activities are responsible for large-scale soil erosion, destruction, and contamination (Hou et al., 2020). A survey report published in 2008 claimed that 2.8 million earthen sites are contaminated soil in the European Union, and 19% of agricultural soil of China possessed toxic pollutants much above the permissible range (Naylor et al., 2020). The soil and permafrost have approx. 4.1 trillion deposits of terrestrial organic carbon. The soil contamination disturbs the carbon cycle and creates a major concern for climate change.

Integrative Strategies for Bioremediation of Environmental Contaminants, Volume Two. https://doi.org/10.1016/B978-0-443-14013-6.00003-2

Table 17.1 The summary of major technological disaster.

Years	Affected region	Polluting agent	Disease and symptoms	Protective measures
1912–65	Japan	Cadmium in Jinzu river Methyl mercury into the local water supply	Itai-itai—painful skin andand bone condition Minamata—paralysis and death	• Government sets new limits on industrial pollutants • Minamata convention on mercury, an international treaty
1958	Niger delta oil pollution in Nigeria	9–13 million barrels of oil spilled in the Niger delta over 50 years since production started in 1958	Extensive damage to fisheries along with the deterioration of water andand soil quality	Settles a lawsuit against the oil producers
1964–90	Ecuador's Amazon degradation	Approx. 400 million barrels of toxic oil waste were released into watersheds on the Amazon rain forest over 30 years	Widespread health problems in the indigenous community	Still determining damages suffered by these technological disasters
1971–96	Peru's Amazon degradation	Estimated a billion barrels of toxic oil wastewater released into Amazonian watersheds	Unexplained diseases, skin ailments, tumors, and miscarriages in Peru's indigenous Achuar community	Settlements outside court, which has not been public
1972–89	Papua New Guinea's Panguna Mine war	Large-scale environmental destruction from rio Tinto's Panguna open-pit copper mine Mining water containing arsenic, sulfur, mercury, cadmium, copper, and zinc was dumped into the local river system of bougainville Island	Approx. 40-mile jaba River was biologically dead	After decade long revolt, the mine closed
1976	Italy's Seveso Dioxin cloud	Release of dioxin chemical in the environment	It affected 2000 people and 80,000 animals' slaughters to check the poison entering the food chain	"Seveso directive (1982)," which regulates the manufacture andand storage of hazardous materials

Table 17.1 The summary of major technological disaster.—cont'd

Years	Affected region	Polluting agent	Disease and symptoms	Protective measures
1978	France's Amoco cadiz tanker spill	Approx. two million barrels of oil spilled off the coast of France	Estimated 200 miles of coastline polluted, harming wildlife	Implementation of revised "International convention for the Preservation of Pollution from Ships (MARPOL)"
1984	Bhopal cyanide gas leak, India	The leak of methyl isocyanide gas from Union carbide chemical Plant	Killed approx. four thousand people directly and sickened approx. half-million people leave survivors with numerous health ailments, including congenital disabilities, blindness, and chronic respiratory troubles	• Environment Protection Act (1986) • Public Liability Insurance Act (1991) implemented by the Indian government
1986	Chernobyl nuclear disaster	A nuclear reactor explosion in the chernobyl nuclear disaster in Soviet Ukraine results in clouds of nuclear waste	It killed 31 people directly though estimates of its long-term death toll vary widely	• The convention on early notification of a nuclear accident was signed by 110 countries • Denmark, Sweden, Italy, and Germany implement nuclear bans
1989	Exxon Valdez oil spill off the Alaskan coast	Approx. 10.8 million gallons of crude oil spilled into Prince William Sound through the Exxon Valdez oil tanker	Pollutes 1300 miles of coastline Killed 250,000 sea birds, 250 bald eagles, 3000 sea otters, and destroyed millions of salmon eggs of Alaska's fishing industry.	• Oil pollution Act (1990) by US congress • Amendment in MARPOL 1992 convention signed over by 155 countries • All tankers must be double-hulled

Continued

Table 17.1 The summary of major technological disaster.—cont'd

Years	Affected region	Polluting agent	Disease and symptoms	Protective measures
2000	Romania's cyanide spill, Romania	Estimated 34 million gallons of cyanide spilled from The baia Mare Gold mine into The Somes, danube, Lupes, and Tisza rivers	• Results in the death of aquatic and plant life of rivers • The impeding access to the drinking water of Hungary, Bulgaria, Serbia, and Romania for several months	• Fine of roughly \$200 million to responsible industry Aurul in 2001 • Bans mining on 85% of the site in 2005
2006	Ivory Coast's Toxic waste dumping	Approx. 400 tons of toxic waste containing petroleum residues and caustic soda dumped into Ivorian city's waste system	The deaths of 17 people and the illness of one hundred thousand people are indirectly linked to waste dumping	Fine of \$195 million for compensation for victims and site cleanup in 2007
2010	Deepwater oil spill in the gulf of Mexico	Spills nearly five million barrels of oil due to explosion on a BP oil ring in the gulf of Mexico The largest oil spill in the US history	• Affected aquatic ecosystem • Killed 11 people • Estimated \$17.2 billion damages to beaches, fisheries, wildlife, and tourism	• Government administration imposes new regulations on deep-water oil drilling • Six-month ban on deep-water oil drilling • \$20 billion fine for site cleanup on responsible authorities
2019–20	Amazon wildfires	Record wildfires in Brazil's Amazon rainforest	Highlighted growing concerns over deforestation and development in Amazon More than 3 million hectares area burned	Brazilian conglomerate JBS S.A., the largest meat processor in the world, pledges to improve its cattle-purchasing practices after investigations linked some of its suppliers to illegal deforestation

The deterioration of worldwide agricultural soils and the consequence of dropping crop yields pose a threat to malnutrition, poverty, and food security for poor people. Moreover, most soil pollutants can reach the food chain that causing severe health issues in animals and humans. The United Nations (UN) has pledged eight sustainable soil development goals to be achieved by 2030. The 68th United Nations General Assembly Session declared 5th December as "World Soil Day in 2013" and 2015 as "International Year of Soils." Following these measures, the United Nations Environment Assembly (UNEA) assembled bodies of the WHO and FAO (Food and Agricultural Organization) to assess soil pollution worldwide and consequent measures and risk by 2020 (Jansson and Hofmockel, 2020).

The extraction of minerals and industrial activities leads to heavy metalloids release. The heavy metalloids are used for various agricultural, industrial, and domestic applications. These metals are so widely adsorbed in the industry that their production continuously increases day by day despite being harmful to the ecosystem. According to a study, the worldwide production of lead and chromium has surged by 232% to 11.3 Mt per year and 514% to 37.5 Mt per year, respectively, in the last half-century. Even new technological advancement relies on heavy metalloids like nickel, lead, cadmium, geranium, solar and electric car batteries, and semiconductor devices. The heavy metalloids possess inherited properties of precipitation, adsorption, and absorption. These characteristics of heavy metalloids made them the most extensively disseminated pollutants of agricultural soils (Borrelli et al., 2017).

The bioremediation process deploys MOs, plants, and enzymes to reduce, degrade, or detoxify harmful matter. The munitions wastes, organic and inorganic compounds, pesticides, halogenated aromatic hydrocarbons, and polyaromatic hydrocarbons are contaminants that are effectively bioremediated by MOs. The UNEP categorized polychlorinated biphenyls (PCBs), PAH, PCPs (pentachlorophenols), BTEX (benzene, toluene, ethylbenzene, and xylene), and DDT (1,1,1-trichloro-2,2-bis(4-chlorophenyl)) ethane as carcinogenic that should be prioritized pollutants for bioremediation.

The terms "bioremediation" and "biodegradation" sound like synonyms, but they have a different context. The bioremediation term is made of "bios" + "remediate" that means "living organism" + "solution of an issue." In simpler words, "bioremediation is the deployment of living creatures' abilities to modify toxic matters in nontoxic and simpler forms." The technologies used to degrade or decompose environment pollutant with the assistance of living organism are part of bioremediation (Gouma et al., 2014). Biodegradation is a technique utilized for the bioremediation process of biodegradable matter. The MOs have been an agent for the transformation and treatment of waste products since the beginning of the 20th century. But bioremediation has emerged as a new "eco-friendly" technology for decontamination of the environment. One of the typical examples of the bioremediation process in municipal wastewater treatment plants is where the metabolic potential of MOs is used for wastewater treatment.

Recent research findings on bioremediation have claimed that bioremediation is the most suitable, sustainable, and relevant process for decontamination of the environment. It is an eco-friendly method to remove organic and metal pollutants from contaminated habitats. Soil pollution is the primary international challenge of the present century in which heavy metal contamination is of significant concern. Heavy metals remain in the soil for a very long period which is hazardous and can cause harmful impacts on human health, soil quality, plant growth, and development. The contaminated soil leads to contaminated food, fodder, fuel, and medicinal products. The heavy metal contamination in soil affects the natural microflora of soil, disturbing vital biochemical processes in soil. The contaminated soil has high microbial metabolic entropy and low carbon fixation potential.

Heavy metalloids act as poison for enzymes. The fertility, crop yield like factors, is also affected by heavy metal contamination.

The primary source of heavy metal contamination in developing countries is the continuous release of untreated industrial waste products and residues directly into the open environment. The other sources include sewage sludge, mine tailings, and surface runoff from polluted water used for irrigation. The mining, smelting, electronic waste processing, and fossil fuel-burning activities result in the release of heavy metalloids in the dust and aerosols. In general, anthropogenic activities are the primary agent behind the elevated levels of lead, copper, mercury, and zinc in the environment. The lead deposition in the soil is majorly by atmospheric deposition.

These primary bioremediation methods have been classified into physical, chemical, and biological, as illustrated in Fig. 17.1.

For many years, the physicochemical techniques used to remediate pollutants in a contaminated environment are inefficient, expensive to operate, and often produce additional residues with pollution potential. Therefore, in recent years, bioremediation methods have received more and more attention due to their low operating costs, best removal efficiency, good performance at low concentrations, and, more importantly, environmental friendliness.

Bioremediation is defined as a process in which, under controlled conditions, organic waste is biologically degraded into an innocuous state below the concentration limits established by regulatory authorities. To improve bioremediation effectiveness, the microorganisms must enzymatically attack the pollutant and convert them into harmless products. Bioremediation can be efficient only in environmental conditions, which will promote microbial growth and activity. The application of bioremediation involves operating environmental parameters to enhance the microbial growth and degradation rate. Bioremediation techniques are typically less expensive than traditional methods such as incineration. Bioremediation is based on natural attenuation people consider it more acceptable than other technologies due to its eco-friendly nature. Most bioremediation systems run under aerobic conditions, but running under anaerobic conditions may permit microbial organisms to degrade otherwise recalcitrant molecules (Science and Sharma, 2012).

The most hazardous pollution now increasing is heavy metalloids these days. The degradation of heavy metalloids is significant as it is highly toxic at high concentration, and it forms complex compound within the cells. Metals appear to affect organic biodegradation by impacting the physiology and ecology of organic degrading MOs. Recent efforts to improving organic biodegradation in the presence of metals include using metal-resistant bacteria, treatment additives, and clay minerals to reduce metal bioavailability. Additional options under consideration include the addition of divalent cations and pH modification.

Microorganisms have evolved the ability to protect themselves from heavy metal poisoning through various mechanisms (such as adsorption, absorption, methylation, oxidation, and reduction). Microorganisms uptake heavy metals actively (bioaccumulation) or passively (adsorption). Microbial methylation plays a vital role in heavy metals bioremediation because methylated compounds are frequently volatile. For example, Hg (II) can be biomethylated by several bacterial species, *Alcaligenes faecalis, Bacillus sp.,* and *Pseudomonas aeruginosa,* to gaseous methyl mercury. The traditional sediment remediation techniques include dredging, capping, and monitored natural attenuation. These new approaches emphasize bioremediation like biostimulation, bioaugmentation, and phytoremediation applied to sediment. The techniques promise to be of lower impact and more cost-effective

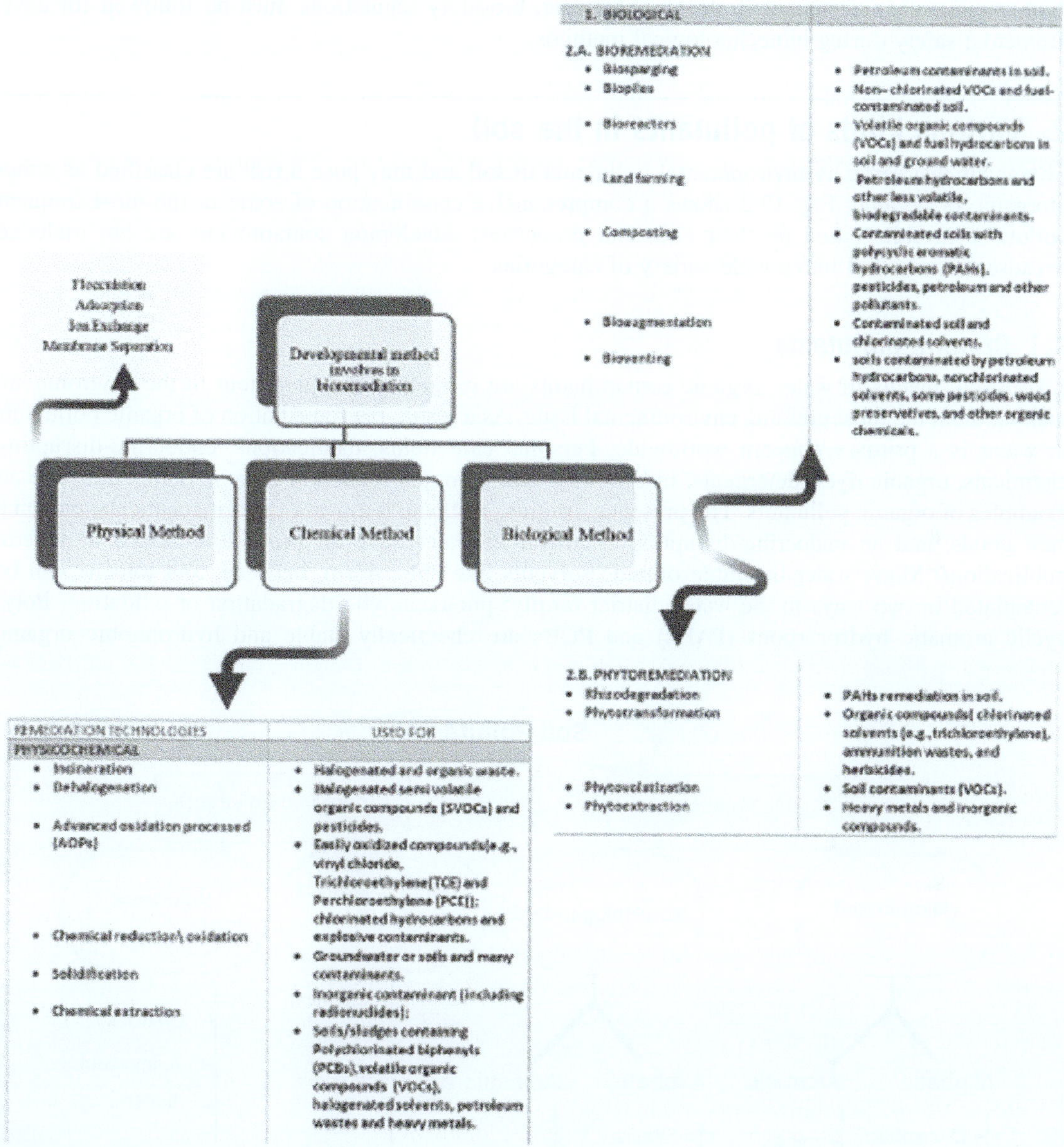

FIGURE 17.1

Different methods of bioremediation.

than traditional management strategies. However, biosafety regulations must be followed for environmental safety during biotechnological methods.

2. Various kinds of pollutants in the soil

Chemicals that serve as environmental pollutants in soil and may pose a risk are classified as either inorganic or organic. Fig. 17.2 shows a comprehensive classification of some of the most frequent pollutants in soils based on their chemical properties; developing contaminants are not included because they can fall into a wide variety of categories.

2.1 Organic pollutants

Like heavy metals in water, organic contaminants are refractory and persistent in the environment, making them the most pressing environmental issue. As a result, the remediation of organic pollutants in water is a primary concern worldwide. Personal care items, medications, endocrine-disrupting chemicals, organic dyes, detergents, insecticides, and common industrial organic compounds are all examples of organic pollutants. The presence, origins, cost, and removal of pharmaceuticals, personal care goods, and an endocrine-disrupting chemical in water have all been summarized in several publications. Many water-insoluble organic solvents like chloroform, benzene, and toluene can be remediated in two ways in the water district family: photocatalytic degradation or oxidation. Polycyclic aromatic hydrocarbons (PAHs) and PCBs are chemically stable and hydrophobic organic

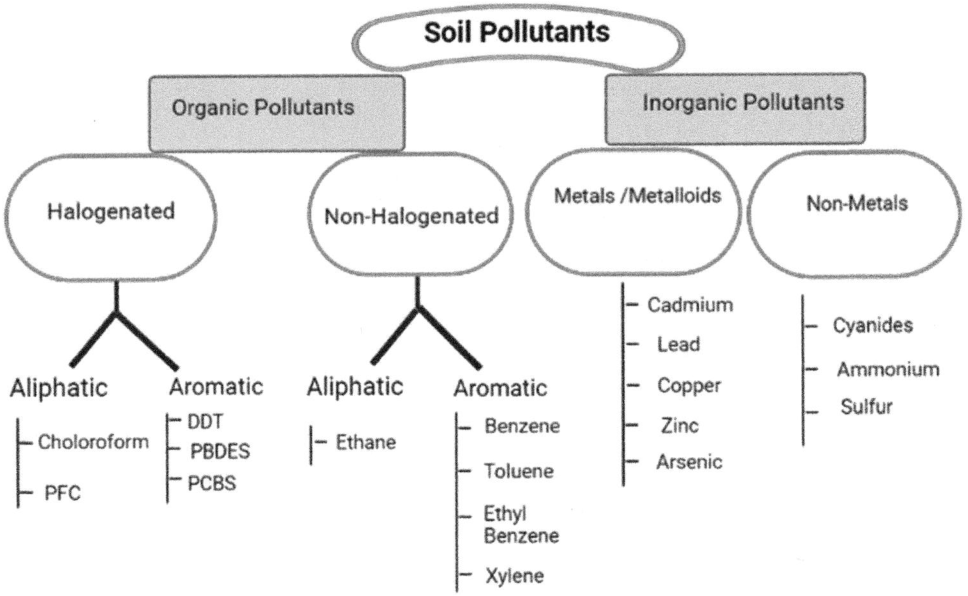

FIGURE 17.2

Various kinds of pollutants present in soil.

molecules abundant in the environment and serve as good indicators of urban activity. PAH can have a natural or anthropogenic origin. If the byproducts primarily originate from incomplete combustion or pyrolysis of organic material, then the origin is biological. These chemicals are produced mainly by anthropogenic activities such as traffic, industry, domestic heating, and incineration. On the other hand, PCBs cannot be formed naturally, and their use is primarily related to the industry. Although their production has ceased, they can still be found in various products such as capacitors, transformers, hydraulic fluid, cutting oils, and so on. They can also be formed byproducts of specific industrial processes or waste incineration. These organic pollutants are discharged mainly into the atmosphere. They accumulate in dry and wet atmospheric depositions after being transported in particle and gaseous forms over short and long distances. As a result, soil contamination can occur as a result of diffuse or point source contamination of organic pollutants, which can become a transboundary environmental issue due to long-range atmospheric transport turns, as evidenced by their accumulation in soils all over the world, even far away from human activity (Lu and Astruc, 2020).

2.1.1 Types of organic pollutants

Pollutants come in a variety of forms and are found in various locations. Some pollutants are resistant to environmental degradation (chemical, biological, and photolytic reactions) and can persist in the environment for lengthy periods. Persistent organic pollutants (POPs) are contaminants that remain in our environment for an extended period. Pesticides, industrial chemicals (PCBs, PBDEs, PFS, etc.), and byproducts of industrial processes (dioxins and furans) are all considered priority pollutants. These contaminants remain in the environment for a long time and travel from one location to another, even crossing international borders. These contaminants have even been found in regions where they have never been utilized, such as the earth's poles. These bioaccumulate in the food chain (animals and humans), causing many well-known health and environmental problems (Cachada et al., 2012). POPs are artificial organic chemicals that can be found almost everywhere on land and in water. These pollutants are among the most long-lasting manmade organic substances that have been released into the environment. Some of these pollutants are incredibly toxic, causing various long-term impacts such as endocrine disruption, mutagenicity, and carcinogenicity. POPs are also chemically stable, making them difficult to break down in the environment or by organisms. Lipophilic in nature, they accumulate in the food chain. PCBs comprise up to 209 distinct chlorinated compounds (known as congeners), of which 113 have been found in the environment. At least half of the PCBs made are still in use or storage, notably in older electrical equipment. As a result, there is still a large stockpile of PCBs that could be released into the environment through spills or leaks from transformers and other equipment.

Furthermore, the migration of these compounds from sediments containing high levels of PCBs to water ensures a continuous supply of the materials in the water phase. Organochlorine pesticides (OCPs) were created to eliminate insects and other pests from agricultural fields, increasing crop yields. Without pesticides, annual crop losses were enormous, accounting for nearly one-third of annual yield. OCPs were created to disrupt and inhibit specific physiological pathways in the organisms they were targeted for. In truth, Paul Hermann Müller, the scientist who invented DDT (dichlorodiphenyltrichloroethane), won the Nobel Prize in 1949 for his work because DDT eliminated the mosquitos that spread malaria and typhus. However, little was known about off-target effects in other organisms, such as invertebrates and vertebrates, when they were released into commerce. OCPs were replaced by other pesticides and removed from the market in the 1970s when knowledge of off-target effects, such as weakening of bird eggshells due to DDT exposure became available. DDT and its metabolites, dichlorodiphenyldichloroethylene (DDE) and dichlorodiphenyldichloroethane (DDD); methoxychlor, dieldrin, toxaphene, lindane, endosulfan, and other agrochemicals are among the OCPs. Table No. 2 POPs are POPs that are commonly found in food (Rios et al, 2007).

Their structures are highly different, with chlorine groups being the only thing they have in common. OCPs are highly lipophilic and resistant to microbial breakdown, making them long-lasting in the environment. Despite this, many OCPs activate the same nuclear receptors and cause similar toxicity. Nontarget organisms have been discovered to concentrate in fatty tissues and are linked to various illnesses, including cancer. As a result, OCPs are at the top of the list of dangerous compounds produced by the United States Department of Health and Human Services Agency for Toxic Substances and Disease Registry (ATSDR) (Martyniuk et al, 2020).

PAHs (polycyclic aromatic hydrocarbons) are organic chemicals found in high air, water, and soil concentrations. PAHs originating from fuels and incomplete combustion processes have recently increased in number. Oil spills in the aquatic environment create PAH pollution and impact marine species. Oil spills are highly correlated with key shipping routes. Furthermore, unintentional oil spills might have a toxicological influence on the marine ecosystem. PAH toxicity and related bioaccumulation features in aquatic species, especially invertebrates, are described here. Although researchers have primarily focused on the carcinogenic toxicity of PAHs, recent studies have revealed the toxicity of PAHs, including endocrine disruption and tissue-specific toxicity.

Furthermore, when evaluating PAH toxicity, the bioaccumulation qualities of PAHs for creatures, including invertebrates, are relevant considerations. Because PAHs are hazardous to aquatic animals, we should prevent PAH pollution in the marine environment. PAHs are classified into four categories in the aquatic environment: those obtained from fuels (petrogenic), those derived from an incomplete combustion process (pyrogenic), those derived from organic metabolism (biogenic), and those derived from the transformation process in sediment (diagenetic). Petrogenic and pyrogenic sources are the most artificial of the four categories of sources, and they are significant contributors to PAH pollution in aquatic ecosystems. Oil spill accidents are among the most significant exposure events for PAH pollution in marine ecosystems. PAHs, saturated aliphatic hydrocarbons, aliphatic unsaturated hydrocarbons, and alicyclic saturated hydrocarbons are all hydrocarbon compounds found in crude oil. Because of their specific toxicity, the impact of these four groups on the ecosystem is particularly concerning. Some essential concerns are discussed below.

2.1.1.1 Effect on human health

POPs are highly lipid-soluble and have semivolatilities. Furthermore, some POPs, such as perfluorooctane sulfonate (PFOS), are water-soluble. These have bioaccumulation tendencies in live creatures with a sluggish metabolism. As a result, these contaminants remain in our bodies for an extended period. POPs that have been halogenated are more resistant to degrading processes. These contaminants can bind to aerosols and travel from one location to another. Their physicochemical features determine POPs' environmental destiny. Direct applications of pesticides in the environment, industrial wastes, and byproducts of industrial operations are all sources of POPs. Agriculture, forestry, horticulture, industries, municipal, medical, and other activities also contribute to the pollution of our environment.

Furthermore, some natural causes, such as volcanic activity and vegetation fires, contribute to introducing these toxins into our ecosystem. POPs are highly toxic because they have the property of bioaccumulation. POPs are well-known and harmful pollutants that are both bioaccumulative and nondegradable. POPs are found in practically everyone's body in the current contaminated environment. POPs have been discovered in fetuses and embryos, which is surprising. In this polluted world, our newborns are born with POPs and have begun to acquire more contaminants. POP concentrations are seen in people of all ages, with older persons having the highest levels. Cardiovascular disease, obesity, hormone disruption, cancer, reproductive and neurological disorders, learning difficulties, and diabetes are major health concerns caused by exposure to these contaminants.

Furthermore, these contaminants cause abnormalities in the female embryo. The bioavailability and dose-response correlations of POPs are detailed for their toxicity profiles. The most hazardous POP, according to the WHO, is 2,3,7,8-tetrachlorodibenzo-p-dioxin (2,3,7,8-TCDD). The following subsections highlight several well-known health issues linked to POPs.

2.1.1.2 Reproductive problems

The functionality disorders of the male and female reproductive systems are at the root of reproductive issues and diseases. Congenital disabilities, premature birth, developmental abnormalities, low birth weight, impotence, diminished fertility, and menstrual irregularities are among them. POP pollution has had an impact on reproductive systems. According to research, some POPs have long been known to have harmful impacts on humans. According to the author, the exposure of POPs reduced sperm quality and quantity. Furthermore, these contaminants influenced sex ratios and the onset of puberty. Pregnancy and endometriosis were both affected by female exposure.

2.1.1.3 Additive and synergistic effects

Individual POPs represent a specific health risk; yet, organisms in the environment are simultaneously exposed to various POPs. As a result, studying the cumulative and synergistic impacts of POPs on creatures, plants, and other environmental identities is exceptionally challenging. It's owing to a scarcity of resources and equipment. POP combinations can sometimes have synergistic effects. The toxicity of POPs can sometimes be exacerbated by one another. As a result, natural POP combinations may be more harmful than predicted in laboratory research. As a result, the additive technique is used to investigate the fatal effects of many POPs simultaneously.

2.1.1.4 Cardiovascular problems

Cardiovascular disorders are the foremost cause of death worldwide. Many people around the world die due to cardiovascular disorders. Hypertension, angina pectoris, cardiac arrhythmias, and other heart disorders are many. Hypertension affects approximately one billion people worldwide. By 2025, it is expected that the number of cardiovascular patients would have climbed to 1.56 billion. Cardiovascular diseases account for around one-third of all mortality, with 17 million people dying each year. Hypertension alone is responsible for 9.4 million fatalities. The proportion of people with hypertension fluctuates depending on their socioeconomic status. In low and middle-income countries (40%), hypertension is more common than in high-income countries (35%). The development of heart illnesses is influenced by a mix of social, behavioral, and metabolic variables and POP pollution. POPs' tainted food has been linked to diabetes, obesity, and other behavioral issues.

POPs are to blame for the onset of these health issues. It is generally recognized that most POPs are lipophilic in nature and, as a result, bioaccumulate in high-density lipoproteins, causing a variety of cardiovascular problems and disorders. Organochlorine (OC) insecticides, PCBs, and polybrominated biphenyls (PBBs) are the POPs that cause cardiovascular problems. The effects of PCBs and OC insecticides on hypertension in Arctic communities were investigated by Valera. Because of the high quantities of these POPs, the authors noticed an increase in hypertension. The authors concluded that these people had a variety of cardiovascular illnesses.

2.1.1.5 Cancers

Cancer is defined as a condition in which abnormal cells develop and spread uncontrollably. The human body is a complex system of interconnected cells and tissues. In cancer, normal cells begin to act differently for various internal and environmental reasons, due to its unpredictable nature and lack of adequate treatment. To date, more than 200 different forms of cancer have been identified. It is the world's second most dangerous disease, accounting for around 21% of deaths each year. Despite advances in medical technology, cancer in its late and chronic stages is incurable.

POP concentrations in low-density lipoproteins are linked to a variety of malignancies. There are some reviews available on the malignancies induced by various POPs. The authors discovered that high levels of these POPs increased the risk of breast cancer. PFOS and perfluorooctanoic acid (PFOA) were found to increase the risk of breast cancer in women.

2.1.1.6 Diabetes

Diabetes mellitus, commonly referred to as diabetes, is a collection of metabolic illnesses marked by high blood sugar levels. Diabetic ketoacidosis and nonketotic hyperosmolar coma are the most severe consequences. Kidney failure, foot ulcers, stroke, heart disease, and eye impairment are all significant long-term consequences. Diabetes has also been linked to POP bioaccumulation in humans. The development of metabolic syndrome has been linked to chlorinated insecticides and PCBs (insulin resistance). POPs identified included polychlorinated dibenzo-p-dioxins, polychlorinated dibenzofurans, and PCBs. POP exposure has been linked to diabetes in several studies.

2.1.1.7 Obesity

Obesity is defined as the accumulation of extra fat in the body, which can lead to various health issues. Obesity is linked to several ailments, including cardiovascular disease, cancer, osteoarthritis, and obstructive sleep apnea. Obesity is on the rise in both adults and children around the world. Bioaccumulation of POPs has also been linked to obesity issues. The significance of POPs in the development of obesity is poorly understood. Obesity as a result of POP buildup has been the subject of some research. In 98 obese and 47 lean people, Dirinck et al. investigated the putative obesogenic effects of OC insecticides and PCBs. The authors came to the conclusion that the diabetogenic result of low-dose POP exposure is more nuanced than a simple obesogenic effect. For two years, Lee et al. investigated the effects of OC, pesticides, PCBs, and PBBs on 90 people. Individuals with high amounts of the reported POPs are more likely to develop obesity and dyslipidemia, according to the author.

2.1.1.8 Endocrine disruption

The endocrine system is in charge of the body's maturation, growth, development, and regulation. Different hormones are secreted by various glands such as the pituitary, pancreas, adrenals, and testes.

All of the body's processes are controlled by these hormones. Disturbances in the endocrine system lead to a variety of diseases and health issues. Endocrine disruptors are chemicals that interfere with the endocrine system, causing problems in the reproductive, neurological, and immune systems and wildlife. Endocrine disruptors impact hormone processes, and all of the dirty dozen POPs are endocrine disruptors. POPs' effects on fetal development have been found to last throughout a person's life. There are some reviews of POP-induced endocrine disruption in the literature. OC pesticides are the most endocrine-disrupting insecticides (DDT, dieldrin, toxaphene and chlordane, mirex, endosulfan, etc.). These act as estrogens, causing the sex organs to change. There were inevitable harmful neurological consequences at five and seven months postterm recognition memory. Furthermore, children with a higher proclivity to startle, as well as reduced motor reflexes, neuromuscular functioning, and decreased responsiveness, had neurobehavioral impairments. Hypoactive reflexes, decreased visual recognitions, and short-term memories were all signs of this at around seven months of age. At the age of four, deficiencies such as weight increase, decreased attentiveness, and poor visual recognition were visible. Dioxin and PCBs have been found to alter neurobehavioral activity in monkey and rat offspring (Alharbi et al., 2018).

2.2 Inorganic pollutants

2.2.1 Heavy metals/metalloids moreover their sources

Metals and metalloids are collectively known as heavy metals, generally with a density greater than 4 ± 1 g cm^{-1}. Heavy metal terms are widely used for the contaminant of freshwater and terrestrial ecosystems, respectively. It naturally occurs in the earth's crust and can easily be found in waters, rocks, soils, sediments, and microorganisms in a natural concentration. The release of these heavy metals by anthropogenic can increase the concentration of these metals in environment pollution that are chromium, mercury, arsenic, boron, platinum, cadmium, lead, and zinc (Aleksandra, 2011). The 17 other heavy metals also affect the environment: vanadium, gold, iron, antimony, cerium, cobalt, bismuth, copper, manganese, tin, nickel, uranium, tellurium, platinum, gallium, silver, and thallium. The toxicity of heavy metals depends on the density of metals or metalloids present in the environment significantly. The increase in toxicity of heavy metals means the concentration of heavy metals deposited on an earth crust, which is unused or unwanted and becomes concentrated by human activities, which cause several damage to the environment and enter the plant, animal, and human tissues. These heavy metals can interfere with the functioning of vital cellular components, and this will happen when humans inhale, diet, or during handling of these heavy metals. The degradation of heavy metals is necessary as it is one of the significant environmental pollutants (Goyal et al., 2020, Table 17.2).

2.2.2 Effects of heavy metals

Living organisms and humans also need metals like copper, molybdenum, manganese, zinc, iron, and cobalt in a small amount for life processes and good health. Metals like zinc, manganese, chromium, copper, potassium, cobalt, magnesium, sodium, nickel, and iron are said to be micronutrients and can be used for:

- Redox processes.
- To stabilize molecule through electroplating.
- Acting as components of various enzyme.
- Regulation of osmotic pressure.

Nonessential metals like gold, lead, silver, mercury, and cadmium are toxic for living organisms and do not have biological functions. It can cause toxicity by displacing essential metals from their

Table 17.2 Examples of heavy metals, it's sources, properties and effects on human body.

S.No.	Heavy metals	Symbol	Properties	Sources	Effects on human
1.	Chromium	Cr	• Density: 7.15 g cm^{-3} • 21st most abundant element in the Earth's surface • Extracted as a chromite ore known as Siberian red lead • Hard • Shiny, steel-gray • Fairly active metal • Reacts with most acids	Mining, industrial coolants, chromium salt manufacturing, and leather tanning	• Gastrointestinal ulceration • Nausea and vomiting • Fever and diarrhea • Vertigo • Toxic nephritis • Liver damage, coma, and death (usually at 1–3g)
2.	Lead	Pb	• Density: 11.3 g cm^{-3}, • 37th most abundant metal, dull silver-gray metal, soft, Easily worked • Found in a mineral ore known as galena, which is made up of lead sulfide and can be combined with silver, zinc, and copper	Lead acid batteries, paints, e-waste, smelting operations, coal based thermal power plant, ceramics, and bangle industry	• Hypertension • Miscarriages • Premature and low births • Stillbirths • Renal impairment In children: • Brain and central nervous system development altered • Reduced intelligence • A decline in educational achievement
3.	Mercury	Hg	• Density: 13.5336 g cm^3 • 66th most abundant metal	Chlor-alkali plants, thermal power plants, fluorescent lamps, hospital waste (damaged	• Inhalation caused: • Lung irritation and eye irritation

Table 17.2 Examples of heavy metals, it's sources, properties and effects on human body.—cont'd

S.No.	Heavy metals	Symbol	Properties	Sources	Effects on human
			• It is found in the ore cinnabar made from mercury sulfide • At room temperature, it is a silvery liquid metal • Rare to find in a natural state	thermometer, barometer, sphygmomanometer, etc.), and electrical appliances	• Genotoxic • It damages the DNA and chromosomes • Mongolism, also known as Down's syndrome
4.	Arsenic	As	• Density: 5.75 g cm^3 • 55th most abundant metal • Found as three allotropic forms The minerals found are: • Arsenopyrite which is made up of iron arsenic sulfide • Realgar made up of arsenic sulfide and is called "ruby of arsenic"	Geogenic/natural process, smelting operations, and fuels	• Gastrointestinal system irritation • Lung irritation • Skin changes • Decreased production of both red blood cells and white blood cells • Increased chances of cancer have been suggested • Infertility and miscarriages
5.	Copper	Cu	• Density: 8.96 g cm^3 • 26th most abundant metal • reddish-gold color • Found in minerals such as chalcopyrite containing copper, iron, and sulfur; bornite also containing copper, iron, and sulfur and known as the peacock ore	Mining, electroplating, and smelting operations	• Flu-like symptoms • Diarrhea • Vomiting • Irritation of the eyes • Dizziness • Irritation caused in the mouth cavity

Continued

Table 17.2 Examples of heavy metals, it's sources, properties and effects on human body.—cont'd					
S.No.	**Heavy metals**	**Symbol**	**Properties**	**Sources**	**Effects on human**
			• Easily worked • Good conductor of heat and electricity • Essential element		
6.	Vanadium	V	• Density: 6.11 g cm^{-3} • 22nd most abundant metal in Earth's crust • Silvery-white • Ductile • Soft • Reacts with hot acids such as hot sulfuric acid	Spent catalyst and sulfuric acid plant	• Severe throat, nose, and eye irritations, and nose bleeds • Throat pain • Dizziness • Headaches • Nausea • Rashes
7.	Nickel	Ni	• Density: 8.9 g cm^{-3} • 22nd most abundant metal • Minerals found are pentlandite which is an iron-nickel sulfide; garnierite which is a green hydrous nickel silicate compound • Silvery metal • Resists corrosion at high temperatures • High amount of Ni came from meteorites	Smelting operators, thermal power plants, and battery industry	• Lung embolisms • Asthma • Allergic reactions (jewelery) • Respiratory failure • Heart disorders • Dizziness (following gas exposure)
8.	Cadmium	Cd	• Density: 8.69 g cm^3 • 64th most abundant metal • Found frequently in combination with zinc	Zinc smelting, waste batteries, e-waste, paint sludge, incinerations, and fuel combustion	• Kidneys primarily affected causing nephrotoxicity

Table 17.2 Examples of heavy metals, it's sources, properties and effects on human body.—cont'd

S.No.	Heavy metals	Symbol	Properties	Sources	Effects on human
			• The only mineral found is greenockite which is made up of cadmium and sulfur • Silvery bluish tint metal		• Infertility caused by a reproductive system failure • Calcium metabolism alterations • Bone fracture • Psychological disorders • Gastrointestinal disorders • Central nervous system complications • Immune system deficiencies • DNA impairment
9.	Molybdenum	Mo	• Density: 11 g cm^3 • 54th most abundant metal • The mineral found is molybdenite which is made up of molybdenum disulphide • The highest melting point from all trace metals • Brilliant alloying agent	Spent catalyst	• A decrease in appetite • Fatigue and weakness • Anorexia • Listlessness • Headaches • Chest pain • Myalgia and arthralgia • Gout and hypochromic microcytic anemia

Continued

Table 17.2 Examples of heavy metals, it's sources, properties and effects on human body.—cont'd

S.No.	Heavy metals	Symbol	Properties	Sources	Effects on human
10.	Zinc	Zn	• Density: 7.134 g cm^3 • 24th most abundant metal, • Silvery-white metal with a blue tinge • Two of the most common ores are zinc blende, made up of zinc sulfide, and calamine made up of zinc silicate • Tarnishes in air • Essential element	Smelting, electroplating	• Nausea and vomiting • Stomach cramps • Anemia • Decrease in high-density lipoprotein (HDL) cholesterol • Pancreatic complications • Fatigue • Epigastric pain

native binding site through ligand interactions. Both essentials as well as nonessentials can cause acute or chronic poisoning (toxicity) when used in large amounts.

They can:

- Damage the structure of DNA at high concentration.
- Damage cell membranes.
- Alter enzyme specificity.
- Disrupt cellular functions.
- Damage central nervous function.
- Brain damage or the reduction of mental processes.
- Lower energy levels.
- Alternate the gene expression.
- Cause skin cancer.
- Muscle disruption.
- Change blood composition.
- Effect lungs when inhaled for a long time.
- Damage kidneys, liver, and heart.

The metal can cause toxicity only when they enter the cells. For this, the organism systems have to be tightly regulated. The system can easily differentiate between very similar metal ions, like oxi-anions chromate, phosphate, etc., and arsenate resembles sulfate structure. The heavy metals can also do severe damage like muscular, physical, and neurological degeneration, which can mimic diseases like multiple sclerosis, Parkinson's disease, Alzheimer's disease, and muscular dystrophy, if exposed for long term. Heavy metals can even cause cancer and allergies on repeated long-term contact (Elliott, 2011).

3. Bioaccumulation

The process of anthropogenic chemicals being taken up by organisms from their environment and nutrition and then absorbed and dispersed into tissues is known as bioaccumulation. As a result, bioaccumulation is a critical concept in ecotoxicology, as it helps define the maximum concentration that an organism can achieve in its tissues about the exposure media and the possible chemical dose/toxicity that an individual may experience. The capacity of MPs to influence organic pollutant bioaccumulation begins with sorption and desorption. MPs have been proven to have considerable "carrier effects" in boosting the bioaccumulation of sorbed organic pollutants and their transfer through the food chain in several studies based on laboratory trials. Furthermore, bioaccumulation is dependent on the capacity of ingested MPs and their sorbed contaminants to reach the tissues and organs of the organism. Even though MPs of less than 10 micrometers have been found in tissue (like muscles) or organs (like gills or liver) of *Danio rerio* (zebrafish), *Carcinus maenas* (European green crab), *Mytilus edulis* (blue mussel), and *Misgurnus anguillicaudatus* (pond loach). Still it is a topic of research that whether MPs can transfer between organs and tissues, or into cells of living organisms or not.

On the other hand, in determining the bioaccumulation of pollutants, the desorption of sorbed chemicals from ingested MPs ("releaser effect") must be taken into account. Surfactants in the gut have been found to speed up the release of hydrophobic organic contaminants from MPs. DDT, phenanthrene, PFOA, and di-2-ethylhexyl phthalate (DEHP) desorption was substantially higher in conditions matching those in the stomach of warm-blooded organisms than in seawater.

On the other hand, MPs may reduce the concentration of dissolved organic pollutants in the environment, particularly in aquatic systems. If animals primarily accumulate organic contaminants through the dissolved phase in their environment, the MPs present may limit their bioaccumulation, even if the ingested MPs are filled with organic pollutants. When a considerable amount of organic pollutants was adsorbed on MPs, the so-called "diluting effect" occurred, resulting in a lower free concentration of organic pollutants. Furthermore, the "diluting effect" is linked to the number of MPs present in the system. Moreover, MPs' ingestion may help remove some of the pollutants ingested from the environment through solid adsorption inside the digestive tract, resulting in less bioaccumulation ("cleaning effect"). The predicted bioaccumulation of pollutants decreased as the ingestion of MPs increased.

The degree of MP intake and subsequent transfer to tissues and organs directly impact the sorbed pollutants' toxicity, bioaccumulation, and even metabolism. Laboratory studies have revealed that the amount of MPs consumed by organisms is precisely proportional to the number of MPs present in the environment. However, in nature, particularly in the open ocean, where MP concentrations are often much lower than in the lab, organism preference and feeding behavior and the morphology and availability of MPs may be more essential determinants in determining ingestion. Many animals, especially those at higher levels, exhibit food preferences based on color, hardness, shape, and surface roughness. The size and type of food that can be eaten and the size of the ingested MPs are determined by the age of the organism.

Furthermore, exposure to MPs has been shown in multiple studies to reduce predatory performance and efficiency, altering animal eating behavior. MPs' size, density, and surface properties are changed by ingestion and excretion, which may affect their subsequent ingestion by benthic animals and, as a result, their interactions with contaminants in the environment and organisms. There is a need for more research into organismal feeding preferences and selectivity for different forms of MPs and the changes in MPs after they are excreted.

It should be noted that, while MPs can collect significant levels of POPs, their content in nature is minimal, even several orders of magnitude lower, when compared to naturally occurring natural particles and dissolved organic matter. Given the impact of natural particles, despite the fact that a wide range of organisms can consume MPs, their contribution to total bioaccumulation in nature (either "carrier/releaser effect" or "diluting/cleaning effect") may be insignificant. The bioaccumulation of DEHP, phenanthrene, and DDT sorbed to MPs in benthic invertebrates, fish, and seabirds was low when compared to ingested food and water (Wang et al., 2020).

4. Bioremediation

Composting and wastewater treatment are two well-known examples of old environmental biotechnologies. Recent research in molecular biology and ecology, on the other hand, suggests that biological processes can be made more efficient. The clean-up of polluted water and land areas is one of these studies' notable accomplishments. Bioremediation is the process of biologically degrading organic wastes under controlled conditions to a benign state or to concentrations below regulatory limits. Bioremediation is the process of degrading environmental pollutants into less hazardous forms using living creatures, typically bacteria. It degrades or detoxifies pollutants that are harmful to human health and/or the environment using naturally occurring bacteria, fungus, or plants. Microorganisms may be native to a contaminated location or isolated from elsewhere and transported to the contaminated site. Living organisms change contaminant substances through reactions that occur as part of their metabolic activities. Multiple species are often involved in the biodegradation of a chemical. Bioaugmentation is a term used to describe the process of introducing microorganisms to a contaminated site in order to improve deterioration.

4.1 Different environmental factors influence bioremediation

The control and optimization of bioremediation processes is a complex system of many factors. These factors include: the existence of a microbial population capable of degrading the pollutants; the availability of contaminants to the microbial population; and the environment factors (type of soil, temperature, pH, the presence of oxygen or other electron acceptors, and nutrients) (Fig. 17.3).

5. Bioremediation of organic pollutants

Bioremediation of organic pollutants can be done through various methods, which are as follows.

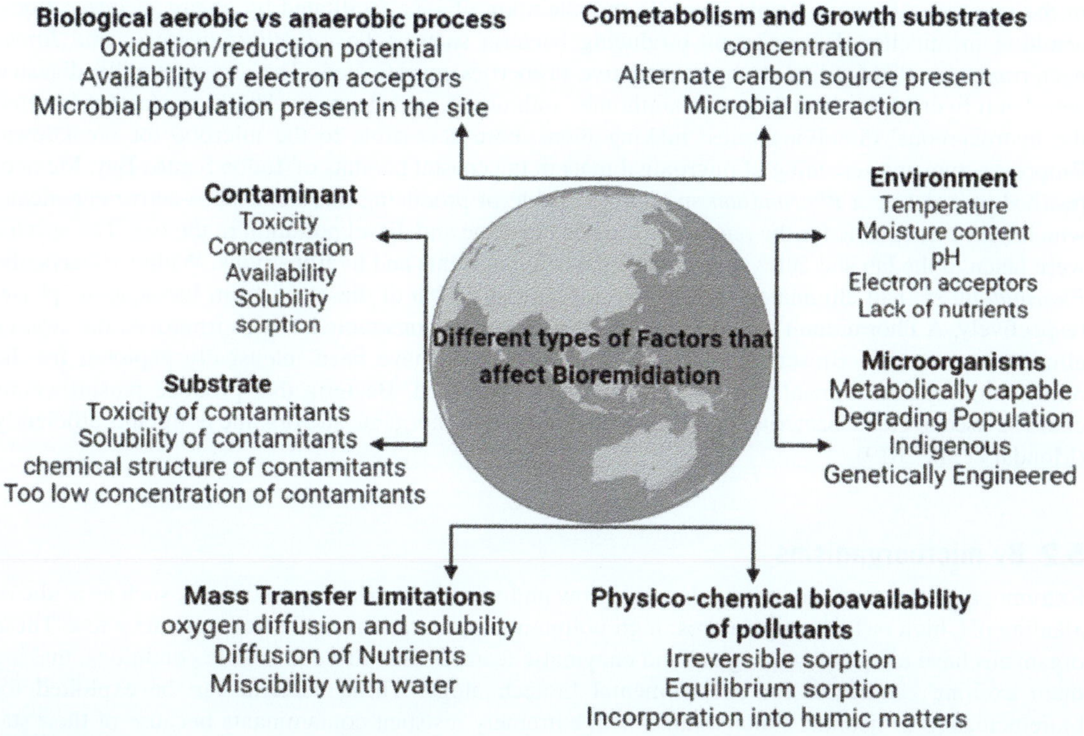

FIGURE 17.3

Different factors influence bioremediation.

5.1 Biosurfactants

Surface-active substances produced by biological systems, primarily microbes, are known as biosurfactants. They contain both hydrophilic and hydrophobic moieties that remain distributed at the interfaces between liquid phases with varying degrees of polarity (oil/water), reducing surface and interfacial tension, which is a crucial feature of lubrication and solubilization. Biosurfactants differ in structure and properties from one organism to the next. They are suitable for bioremediation techniques because bacteria quickly destroy them in water and soil. Biosurfactants are divided into five categories:

1. Lipopolysaccharides, from *Acinetobacter calcoaceticus.*
2. Fatty acids and neutral lipids.
3. Glycolipids, like rhamnolipids produced by *P. aeruginosa* and sophorolipids from *Candida spp.*
4. Lipopeptides produced by *Bacillus subtilis.*
5. Phospholipids obtained from *Corynebacterium lepus.*

Biosurfactants increase bioavailability and remove pollutants by pseudo-solubilization and emulsification during the decontamination of organic pollutants. In contrast, the use of biosurfactants in the recovery of heavy metals involves the chelation of ions facilitated by chemical interactions, resulting in micelles. Biosurfactant-producing bacteria such as *Lactobacillus pentosus* and *Brevibacterium spp.* PDM-3 had its biodegradative properties investigated. The strains readily digested petroleum hydrocarbons such as phenanthrene, anthracene, and fluorene. The biosurfactant boosted the hydrocarbons' dissolving rates, making them more accessible to the microbe for breakdown. Bioprospecting and screening of microalgal mats in the coastal habitats of Todoa Santos Bay, Mexico, resulted in isolating a *Phormidium* sp. strain capable of producing marine surface-active chemicals, which were responsible for the remediation of hexadecane and diesel pollution in the bay. The species were taken to the lab and cultivated in seawater with nutrients and hydrocarbons. Within 10 days, the *Phormidium* sp. had eliminated 45% of hexadecane and 37% of diesel oil from the aqueous phase, respectively. A Phormidium structure in a biodisk was also constructed, which improved the strain's elimination capacity. Biosurfactants derived from bacteria have been intensively explored for the remediation of contaminants, and some have been proposed. Bacteria that produce biosurfactants could be used to create bacteria-algal consortia that remove inorganic and organic pollutants efficiently (Mondal et al., 2019).

5.2 By microorganisms

Extremophiles are microorganisms that can grow and survive in extreme conditions, such as acidic or alkaline pH, high or low temperatures, high pollutant concentrations, and salts, to name a few. These organisms have unique physiological and enzymatic features that thrive in hostile conditions, making them exciting candidates for environmental biotechnology. These bacteria can be exploited for bioremediation of habitats contaminated with extremely resistant contaminants because of their stability and durability under harsh environmental circumstances. Extremophile fungi, such as black yeasts, can adapt to different levels of environmental stress and tolerate a wide range of pH, temperature, and xenobiotic compounds, making extremophile treatment of industrial wastes a versatile approach that can lead to novel environmental recovery technologies.

Some filamentous fungi and yeasts are isolated from substrates collected in marine sediments and terrestrial soils. Antarctic habitats were shown to be able to manufacture ligninolytic enzymes. The activities of protease and lipase in yeasts and filamentous fungi, which could be effective in treating protein- and oil-based wastes and effluents, respectively.

Bacteria that have evolved to the cold have been proven effective at degrading petroleum hydrocarbons. Archaea from harsh conditions have also been promising bioremediation resources for hydrocarbons. Scientist used *Bacillus licheniformis M2-7*, a thermophilic bacteria in the biotransformation of benzo[a]pyrene. Using gas chromatography, the authors identified an intermediate molecule matching to phthalic acid, which had a low chemical structural complexity and was likely less harmful than benzo[a]pyrene. Despite the authors' emphasis on determining whether the biotransformation of this PAH can produce harmful intermediate molecules, bioindicators were not used to assess toxicity (Giovanella et al., 2020).

5.3 By enzymes

The capacity of microbes to detoxify or change organic contaminants is determined by their accessibility, bioavailability, and metabolic potential. Several enzymes catalyzed metabolic reactions, like hydrolysis, oxidation, the addition of oxygen to double bonds, oxidation of amino group to nitro group, dehalogenation, reduction of the nitro group to the amino group, replacement of sulfur with oxygen, ring cleavage, and addition of hydroxyl group to benzene ring occur during degradation.

Hydrolases are a class of enzymes that work without a redox cofactor and are involved in the breakdown of pesticides with ester, amide, carbon-halide, or phosphate linkages. Similarly, several hydrolyzing enzymes exist, which aid in pesticide breakdown.

Phosphotriesterases (PTEs) are homodimeric proteins with a molecular weight of 36Kda (monomeric) that detoxify and hydrolyze organophosphate insecticides. The OPD (organophosphate-degrading) gene in the bacterium that makes PTEs reduce the capacity of organic pollutants to inactivate acetylcholinesterase (AchE).

Esterases are a very diverse class of hydrolases that have been found to be particularly significant in the metabolism of organic contaminants. There are several esterases, but two are particularly important: esterases A and B. The esterases A contain the Cys residue, and the organophosphate interacts with the –SH functional group, allowing them to be easily hydrolyzed by water. Furthermore, the organophosphate reacts with the Ser-OH bond to generate a P = O bond, which is then hydrolyzed by water. In contrast, esterases B contain the Ser residue and are known as carboxiestrases, which hydrolyze endogenous compounds, organic pollutants, and acid anhydrides in mammals.

Cytochrome P450 and NADPH-cytochrome P450 reductase are two membrane proteins that make up the mixed-function oxidases (MFOs) enzyme system. The MFO is activated by NADPH and O_2, with one molecule being reduced to water and the other being incorporated into the substrate. Because it utilizes molecular oxygen for hydrolysis and oxidation of the substrate, the cytochrome P450 family is a large and well-characterized group of monooxygenase enzymes. P450 also requires a non-covalently bound cofactor to recycle their redox center, which is required for pesticide cleanup. MFO is also found in insects' endoplasmic reticulum and mitochondria, where it aids in growth, development, and reproduction, among other things. Glutathione-S-transferase (GST) is a transferase system that catalyzes the conjugation of glutathione (GSH) to an electrophilic site on the xenobiotics being introduced. It is necessary for the detoxification of contaminants. Glutathione is made up of glycine, cysteine, and glutamic acid. Oxidoreductases are another set of enzymes that catalyze the oxidation/reduction reaction by using molecular oxygen as an electron acceptor.

According to cleanup regulations, PAH must be removed from contaminated soil during rehabilitation. PAHs are broken down by dioxygenase, ligninolytic enzymes, and cytochrome P450 monooxygenases. Bacteria destroy low molecular weight PAHs such as naphthalene, phenanthrene, and anthracene in soil. Several filamentous fungi, basidiomycetes, and white-rot fungi have been discovered to be more efficient at utilizing PAHs (naphthalene, phenanthrene, pyrene, catechol, and others) than bacteria (Gaur et al., 2018).

5.4 By PGPR

Although PGPR was first used to stimulate plant growth and biocontrol of plant diseases, bioremediation with PGPR has recently received much attention. Microorganisms can degrade and even mineralize organic chemicals in the presence of plants, unlike inorganic compounds. As a result, finding effective mechanisms for the breakdown and mineralization of organic substances could be crucial in the future. So far, bacteria capable of degrading organic pollutants like PCBs have been isolated from various locations, and the processes and encoding genes have been thoroughly investigated. However, most of these bacteria cannot thrive in soils, particularly the rhizosphere, due to the near-starvation circumstances. Several successful techniques for improving the degradation efficiency and bacterial tolerance to soil pollutants have been discovered.

Increasing plant growth phytoremediation has various advantages over physical and chemical remediation.

- It is possible to achieve high levels of microbial biomass in the rhizosphere.
- It is inexpensive.
- It has the potential to be quick.
- It preserves the natural qualities of soil.
- It obtains energy primarily from sunshine.

Despite these benefits, certain plants have a low tolerance for soil pollutants, limiting the degradation efficiency to a level that is insufficient for substantial soil rehabilitation. The addition of PGPR boosted organic pollutants (polycyclic aromatic hydrocarbon and creosote) elimination, most likely through improving plant germination and survival is strongly contaminated soils, as well as increasing plant growth and root biomass accumulation. Ethylene is necessary for plant growth, but too much of it, caused by stress, might stifle growth. PGPR improves plant development by absorbing aminocyclopropane carboxylic acid (ACC), the immediate precursor to ethylene, and decreasing ethylene production in stressed plants by creating 1-aminocyclopropane-1-carboxylate deaminase (ACC deaminase). Even with both PGPR and plants, remediation technologies for several environmental toxins such as total petroleum hydrocarbons (TPHs) may be inefficient. It was discovered that combining PGPR with certain contaminant-degrading bacteria proved efficient. As a result, a multiprocess phytoremediation system (MPPS) was created. To treat TPHs, they used PGPR as well as specialized contaminant-degrading bacteria. Specific contaminant-degrading bacteria can be chosen in this system based on the qualities of the pollutants. They can quickly digest some readily available chemicals, but their primary function is to stimulate plant growth and increase plant tolerance to contaminants. (Zhuang et al., 2007).

6. Bioremediation of heavy metals

Heavy metals can be degraded using various microorganisms, enzymes, and plants. Some of the examples are given below.

6.1 By microorganisms

The most common methods used to degrade heavy metals are ultrafiltration, coagulation, floatation, chemical precipitation, ion exchange, evaporative recovery, flocculation, nanofiltration, reverse osmosis, electrodialysis, etc., and some physiochemical methods are also used like soil washing, extraction, immobilizations, and stabilization. Still, all the above techniques are expensive, and the high energy and chemical required. On the other hand, the end products are harmful and poisonous (Kulshreshtha et al., 2014).

The best and efficient way is the use of indigenous MOs or genetically modified microorganism, which have the capability of removing toxins from the environment and stabilizing the ecosystem by converting them into nontoxic secondary compounds. The organism used should have the capacity to tolerate the toxicity of heavy metals, and then only bioremediation will succeed (Ojuederie and Babalola,2017).

The examples of some bacteria which can remediate the heavy metals like cadmium, lead, and copper are *Enterobacter cloacae KJ-46, Viridibacillus arenosi B-21, E. cloacae KJ-47, and Sporosarcina soli B-22.* The experimental data show the efficiency of using the above bacteria in combination can degrade up to 5.6% of cadmium, 98% of lead, and 85.4% of copper collected from the contaminated soil. The mechanism which the researcher follows to degrade the heavy metals are:

- Sequestration of toxic metals by cell wall components or by intracellular metal-binding proteins and peptides such as metallothioneins (MTs) and phytochelatins along with compounds such as bacterial siderophores, which are primarily catecholates, compared to fungi that produce hydroxamate siderophores.
- The biochemical pathway is altered to block the uptake of metals.
- The metal is converted into a harmless form by using enzymes.
- By theuse of a precise efflux system, the intracellular concentration will reduce (Table 17.3).

6.2 By enzymes

The degradation of heavy metals depends on the enzyme present in the plant or microbes. Plant or microbial enzymes are the best methods to degrade heavy metals into nonhazardous form. They catalyze the heavy metals by using polymerization, reduction, oxidation, hydrogen ion abstraction, etc.

For example, the oxidoreductases enzyme uses the energy released during the biochemical reaction, cleaves the bond, and also helps in transferring the electron from one organic compound, which is reduced to another chemical compound. The harmful contamination is oxidizing into harmless compounds by using the above reaction. Oxidoreductases also help humify various phenolic compounds deposited in soil due to lignin decomposition. It also detoxifies various other compounds like xenobiotics, for example, anilic or phenolic compounds, by using the reaction like the association with humic compound, polymerization, or copolymerization of different substrates. The microbial enzymes degrade the azo dye and help decolorization (Saratale et al., 2016).

6.3 By plants

The remediation or degradation by plants is known as phytoremediation. This plant is used as in situ, and the microorganisms associated with that plant degrade, contain, or render harmless contaminants in the environment. Phytoremediation is also used to degrade organic compounds like pesticides, PCBs, polynuclear aromatic hydrocarbons, and inorganic compounds like heavy metals (Rezania et al., 2015)—the advantage of using this technique as it is an autotropic process system. Hence, the requirement of nutrients is minimal for the large biomass, requires simple management, and is also acceptable by society because of its environmental sustainability and aesthetic view. The plant also accumulates toxic heavy metals in its vegetative and reproductive parts. The plants help maintain soil

Table 17.3 Different organisms name with remediation of different types of heavy metals.

Microorganism	Name of organism	Remediated heavy metals	References
Bacteria	*Escherichia coli* strain	As and Hg	Zhao et al. (2005)
	Methylococcus capsulatus	Cr^{6+}	Zhao et al. (2005)
	Pseudomonas fluorescens 4F39	Ni	Zhao et al. (2005)
	Acinetobacter species	Zn	Tabaraki et al. (2013)
	Desulfovibrio desulfuricans	Cd, Ni, and Cr	Joo et al. (2015)
	Enterobacter cloacae	Pb, cd, and Ni	Banerjee et al. (2015)
	Pseudomonas aeruginosa	Pb	Ahmady-Asbchin et al. (2015)
	Enterobacter species	Cu	Bestawy et al. (2013)
	Kocuria flava	Cu	Achal et al. (2011)
	Stenotrophomonas species	Cu and Cd	Zaki and Farag (2010)
Fungus	*Aspergillus lentulus*	Cr and Pb	Iskandar et al. (2011)
	Aspergillus niger	Cu(II) and Pb(II)	Kumar and Mehta (2014)
	Aspergillus nidulans	Pb and Cr	Fazli et al. (2015)
	Aspergillus versicolor	Cd	Zhang et al. (2016)
	Pleurotus ostreatus	Pb	Mezaguer et al. (2013)
	Pleurotus mutilus	U	Şahan et al. (2016)
	Trametes versicolor	Zn	Iskandar et al. (2011)
	Termitomyces clypeatus	Cr	Iskandar et al. (2011)

fertility without affecting the topsoil by changing the availability of heavy metals (Ali et al., 2013). The phytoremediation technique is an eco-friendly and economical method over other techniques used to degrade heavy metals. It also helps in improving soil fertility by releasing various organic matters. The plant has multiple mechanisms like rhizodegradation, phytoextraction, phytovolatilization, phytostabilization, and phytofiltration to consume heavy metals from the environment.

6.4 By nanoparticles

In recent years, nanotechnology has been rising in all areas of human life. Due to its high surface-area-to-volume ratio, nanoparticles are widely used in water treatment, catalysis, biosensors, and pollutant decomposition. With the flow of these advancements, scientists working on bioremediation have also reported on various combination approaches that integrate nanotechnology with bioremediation and ultimately improve bioremediation efficiency.

Joint efforts have paved the way for the term "nanoremediation," the involvement of nanoparticles in repairing heavy metal toxicity. In general, microbial bioremediation has been found to be ineffective at high metal concentrations, as these high concentrations are fatal to restoring organisms (Mukherjee et al., 2017).

In this scenario, the integration of nanoparticles into microbial cells offers immense potential for cleaning contaminated environments due to its strong reaction sites that can contact the contaminants. And this leads to rapid detoxification or immobilization. The literature suggests that chemically synthesized metal nanoparticles have been integrated into biological agents such as enzymes and whole cells in the laboratory to support the natural repair of various contaminants. Incorporating bioremediation enzymes such as peroxidase into iron/magnetite nanoparticles has been reported to improve enzyme stability and activity against contaminant substrates. In addition, the enzyme embedded in magnetite can be easily separated from the treatment site for effective reuse. Furthermore, in another report, palladium nanoparticles were deposited on the cell wall of *Shewanella oneidensis* to purify polychlorophenol residue effectively. Nanoparticles can also be used to immobilize microbial cells such as Pseudomonas. Pseudomonas can decompose or biologically regenerate certain chlorinated chemicals. It is also noteworthy that various dyes such as methylene blue and methyl orange were degraded by biologically synthesized gold and silver nanoparticles (Meenakumari and Philip, 2015). Therefore, the advantage of nanoremediation is to reduce the toxic effects of heavy metals on microorganisms, improve the activity of microorganisms on specific contaminants, and ultimately reduce the time required for repair, and thus the overall cost.

7. Advantages of bioremediation

- Bioremediation is a natural process and is therefore perceived by the public as an acceptable waste treatment process for contaminated material such as soil. Microbes can degrade the contaminant increase in numbers when the contaminant is present; when the contaminant is degraded, the biodegradative population declines. The residues for the treatment are usually harmless products and include carbon dioxide, water, and cell biomass.
- Theoretically, bioremediation is helpful for the complete destruction of a wide variety of contaminants. Many compounds that are legally considered to be hazardous can be transformed into harmless products. This eliminates the chance of future liability associated with the treatment and disposal of contaminated material.
- Instead of transferring contaminants from one environmental medium to another, such as land to water or air, the destruction of target pollutants is possible.
- Bioremediation can often be carried out on-site, often without causing a significant disruption of everyday activities. This also eliminates the need to transport quantities of waste off-site and the potential threats to human health and the environment that can arise during transportation.
- Bioremediation can prove less expensive than other technologies used to clean up hazardous waste.

8. Disadvantages of bioremediation

- Bioremediation is limited to those compounds that are biodegradable. Not all combinations are susceptible to rapid and complete degradation.
- There are some concerns that biodegradation products may be more persistent or toxic than the parent compound.

- Biological processes are often peculiar. Important site factors required for success include the presence of metabolically capable microbial populations, suitable environmental growth conditions, and appropriate levels of nutrients and contaminants.
- It is challenging to extrapolate from the bench and pilot-scale studies to full-scale field operations.
- Research is needed to develop and engineer bioremediation technologies appropriate for sites with complex mixtures of contaminants that are not evenly dispersed in the environment. Contaminants may be present as solids, liquids, and gases.
- Bioremediation often takes longer than other treatment options, such as excavation and removal of soil or incineration.
- Regulatory uncertainty remains regarding acceptable performance criteria for bioremediation. There is no accepted definition of "clean," evaluating the performance of bioremediation is difficult, and there are no acceptable endpoints for bioremediation treatments (Kulshreshtha et al., 2014).

9. Conclusion and future aspects

Bioremediation is the most suitable technology to remove pollutants from sewage water or contaminated soil. Bioremediation of organic and metal pollutants in contaminated soil and water can be done by many methods using fungal and bacterial microorganisms or by in situ and ex situ bioremediation methods. Fungal and bacterial bioremediation are eco-friendly and less economical but time-consuming. Bioremediation of pollutants is a need of society due to the hazardous effect of pollutants on human health (carcinogenicity, toxicity, and bioaccumulation). Hence, toxic contaminants like hexavalent chromium need to be remediated. Bioremediation preserves the natural properties of soil and water, is less expensive and environmentally friendly, but on the other side, it is also a time-taking process. There are more merits of bioremediation than its demerits; hence, it is the best method to resolve the problem of metal and organic pollutants contamination in the contaminated area.

References

Achal, V., Pan, X., Zhang, D., 2011. Remediation of copper-contaminated soil by kocuria flava CR1, based on microbially induced calcite precipitation. Ecol. Eng. 37 (10), 1601–1605. https://doi.org/10.1016/j.ecoleng.2011.06.008.

Ahmady-Asbchin, S., Safari, M., Tabaraki, R., 2015. Biosorption of Zn (II) by Pseudomonas aeruginosa isolated from a site contaminated with petroleum. Desalination Water Treat. 54 (12), 3372–3379. https://doi.org/10.1080/19443994.2014.913202.

Aleksandra, S.-N., 2011. Phytotechnologies: Importance in Remediation of Heavy Metal–Contaminated Soils, pp. 277–295. https://doi.org/10.1007/978-94-007-1914-9_12.

Alharbi, O.M.L., Basheer, Al A., Khattab, R.A., Imran, A., 2018. Health and Environmental Effects of Persistent Organic Pollutants, vol 263. Elsevier B.V.

Ali, H., Khan, E., Anwar Sajad, M., 2013. Phytoremediation of heavy metals-concepts and applications. Chemosphere 91 (7), 869–881. https://doi.org/10.1016/j.chemosphere.2013.01.075.

Banerjee, G., Pandey, S., Kumar Ray, A., Kumar, R., 2015. Bioremediation of heavy metals by a novel bacterial strain Enterobacter cloacae and its antioxidant enzyme activity, flocculant production, and protein expression in presence of lead, cadmium, and nickel. Water Air Soil Pollut. 226 (4). https://doi.org/10.1007/s11270-015-2359-9.

Bestawy, E.El, Helmy, S., Hussien, H., Fahmy, M., Amer, R., 2013. Bioremediation of heavy metal-contaminated effluent using optimized activated sludge bacteria. Appl. Water Sci. 3 (1), 181–192. https://doi.org/10.1007/s13201-012-0071-0.

Borrelli, P., Robinson, D.A., Fleischer, L.R., Lugato, E., Ballabio, C., Alewell, C., Meusburger, K., Modugno, S., Schütt, B., Ferro, V., Bagarello, V., Oost, K.Van, Montanarella, L., Panagos, P., 2017. An assessment of the global impact of 21st century land use change on soil erosion. Nat. Commun. 8 (1). https://doi.org/10.1038/s41467-017-02142-7.

Cachada, A., Pato, P., Rocha-Santos, T., Ferreira da Silva, E., Duarte, A.C., 2012. Levels, sources and potential human health risks of organic pollutants in urban soils. Sci. Total Environ. 430, 184–192. https://doi.org/10.1016/j.scitotenv.2012.04.075.

Elliott, K.C., 2011. Nanomaterials and the precautionary principle. Environ. Health Perspect. 119 (6), 240–243. https://doi.org/10.1289/ehp.1103687.

Fazli, M.M., Soleimani, N., Mehrasbi, M., Darabian, S., Mohammadi, J., Ali, R., 2015. Highly cadmium tolerant fungi: their tolerance and removal potential. J. Environ. Health Sci. Eng. 13 (1), 1–9. https://doi.org/10.1186/s40201-015-0176-0.

Gaur, N., Narasimhulu, K., PydiSetty, Y., 2018. Recent advances in the bio-remediation of persistent organic pollutants and its effect on environment. J. Clean. Prod. 198, 1602–1631. https://doi.org/10.1016/j.jclepro.2018.07.076.

Giovanella, P., Vieira, G.A.L., Otero, I.V.R., Pais Pellizzer, E., de Jesus Fontes, B., Sette, L.D., 2020. Metal and organic pollutants bioremediation by extremophile microorganisms. J. Hazard Mater. 382, 121024. https://doi.org/10.1016/j.jhazmat.2019.121024.

Gouma, S., Fragoeiro, S., Bastos, A.C., Magan, N., 2014. Bacterial and Fungal Bioremediation Strategies. Elsevier Inc.

Goyal, D., Yadav, A., Prasad, M., Singh, T.B., Shrivastav, P., Ali, A., Kumar Dantu, P., Mishra, S., 2020. Effect of heavy metals on plant growth: an overview. In: Contaminants in Agriculture: Sources, Impacts and Management, March 2016. pp. 79–101. https://doi.org/10.1007/978-3-030-41552-5_4.

Hou, D., O'Connor, D., Igalavithana, A.D., Alessi, D.S., Luo, J., Tsang, D.C.W., Sparks, D.L., Yamauchi, Y., Rinklebe, J., Yong Sik, O., 2020. Metal contamination and bioremediation of agricultural soils for food safety and sustainability. Nat. Rev. Earth Environ. 1 (7), 366–381. https://doi.org/10.1038/s43017-020-0061-y.

Iskandar, N.L., Zainudin, N.A.I.M., Tan, S.G., 2011. Tolerance and biosorption of copper (Cu) and lead (Pb) by filamentous fungi isolated from a freshwater ecosystem. J. Environ. Sci. 23 (5), 824–830. https://doi.org/10.1016/S1001-0742(10)60475-5.

Jansson, J.K., Hofmockel, K.S., 2020. Soil microbiomes and climate change. Nat. Rev. Microbiol. 18 (1), 35–46. https://doi.org/10.1038/s41579-019-0265-7.

Joo, J.ock, Choi, J.H., Kim, I.H., Kim, Y.K., Oh, B.K., 2015. Effective bioremediation of cadmium (II), nickel (II), and chromium (VI) in a marine environment by using desulfovibrio desulfuricans. Biotechnol. Bioproc. Eng. 20 (5), 937–941. https://doi.org/10.1007/s12257-015-0287-6.

Kulshreshtha, A., Agrawal, R., Barar, M., Saxena, S., 2014. A review on bioremediation of heavy metals in contaminated water. IOSR J. Environ. Sci. Toxicol. Food Technol. 8 (7), 44–50. https://doi.org/10.9790/2402-08714450.

Kumar, S., Mehta, D., 2014. Elemental analysis of soil samples using thick target-particle induced X-ray emission (TT-PIXE) technique. J. Nuclear Phys., Mater. Sci., Radiat. Appl. 1 (2), 225–237. https://doi.org/10.15415/jnp.2014.12018.

Lu, F., Astruc, D., 2020. Nanocatalysts and other nanomaterials for water remediation from organic pollutants. Coord. Chem. Rev. 408, 213180. https://doi.org/10.1016/j.ccr.2020.213180.

Martyniuk, C.J., Mehinto, A.C., Denslow, N.D., 2020. Organochlorine pesticides: agrochemicals with potent endocrine-disrupting properties in fish. Mol. Cell. Endocrinol. 507, 110764. https://doi.org/10.1016/j.mce.2020.110764.

Meenakumari, M., Philip, D., 2015. Degradation of environment pollutant dyes using phytosynthesized metal nanocatalysts. Spectrochim. Acta Mol. Biomol. Spectrosc. 135, 632–638. https://doi.org/10.1016/j.saa.2014.07.037.

Mezaguer, M., Kamel, N.El H., Lounici, H., Kamel, Z., 2013. Characterization and properties of pleurotus mutilus fungal biomass as adsorbent of the removal of Uranium(VI) from Uranium leachate. J. Radioanal. Nucl. Chem. 295 (1), 393–403. https://doi.org/10.1007/s10967-012-1911-y.

Mondal, M., Halder, G., Oinam, G., Indrama, T., Tiwari, O.N., 2019. Bioremediation of Organic and Inorganic Pollutants Using Microalgae. Elsevier B.V.

Mukherjee, T., Chakraborty, S., Biswas, A.A., Das, T.K., 2017. Bioremediation potential of arsenic by non-enzymatically biofabricated silver nanoparticles adhered to the mesoporous carbonized fungal cell surface of Aspergillus foetidus MTCC8876. J. Environ. Manag. 201, 435–446. https://doi.org/10.1016/j.jenvman.2017.06.030.

Naylor, D., Sadler, N., Bhattacharjee, A., Graham, E.B., Anderton, C.R., Ryan, M.C., Lipton, M., Hofmockel, K.S., Jansson, J.K., 2020. Soil microbiomes under climate change and implications for carbon cycling. Annu. Rev. Environ. Resour. 45, 29–59. https://doi.org/10.1146/annurev-environ-012320-082720.

Ojuederie, O.B., Babalola, O.O., 2017. Microbial and plant-assisted bioremediation of heavy metal polluted environments: a review. Int. J. Environ. Res. Publ. Health 14 (12). https://doi.org/10.3390/ijerph14121504.

Rezania, S., Ponraj, M., Talaiekhozani, A., Mohamad, S.E., Md Din, M.F., Taib, S.M., Sabbagh, F., Sairan, F.Md, 2015. Perspectives of phytoremediation using water hyacinth for removal of heavy metals, organic and inorganic pollutants in wastewater. J. Environ. Manag. 163, 125–133. https://doi.org/10.1016/j.jenvman.2015.08.018.

Rios, L.M., Moore, C., Jones, P.R., 2007. Persistent organic pollutants carried by synthetic polymers in the ocean environment. Mar. Pollut. Bull. 54 (8), 1230–1237. https://doi.org/10.1016/j.marpolbul.2007.03.022.

Şahan, T., Ceylan, H., Aktaş, N., 2016. Optimization of biosorption of Zn(II) ions from aqueous solutions with low-cost biomass trametes versicolor and the evaluation of kinetic and thermodynamic parameters. Desalination Water Treat. 57 (26), 12156–12167. https://doi.org/10.1080/19443994.2015.1048304.

Saratale, G.D., Jung, M.Y., Oh, M.K., 2016. Reutilization of green liquor chemicals for pretreatment of whole rice waste biomass and its application to 2,3-butanediol production. Bioresour. Technol. 205, 90–96. https://doi.org/10.1016/j.biortech.2016.01.028.

Science, L., Sharma, S., 2012. Bioremediation: features, strategies and applications. Asian J. Pharm. Life Sci. 2 (2), 202–213.

Tabaraki, R., Ahmady-Asbchin, S., Abdi, O., 2013. Biosorption of Zn(II) from aqueous solutions by acinetobacter sp. Isolated from petroleum spilled soil. J. Environ. Chem. Eng. 1 (3), 604–608. https://doi.org/10.1016/j.jece.2013.06.024.

Wang, T., Wang, L., Chen, Q., Kalogerakis, N., Ji, R., Ma, Y., 2020. Interactions between microplastics and organic pollutants: effects on toxicity, bioaccumulation, degradation, and transport. Sci. Total Environ. 748, 142427. https://doi.org/10.1016/j.scitotenv.2020.142427.

Zaki, S., Farag, S., 2010. Isolation and molecular characterization of some copper biosorped strains. Int. J. Environ. Sci. Technol. 7 (3), 553–560. https://doi.org/10.1007/BF03326164.

Zhang, S., Zhang, X., Chang, C., Yuan, Z., Wang, T., Zhao, Y., Yang, X., Zhang, Y., Guixiao, La, Wu, K., Zhang, Z., Li, X., 2016. Improvement of tolerance to lead by filamentous fungus pleurotus ostreatus HAU-2 and its oxidative responses. Chemosphere 150, 33–39. https://doi.org/10.1016/j.chemosphere.2016.02.003.

Zhao, X.W., Zhou, M.H., Li, Q.B., Lu, Y.H., He, N., Sun, D.H., Deng, X., 2005. Simultaneous mercury bioaccumulation and cell propagation by genetically engineered Escherichia coli. Process Biochem. 40 (5), 1611–1616. https://doi.org/10.1016/j.procbio.2004.06.014.

Zhuang, X., Chen, J., Shim, H., Bai, Z., 2007. New advances in plant growth-promoting rhizobacteria for bioremediation. Environ. Int. 33 (3), 406–413. https://doi.org/10.1016/j.envint.2006.12.005.

Index

Note: 'Page numbers followed by "*f*" indicate figures and "*t*" indicate tables'.

C

D

E

I

R

S

CPI Antony Rowe
Eastbourne, UK
August 07, 2023